The Lost History of the New Madrid Earthquakes

CONEVERY BOLTON VALENCIUS

The Lost History
of the New Madrid
Earthquakes

The University of Chicago Press ★ Chicago and London

The University of Chicago Press, Chicago 60637
The University of Chicago Press, Ltd., London
© 2013 by Conevery Bolton Valencius
All rights reserved. Published 2013.
Paperback edition 2015
Printed in the United States of America

22 21 20 19 18 17 16 15   3 4 5 6 7

This book was supported by a 2008–2009 research fellowship
from the National Endowment for the Humanities, a vital
tax-supported source of intellectual inquiry in the United
States. Any views, findings, conclusions, or recommendations
expressed in this publication do not necessarily reflect those of
the National Endowment for the Humanities.

I am grateful to Cambridge University Press for permission
to use material first published in Conevery Bolton Valencius,
"Accounts of the New Madrid Earthquakes: Personal Narratives
and Seismology over the Last Two Centuries," in Deborah R.
Coen, ed., *Witness to Disaster: Earthquakes and Expertise in
Comparative Perspective,* special issue of *Science in Context* 25,
no. 1 (February 2012): 17–48.

ISBN-13: 978-0-226-05389-9 (cloth)
ISBN-13: 978-0-226-27375-4 (paper)
ISBN-13: 978-0-226-05392-9 (e-book)
DOI: 10.7208/chicago/9780226053929.001.0001

Library of Congress Cataloging-in-Publication Data

Valencius, Conevery Bolton, 1969–
   The lost history of the New Madrid earthquakes / Conevery
Bolton Valencius.
      pages ; cm.
   Includes bibliographical references and index.
   ISBN 978-0-226-05389-9 (cloth : alkaline paper) —
ISBN 978-0-226-05392-9
(e-book) 1. New Madrid Earthquakes, 1811–1812. 2. Earth-
quakes—Mississippi River Valley—History. I. Title.
   QE535.2.U6V354 2013
   363.34'95097709034—dc23

                                        2013007013

♾ This paper meets the requirements of ANSI/NISO Z39.48-1992
(Permanence of Paper).

TO MATTHEW

CONTENTS

ILLUSTRATIONS

Introduction: Earthquake Cracks

On a cold night early in 1826, Davy Crockett and his hunting dogs chased a bear through the rough terrain of west Tennessee. Crockett was above all a practical frontiersman. As he ran through the woods, his main concern was for his gun—he worried that he might break it as he stumbled on the "earthquake cracks" that ran through the ground beneath him. Eventually, the bear he was chasing wedged itself down inside a larger crack, about four feet deep, where Crockett's hunting dogs could not surround it. Undeterred, Crockett crawled into the crevasse himself and knifed the bear from the side. The next morning, after Crockett's hunting partner had caught up with him, the two men butchered the bear and salted and packed its meat. As they did so, his partner examined the crack and "said he wouldn't have gone into it . . . for all the bears in the woods." Sure enough, as Crockett recounted in his autobiography, that night "a most terrible earthquake . . . shook the earth so, that we were rocked about like we had been in a cradle." "We were very much alarmed," he explained, "for though we were accustomed to feel earthquakes, we were now right in the region which had been torn to pieces by them in 1812, and we thought it might take a notion and swallow us up, like the big fish did Jonah."[1]

Davy Crockett was not overly frightened to be shoulder-to-shoulder with a bear, but despite his bravado in crawling down into the earthquake crack, he admitted to being "very much alarmed" by the tremors that terrified his partner. At the same time, the earthquakes that periodically shook his favorite hunting grounds were for Crockett a recognizable feature of his environment. He knew them as reverberations of the great Mississippi Valley earthquakes that had rocked and reshaped his region in the winter of 1811–12. Those "great shakes" continued—and continue today—in smaller tremors felt by hunters, bears, dogs, and other inhabitants of the American heartland.

The New Madrid Earthquakes

The earthquakes Crockett remembered as having "torn to pieces" his hunting grounds in 1812 were what we now call the New Madrid earthquakes, a series of powerful tremors that rent the midcontinent and were felt across North America. From December 1811 to February 1812, three large earthquakes—and numerous others—shook an area centered on what is now the small jut of southeast Missouri known as the Missouri "Bootheel" (site of New Madrid, a small Mississippi River trading port named for the Spanish capital but now pronounced, defiantly and definitively, "new MAD-rid").[2]

Around the epicenters of the 1811–12 quakes, the heart of the central continent where the present-day states of Missouri, Arkansas, Tennessee, Kentucky, and Illinois draw near each other, these earthquakes' effects were terrifyingly intense. The air filled with loud noise, foul stench, and mysterious flashes of light. Large areas were covered with rising warm water that threatened to drown many in what is now southeast Missouri. The surface of the earth rolled in visible ground waves so powerful that sections of forest snapped off midtrunk as the trees recoiled.

Ground shaking caused the deep, soft soil of this vast floodplain region to separate ("like a runny egg casserole if you took it out of the oven and shook it side to side," as New Madrid researcher Tish Tuttle described it),[3] causing the earthquake cracks that could run for almost a mile and could be anywhere from centimeters to many meters deep. In areas near southeast Missouri and eastern Arkansas, water just under the surface of the earth was put under tremendous pressure by seismic movement. The water and sand under the soil in much of the Mississippi floodplain responded by liquefying—turning from solid foundation into volatile quicksand.*[4] Seeking an outlet, this liquefied sand shot upward through the clay underlayer and dirt of the surface into fountains of liquefied sand which created immense, round "sand blows," volcano-like, rounded cones of white sand fed

---

*Liquefaction is a surprisingly complicated process to model in a rigorous way, but anyone who has ever wriggled bare feet on a beach is familiar with it: smush wet sand vigorously between your feet, and the formerly solid beach you have been walking on will spurt up between your toes as a liquid. The seismic threat of liquefaction is also well captured in a tabletop demonstration: pour dry sand into a glass baking dish; place a few building blocks, or better yet toy houses, on top; then slowly pour in water until about two-thirds of the thickness of the sand is saturated. The houses or blocks will rest on this wet sand without problem. Then, accompanied perhaps with dramatic sound effects, gently shake the table or the pan: water will work its way to the surface to flood your toy homes, which will lean over and start sinking in a way that bodes ill for people living in real, life-size liquefaction zones.

by a central channel or "sand dike" of white sand rising up through red clay or black dirt.[†5]

Because of the seismic commotion, a portion of western Kentucky lurched upward, raising itself up into an enormous dome. This uplift is still visible today. It does not seem very dramatic at first, and in fact most people who live in the area or travel through pass by this uplift without a second thought—much as we can travel right on by the evidence of sand blows, fault scarps, and topographic change. Only in the context of the flat, broad expanse of the Mississippi floodplain does this seemingly gentle bulge in the terrain seem peculiar—and to the eyes of a seismologist, ominous indeed. The sudden uplift also turned the continent's central river back on itself: the mighty Mississippi ran briefly but dramatically backward before the current wore down the new-made falls.

Even far from the epicenters, the New Madrid earthquakes were powerful. People felt them in southern New Hampshire. They terrified settlers in the Ohio Valley, woke sleepers on the lower Eastern Seaboard, and crashed furniture in the nation's capital. In southwestern Kentucky, the tremors revealed a brutal murder by two slaveholders: nephews of Thomas Jefferson had in a drunken rage killed and dismembered a young slave and thrown his body parts into a fire. Only tremors that knocked over the chimney revealed their grisly crime. Further east in Kentucky, those same tremors badly frightened John James Audubon's horse. In South Carolina, wells went dry afterward. For North Americans of the 1820s—Illinois settlers and Cherokee farmers, slaveholders and those living in slavery, East Coast city folks and backwoods hunters such as Davy Crockett and his partner—these disruptions were an obvious and well-remembered part of local history, a set of disturbances whose environmental traces were all too apparent and all too frightening.[6]

[†] Despite decades of plowing, sand blows are still readily visible in some locations near New Madrid and even south into Arkansas. When I followed a team of researchers out on fieldwork in eastern Arkansas in the early fall of 2003, they showed me the white circles of sand identifiable in growing season by how scraggly the cotton grows—sand doesn't hold water in the hot Arkansas sun. From one step to another you pass from thriving plants to dry brown scrubby ones, stepping from soil onto the remaining evidence of an earthquake from two hundred years ago. Sand blows are characteristic of soil liquefaction in many parts of the world, but generally they are the size of dinner plates, maybe the occasional serving platter. In the New Madrid region, they are so large and plentiful—"world class" in the charming technical term employed in seismological journals—that experienced seismologists can be standing in the middle of one and not realize it: the edge might be over in an adjacent field.

Explanations

A key question about all this disruption, for people of Davy Crockett's time—and for many people today—is, what happened? And why did it happen? What explains such sudden tremors in the earth, much less the widespread reports of flashing lights, bad smells, ominous noises, and spouts of liquefied sand?

To a very large extent, *we still do not know.*

We know that there was a series of large earthquakes in 1811–12. Estimates of their magnitude have ranged from 7.0 to 8.7, though most current estimates are closer to 7 than 8.‡ Seismologists have mapped the intersecting faults around the New Madrid area and debate ongoing seismic movement in the region.[7]

But we still in some essential sense do not know why these ongoing quakes occur. We know—at least in the big picture—a fair amount about how earthquakes work. The framework of plate tectonics explains that our present-day continents are parts of what was once one enormous supercontinent, ripped apart as pieces of earth's crust move apart from each other on flows of immensely slow-moving beds of mantle. This is why, as most

---

‡ Ask earth scientists about the magnitude of these earthquakes, and they start to sigh and look pained. There are almost a dozen different ways to express the size of earthquakes (different scales work better for relatively small quakes than for relatively large ones, and there has been significant change in how earthquakes are measured since the first scales were developed), and almost no one outside the earth sciences community really understands the differences in these scales. Add to this the fact that the size of the New Madrid events is one of the hottest areas of debate about them, and you get scientists who have to give very long answers when they know that everyone really wants a simple, uncomplicated number. The best way to compare large quakes worldwide is with "moment magnitude," a measure that incorporates the force released at the source of an earthquake, the size of the ruptured fault, and the slip displacement (most of us can picture slip displacement as the "overhang" or side-to-side shift created by a quake). Overall, the current upper estimates of moment magnitude for the biggest three shocks are in the range of 7.8 to 8.1, while the lower estimates are around 6.7. Now, since these scales are logarithmic—an earthquake of 6.0 moment magnitude has *ten times more* amplitude on a seismogram than an earthquake of 5.0, which means it releases a whopping *thirty-two times* more energy—that is one whale of a difference. Saying to an earth scientist that the range is "upper 6s to low 8s" is an almost meaningless statement. But for most of the rest of us, such a range at least expresses the idea that these are indeed sizable, powerful quakes—easily in the league of the 6.9 Loma Prieta "World Series" quake that knocked out the Bay Bridge between San Francisco and Oakland in 1989 or the 6.7 Northridge quake northwest of Los Angeles in 1994, but not anywhere near the immense and devastating 9.0 Sumatra-Andaman Islands earthquake of December 2004.

schoolchildren with access to globes have noticed at one point or another, all the continents have a rather jigsaw-y look to them—a jigsaw puzzle long played with, the pieces worn and rounded into hazy approximation.[8]

Most of the earthquakes we are familiar with occur when one huge piece of earth's crust grinds up against another on its travels (think of Los Angeles's slow, jerky, and inexorable movement northward toward San Francisco on the San Andreas fault), or when one piece of crust dives down like a sounding whale, being subsumed in the process by another massive plate (such subduction can produce massive waves—as happened, tragically, in the December 2004 Sumatra-Andaman Islands earthquake and tsunami that killed over a quarter-million people and caused devastation throughout southern and eastern Asia). If you hold your two hands flat in front of you, with their edges touching, and move them up and down next to each other, or move one under the other, you have a rough but workable model of why most earthquakes happen. For people worried about earthquakes in California—or Japan, or Indonesia, or anywhere else on a plate boundary—this basic model of crashing, scraping, and subducting plates works very well.

But this model does not explain why or how earthquakes occur in the *middle* of tectonic plates. New Madrid, Missouri, is roughly 1,500 miles from the San Andreas fault. So-called intraplate quakes are not well explained by plate tectonics, nor by any model in seismology. Earthquakes can occur for many reasons besides crunching tectonic plates: small local tremors are frequently produced by deep drilling, for instance, and significant seismicity may be attributable to the "glacial rebound" of continents relaxing back upward after being relieved of their heavy load of glacial ice after the last ice age—in fact, glacial rebound may be one main reason for the New Madrid earthquakes. Yet none of the current forms of explanation alone would seem to account for the scale of events that occurred in the Mississippi Valley in 1811–12. Something big enough to make the Mississippi River run backward has to have an awful lot of motive force behind it. Just where did all that power come from?[9]

More than nineteenth-century history is at stake in the answer to that question. Recent research on New Madrid area seismicity has demonstrated that the 1811–12 earthquakes were not simply an anomaly, but part of a continuing series of earthquakes that goes back several thousand years. What is more, they seem to have occurred relatively frequently and regularly—possibly every two to six hundred years. Intraplate quakes can clearly pack a wallop. Two of the most lethal earthquakes in recorded history were intraplate quakes, both in China: the 1556 earthquake in Shaanxi Province, which buried at least 830,000 people, and the July 1976 magnitude 7.7 earthquake which leveled the northeast China city of Tangshan, with an estimated death toll as high as 650,000. Figuring out more about the mechanisms and likelihood of future similar shocks in the New Madrid area is a clear public imperative.[10]

A Lost History

For much of the last two hundred years, Americans and residents of the middle Mississippi Valley have forgotten the earthquake history that presented such obvious pragmatic challenges for Davy Crockett. Accounts like Crockett's are part of the reason why this story has been lost. Stories of earthquakes powerful enough to remake the topography of the midcontinent became just another part of the tradition of American frontier tall tales told by larger-than-life figures like Davy Crockett.[§] Down-home narratives like this hunting tale might strike most of us as quaint Americana, not as evidence for serious and sober analysis. Narratives like Crockett's cast doubt upon the very events they chronicle.[11]

In the years, decades, and two centuries since, human enterprise and the effects of nature itself worked to wipe away the traces of earth and memory linked to the New Madrid quakes. The 1811–12 quakes damaged the local landscape in dramatic and visible ways, rending the surface of the earth, spewing forth material from underground, and scarring the landscape. Yet such changes to soft-soiled floodplain were subject to rapid erosion. Only the deepest of earthquake cracks are now recognizable as earthquake evidence. Sand blows easily visible even in the early twentieth century as "barren spots upon which little will grow" became harder to identify by the early twenty-first century. Generations of plowing and grading blurred earthquake traces in the soft soil of the middle Mississippi Valley. In some of the open fields, when the cotton or soybeans are just starting to grow in early spring, sand blows become visible after spring rains: the surrounding

---

[§] Indeed, many people today may not be at all sure that Davy Crockett was even a real person—his very existence has entered the realm of folk story. Books like his 1834 autobiography framed him for Americans of the nineteenth century as an early frontier celebrity alongside Daniel Boone. The popular 1955 ABC Disneyland five-part TV miniseries *Davy Crockett* did pretty much the same for Americans of the mid-twentieth century. Lost in this mythology is that Crockett—a formal "David" to his friends—was also a politician of some savvy, who when not on a hunting expedition dressed, as one contemporary noted, "like a gentleman and not a backwoodsman," argued national policy, and opposed president Andrew Jackson's unjust Indian policies even though that opposition ended his legislative career.

FIG. I.1. A challenging moment in the Crockett household. This kind of image made Davy Crockett a well-known figure in nineteenth-century America. Pale East Coast shop clerks pored over woodcuts like this and dreamed of life in the backwoods. This kind of image also made Crockett's stories of earthquakes—and those of many of his contemporaries—easily to dismiss and ignore. Tales of huge cracks in the earth, shooting sand blows, and retrograde rivers seemed exactly as credible as this bear-versus-family battle. (*The Crockett Almanac*, vol. 1, no. 3 (1837), RB 95164, reproduced by permission of the Huntington Library, San Marino, CA)

soil darkens with water that drains swiftly through the white sand of the blows. Yet such ghostly white circles might strike most casual observers as an agricultural problem, not seismic testimony. Unlike fault lines that tear rocky outcrops, the New Madrid earthquakes left comparatively few unambiguous environmental traces.[12]

Over the twentieth century, earth scientists and casual observers alike have become fascinated by the more frequent and more easily explained kind of earthquakes, at the borders of the planet's massive tectonic plates. Contemporary towns and cities of the Mississippi and Ohio Valleys—Memphis, St. Louis, Louisville—do not figure in contemporary culture as earthquake zones in the way that Tokyo or San Francisco do. We walk along the deceptively smooth surfaces of the New Madrid seismic zone in ignorance and unconcern.

Every once in a while, though, someone stumbles into cracks left by these quakes. People living in the small, sleepy towns of eastern Arkansas

FIG. I.2. Sand blows in Mississippi County, Arkansas, two centuries after the New Madrid earthquakes. This photograph from New Madrid researcher Tish Tuttle shows the dramatic size and extent of earthquake "sand blows" (notice the nearby trees). Paradoxically, this image also shows the relative invisibility of such dramatic events: if most of us driving by even noticed white splotches in a floodplain field, we might assume they had something to do with irrigation or local soils, not huge long-ago earthquakes. (Courtesy of Martitia Tuttle, M. Tuttle and Associates)

will awaken in the middle of the night to hear a roaring, rushing noise, feel their beds shaking, and see pictures rattling against the walls. Paddlers or hunters out on the region's many streams and sloughs will notice white arms of sand branching upward through the dark color of a sharply cut stream bank, remnants of "sand dikes" that once spewed hot liquefied sand and organic matter high into the air. Historians in North Carolina, in Ohio, in Georgia, gently unfolding stiff sheets of correspondence from the early nineteenth century, will come across startled news of "the late awful visitation of Providence."[13]

Like Davy Crockett, *The Lost History of the New Madrid Earthquakes* crawls down into the cracks left by these dramatic New Madrid quakes. In dim crevices, we can find the traces of environmental and social upheaval. We can reconstruct broken shards of long-neglected history and bring into the light this story, once well known and now virtually forgotten.

This book answers many of the mysteries surrounding these dramatic early nineteenth-century earthquakes. First, how did these frightening events matter at the time? Careful investigation reveals that they were key events in the social, political, religious, and territorial upheavals of the moment. They were well known not only to those whose lives and livelihoods were transformed by them, but by people across the country, who gave voice to a uniquely American vernacular science as they debated the import of the quakes. Second, how could earthquakes that mattered so much, to so many, be almost completely forgotten in the decades and centuries following, especially when they carried the threat of a repeat performance? No one factor could erase these earthquakes: rather, a combination of changes—social, environmental, and scientific—combined to submerge knowledge of the New Madrid earthquakes for much of the modernizing twentieth century. It took the American Civil War and its aftermath, new racial and social tensions of the twentieth century, environmental transformations wrought by swamp drainage, timbering, and farming, and a radical shift in the way seismology was done to erase the signs and memories of such a dramatic set of events. Third, how did scientists come to rediscover these earthquakes after centuries of neglect? Even as seismology became a science of instruments and careful measurement, old narratives of long-past quakes turn out to have surprising salience for contemporary investigation. And finally, what should we make of the threat of future earthquakes in the New Madrid seismic zone, given this history? What might seem a set of questions about the past become challenges still unresolved about priorities for the future.[14]

To explore how people first made sense of the New Madrid earthquakes, we begin with an account by a literate Mississippi River traveler named William Leigh Pierce. His account serves both to introduce the exciting events of the earthquakes and to show us how people of his era investigated and came to understand puzzling natural phenomena: through sometimes chatty first-person narratives that included evaluation of trading routes, measurements of sand blows, and commentary on religion as equal ingredients. Making knowledge about the natural world was not separate from other kinds of reporting and conversation. Rather, early Americans folded the creation of knowledge into storytelling and the work of building commercial networks. Reading carefully such documents of the past, we learn how people of earlier times thought and put together their picture of the world—about earthquakes and about all manner of phenomena.

One reason the New Madrid earthquakes have been so effectively forgotten is that they do not seem to have mattered much—at least not in the histories written by those who gained social and economic power after the quakes. Nothing much was happening in that obscure part of the world before the quakes, goes the conventional story, and nothing much of importance happened there afterward. These may have been physically dramatic events, but they did not have much impact on human history.

Reading carefully the records of the New Madrid quakes shows how very wrong that conventional story has been. The area right around the earthquakes' epicenters was in fact a hotbed of cultural change and settlement in the tumultuous period of the beginning of the nineteenth century, especially settlement by Indian groups from east of the Mississippi. The earthquakes transformed the terrain by creating swampy "sunk lands" and effectively erased Indian settlement in what was once a thriving New Madrid hinterland.

The New Madrid earthquakes also mattered a great deal to Indian communities across eastern North America. For Cherokees in the midst of cultural upheaval, the earthquakes gave fire to a movement of apocalyptic prophecy. For Creeks resisting American takeover, the earthquakes became part of a war of resistance. For Indian people throughout eastern North America, the spiritual symbolism of the quakes served as a call to ally with the Shawnee leaders and prophets Tecumseh and Tenskwatawa in a movement of cultural and military resistance.

The New Madrid earthquakes similarly became an element of American spiritual movement, as they underscored the physically demonstrative spirituality of the Great Revival. People felt the quakes in their bodies, feeling odd, ill, and off-balance. Just so, they worked through the spiritual meaning of the earthquakes as physical manifestations of Christian faith. The New Madrid earthquakes jolted early Americans into belief—and also, in surprising ways, into scientific questioning. The shocks of earthquake felt like the shocks of the Holy Spirit, and also like the shocks of the fascinating new science of electricity. Musing about the jolts of earthquakes, Americans thought through profound questions of spirit and of causation.

Historians have dismissed early American science as derivative and inconsequential, struggling unsuccessfully to catch up to the breakthroughs and insights of European courts and laboratories. Looking at the New Madrid earthquakes—obscure events of an obscure cultural frontier—reveals how mistaken this view is. Across North America, people rushed to exchange information, questions, theories, and arguments about the causes of these earthquakes. Scientific inquiry, scientific conversation, and scientific thinking existed throughout early American society. The surprise is that this everyday science of early America was not just performed in labs or associated with those we might recognize as scientists: it was connected with almost all forms of society.

Listening to early Americans discuss and debate these earthquakes shows us what science really was in the early United States: a set of questions and debates in which many people, from a wide range of geographic and social places, regarded themselves as engaged participants. We have let these conversations lie invisible because they do not in any way resemble our usual histories of science. To recognize how American vernacular science was constructed through the New Madrid earthquakes is to recognize how fundamentally scientific thinking and scientific questions have shaped many aspects of early American civic and intellectual life.

But this story of how people came to grapple with the New Madrid earthquakes does not stop with William Leigh Pierce, Davy Crockett, and their early American contemporaries. Once well known, the New Madrid earthquakes were gradually and inexorably forgotten, the memory of them dormant for over a century. Huge dredging machines remade earthquake terrain into farmland; photos of African American sharecroppers protesting deep social injustice displaced older images of early nineteenth-century Native settlers; and the river around New Madrid became known as the site of an exciting battle rather than alarming seismic upheaval. In the extended process of forgetting, narratives of the New Madrid earthquakes, from folksy stories like Crockett's to early scientific reports and Native American oral histories, were all rendered virtually invisible—just as the Cherokees and other Indians settling the middle Mississippi Valley, who reacted so clearly and forcefully to the earthquakes, were themselves rendered invisible, their own history of resettlement in the region buried and denied.

To early twentieth-century boosters of the area trying to encourage new settlement, to hardscrabble tenant farmers just trying to get by, to scientists enthralled by the new techniques of instrumental seismology, stories of wide-eyed river men and awestruck Indians seemed quaint, irrelevant, even ridiculous. Until recently, historians also forgot these narratives, conventionally regarding the history of American seismology as beginning when experts invented modern devices to measure the earth. Because the New Madrid quakes occurred in the preinstrumental era of the earth sciences, before the invention of seismometers, there are no instrumental records of their movements. Unrecorded, unregistered, the New Madrid earthquakes submerged into a haze of inexact hearsay—the subject of a few novels, but little mentioned in our history textbooks. For the better part of a century and a half, the New Madrid earthquakes became the subject of derision and doubt.[15]

Then, because of late twentieth-century changes within seismology, accounts like Davy Crockett's moved once again into the center of discussion. Eyewitness accounts, so long doubted and derided, have become once again a crucial source of evidence. Present-day seismologists compile, quote, and argue about personal letters from 1812; they debate settlement patterns of the Ohio Valley; they pore over census data from the young United States. They act, in other words, like historians.

Modern researchers engage with other forms of early nineteenth-century knowledge as well. The local perspective of people working the land has begun, in small but consequential ways, to figure in scientific reassessments of the New Madrid earthquakes. Bodily knowledge, once a common form of information about earthquakes and then utterly rejected as unscientific, is once again a form of information about the severity and extent of seismic shocks. Such reassessments suggest further ways that once-valued and subsequently rejected information about the earthquakes, especially earthquake reports from Indian communities, information about the response of animals, or testimony about the lights and weather associated with tremors, might usefully inform some of the questions asked in our modern sciences of the earth.

As the New Madrid earthquakes have come back to light, the science and the history of the quakes have become the subject of contentious public policy debate. Different scientific theories have practical implications for infrastructure regulations. Is the New Madrid seismic zone "cooling down" in ways that present decreasing risk for the future? Or are the 1811–12 events fair warning of similar seismic tumult likely to recur beneath cities and farms of the American heartland? The financial burden of earthquake preparation and seismic building codes would be felt heavily in the affected areas, so scientific theories and risk assessments matter a great deal in practical terms. The facts and forecasts of the New Madrid earthquakes are presently debated in city council meetings, insurance industry websites, and the pages of local newspapers, as communities try to create appropriate responses to earthquake threats about which prominent scientists still struggle to come to consensus.

How do we know what we know—or what we think we know—about the New Madrid earthquakes of 1811–12, and possibly of the future? The smelly barking dogs of a famous American backwoods bear hunter might seem a strange setting for the search for thinking about natural events in the United States. Yet in just such places, and through just such records, was knowledge about natural processes created in early American life. Later scientists would come back to reports like Crockett's to try to figure out what had actually happened. Accounts of the New Madrid earthquakes are not just a colorful curiosity, but a fundamental ingredient of modern scientific analysis of the puzzling midcontinent.

For North Americans of the early nineteenth century, it was clear that these disruptions—events only slowly even named as earthquakes—had created important material changes. The New Madrid earthquakes marked the land and the people on it. Digging into a past landscape, a landscape of physical terrain and imaginative questioning, brings to the surface evidence of earthquakes with possible implications for the American future.

1 ✳ A Great Commotion: The Experience of the New Madrid Earthquakes

EARTHQUAKE

We have postponed a number of articles prepared for this evening's paper, to make room for the following very interesting communication from an intelligent friend at New Orleans.

It is we presume the most particular and satisfactory account of the earthquakes on the Mississippi which has yet been published: And Mr. Pierce being an ear and eye witness to the scenes he describes, the authenticity of his narrative cannot be doubted.[1]

Hampshire Federalist, *Springfield, Massachusetts, February 11, [1812]*

✳

By February 1812, many people in North America were unsettlingly aware of a series of disruptions that had occurred somewhere along the edge of the growing American nation. From the upper Missouri River to Upper Canada, from Georgia to Connecticut, many people over the previous three months had felt rocking and shaking, heard strange sounds, witnessed the movement of objects and the alarm of animals and birds, or felt disturbing symptoms in their own bodies. These puzzling and sometimes frightening events were big news in 1812.

In widely circulated journals and idiosyncratic local papers, from substantial communities and tiny settlements, people shared their impressions and sought others' knowledge about these natural disturbances. Some reports were just a few snippets of description—questions, really: had anyone else felt the shakes? heard all the noise?—while other accounts went on for long passages comparing many neighbors' experiences or offering parallels from historical tremors. Surveying the abundant printed evidence of the New Madrid earthquakes is like looking through the pieces of a massive

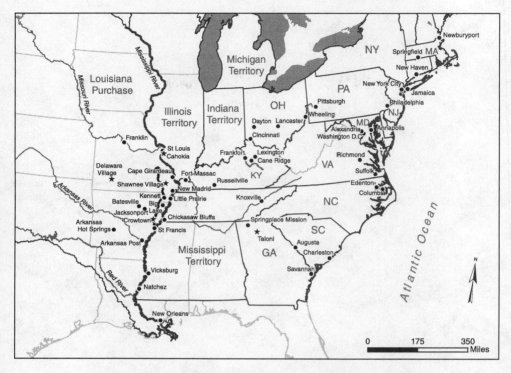

FIG. 1.1. United States, 1811–12, at the time of the New Madrid earthquakes. (Geographer: Bill Keegan)

puzzle spilled from a box onto a table top. The brightly colored pieces are hard to sort for order and connection.

One particular New Madrid account, published by the *Hampshire Federalist* on 11 February 1812, brings the larger pattern into clear focus. Written by river traveler William Leigh Pierce, this "interesting communication" was among the longest of the immediate eyewitness reports; it appeared in scores of different newspapers, and it addressed events and fears that appear common to many other accounts. His narrative thus provides a guide for modern people interested in the quakes just as it did for readers of many newspapers in 1812. To follow Pierce in his downriver journey is to read how the New Madrid earthquakes would enter American experience.

Pierce's account held great interest in 1812, and it remains fascinating in the early twenty-first century. In the 1960s and '70s, during the beginnings of seismological reassessment of the New Madrid earthquakes, researchers lamented their reliance upon "often exaggerated, inaccurate, and sometimes fanciful narratives prepared at the time."[2] More recently, researchers

A MISSISSIPPI FLAT-BOAT.

FIG. 1.2. "A Mississippi flat-boat." William Leigh Pierce was traveling the Mississippi in a boat much like this when the first shocks of the New Madrid earthquakes hit in December 1811. Earthquakes that disrupted American access to the commercial artery of the Mississippi were of national interest, as the many reprintings of Pierce's earthquake account make clear. (Samuel Adams Drake, *The Making of the Great West, 1512–1883* (1891), 164, image 2003-0517, courtesy of the State Historical Society of Missouri–Columbia)

have returned to reassess accounts like his, to piece together a modern scientific explanation for the events he chronicles. Pierce's narrative thus offers a window not simply into the experience of the earthquakes when they happened, but into how successive generations of researchers have come to comprehend them.

Public interest in New Madrid accounts like Pierce's reveals the tensions of 1812, when new commercial networks were spreading further across North America, explorers sought to survey western territories only recently claimed by the United States, and war over commerce, territory, and culture was brewing between beleaguered Native nations and a young American republic and between the United States and Britain. As we follow Pierce on his journey downriver, we can discern with vivid clarity the scenes of destruction he describes, the lens of natural inquiry through which he registers and describes the devastation, and his assessment of its social and political consequences in the national and environmental context of early nineteenth-century America. Reading Pierce, we can see how knowledge was made in early America.

William Leigh Pierce and His Account of the Earthquakes

At the end of 1811, William Leigh Pierce was part of a convoy of shallow-bottomed boats taking goods from the Ohio River down the Mississippi to the international trading port of New Orleans. Pierce was part of the United States' second national generation. His mother had been a Carolina planter and his father (also a writer and also named William Leigh Pierce) fought in the American Revolution and was a delegate to the Constitutional Convention that had formed the framework for the country.[3]

On the night the first of the "hard shocks" hit, Pierce's convoy happened to be lying anchored off the small community of Big Prairie, a little below the trading town of New Madrid, Missouri, in the Mississippi River below the confluence of the Ohio and Missouri. He mailed off a long narrative of his journey once he reached New Orleans. His account thus traced the earthquakes' effects along the middle and lower Mississippi. The *Hampshire Federalist*'s staff felt evident relish at being able to relay Pierce's report of what he saw and heard. It promised to be a good one.

The reading public at the time evidently agreed: Pierce's account of the earthquakes—which eventually included several follow-up reports as well—was excerpted, quoted, and reprinted throughout the American popular press in 1812. Such multiple and overlapping publishing was typical of the conversations about the New Madrid earthquakes: accounts frequently appeared in multiple forms, quoted and re-cited, their information hashed through repeatedly in various abridgements or compilations. Newspapers such as the *Hampshire Federalist* were where people got news and argued opinions in early America: almost every town had at least one. They were stridently partisan (just like the *Federalist*), and they were passed along and shared long after their particular issue date. More literary readers could also have encountered Pierce's report in several books and pamphlets on the recent quakes, as well as in his own narrative poem, *The Year*, published in 1814 to commemorate the war and earthquakes of 1812.* Collections of earthquake accounts were soon rushed into wider circulation: Robert Smith, a Philadelphia printer, published a pamphlet compiling reports on the quakes in 1812, as did an anonymous printer in Newburyport, Massachusetts.[4]

Accounts of the New Madrid earthquakes were generally written as letters, and their form bears witness to the close connections between letter writing and newspaper reporting in the early nineteenth century. Like reports of new settlement sites, relations with Native groups, or recent epidemics, many accounts of the quakes began as letters from one individual to another person or family, with the explicit understanding that they were for common consumption and might be forwarded to a local paper. Particularly vivid recountings of the New Madrid earthquakes were reprinted several times over, appearing often in newspapers located successively further East.[5][†]

Little can be gleaned from many authors, who followed nineteenth-century conventions of letter writing and publication in signing many reports only with an initial or pseudonym—"J.S." or "Dr. M." or "A Concerned Resident." Such conventions may have allowed for the participation of female writers at the time, but hide them now. Women also may have sent their own accounts through male family members, as was common practice.[6]

Of course, not all of the discussion in the early nineteenth century about New Madrid—or anything else—was written down. People discussed the New Madrid earthquakes in trading houses, around campfires, in Indian lodges, in sewing parlors, and in big-city coffeehouses. At two centuries' distance, what remains are the parts that got written down. Only a few snatches of conversation and discussion remain caught in diaries or letters

---

*In the nineteenth century, an event possessed real importance only if someone wrote a long poem about it. Henry Rowe Schoolcraft, later an authority on American Indian tribes, likewise incorporated the New Madrid earthquakes into his mock-serious epic poem *Transallegania*, in which the earthquakes occur because of a fistfight between personified metals beneath the earth. As the metals quarrel, "The rivers they boiled like a pot over coals, / And mortals fell prostrate and prayed for their souls: / Every rock on our borders cracked, quivered, and shrunk, / and Nackitosh tumbled, and New Madrid sunk."

†The Center for Earthquake Research and Information, which connects the USGS and the University of Memphis, explains in their online database of earthquake accounts that such excerpting and reprinting is likely responsible for the widespread myth that the New Madrid earthquakes rang church bells in Boston. They did not, but the quakes *did* ring bells in Charleston—and an article to that effect was clipped and reprinted in Boston papers in a way that is easy to misread.

like tufts of fur caught on barbed wire, small but telling clues to a larger whole.[7] Through this written record emerges a story of the gradual creation of public knowledge and public memory about these earthquakes—knowledge and memory that would all but disappear by the end of the nineteenth century, only to re-emerge through close consideration of accounts like Pierce's as the twentieth century gave way to the twenty-first.

> TO THE EDITOR OF THE NEW-YORK EVENING POST.
> BIG PRAIRIE, (on the Mississippi, 761 miles from New Orleans,) Dec. 25th, 1811.
> DEAR SIR,
> Desirous of offering the most correct information to society at large, and of contributing in some degree to the speculations of the philosopher, I am induced to give publicity to a few remarks concerning a Phenomenon of the most alarming nature. Through you, therefore, I take the liberty of addressing the world, and describing, as far as the inadequacy of my means at present will permit, the most prominent and interesting features of the events, which have recently occurred upon this portion of our Western Waters.

To begin his account, Pierce situated himself in terms everyone of his day would understand: 761 miles north of New Orleans, a vast distance by land or upstream, but a distance traversable in less than three weeks downstream. He had entered the Mississippi from the Ohio, a central river route of the eastern American country; to his north was St. Louis, a small but important and growing trading town that anchored the confluence of the Missouri and the Mississippi. He was on the "Western Waters," that is, on the flowing network of rivers that Americans envisioned as soon connecting communities and settlements of east and west.[8]

Pierce also situated himself intellectually, as part of a community of observers responsible for contributing usefully both to "society at large" and to "the speculations of the philosopher." Far from implying some esoteric or removed group, in Pierce's intellectual world this was a typically ornate way of talking about the widespread process of figuring out why the world worked as it did. This was the job not of remote, long-bearded elites in a high tower, but of men of education and substance casting keen eyes upon the world—among whom Pierce clearly included himself.

> Proceeding on a tour from Pittsburg to New Orleans, I entered the Mississippi where it receives the waters of the Ohio, on Friday, the 13th day of this month, and on the 15th, in the evening, landed on the left bank of this river about 116 miles from the mouth of the Ohio. The night was extremely dark and cloudy; not a star appeared in the Heavens and there was every indication of a severe

rain—for the three last days indeed, the sky had been continually overcast, and the weather unusually thick and hazy.

Unusual weather was no small side comment. In the natural philosophy of Pierce's day, processes of earth and sky were closely related, and many theorists posited that changes in weather might cause or at least might indicate earthquakes. Such clues would prove important to those attempting to tease out the web of connections that bound and made sense of such an overwhelming phenomenon of the earth.

A trip from Pittsburgh to New Orleans, similarly, was no small undertaking: it meant venturing across what American imagination viewed as the heart of the continent, to the confluence region where the Ohio, the Mississippi, and the vast Missouri Rivers all came together in an area of longstanding commercial and cultural importance. Such a trip encompassed the middle Mississippi Valley—the area roughly from the coming together of those three rivers to the mouth of the Arkansas, pulling together the modern American states of Missouri and Arkansas, western Tennessee, Kentucky, and Mississippi, and southern Illinois.[9]

The region where Pierce anchored during that first "hard shock" was at the northern end of what is known as the Mississippi Delta or the Mississippi embayment: the deep, gradually broadening and widening, roughly conical trough formed by the gradual deposition of organic sediment by the ancient grandparent of the Mississippi River. Over vast time, this long-flowing ancient river formed all of the Mississippi Delta, from just below the mouth of the Ohio down to the Gulf of Mexico. The rich sediment deposited by the ancient Mississippi and lack of underlying rock structure endow the Mississippi floodplains with their astonishing fertility: corn and cotton grow extremely well in soils hundreds of meters thick. The loosely deposited soil (in geological terms, the unconsolidated sediment) also endowed the region with the propensity to shake astoundingly and alarmingly when subjected to seismic waves.‡ The entire Mississippi basin is the large-scale equivalent of fill land, much like the portions of the San Francisco Marina neighborhood that turned to liquefied morass during the 1989 Loma Prieta quake. In geological terms, the structure and history of the Mississippi embayment would later become part of what was under debate as researchers struggled to understand the events that Pierce was about to relate. In political and cultural terms, the Mississippi embayment was an area of conflict and struggle in December 1811.[10]

---

‡ "Like a big bowl of jelly," in the alarming phrase that kept recurring in my conversations with area geologists.

The west bank of the Mississippi had just become legally American a few short years before, with the Louisiana Purchase of 1803. Beginning in the late 1600s, the French had worked with Quapaws along the Arkansas River and with Osages on the branches of the Missouri to negotiate trade, mining, and small-scale French settlements. After the multinational conflict known in Europe as the Seven Years' War and in North America as the French and Indian War, France ceded international control to Spain in 1763. Spanish control was even weaker than France's had been, and Spain's diplomacy less adept at the familial metaphors and elaborate exchanges of mutual obligation by which Native groups like the Quapaws and Osages were accustomed to maintain peace or wage war. In 1800, because of European financial and political pressure, Spain ceded Louisiana (much of the area between the Mississippi and the Rocky Mountains) to the French under secret agreement. Soon thereafter, American negotiators found themselves somewhat surprised to be negotiating with a weary Napoleon a purchase price for the mid-American continent between the Mississippi and the "Stony Mountains."[11]

The Louisiana Purchase is justly famous in American history, but it was somewhat puzzling to leaders at the time. President Thomas Jefferson fretted that his purchase was expedient but constitutionally unsanctioned, and no one had a very precise idea of what territory had in fact changed hands. But what was clear to most people—Indian, French, American, and various others—was that the middle Mississippi, the area that opened out onto the Purchase, was a crucial area for shaping the diplomacy, trade, and politics of future settlement.§[12]

§ The bramble of names and organizations given the middle and lower Mississippi Valley in the first decades of American ownership accurately reflects the geographic confusion and political turmoil surrounding them. In 1804, Congress split the Louisiana Purchase into two portions: the southern Territory of Orleans and the more northerly District of Louisiana, administered as part of the Territory of Indiana. In 1805 Congress moved the Territory of Orleans to the second stage of territorial government and turned the District of Louisiana into Louisiana Territory, with a capital at St. Louis (Louisiana Territory was therefore north of the present-day state of Louisiana). In 1806, territorial governor and international conniver James Wilkinson took time out from his plots with the Spanish to subdivide the Territory of Louisiana to produce the District of Arkansas. When the earthquakes hit, New Madrid was thus in the New Madrid District of the Louisiana Territory, just north of the District of Arkansas. Shortly after the New Madrid earthquakes, in 1812, the Territory of Louisiana became the Territory of Missouri and the former Territory of Orleans was admitted to the Union as the state of Louisiana. Debates over slavery would complicate the admission of the state of Missouri, until the infamous "Missouri Compromise" allowed Missouri to join the Union, but as a slave state—provided that any subsequent slave states were south of its southern border, and that

An important concern for Americans, both in the Louisiana Purchase and in later conflicts with European powers that resulted in the War of 1812, was keeping Mississippi commerce flowing. One of the first acts of American possession of Louisiana was to lift trade limits imposed by the Spanish. By the time William Leigh Pierce's convoy was descending the river, many people on both sides of the Atlantic looked with anticipation or anxiety to a British assertion of control over the Mississippi Valley and its commerce, perhaps beginning with a military strike on the vulnerable port of New Orleans.[13]

> It would not be improper to observe, that these waters are descended in a variety of small craft, but most generally in flat bottomed boats, built to serve a temporary purpose, and intended to float with the current, being supplied with oars, not so much to accelerate progress as to assist in navigating the boats, and avoiding the numerous bars, trees and timber, which greatly impede the navigation of this river. In one of these boats I had embarked—and the more effectually to guard against the Savages, who are said to be at present much exasperated against the whites, several boats had proceeded in company.

Merchants like Pierce might be taking a boatload of commodities—cotton, sorghum, whiskey—down to the port at New Orleans, or they might descend the Mississippi carrying the trade goods Indian and American settlers wanted, such as cloth, buttons, tools such as needles or awls, salt, jewelry, tobacco, specialized devices for a variety of trades, weapons and ammunition, books and newspapers, and news. They bartered sometimes for money, but often for trade goods: animal fat, crops, lead from the open-pit mines of Missouri. Farmers and traders plying the frontier often traveled in rough craft constructed for one journey: on arrival in New Orleans, the boats would be torn up and sold for the lumber while the merchant might simply walk home. In his introduction, Pierce makes clear that his trip was like many such trading voyages, leaving laden with goods from the eastern entrepôt of Pittsburgh.[14]

Pierce's trip was also typical in his fear of interactions with the Indian groups along the river. War seemed likely with Great Britain, as well as with confederacies of Native peoples, especially the Creeks in the Southeast and

---

they would enter the United States only if paired with a free state in order to preserve balance in the US Senate. Missouri was paired with Maine to join the Union in 1821. This tenuous "balance" of unfreedom and freedom held, more or less, until bloody debate over the entrance of Kansas as a free or slave state helped precipitate the Civil War of 1861–65. Arkansas grew more slowly—in part because of the difficulty of traversing the earthquake-riven sunk lands—only gaining statehood in 1836.

the Shawnee-led coalition headed by Tecumseh and Tenskwatawa. In the spring of 1811, there were several killings and kidnappings along the Mississippi River and in Illinois Territory, and fear of many more to come. Rumors flew up and down the Ohio and the Mississippi. The same newspaper editions that brought news of earthquakes in the Mississippi Valley to American cities and towns also reported fears of "formidable combinations of the savages." Early in 1812, confused (and erroneous) reports from St. Louis appeared in national newspapers of attacks on the settlement of St. Francis by four or five hundred Osages. Spies reported back that Indians had stolen horses and killed stock. American settlers along the upper Mississippi abandoned their farms for fear of attack.[15]

> *Precisely at two o'clock on Monday morning the 16th instant, we were all alarmed by the violent and convulsive agitation of the boats accompanied by a noise similar to that which would have been produced by running over a sand bar—every man was immediately roused and rushed upon deck.—We were first of opinion that the Indians, studious of some mischief, had loosed our cables, and thus situated we were foundering. Upon examination however, we discovered that we were yet safely and securely moored. The idea of an Earthquake then suggested itself to my mind, and this idea was confirmed by a second shock, and two others in immediate succession. These continued for the space of eight minutes. So complete and general had been the convulsion, that tremulous motion was communicated to the very leaves on the surface of the earth. A few yards from the spot, where we lay, the body of a large oak was snapped in two, and the falling part precipitated to the margin of the river; the trees in the forest shook like rushes: the alarming clattering of their branches, may be compared to the effect which would be produced by a severe wind passing through a large cane brake.*

Many people experienced an ordinary world in commotion. Since the big December quakes hit late at night, those who were awake were often engaged in quiet activities like reading (or at least many of the people who later reported that they felt the quakes while awake in their beds late at night *said* that what they were doing was quietly reading). Many people—especially those further off—noticed paintings, mirrors, or curtains swaying and reached to close windows against a stray breeze. Bells and clocks tolled unaccountably. Some people felt lurching, jerky sensations, while others described sounds that were not so much heard as felt. A householder in the Creek nation awoke to horrible squawking and thought someone was stealing all her turkeys and chickens in the middle of the night. A man named Major Burlison, who lived along the "great road" leading into the Mississippi River port of New Madrid, sprang from bed at the first shocks, "between sleep and awake." "He never thought of an earthquake," he recounted, "but concluded that either the house was haunted, or the end of the world was at hand." As they felt the tumult of the hard shocks near the epicenters, many people struggled to figure out what was happening—a struggle that would continue as people tried to piece together why earthquakes happen at all.**[16]

Ordinary objects served to register the strangeness of the earth. Water in a small brook, reported one New York man, was shaken by the tremors. Tremors set cradles rocking, threw books down, and upset the livestock. In Annapolis, during the January shock, people who were "skaiting" on the inlets of the Chesapeake Bay fled in terror for the shore, while the steeple of the historic Maryland statehouse was seen to vibrate six or eight feet at the top of its 181-foot height. In Dayton, a surveyor attempting to lay a straight road was unable to get his magnetic needle to settle for several days after the December shakes.[17]

Confronted with such strangeness, people reached for familiar metaphors. A traveler through the Creek nation felt the earthquakes' motion as "similar to a cradle rocking from one side to the other." John J. Audubon noted in Kentucky that "the earth waved like a field of corn before the breeze." Others felt themselves buffeted by waves on a ship at sea.[18]

For people who felt only mild effects like the tinkling and swaying of apothecary bottles in a shop in New Haven, the recognition that they were caused by a far-off earthquake might come only as they read continuing and more detailed accounts from others in places nearer the epicenters. Politician and planter Winthrop Sargent observed that "in general the effects were more like those of a tornado or whirlwind, than of an earthquake." Discussion of the New Madrid earthquakes was in large measure a project of understanding them *as* earthquakes. For some like Pierce, the movement of the earth was unmistakable—he very quickly identified the event as an

---

** Even in earthquake zones, people who feel quakes often do not have any clear sense at first of what is happening to them. Riding a clunky bike across the Stanford campus in the late 1980s, I felt the bike shift suddenly underneath me and fell sprawling to the ground. I was embarrassed at my clumsiness—especially on that fit, athletic campus. Only when I rode home to my residential co-op to find that day's cooks in tears in the kitchen, surrounded by shards of broken plates, did I realize that we had just experienced a moderate earthquake. Another survivor of a far more serious California quake reported thinking that a bulldozer had driven into his house. People who feel quakes often think that a car or truck has hit their buildings, or that large AC units have just turned on, or that a "herd of possums" is stampeding across the roof. It is apparently easier to posit wild or far-fetched calamities than to accept the reality that our stable earth is often anything but.

earthquake. What marked all of the New Madrid accounts, however, was a sense of events as tumult in every plane, not restricted to the movement of earth—what one writer from Lancaster Ohio called a "conflict of the elements."[19]

What was not surprising was that William Leigh Pierce and his crew would think first that they were being attacked. International and cross-cultural tensions made that all too likely. In the winter of 1811, the United States and Britain growled and raised hackles at each other, not least over control of the waterways on which Pierce traveled. At the same time, a Native confederacy based in the upper Ohio Valley, led by brothers Tecumseh and Tenskwatawa, threatened to ally with the British to retake control of key lands of American commerce and settlement. Similar trouble seemed brewing in the southeastern territory of the Creeks. Few Americans had any clear sense of what might be happening in the northern and western territories along the upper Missouri, or indeed in the southwestern portions of the Louisiana Purchase, but recent reconnaissance like that of President Jefferson's Corps of Discovery under William Clark and Meriwether Lewis had made clear that French traders and Native groups skilled with horses, guns, and trading languages had well-established networks that would be challenged only with difficulty by Americans pushing west with trade or territorial claims. Along the Mississippi, recent harassment of river traffic heightened tensions and suspicions for people with valuable cargo under their care. American control of the Mississippi was shaky, contested not only by the sovereignty of Indian groups but by brigands and river pirates. Pierce had good reason to fear attack by Indians or marauders.[20]

Exposed to a most unpleasant alternative we were compelled to remain where we were for the night, or subject ourselves to imminent hazard in navigating through the innumerable obstructions in the river, considering the danger of running two fold, we concluded to remain. At the dawn of day I went on shore to examine the effects of the shocks; the earth about 20 feet from the water's edge was deeply cracked, but no visible injury of moment had been sustained, fearing, however, to remain longer where we were, it was thought most advisable to leave our landing as expeditiously as possible: this was immediately done—at a few rods distance from the shore, we experienced a last shock more severe than either of the preceding. I had expected this from the lowering appearances of the weather; it was indeed most providential that we had started, for such was the strength of this last shock that the bank to which we were (but a few moments since) attached, was rent and fell into the river, whilst the trees rushed from the forests, precipitating themselves into the water with force sufficient to have dashed us into a thousand atoms.

Pierce did not know this when writing his letter, but he was describing only the first of what would ultimately be three series of shocks—a pattern of serial tremors very different from the foreshock-main shock-aftershock sequence more typical of earthquakes on plate boundaries. The dawn aftershock he described is often discussed as a major tremor on par with the "smart shock" of the night before. Subsequent main shocks followed on 23 January and 7 February. Naming such upheaval proved challenging at the time and proves challenging still. A local trapper describing events to visiting scientific tourist Charles Lyell in 1846 referred to "the great shake." Others described "terrible heavings of the earth" or simply, the *Shakes.* Even the unclarity of naming reflects uneasiness about how to structure understanding of these events.[21]

> It was now light, and we had an opportunity of beholding in full extent all the horrors of our situation. During the first four shocks tremendous and uninterrupted explosions, resembling a discharge of artillery, was heard from the opposite shore, at that time I imputed them to the falling of the river banks.

Pierce was not alone in registering the huge noises of weapons and warfare. Others similarly felt earthquake shocks followed by "distinct reports like cannon." Many other witnesses emphasized how colossally noisy the earthquakes were, like carriages running over cobblestoned streets, like a "violent tornado," "like the burning out of a chimney," or like "the noise made by emptying loads of small stones." In the sensory world of the early nineteenth century, sounds were meaningful, a significant part of how people apprehended an environment. For many who heard as well as felt the New Madrid quakes, the bewildering noise of the events symbolized disruption and confusion. In Lancaster, Ohio, "The ringing of the bells occasioned by the violent agitation of the earth" was matched by "the howling of dogs; the bellowing of cattle, and the running to and fro of horses."[22]

The ways in which people described the disturbing noises of the New Madrid tremors hint at larger frameworks for how they understood the world. The noise of earthquakes was associated with the sounds of air and fire: tremors sounded like the blowing of bellows, like "steam escaping from a boiler," like "a blaze of fire acted upon by wind." Such reports indicate the continuing currency of essentially Aristotelian geological theories about the earth as a hollow sphere or as a partial solid containing underground passages: much theorizing about earthquakes argued that they were the surface manifestation of explosions in vast underground cavities.[23]

Sounds could also indicate that earthquakes of the Mississippi Valley were linked with other phenomena spanning the globe. An 1869 geography of the Mississippi Valley echoed early nineteenth-century concerns by situ-ating the events of New Madrid in a global context. "The telluric[††] activ-ity of which these events were a part," argued the author, "extended over half a hemisphere": the effects of this larger convulsion—felt in the Mis-sissippi Valley as dramatic quakes—included the elevation of an island in the Azores, earthquakes in Caracas in the spring of 1812, a volcano in St. Vincent, and "the fearful subterranean noises which were heard on the Lla-nos of Calabazo, and at the mouth of the Rio Apure [both in Venezuela], and even far out at sea." If the groaning and roaring sounds beneath foreign seas were like those of the Mississippi Valley, then those utterly distant locations could in fact be related through as-yet-unknown geologic cause. Writers on the nineteenth-century borderlands of the United States regarded them-selves as connected to world events and distant places not only through networks of communication and trade, but through the resonance of the earth itself.[24]

> *This fifth shock explained the real cause. Wherever the veins of the earthquake ran; there was a volcanic discharge of combustible matter to great heights, an incessant rumbling was heard below, and the bed of the river was excessively agitated, whilst the water assumed a turbid and boiling appearance—near our boat a spout of confined air breaking its way through the waters, burst forth, and with a loud report discharged mud, sticks, &c. from the river's bed at least 30 feet above the surface. These spoutings were frequent, and in many places appeared to rise to the very heavens. Large trees which had lain for ages at the bottom of the river were shot up in thousands of instances, some with their roots uppermost and their tops planted; others were hurled into the air; many again were only loosened, and floated upon the surface. Never was a scene more re-plete with terrific threatenings of death; with the most lively sense of this awful crisis we contemplated in more astonishment a scene which completely beggars description, and of which the most glowing imagination is inadequate to form a picture. Here the earth, river, &c. torn with furious convulsions, opened in huge trenches, whose deep jaws were instantaneously closed; there through a thousand vents sulphureous streams gushed from its very bowels leaving vast and almost unfathomable caverns.—Every where Nature itself seemed tottering on the verge of dissolution. Encompassed with the most alarming dangers, the manly presence of mind and heroic fortitude of the men were all that saved them. It was a struggle for existence, and the need to be purchased was our lives.*

[††] *Telluric*, for those not enmeshed in the often curious formalism of nineteenth-century writing, indicates something that comes from or arises from the earth or soil. We might instead use *terrestrial* or *earthly*.

Pierce's descriptions indicate some of the central geologic concerns and frameworks of his era. To Pierce, an earthquake could extend or propagate beneath the earth in "veins" like those of precious metal—or perhaps even more like the vital pulsing blood vessels of a living body. The shaking of the quakes would then produce what he called in highly somatic vocabulary "volcanic discharge" from the river itself and, as he explained elsewhere, in sand blows on land. Absent the geological markers, this could be the description of a person wracked with illness. Reports about the tumultuous world shared the language, convention, and expectations of narratives of the sick or vulnerable human form.

Yet as grounded as Pierce was in the frameworks of his own era, the physical phenomena he described would also become central to later recovery of New Madrid earthquake history. What Pierce experienced as "spoutings" from the bottom of the Mississippi River also occurred on land, as dramatic expulsions of liquefied sand known as sand blows. Sand blows or "sand slews" were and are a main form of geological evidence of the New Madrid earthquakes.‡‡ The most extensive and terrifying sand blows occurred around the epicenters in present-day southeast Missouri and eastern Arkansas. Such liquefaction features, as they are known, are a signal feature of the New Madrid region. Sand blows and other liquefaction features generally occur only in relatively big tremors, magnitude 5 and above. In many of the world's earthquake zones they are the size of dinner plates, perhaps bicycle wheels; near the epicenters of the New Madrid shocks sand blows are commonly a meter to a meter and half thick and ten to thirty meters in diameter. In the southern jut of the Missouri Bootheel, a remaining sand blow is so large that local geology students call it "the Beach." In parts of the over ten thousand square kilometers of the New Madrid seismic zone, they cover up to a quarter of the soil surface. Recent seismological investigation has discovered them near the mouth of the Arkansas River, over 250 kilometers south of the epicenters. The quakes also caused similar phenomena at quite a distance. Moravian missionaries reported that in the Cherokee town of Taloni, in what is

---

‡‡ Sand blows are the surface manifestation of liquefaction, and in what we have learned to call the New Madrid seismic zone they are visible as round white splotches because of the light color of the subsurface sand. Sand dikes are the sediment-filled cracks in the surface layer through which this liquefied soil spurts to the surface. These upward-reaching, white veins can sometimes be revealed by erosion, as in a stream bank, or they can be intentionally revealed by the process of geological trenching, in which researchers (as may perhaps be obvious) dig a trench and read the earth's history on its walls. Sand sills are a related phenomenon: lenses of subterranean deposit formed essentially the same way as sand blows, but which do not reach the surface.

FIG. 1.3. Side view of an earthquake fissure filled with intruded sand in Mississippi County, Missouri. A hundred years after the quakes, USGS geologist Myron Fuller documented this fissure and similar earthquake evidence—liquefaction features, they would be called a century later. Fuller's report provided evidence for significant effects of the New Madrid earthquakes, as well as the pre-1811 earthquake history of the region. Yet his work lay neglected for decades, as other forces, environmental, social, and technological, pushed the New Madrid earthquakes out of popular attention and professional research. (Myron Fuller, *The New Madrid Earthquake*, 1912, courtesy of the US Geological Survey Photographic Library)

now northeast Georgia, the New Madrid earthquakes were reported to have caused sizable sinkholes, some of which filled with water.[25]

As William Pierce reported during the period of upheaval, along with sand blows were the earthquake cracks ranging from smallish fissures to large chasms. Stephen Austin, later a founder of the American state of Texas but at the time, like Pierce, a young man traveling the Mississippi, stopped by the New Madrid region shortly after the quakes to find the earth's surface "very much crack[d] . . . and perforated." A river traveler near New Madrid surveyed widespread cracking open of the earth, reporting that he could barely go onshore, "for the ground is cracked and torn to pieces in such a way as makes it truly alarming." Equally alarming was the wide area over which such damage was reported: crevices and cracks were reported as far as Fort Massac on the Ohio River.[26]

Most of these cracks have since been softened by erosion, destroyed by plows, or dug out by drainage work. Into the mid-nineteenth century, though, they were still part of the landscape of the epicenters. In 1853 a Missouri state surveyor working in Pemiscot County in southeast Missouri downriver of New Madrid noted "a number of ravin[e]s, called 'Earthquake Cracks,' bearing N. and South." Geologist Myron Leslie Fuller reported in 1912 that certain fissures were still deep enough to hide a man riding on horseback.[27]

Earthquake cracks represented some of the most threatening aspects of the New Madrid earthquakes for the people who experienced them. One old trapper reported that his grandfather had received a boatload of castings, or molds, from Philadelphia, which he stored in the cellar of his house near the epicentral region. "During one of the shocks," he recounted in the 1860s, "the ground opened immediately under the house, and [the castings] were swallowed up, and no trace of them was afterwards obtained." Other stories reported a cow near New Madrid completely "engulphed."[28]

During the day there was with very little intermission, a continued series of shocks, attended with innumerable explosions like the rolling of thunder; the bed of the river was incessantly disturbed, and the water boiled severely in every part. I consider ourselves as having been in the greatest danger from the numerous instances of boiling directly under our boat; fortunately for us, however, they were not attended with eruptions. One of the spouts which we had seen rising under the boat would inevitable have sunk it, and probably have blown it into a thousand fragments; our ears were constantly assaulted with the crashing of timber, the banks were instantaneously crushed down, and fell with all their growth into the water. It was no less astonishing than alarming to behold the oldest trees of the forest, whose firm roots had withstood a thousand storms and weathered the sternest tempests, quivering and shaking with the violence of the

FIG. 1.4. Mississippi River, mid-quake. During the three "hard shocks" of the 1811–12 New Madrid earthquakes, the Mississippi River became a maelstrom. Huge waves sloshed small boats, undermined banks collapsed, long-buried trees floated to the water's surface, whole islands disappeared. For weeks afterward, travelers reported seeing the debris of wrecked river craft float downstream. Yet the twentieth century, the death toll of the New Madrid earthquakes was minimized or forgotten. (Richard Miller Devens, *Our First Century* (1877), p. 220, image 023992, courtesy of the State Historical Society of Missouri-Columbia)

> *shocks, whilst their heads were whipped together with a quick and rapid motion; many were torn from their native soil, and hurled with tremendous force into the river, one of these whose huge trunk (at least 3 feet in diameter) had been much shattered, was thrown better than a hundred yards from the bank, where it is planted in the end of the river, there to stand a terror to future navigators.*

The earthquakes intensified the deceptive threats of an economically vital and politically contested river. In particular, the dislodging of vast stands of trees greatly increased the numbers of "planters" and "sawyers" which endangered river travelers. These were large trees which obstructed the river, "planted" either as stationary obstructions or with their roots anchored by their weight on the bottom and their top branches "sawing" up and down in the current where they could puncture and chew unwary boats. These trees could dash a boat to pieces or sink it: as one river traveler reported, boats carrying lead from the mining districts of southeast

Missouri down the river to New Orleans were particularly vulnerable—their valuable cargo was so heavy that the boats sank almost instantly when broached by the errant limb of a planter or sawyer.[29]

Earthquakes caused gigantic sloshing in the river, which ate away and undermined riverbank. This essential process was familiar from seasonal flooding. On occasion, large stretches of bank would simultaneously collapse into the Mississippi in "booming floods" that were at once hugely noisy and hugely dangerous: those booming collapses themselves created waves in the river that could capsize and sink small craft. Earthquakes thus had some effects similar to those of usual seasonal rhythms of flood and storm, but more intense and clustered. This similarity to usual rhythms would also contribute to the ambiguity over forms of environmental evidence: whether a sawyer was created by a spring rise or by seismic tremor would not be apparent from the sawyer itself.[30]

> Several small islands have been already annihilated, and from appearances many others must suffer the same fate. To one of these, I ventured in a skiff, but it was impossible to examine it, for the ground sunk from my tread, and the least force applied to any part of it seemed to shake the whole.

The seismic upheaval of that winter changed the map of the river, causing several Mississippi River islands to disappear. The river's islands were no more stable than outer banks of an ocean harbor: though some were large and well timbered, they were constantly being eroded or added to by the changing currents of seasonal floods.

During the night of upheaval described by Pierce, some of the smaller islands in the middle Mississippi slid back into the river or were belched entirely into the air by sand boils. As one letter writer described from New Madrid in late February: "There is a certainty that a bar composed of stone coal, burnt substance, &c. has been thrown directly under the bed of the river. Island No 8, from the junction of the Ohio to the Mississippi is entirely sunk."§§ Others were sunk so as to be visible but no longer usable as landing places. By the late 1810s, the "sunken islands" were still visible underwater as eerie reminders of seismic upheaval. Island disappearance was thus the subject of great interest at the time, discussed in wondering tones in letters and accounts of the great quakes.[31]

As with the implanting of new planters and sawyers, such changes were not different in kind from what might happen in a spring "fresh," or flood—

---

§§ In the early nineteenth century, islands in the Mississippi were numbered sequentially from the mouth of the Ohio in an attempt to standardize knowledge of these changeable features. Island Number 8 was thus the eighth island south of that confluence.

but they were significantly different in extent and swiftness. Where generally in the course of a seasonal flood a few trees or an area of undermined bank might fall into the river, now whole swaths of riverbank had. In any given spring the Mississippi might extend a loop by eroding banks—such erosion had already pushed the town of New Madrid more than a mile from its original site by 1811—but the disruption of the earthquake shocks created closely spaced and enormous disturbances that did in a short time what otherwise might have happened over the course of years or decades. These sudden changes were hard to assess and hard to assimilate—rivermen accustomed to continually "learning" the river, adding to their knowledge after every season of flood or storm, were confronted with a perilously complete transformation of the river they thought they knew.***[32]

> Anxious to obtain landing, and dreading the high banks, we made for an Island which evidenced sensible marks of the earthquake: here we fastened to some willows, at the extremity of a sunken piece of land, and continued two days, hoping that this scene of horrors was now over—still however the shocks continued, though not with the like frequency as before.
>
> On Wednesday in the afternoon I visited every part of the Island where we lay, it was extensive and partially covered with willow. The Earthquake had rent the ground in large and numerous gaps; vast quantities of burnt wood in every stage of alteration, from its primitive nature to stove coal had been spread over the ground to very considerable distances; frightful and hideous caverns yawned on every side, and earth's bowels appeared to have felt the tremendous force of the shocks which had thus riven the surface. I was gratified with seeing several places where those spouts which had so much attracted our wonder and admiration had arisen, they were generally on the beach and have left large circular holes in the sand formed much like a funnel. For a great distance around the orifice vast quantities of coal have been scattered, many pieces weighing from 15 to 20 pounds were discharged 160 measured paces.—These holes were of various dimensions, one of them I observed most particularly, it was 16 feet in perpendicular depth and 63 feet in circumference at the mouth.

*** In his memoir *Life on the Mississippi*, Mark Twain describes the knowledge of the Mississippi's course hard-gained by steamboat pilots of midcentury through careful observation and memorization of the river's many twists and turns, as well as through communication among the network of river pilots who kept each other abreast of new developments, passing along updates about particular sawyers or specific changes to channel with the river's rise or fall. Though steamboats were much larger and thus more challenging to pilot than the smaller flatboats of William Leigh Pierce's era, the process of river learning was essentially similar: to mistake a cutoff for a main channel might cause a costly delay—or lead to loss of life and valuable cargo.

In a juxtaposition jarring to modern sensibilities, many New Madrid earthquake narratives similarly combine romantic description and precise accounting, describing a "scene of horrors" alongside careful measurement of sand blow circumference and the weight of matter thrown from sand blows. Wonder and analysis together characterize early nineteenth-century approaches to understanding these quakes. These forms of understanding would only be cleaved in the latter decades of the century, as leading researchers in Europe and Japan articulated and developed the science of seismology in large part through global tools of sensing and measurement. Only then would paced-off measurements of sand blow holes and expulsions become regarded as obsolete and unscientific. In William Leigh Pierce's era, his careful pacings and his horrified astonishment were of a piece, alike valuable elements of comprehending vastly challenging tumult.[33]

As Pierce's astonished account makes clear, a particularly striking aspect of the seismic disruption for many observers was the ground's violent expulsion of a wide and disturbing variety of substances. The sand blows and associated "earthquake cracks" that rent the earth spewed out warm water, awful smells, and organic detritus; observers reported trees being swallowed and thrown back out into the air. Naturalist and traveler Louis Bringier recorded explosions of ten to fifteen feet and "black showers" of "carbonized wood." Many observers called this "stove coal" or "channel coal," a form of coarse coal known as lignite. Traveler John Shaw described this as a "hard, jet black substance, which appeared very smooth, as though worn by friction," unlike anthracite or bituminous coal. Unlike erodible sand and soil, the harder chunks of lignite provided proof for many of what they had actually seen. Geologist Edward Shaler reported almost a century after the quakes that lignite could be found in castings of earthworms throughout what locals called the "sand slew" country. Some fissures were reported to still emit gases or air into the late nineteenth century.[34][†††]

[†††] Chunks of ejected lignite are still found along the New Madrid seismic zone, remarkable but usually unnoticed testimony to huge environmental upheaval of two centuries past. In the course of this research I found myself staring in honest-to-gosh open-mouthed fascination at a piece of lignite at the Arkansas Archeological Survey research station in Blytheville, flabbergasted to imagine how this very ordinary-looking, rough-surfaced piece of black coalish rock came to be spewed out of the earth in the midst of huge seismic tremors, and then be collected and come to rest so innocuously on the table in front of me.

On Thursday morning the 19th, we loosed our cables with hearts filled with fervent gratitude to Providence, whose protection had supported us through the perils to which we had been exposed.

By the early nineteenth century, earthquakes were both spiritually resonant and scientifically intriguing, a legitimate object of reflection in each
of those related realms. Pierce's quiet acknowledgment of the protection of
God speaks to the changing theological understanding of natural disaster
in his era. Certainly many looked to discern the voice of their creator in the
disruption of the New Madrid temblors. When God speaks, sang one hymn
of the period, "The mountains tremble at the noise, / The valleys roar, the
deserts quake." The events caused many to acknowledge what one politician
termed "weakness and dependence on the everlasting God." Such spiritual
meanings are historically rooted in the experience of earlier earthquakes
in Europe and North America, to which many responded with relatively
straightforward theological explanations: God causes earthquakes to punish sin or warn against present wrongdoing. The Bible that people of Pierce's
era knew as an intimate and familiar guide gave many examples of earthquakes as God's punishment. In the Christian Old Testament, God sends
earthquakes to wreak judgment: like thunder, lightning, or whirlwind, they
are earthly expression of the terror and majesty of God's voice.[35]

Yet for many of their contemporaries, as for Pierce, the earthquakes held
spiritual resonance but not necessarily simple or straightforward judgment. Stephen Austin expressed a widely shared sense of emotional and
philosophical dismay. Surveying the devastation of New Madrid, Austin
wrote of "the wise Dispensations of Providence ... puting in motion the terrible engines of his Power and by some extraordinary convulsions throwing
a hitherto fertile country into dessolation and plunging such of the unfortunate wretches who survive the ruin, into Misery and dispair." Austin's
sorrow reflected a widespread and somewhat modulated view that earthquakes reveal the "terrible engines" of God's power, but not necessarily in a
form directed against any one particular community's actions.[36]

Such a spiritually thoughtful but slightly distanced tone marks the New
Madrid earthquakes as occurring in the aftermath of several powerful
eighteenth-century earthquakes. These had focused European debates over
naturalistic explanations and earthquake theology. On the first of November, 1755, an earthquake and its subsequent tsunami destroyed Lisbon, killing roughly 12,000 people. The catastrophe also shook up earthquake theorizing, causing widespread reflection among European thinkers on the cause
and moral significance of lethal earthquakes. Voltaire used that earthquake
as a primary satiric target for skewering a simplified optimism that all turns

out well in the end: an earthquake that buries people at prayer is hard to imagine as Godly providence. (That same month, the Cape Ann earthquake off the coast of New England likewise amplified British colonists' interests in the subject). Yet even as they debated the spiritual meaning of the earthquake, many who wrote about Lisbon also made careful observations about specific seismic effects—changes to hot fountains near Prague, for instance—adding to contemporary European scientific knowledge. Through the later eighteenth century seismic events continued to excite scientific interest as well as bring tragedy. Savants had eagerly discussed the minor tremors of the "year of earthquakes" in 1750 in Britain, and they investigated the horribly destructive quakes in Sicily and Calabria in 1783. Discussion of the New Madrid quakes continued this tradition of observation and discussion of the philosophical as well as physical meaning of earthquakes.[37]

Pierce continued to describe his reactions as he and his crews continued down the river. They traveled through "a scene of ruin and devastation," where trees and "whole brakes of cane," the native bamboo that thrived throughout the middle Mississippi Valley, tumbled into the river to form eerie and dangerous "subterranean forests."

> The obstructions in this river which have always been quite numerous are now so considerably increased as to demand the utmost prudence and caution from subsequent navigators, indeed I am very apprehensive that it will be almost impassable in flood water, for until such time it will be impossible to say where the currents will hereafter run, what portion (if any) of the present embarrassments [that is, obstructions] will be destroyed and what new sand bars, & c. may yet be caused by this portentious phenomenon. Many poor fellows are undoubtedly wrecked, or buried under the ruin of the banks. Of the loss of four boats I am certain.

Pierce was among the first of many to comment on how the quakes changed the channel—and in some places the course—of the Mississippi. He was one of the few commentators to so perceptively understand the earthquakes' toll on those who worked the continent's mighty river. Most reports on the quakes, in Pierce's day and for the following two centuries, have insisted that almost no one died or was hurt, since the middle Mississippi Valley was so little settled. Such assessments, however, take into account only the very recent American settlements or the several-generations-old French villages, not the Indian communities both new and of long standing. Such assessments also left unacknowledged the French and mixed-race boatmen who poled and oared the Mississippi. These Métis or Creole river people were often children of French fathers and Indian mothers who encouraged their sons to manly river work rather than womanly farming. Many of these mixed-culture men comprised the crew of Lewis and Clark's expedition; few of them were much in the minds of most American observers. After the New Madrid earthquakes, some travelers took note of the remains of flatboats and debris from many wrecks which drifted downstream for weeks after each of the main shocks, but many left unremarked the evident death upon the river. The total death toll in 1811–12 is undoubtedly small compared to many of the world's earthquakes, and it will almost certainly never be ascertained with precision—but records such as Pierce's make clear that people, perhaps many people, died as a result of these tremors.[38]

> *It is almost impossible to trace at present the exact course of this earthquake or where the greatest injuries have happened, from numerous inquiries however, which I have made of persons above and below us at the time of the first shock, I am induced to believe, that we were very nearly in the height of it, the ruin immediately in the vicinity of the river, is most extensive on the right side in descending: For the first two days the veins appeared to run in due course from W. to E. afterwards they became more variable, and generally took a N. W. direction.*

As readers, writers, gossipers, and commentators compared accounts and news of this "great commotion," it soon became clear that the earthquakes were most intense around the bend of Mississippi near New Madrid and Little Prairie. Pierce was indeed "in the height of it." The most affected area was around what is now termed the epicenters of the main shocks, the multistate area around the town of New Madrid.

Yet the geology of midcontinent—the tendency of soft alluvial soils near rivers to shake and the ability of old rocks of eastern North America to transmit shaking—meant that the quakes were felt at remarkably far distances. Other areas of intense disturbance extended up through the bottomland of the middle and even upper Ohio River, where people in Kentucky and Ohio experienced the earthquakes dramatically, and included a few eastern port cities. In Columbia, South Carolina, as the local newspaper reported in late January, "The violence of the jar produced by the first shock, bent the lightning rod of the south Carolina College, threw down the plastering, and cracked the chimnies of some of the houses, and stopped the clocks in others." Less affected places felt shaking and sometimes experienced other effects like noises and flashing lights. In a practical sense, this spanned the entire United States at the time of the quakes.[39]

> *At New-Madrid, 70 miles from the confluence of the Ohio and on the right hand, the utmost consternation prevailed amongst the inhabitants, confusion, terror and uproar presided, those in the town were seen running for refuge to the country, whilst those in the country fled with like purpose towards the town. I am happy however to observe that no material injury has been sustained.*

THE GREAT EARTHQUAKE AT NEW MADRID

FIG. 1.5. "The Great Earthquake at New Madrid." This crude woodcut from almost four decades after the earthquakes seems dramatic and overblown. It *is* dramatic and overblown. Yet is it also a fair representation of phenomena widely reported at the time: cracks in the earth gaped open as structures shook, trees bent and snapped, and lightning flashed in the sky. The lightning indicates scientific explanations from that period: perhaps earthquakes were caused by some form of electricity? Today, modern science of New Madrid does not recognize lightning as related to earthquakes—but earthquake cracks, damaged dwellings, and bent trees are all amply documented. (Henry Howe, *Historical Collections of the Great West* (1857) vol. 2, p. 237, image 023993, courtesy of the State Historical Society of Missouri-Columbia)

> *At the Little Prairie, 103 miles from the same point, the shocks appeared to*
> *have been more violent, and were attended with severe apprehensions, the towns*
> *were deserted by its inhabitants, and not a single person was left but an old negro*
> *man, probably too infirm to fly, every one appeared to consider the woods and*
> *hills most safe and in these confidence was reposed, distressing however as are*
> *the outlines of such a picture, the latest accounts are not calculated to increase*
> *apprehensions, several chimneys were destroyed and much land sunk, no lives*
> *however have been lost.*

Pierce's account indicates both the physical and the social geography at the time of his river voyage. His accounts trace for his readers the changes to the topography of land and river, but he also captures the human consequences of the seismic events. Pierce's report of the abandoned man and the apparent insignificance of his probable death indicates the historical disregard of black residents of the area.

The man Pierce mentions was likely a slave, as were most African Americans in the American territories along the west bank of the Mississippi in 1811. It is certainly possible, though, that he was a free black man, heir to a French legal and social legacy along the middle Mississippi River in which white, black, and Native people coexisted uneasily but, in comparison to American social structures, with some degree of social flexibility. The French villages of Upper Louisiana had slaves, but they also had free black residents. The "old negro man" of Little Prairie was living within a changing reality, in which racial topographies were hardening. Over the half century following the quakes, the very possibility of being a free black person was limited by increasingly restrictive laws and threatened by violence: an "old negro man" in 1850s New Madrid would certainly be living in slavery. At the time of Pierce's voyage some tenuous ambiguity remained. Yet the man's social status is all too clear: he may have been too frail to take flight, but he was also too little valued to be carried along by anyone else.[40]

Scandalized traveler John Shaw recounted that a seventeen-year-old girl named Betsey Masters, who lived about twelve miles from New Madrid, had her leg crushed and pinned by a falling beam during the quakes. Her family could not rescue her, so they left her with some food and water nearby while they fled. Though a "total stranger," Shaw was the only person, he reported, willing to go back and check on her: he brought her more provisions but left deeply troubled by her ultimate fate. Happily, he reported, "Miss Masters eventually recovered." No similar coda exists for the old, "infirm" man: Little Prairie was completely abandoned with the devastation of the quakes. He may simply have perished.[41]

A little below Bayou River, 130 miles from the same point and 13 miles from the spot where we lay, the ruin begins extensive and general.

At Long Reach, 146 miles, there is one continued forest of roots and trees which have been ejected from the bed of the river.

At and near Flour Island, 174 miles, the destruction has been very great and the impediments in the river much increased.

At the Devil's Race ground, 193 miles, an immense number of very large trees have been thrown up, and the river is nearly impassable. The Devil's Elbow, 214 miles, is in the same predicament, below this, the ruin is much less, and indeed no material traces of the earthquake are discoverable.

What did it mean that "the ruin begins extensive and general?" Pierce had already recounted a number of alarming changes to places around the epicenters—sand blows, caved-in banks, sheared-off forests—and his commentary as he continued below New Madrid signaled further environmental tumult. The New Madrid quakes caused extensive change to environments and even topography, though the nature, extent, and permanence of these changes became subject to debate by the late nineteenth century and have remained debated since.

In the spring of 1846, noted British geologist Charles Lyell—whose *Principles of Geology* had impressed upon the reading public and the young Charles Darwin the power of slowly summed processes of deep time—visited the New Madrid region to see evidence of the earthquakes. He noted the depressed topography near New Madrid. He also explored other earthquake evidence, observing that the seismic shocks destroyed nearby forest.[42]

Many observers at the time noted extensive topographic change. A Methodist preacher near the epicenters described a night of horrors: "One shock following another; the solid ground running in waves, resembling those on the face of the lake." He wrote that in the area near New Madrid "shocks tore the earth asunder, changing the face of the country, raising lakes to dry land and timbered lands becoming wet and swamp lands, level beautiful prairies thrown into mounds and deep chasms." Later, traveler Louis Bringier noted more dispassionately that "the country here was formerly perfectly level, and covered with numerous small prairies of various sizes, dispersed through the woods. Now it is covered with slaches (ponds) and sand hills and mounticules, which are found principally where the earth was formerly the lowest." Researchers in subsequent centuries would identify landslides along eastern Mississippi Valley bluffs from the confluence of the Ohio down through Memphis.[43]

Areas in the town of New Madrid were what the French residents termed *callé*, sunk down by the settling and radical shifting of earth during main shocks of the quakes. Likewise subsided were areas to the south, along the St. Francis River, where the earthquakes created the "sunk lands" of northeastern Arkansas, and in scattered locations in west Tennessee. In the early nineteenth century, such subsidence was widely reported as evidence of earthquake. The "sunk lands" would later be reinterpreted by geologists and others as ordinary processes of flooding and swamp overflow, and such reports would be dismissed. Matter-of-fact assertions in 1812 would become by the early twentieth century evidence in pointed debates.[44]

The earthquakes formed extensive lakes in the areas near their epicenters. Mississippi River booster and missionary Timothy Flint explained both the extent and the confusion of this topographic transformation: "The whole country, to the mouth of the Ohio in one direction, and to the St. Francis in the other, including a front of three hundred miles, was convulsed to such a degree as to create lakes and islands, the number of which is not yet known." The number of such lakes is still not well known. Many maps of what is now the state of Missouri from the early nineteenth century indicate large lakes in the southeastern corner that do not appear on later mappings. Like the sand blows, some of these earthquake lakes apparently became less visible and present because of erosion by water and filling in by soil and wind erosion, as well as the overgrowth of vegetation.[45]

Other lakes were drained in the extensive swampland reclamation projects that remade the area south of New Madrid in the late nineteenth and early twentieth centuries. One significant and lasting body of water resulted from the "hard shock" of 7 February 1812. This tremor created the "Tiptonville Dome," an uplift of formerly flat alluvial plain that stretched from western Tennessee across the Mississippi. It is still visible as a gentle rise adjoining the Mississippi, unremarkable except in comparison to the flat alluvial plain surrounding it. The rising land impounded Reelfoot Creek and formed Reelfoot Lake. Reelfoot Lake is a shallow "flooded forest," in the words of the Tennessee State Parks Department. It is also a main fishing attraction in western Tennessee (as well as a place to see overwintering bald eagles). The spiky trunks of trees submerged in 1812 would tear up fiberglass hulls, so special V-shaped fishing boats or graphite-coated metal or wood jon boats float quietly over the submerged stumps of forests killed by the New Madrid earthquakes.[46]

The western country must suffer much from this dreadful scourge, its effects will I fear be more lasting than the fond hopes of the inhabitants in this section of the union may at present conceive. What have already been the interior injuries I cannot say. My opinion is that they are inferior in extent and effect.

FIG. 1.6. Reelfoot Lake, 2004. In the summer of 1811, Reelfoot River was a largish creek. That winter, earthquakes dammed it to create a shallow but extensive body of water. Two hundred years later, Reelfoot Lake is a popular recreational fishing site where companies rent special shallow-hull fishing boats adapted well to the cypress trees that stick up to the surface of the water—and like these trees in the background, sometimes grow past it. (Conevery Bolton Valencius)

Around New Madrid, everyone who could flee, did. River boatman John Vettner and his crew reportedly "offered, at New Madrid, half their loading for a boat to save it, but no price was sufficient for the hire of a boat. Mrs. Walker offered a likely negro fellow for the use of a boat for a few hours, but could not get it." In New Madrid, residents were said to have fled to the nearest high ground, a small knoll in town or a hill some miles off, hoping to escape from the rising waters. Meanwhile, two hundred refugees from Little Prairie escaped to New Madrid, wading up to their waists in water.[47]

The abandonment of New Madrid and Little Prairie was sudden and nearly complete. In 1808, there were a hundred houses in New Madrid and twenty-four log cabins in the trading post of Little Prairie. As late as 1817, only twenty log houses remained in New Madrid, and Little Prairie was abandoned. To the south, most of those who lived through the quakes in the region of the "sunk country" abandoned their homes—or what was left of them.[48]

Recognizing the plight of Americans caught most centrally in the disruption of the earthquakes, Congress passed legislation in 1815 for the relief of landowners. This act was one of a small number of early congressional disaster relief measures that undergirded the much later expansion of national welfare programs. Yet this disaster relief program was itself a disaster. The act granted landowners whose property was "materially damaged by the earthquakes" a certificate for an equivalent amount of land elsewhere. The resultant "New Madrid certificates" became a form of almost uncontrollably multiplying currency. After the announcement of the legislation, speculators rushed into the area to buy up claims from residents with no idea that their land was about to become much more valuable (a few savvy residents likewise sold their land claims several times over to guileless would-be speculators). Shysters waving New Madrid claims filed for choice land in central Missouri, regardless of its prior settlement by Native people or even other Americans. Enterprising Missourians attempted to claim the medicinal Arkansas Hot Springs (only Supreme Court decision in 1876 ultimately prevented them). A collection of landowners wielding New Madrid claims successfully finagled a determination of Little Rock as the territorial capital of Arkansas. Widespread fraud plagued the whole process. Litigation clogged courts and gave headaches to attorneys general for decades: one New Madrid case was not decided by the US Supreme Court until 1851. Claims made under New Madrid certificates settled areas of both Arkansas and Missouri, but the US effort to help those in the New Madrid and Little Prairie area did not do so.[49]

Yet before these legal tangles were even midway through the courts, New Madrid had recovered in population. By 1821 surveyor William Rector advised the commissioner of the General Land Office that "much of the Lands that have been abandoned by the claimants on account of alledged injury by the Earthquakes, will sell for a better price, then the Lands that have been located in virtue of the claims would now sell for ... Because those New Madrid Lands are generally exceedingly rich and well suited to the produce of Cotton." While Little Prairie remained abandoned, by 1830 the formal New Madrid population again equaled that of 1810—mostly new American settlers for whom good, cheap land was invaluable, whatever its seismic history. Subsequently, many who have surveyed the history of the region argue that since the New Madrid population rebounded so robustly, the earthquakes had very little lasting effect.[50]

At the same time, the repopulation of New Madrid fails to register a broader truth of Pierce's pessimism about the terrible impact of the earthquakes. The area surrounding New Madrid had occupied a significant role in Native American and French settlements in the region before the quakes.

After the quakes, New Madrid was not a very significant town for American settlement or much of anything else (only in the Civil War would its position atop a bend in the Mississippi River again prove crucial). The tremors were a main factor in diminishing that stretch of the middle Mississippi to the sleepy, insignificant settlements they soon became in American life and American imagination.

> The continuance of this earthquake must render it conspicuous in the pages of the historian as one of the longest that has ever occurred: from the time that the first shock was felt at 2 o'clock in the morning of the 16th, until the last shock at the same time in the morning of the 23d, was 168 hours.

Because each episode occurred as a long series of shocks, they were hard to get past, to get through—people stayed in the midst of upheaval for a very long time. Earthquakes continued to occur and overwhelm: one correspondent noted in late December that "there has been in all forty-one shocks ... the last one at eleven o'clock this morning (20th) since I commenced writing this letter." For years after the main shocks, even people far off felt tremors. By the early 1820s, the tremors settled down, becoming less sensible to those further afield, but continuing to be felt and heard in the middle Mississippi Valley.[51]

> Nothing could have exceeded the alarm of the aquatic fowl, they were extremely noisy and confused, flying in every direction, without pursuing any determinate course, the few Indians who were on the banks of the river have been excessively alarmed and terrified.

Pierce's account was typical in its lack of detail about Native experience of the quakes. Many American observers recorded that local Indians reacted with alarm, terror, and flight, but few recorded careful observations or thorough conversations. Pierce, like many American witnesses, treated the distress of Indians as parallel to that of surrounding animals, simply another marker indicating disruption to the natural world.

Early nineteenth-century life was full of creatures. In cities as well as out in the countryside, pigs and horses, mules and oxen, mice and poultry—as well as the dung heaps and smells and flies that went with them—were constantly around and underfoot, pulling loads, eating trash, rooting through slops. Cats chased chickens, hunting dogs snoozed on porches, long clouds of migrating birds filled the skies every spring and fall. The unrest of birds and animals thus provided a form of standardization easily conveyed and widely understood to express the upheaval of the natural world.[52]

Pierce's account reflects a holistic knowledge making, in which many realms of experience were connected: earth, sky, Indians, animals. In

Pierce's writing as in many New Madrid accounts, earthquakes were a phenomenon not only of the earth, but of a commotion of "all nature." In particular, they were widely associated with the comet of 1811.

The comet of 1811 was noted by people throughout the world: it was one of the brightest and longest comets ever recorded, appearing in many skies for the better part of a year. Over North America it became visible in the sky at the end of August and increased in brightness through September and October. Many people later remembered that it disappeared from view right about the same time as the first December shock. By the time of the last "hard shock" in February 1812 it was no longer visible to most observers. Just as the comet of 1066 was associated with the successful English invasion of William the Conqueror, the movement and brightening of the great comet of 1811 was widely taken as a harbinger of invasion, fighting, or coming judgment. In Europe it was called "Napoleon's comet" (it appears as a significant omen in Leo Tolstoy's *War and Peace*).[53]

For observers across North America, the comet seemed to be closely connected with the earthquakes: together, they seemed to be an ominous call or warning.‡‡‡ Politician, physician, and natural science authority Samuel Latham Mitchill led off his extensive compilation and report on the New Madrid earthquakes with mention of the comet. Indeed, such celestial and natural events spanned the areas of Mitchill's expertise. Comets, like earthquakes, were widely feared as bringing epidemics in the nineteenth century. The memoir of an early Methodist preacher from the Ohio Valley sums the experience of many in 1811: "War, Indians, Comets, Earthquakes."[54]

Yet in William Leigh Pierce's account, as in many similar narratives, concern for natural portents coexisted with specific, careful, timed observations. Directly after his passage about the comet, Pierce included a "table of the shocks, with the exact order of the time in which they occurred." Like quite a few people who experienced the quakes, he carefully noted what had happened, repeating precisely what happened when—recording, for instance, after the dawn shock on 16 December at 8:00 a.m., "nine shocks in quick succession," and on 17 December, among many tremors, "5 in the morning a great and awful shock followed, with three others." His listing of the shocks just in the first two days of the quakes formed a long paragraph: the shocks were nearly constant, subsiding only slowly (and temporarily). By the following week his listing tapered down to end with "22d Dec. 11

‡‡‡ Celestial events like eclipses and meteor showers were widely taken as meaningful by people in slavery: Virginia preacher and field hand Nat Turner understood an eclipse in February 1831 as a signal to launch his slave rebellion. Similarly, in a traditional African American spiritual, "when the stars begin to fall" is the time of a changed world, offering redemption and freedom.

o'clock A. M. a slight shock, 23d Dec. at 2 in the morning a very severe shock." This comfortable proximity of description and tabulation, of symbol and quantification, speaks to the holistic understanding of the natural world in the early American nineteenth century: symbolic portents and careful natural observation could work together to reveal aspects of the truth of an event.[55]

> *Thus we observe that there were in the space of time mentioned before eighty nine shocks—it is hardly possible to conceive the convulsion which they created, and I assure you I believe that there were many of these shocks which had they followed in quick succession, were sufficient to shake into atoms the firmest edifices which art ever devised.*

Pierce was an astute observer: seismic shocks in series do severely damage buildings and other structures. Whether such shocks would be "sufficient to shake into atoms the firmest edifices which art ever devised" would some hundred and ninety years later become hotly debated in Memphis, Tennessee, as seismologists attempted to interpret evidence like Pierce's.

> *I landed often, and on the main shore as well as on several islands found evident traces of prior eruptions, all which seem corroborative of an opinion that the river was formed by some great earthquake—to me indeed the bed appears to possess every necessary ingredient. Nor have I a doubt but that there are at the bottom of the river strata upon strata of volcanic matter. The great quantity of combustible materials which are undoubtedly there deposited, tend to render a convulsion of this kind extremely alarming, at least however, the beds of timber and trees interwoven and firmly matted together at the bottom of the Mississippi are tolerably correct data from which may be presumed the prior nature, &c, of the land. The trees are similar to the growth upon the banks, and why may not an inference be drawn that some tremendous agitation of nature has rent this once a continued forest, and given birth to a great and noble stream. There are many direct and collateral facts which may be adduced to establish the point, and which require time and investigation to collect and apply.*

Pierce was one of the first to assert that the middle Mississippi Valley showed environmental signs of massive earthquakes in the past, but his assertions about volcanoes would not survive changes in geological thinking. Though in the early nineteenth century earthquakes were often understood as the surface manifestation of deep-seated, subterranean explosion, a kind of subterranean volcano, by the late nineteenth century this notion of subterranean explosion seemed dated and wrongly directed. For late nineteenth- and twentieth-century scientists, such musings about volcanoes in the Mississippi Valley were simply yet more reason to view accounts like Pierce's as mere tall tale.

Yet more recently seismologists have validated the broad strokes of Pierce's assessment about the region's earthquake history. The Mississippi Valley has been the site of very large seismicity in the deeper past. Pierce's observations made sense in his era and were widely sought and shared. Ironically, when scientific theories changed in the late nineteenth century, his entire account was minimized or dismissed outright precisely because he had attempted to theorize about the geologic implications of his experiences in terms which seemed utterly wrongheaded by late nineteenth- and twentieth-century scientific standards. In yet another ironic twist, in the past few decades, not only have testimonies from the time regained credence, these passages from Pierce's account have been key to the return of scientific interest, as researchers reread them to find details once dismissed as implausible and now considered as possible evidence for new seismological conclusions.

> It is a circumstance well worthy of remark that during the late convulsions the current of the river was almost instantaneously and rapidly increased. In times of the highest floods it rates at from 4 to 5 knots per hour. The water is now low and when we stopped on the 16th inst. at half after 4 P. M. we had then run from that morning 52 miles, rating at 6 knots generally. This current was increased for two days and then fell to its usual force. It is also singular that the water has fallen with astonishing rapidity. The most probable and easy solution of this fact, which presented itself to my mind, was, that the strength of the Mississippi current was greater than the tributary streams could support. Either this must have been the case or some division of waters above has occurred, or destruction below has created some great basin or reservoir for the disembarging of the main body of water. The latter presumption I apprehend cannot be correct as our progress towards the mouth of this river, is marked with little or no injury.

In one of the earthquakes' most startling and alarming effects at the time, the February shock did actually force the Mississippi backward. All the main shocks affected the river's current: the Mississippi in 1811–12 was a broad river, and the powerful tremors underneath, around, and through its waters caused waves to move back and forth, amplifying and interacting as they bounced off the river's banks. This large-scale sloshing destroyed riverbanks and smashed many river boats to pieces.

But in an even more dramatic way, the earthquakes affected the Mississippi by creating temporary falls. A pair of travelers who passed through New Madrid and then Lexington gave an account of the 7 February shock, reporting that "some obstruction had presented itself in the river something like a rapids or falls, which greatly endanger the navigation."[56] This Tiptonville Dome disruption not only created Reelfoot Lake but disrupted the Mississippi. The uplift blocked the clear flow of the enormous river and forced much of its water back upstream. The flow that did make it over the uplift swiftly eroded the soft soil, so that the main current continued downstream within a very short time, but for those who experienced the tumultuous and backward current the power of the earth's movement was spectacular.

The river traveler who experienced this reported that "there was a back current in the river which drove the boat several miles up a small bayou, and during the convulsion, the motion of the boat was so violent as to stave many of the barrels of flour in the boat." After three days of quaking, he abandoned his goods and left the area. He and other dazed travelers reported these events back east, where they were regarded as true witness. In succeeding years and certainly by the late nineteenth century, when no obstruction was visible in the Mississippi, such tales seemed fanciful accounts born of fear rather than any observation. Only over the course of the late twentieth century, with evidence elsewhere in the world of exactly this sort of earthquake-induced waterfall and river blockage (including a similarly spectacular and short-lived waterfall in 1996 on the Tachia River after a 7.6 earthquake in Taiwan), did such accounts appear again as an accurate description of seismic events.[57]

> *Thus, my dear sir, I have given a superficial view of this awful Phenomenon; not so much to convey instruction upon a very interesting subject as to satisfy the curiosity of the public relative to so remarkable an event. At some more convenient season it is my intention from facts which I had the opportunity of collecting, to canvas the subject more in detail, you are therefore at liberty to make whatever use you please of this brief sketch and publish the whole or extract such parts as you may deem best adapted.*
>
> *Should other interesting circumstances occur relative to this Phenomenon, I will do myself the pleasure of making you another communication.*
>
> *With much respect, I am, sir, your obedient servant.*
>
> WILLIAM LEIGH PIERCE.

Pierce finished his first, main account of the New Madrid earthquakes by indicating their place in continuing conversations. He and others would continue to collect observations and "canvas the subject more in detail." Many compilers and newspaper editors would indeed "publish the whole" and "extract such parts" as they pleased. Pierce's would be one element in what became a small newspaper and publishing boom.

William Pierce too continued to work on his own understanding of these baffling events. In a subsequent letter, which he wrote on 13 January 1812 from New Orleans, Pierce elaborated on some aspects of his description of "the late Earthquake." He also clarified its extent, based on his questioning of others arriving in New Orleans:

> *Its range appears to have been by no means confined to the Mississippi. It was felt in some degree throughout the Indiana Territory, and the states of Ohio, Kentucky and Tennessee. I have conversed with gentlemen from Louisville and Lexington, (Ken.) who state that it was severe in both of those places. At the latter indeed it continued for twelve days, and did some inconsiderable injuries to several dwellings. From thence it ranged the Ohio River, encreasing in force until it entered the Mississippi, and extended down that river to Natchez, and probably a little lower. Beyond this it was not perceived.[58]*

A main interest in the New Madrid accounts was to figure out not only what had happened, but *where.* Writers sought to compare reports to estimate the extent of disturbance. *Niles' Register*, for instance, reported that "in Georgia the effect was much greater than in Virginia." Many accounts represent a narrative mapping. One report from a Lexington paper in late January noted that "from the accounts already received through the medium of the Newspapers, it seems, that the shocks of the earthquake have been very sensibly felt throughout this state,—the Indiana Territory-Tennessee-Western parts of Pennsylvania-Ohio-Maryland-N. & S Carolina-In Philadelphia-Washington City-Norfolk-Alexandria and Richmond, Va.-" Such breathless listing helped map extent through a record of physical sensation, recording where shocks were "sensibly felt" and subsequently discussed in local newspapers. In such narratives, readers mapped for themselves the connection of felt effects that bound together portions of their nation both old and new.[59]

In this context, discussing earthquakes was a main way of discussing the West. Much was unknown about western North America and the puzzles posed by the natural history of western lands. Those interested in natural history pored over reports of mammoth bones and teeth, of reptiles with strange bodies that older Indians had long thought to have been extinct. Southwestern territories were said by reliable authorities in the first decade of the 1800s to contain unicorns and giant water serpents. Tales of huge surface silver deposits—likely based on massive and massively fascinating meteors—fueled decades of mining explorations to desert regions of the Southwest. The wilds of the West featured in an 1812 history: domesticate a deer, it advised, by discovering a sleeping fawn, carrying it some distance, then waking it up: it will eat out of a hunter's hand. "Paroquets," now-extinct

Carolina parakeets, have harsh voices but beautiful bright colors: green with yellow crowns. They sometimes overwinter in large groups in holes in trees, where occasionally they will be discovered in large groups frozen to death. Such detail fascinated early settlers and glittered before western boosters, who avidly read traveler's reports and rushed to seek the riches of the "far west."[60]

To explore new and potential American lands, in the first decade of the nineteenth century president Thomas Jefferson sent a series of missions out to areas Americans did not know. The most famous is the Corps of Discovery led by Meriwether Lewis and William Clark to the Pacific Northwest and the upper Missouri, but this was only one front of a whole set of endeavors: Thomas Freeman and Peter Custis went up the Red River in 1806; George Hunter and William Dunbar talked with hunters and tested the Hot Springs of the Arkansas Ouachita River in 1804–5; and Missouri territorial governor James Wilkinson authorized Zebulon Pike to identify for Americans the source of the Mississippi and the southern Rockies in 1806–7. The energy and effort expended in these voyages of exploration across grassland seas and thick-grown swamplands indicate the intense interest in finding and capitalizing upon the riches of still-unknown western lands.[61]

Slow communication and uncertain mails meant that early nineteenth-century Americans were hungry for accurate information about far-distant places and events. In 1807, a frustrated federal agent in Arkansas Territory was reduced to piecing together recent acts of Congress from torn scraps of newspaper used as a wrapper for commodities shipped from New Orleans. Mail was iffy at best. A territorial judge in 1813 found difficulty in resigning from the bench in Kentucky because his letters of resignation had such trouble reaching Washington. In this context of intense curiosity and sparse knowledge about events in the "far west," reports like Pierce's were central and valuable.[62]

Pierce's account was not simply storytelling, but knowledge making. He told readers about territory new to them, and to their nation, and he helped share ways of structuring that knowledge through his melding of spirituality and exact observation, his emphasis on careful quantification, and his blending of the marvelous and the prosaic, the practical and the sublime. In making known the wilds of the west, Pierce helped make known the ways in which people could find out more for themselves, how they could discuss and share what they learned. His assertions about the American future of the region also made new places comprehensible as parts of the American nation. His account was part of an establishment of claims to know as well as own territories beyond the Mississippi.

William Leigh Pierce evocatively described what he saw, smelled, heard, felt, and feared. He demonstrated to his readers how all these forms of experience were part of creating knowledge about them. In his ways of coming to terms with the earthquakes, Pierce represents many other writers who were shaken and frightened by the tremors and wrote down similar accounts in many similar media. His account brings into focus a diffuse and important element of early American culture that usually lies invisible beneath conventional histories.

Yet Pierce's account of the New Madrid earthquakes, particularly his pessimistic assessment that they would long limit immigration into the middle Mississippi Valley, was already being overwritten by the time his final letters were being published in eastern newspapers. In the beginnings of a process of reassessment and rewriting that continued over the next two centuries, knowledge of the New Madrid earthquakes was being reworked even as it began. Another voyage on the Mississippi, upstream from Pierce, ultimately proved much more influential in shaping how Americans assessed, claimed, and spread onto the territory of midcontinent.

The Steamboat *New Orleans* and a New Nineteenth-Century View of the Mississippi

Among the many river craft that suffered though the New Madrid earthquakes was one unlike all the rest: upriver of New Madrid, the first steamboat to navigate the "western waters" of the Ohio and Mississippi River shuddered with the December tremors as its passengers lurched in terror. Like Pierce's boat, the *New Orleans* and its travelers avoided foundering amid the river's tumult. Like his, the *New Orleans* docked safely at its goal. Unlike his entirely typical craft and crew, the *New Orleans* and its travelers amazed all they passed. The success of that voyage was part of a reshaping of commerce and communication that would transform nineteenth-century life, first and especially along the rivers where the New Madrid earthquakes had been felt so strongly.[63]

On board the *New Orleans* were Nicholas Roosevelt and Lydia Latrobe Roosevelt, their toddler and large Newfoundland dog, and their crew and domestic staff. The Roosevelts were a romantic couple: she was seventeen and he was her father's forty-one-year-old business associate when they fell in love and were married.§§§ They were both passionately committed to

---

§§§ Nicholas Roosevelt was indeed one of *those* Roosevelts: his brother eventually became the great-grandfather of president Theodore Roosevelt. Lydia Latrobe Roosevelt, meanwhile, was part of a family who created the civic spaces and technologies for much of

FIRST BOAT BUILT ON THE WESTERN WATERS, 1812.

FIG. 1.7. In a coincidence that would seem implausible in a Hollywood blockbuster, the first steamboat to travel the Mississippi, the *New Orleans*, was underway when the first of the New Madrid earthquakes struck in December 1811. The sheer technological and commercial power represented by the *New Orleans* meant that the earthquakes did not in the end hold back American settlement of the "far west," as people like flatboat merchant William Leigh Pierce feared they would. (*Lloyd's Steamboat Directory, and Disasters on the Western Waters* (1856), 42, image Ref103-1, courtesy of the State Historical Society of Missouri-Columbia)

modern America. Her father, architect Benjamin Henry Latrobe, was a main influence on early American design; her younger brother John Hazlehurst Boneval Latrobe helped secure the landmark agreement to run the first telegraph lines in the United States down the Baltimore & Ohio Railroad; and another brother, Benjamin Henry Latrobe, was a civil engineer who worked after the Civil War as consultant for the Brooklyn Bridge. Lydia Roosevelt was apparently a strong force within her family's many projects but, like many nineteenth-century women, left little of her own historical record: we know her story mostly through an 1871 book about her steamboat pioneering written by younger brother John H. B. Latrobe and an 1835 account by cousin Charles Joseph Latrobe, an adventurer who traveled the United States with Washington Irving, tutored European nobility, became a noted Alpine mountain climber, and ended his career as colonial governor of Melbourne, Australia, during that region's gold rush. Just as William Leigh Pierce

steam as a new form of transit for their modern century. Nicholas Roosevelt was an inventor and engineer: with the help of influential backers, he had designed and manufactured the steamboat *New Orleans* to demonstrate the practical and commercial possibilities of steam-powered shipping on the Mississippi. The boat was a sky-blue, 116-foot side-wheel steamboat with a 7-foot draft. It was powered by a 34-inch-cylinder low-pressure engine that developed less than 100 horsepower (as well as two masts, as backup). The Roosevelts left Pittsburgh in their daring new craft late in 1811. Nicholas was widely condemned for his foolhardiness; Lydia, visibly pregnant, was widely admired for pluck in braving not only the dangers of the river but the likelihood of being blown to bits by this new and seemingly outlandish form of travel.[64]

The steamboat was hugely noisy and engulfed in commotion: the furnace roared, its crew worked and sweated to keep it fed, its engine belched smoke, the smokestacks trailed sparks through the night, and the great wheel churned a huge path through the water. Families along the river in Kentucky gave alarm that British and Indian allies were sweeping downriver. Others concluded that some new sort of sawmill must be headed west.[65]

During their stay in Louisville, Lydia gave birth to the second of the nine children the couple would raise. Later, as crowds watched and Lydia Roosevelt stood dramatically by the prow, having refused to go downstream by land to rejoin the boat safely below the dangerous rocky stretch of white water known as the Falls of the Ohio, the *New Orleans* gathered its maximum steam possible—it had steerage way only if it was running faster than the swift current—and succeeded in passing successfully over the boulder obstructions that marked the gradual but treacherous falls.****[66]

One night about two weeks later, while the boat was anchored on the Ohio below the falls, those on board felt a tremendous shock as if they had suddenly run aground. All that night they felt unusual shocks and tremors

---

represented many Mississippi River travelers, Lydia Latrobe Roosevelt and her sprawling and manifestly talented extended family were perhaps the quintessential American improvers in an era deeply committed to personal and national improvement.

**** While in Louisville, the Roosevelts had also engaged in a piece of demonstration long savored by contemporaries in an era that appreciated good theater. Confronted with widespread skepticism about the boat's power and prospects, they invited community leaders in Louisville to a formal dinner on board the *New Orleans*. Midmeal, guests heard sudden rumbling, felt motion, and rushed out on deck, convinced the boat had slipped anchor and was drifting fatally toward the dangerous falls. Instead, a satisfied Nicholas Roosevelt pointed out that the boat was instead confounding critics by making headway *up* the river as the guests enjoyed their dinner.

that they soon understood to be massive earthquake tremors. For the rest of their journey, these travelers, just like William Leigh Pierce, voyaged across a landscape flooded and transformed by the earthquake shocks. They must have passed his convoy as they traveled downstream. One night those on board the *New Orleans* were kept awake by rasping scrapes of large masses passing over the sides of the ship. The sounds were of an island disappearing: the one they had anchored to. Villagers at Little Prairie called out to the passing steamboat for rescue, fearing that the ground was "gradually sinking" beneath them. Taking stock of their own dwindling provisions, the crew of the *New Orleans* instead steamed by without stopping.[67]

From the beginning of its journey, the tumult of the steamboat was so astonishing that people thought it must be some sort of natural event. When it hove into view in Louisville, Kentucky, on the night of 1 October, the roar of escaping steam as the boat turned to drop anchor disturbed people throughout the town: some people insisted that the comet had dropped into the river and produced the "hubbub." Along the Mississippi, Chickasaws told the crew that the boat was called the "Penelore," or "fire Canoe," and its noise and sparks were thought to be connected to the celestial comet. Perhaps in echo, the next steamboat to travel the western waters after the *New Orleans* was named the *Comet*.[68]

Earthquake shocks multiplied such associations. As it steamed past this earthquake terrain, the *New Orleans* became connected in popular imagination with the tumult of the earthquake. The "smoky atmosphere" during the earthquakes was associated with the belching smoke of the steamboat, and the noise and deep rumbling of the earthquakes with the loud, deep beating of the steamboat paddles.[69]

The *New Orleans* did not in the end founder in the seismic waves created on the Mississippi, but successfully docked in the city for which it was named. For sheer romantic storytelling, the drama of proud inventor and young wife triumphantly surveying New Orleans from aboard the new mode of Mississippi River travel, newborn son in arms, is hard to beat. But one element of dénouement comes close: during the eventful voyage, the captain of the boat and one of the Roosevelts' maids fell in love and were married in the safely docked *New Orleans*: the world's first steamboat wedding.[70]

The *New Orleans* heralded a profound change in the world of Mississippi River commerce so recently chronicled by William Pierce. A merchant writing back to his trading firm in Rhode Island from Natchez reported on the third of January 1812 that the craft had just "arrived & Delivered a Cargo at this Place" and was headed down for New Orleans. "It is thought She will answer well," he mused, "and will very much Improve the navigation in this River both for Safety & Expediency."[71]

With explosive speed as well as frequently explosive boilers, steamboats remade western commerce. On 2 August 1817, a large crowd gathered to welcome the *Zebulon M. Pike* as it made the first steamboat ascent of the Mississippi past the mouth of the Ohio. The technology was still being developed—the *Zebulon M. Pike* sometimes needed crews with long poles to push it against the current—but the boat's belching engine was nonetheless noisily impressive. Local Indians fled from it, as others had from the *New Orleans*, as white American crowds jeered. Within two years the *Independence* headed up the Missouri to Franklin—site of New Madrid certificate claims—to inaugurate steam transit on that major river. By 1819, St. Louis was a paddlewheel port, its harbor bristling with the tall chimney stacks from boats sometimes three deep at the dock, and riverside areas were soon substantially deforested by the engines' insatiable need for combustible fuel. By the late 1820s and early 1830s more remote streams like the White River were pioneered by steamboats: Jacksonport, Arkansas, which had been explored and recorded by an American and Native American pair of trapping partners only in 1822, was reached in 1833 by an early steamer on the White River.[72]

Steam travel helped create the nineteenth-century revolution in communication and travel. A young Abraham Lincoln patented an inflating apparatus to allow steamboats to free themselves from shoals like the Falls of the Ohio that had threatened the *New Orleans*. The frequent, murderous explosions of steamboat boilers punctuated but did not halt this flood of commercial development. (The *New Orleans* itself was soon wrecked on an upriver journey, near Baton Rouge.)[73]

William Leigh Pierce maintained that the earthquakes he experienced would devastate the area for the foreseeable future. His perspective was understandable: he recorded what he saw from the deck of a vulnerable flatboat, placed in peril by the river's tumult and virtually powerless to move against it, looking out at a devastated landscape. From the point of view of a steamboat pushing through the river, the prospects for the western Mississippi Valley looked considerably different. The *New Orleans*, like Pierce's smaller craft, had to watch for planters and sawyers, but it could power over the current that constrained Pierce's convoy and all prior craft on the river. Lydia Roosevelt could stand on a 116-foot-long deck and look out on a different future for the West.

Just as in the complicated resolution of the land claims made under New Madrid certificates, local consequences of this journey were complex. The family so emblematic of the new steam power of the western rivers became tangled in extensive disputes over priority of invention and the legality of official monopolies. Nicholas Roosevelt lost several legal and economic skirmishes with his former partners. But the family's journey had helped remake the Mississippi River as a zone of commerce powered by wood and coal, not raw strength of arm, a commerce that could now push upstream against a powerful current, a commerce that could transport goods and people and news much more rapidly than before, in the world the earthquake had unmade. Pierce's era had come to an end, not because of the physical damage done by the temblors, but because of the dramatic change in communication and transportation heralded by the "fire Canoe" just upstream of his flatboat convoy.[74]

The Remaking of Memory along the Mississippi

Soon after the earthquakes, Kentucky resident Joseph Ficklin observed that "the Indians cannot have suffered much in their tents and bark homes. But the United States will suffer in the sales of their public lands west of the Mississippi for an age. At least the present generation must be buried before the spirit of wandering in that direction, revives." Ficklin had ample reason for his assessment, but he was profoundly wrong.[75]

American settlement of the middle Mississippi Valley paused for a few years, but certainly not for a generation, after the environmental tumult of 1811–12. The decline in newcomers represented merely a blip in the torrent. In the end, steam and commerce and land proved far more powerful in promoting settlement of the West than earthquakes were in stopping it.[76]

In the spring of 1817, on his last day in office, president James Madison vetoed a bill that would have funded federal support for roads and canals. Advocates of such improvements to communication and transportation, including congressional representatives John C. Calhoun of South Carolina and Speaker of the House Henry Clay of Kentucky, were dumbfounded. The role of the federal government, Calhoun had argued, was to "counteract every tendency to disunion," to "bind the republic together" with improvements such as those Madison had just vetoed. After Madison's veto, Clay commented ruefully that "no circumstance, not even an earthquake that should have swallowed up one half of this city, could have excited more surprise." Clay spoke with the dismay of those who understood the importance of roads and canals to the unfolding nation he and Calhoun helped lead. He spoke also as one acquainted with stories of earthquake and upheaval: his was a metaphor closely connected to the recent experience of his Kentucky constituents, applied to the changes in communication and transportation that would soon come to transform their nation and their world—although largely under local and private initiatives, rather than federal direction.[77]

The irony was that for all this evident upheaval and drama, the New Madrid earthquakes mark a substantial division in the history not of the river, but of the people on it. Nowhere was this more clear than in the transformation of the region south of New Madrid. What had been an active trading hinterland became a swampy backwater. As Americans flooded into what soon became the states of Missouri and Arkansas, they effaced a prior history until it became as if it had never been. In the early nineteenth century New Madrid was no boom town, but one of the first frontier towns to go under. William Leigh Pierce may have glimpsed the end of the New Madrid hinterland from his flatboat; Nicholas and Lydia Roosevelt helped bring it about from the deck of their churning steamboat.[78]

2 * Earthquakes and the End of the New Madrid Hinterland

In 1818, an aspiring scientific writer named Louis Bringier described how the New Madrid earthquakes had disrupted recent Cherokee settlements along the St. Francis River and caused the emigration of the entire community. Unlike the reports from William Leigh Pierce, Bringier's observations have been dismissed by experts then and now as scattered and inconsequential—just as the Native American settlements he describes have been dismissed in the same terms. Yet much of what Bringier reported is supported by other historical records. He may have been a disorganized writer, but he was an accurate one.[1]

Not empty territory, the area south of New Madrid, particularly along the St. Francis River, had emerged in the late eighteenth and early nineteenth centuries as a region of settlement, exchange, and cultural interaction. The New Madrid hinterland, a zone of encounter and negotiation with New Madrid as its northern focus, stretched down the St. Francis River into the Native, French, and American settlements of what would become Arkansas. Cherokee and other Native American settlements along the St. Francis were central to the diplomacy, trade, and exchange of the region.[2]

Louis Bringier visited one of those St. Francis River Indian villages in June 1812. He was one of many to describe large, well-recognized communities along the river. Yet in American history making, the New Madrid earthquakes have been regarded as the inevitable postscript to a small, fading French colonial outpost about to be eclipsed by energetic American settlements. Bringier's account shows how well the people of the very early nineteenth century understood the Native American networks of the middle Mississippi Valley, in contrast to how profoundly later history writing has effaced all memory of this generation of Native emigration and settlement. The New Madrid earthquakes were no inconsequential blip, but part of a convulsive political, cultural, and environmental disruption that brought a sudden end to a thriving economic and cultural zone.[3]

Much was changing in 1812. War, negotiation, and emigration were re-drawing maps over much of lower North America. But in the territory of the New Madrid earthquakes, the region shook with more than the tumult of political and social change. The environmental tumult of the quakes that rocked the area thoroughly and repeatedly in the winter of 1811–12 altered the physical and social landscape forever. The New Madrid earthquakes, and the human responses to them, destroyed the New Madrid hinterland. Rivers became swamps, and as a result, flourishing trade routes became brambly hideouts. Native communities who had pioneered the area over the previous decades fled the devastation, turning into refugees and in-terlopers on lands further west and effectively accomplishing American aims of Indian removal. Environmental stress added to military and cul-tural conflicts as Cherokees fleeing the earthquakes pushed against Osage claims. Seismic forces are a forgotten but important added pressure on top of population disparities, overhunted environments, asymmetrical military force, and a tragically uneven burden of disease, forces pushing Indians out of lands that Americans wanted.

Understanding the New Madrid hinterland—and especially the role of environmental change in disrupting these little-known early settlements along the St. Francis—is crucial to tracing the impact of Native American groups on the emerging economy and society of early Arkansas, Missouri, Tennessee, and the Mississippi Valley generally, and equally crucial to rec-ognizing what the New Madrid earthquakes did and how they were under-stood. Historians have long recognized the permanent scars on American communities caused by the social tumult and ethnic conflicts of the late eighteenth and nineteenth centuries. But in the middle Mississippi Valley these larger social shifts were located in a specific topography, one that was scarred permanently as well, by the earthly tumult of the quakes of 1811–12. Topographic changes framed social changes. Environmental forces made possible the creation and dissolution of communities, and often spurred the movement of individuals and whole settlements. In the New Madrid hinterland, such environmental change is part of a history lost along with the memory of the early Indian settlers recorded by a few Americans such as the earnest Louis Bringier.

Earthquake Visions and Cherokee Communities

Louis Bringier was like William Leigh Pierce: a man of education and family connections, down on his luck, traveling the borderlands of middle North America in the late eighteenth and early nineteenth centuries. Bringier was the eldest son of a wealthy Louisiana planter family, but like other

Creole princelings, he lost much of his fortune gambling. When he was in his late thirties, he engaged in a series of travels in the midcontinent and Spanish territory. At some point he went to Mexico to mine silver and gained what was reputed to be a fortune, only to have it confiscated during political upheaval. Around 1810, he began a series of trips through the middle Mississippi Valley, interacting extensively with Native settlers and purchasing land near the lower St. Francis River Cherokee towns. In 1816 he traveled through Arkansas as a member—possibly the interpreter—of a band of spies led by famous pirate Jean Lafitte, scouting information for Spanish authorities interested in the expansion or defense of Texas. His motley travels and ambitions attest to the foment of regimes, loyalties, peoples, and economic possibilities of tumultuous early nineteenth-century North America.[4]

After his second Arkansas-area trip, a colleague asked him to systematize and write down his observations about the Mississippi Valley for the fledgling *American Journal of Science*. Bringier responded with a series of enthusiastic observations that leapt from lead mines to hot springs to topography to Native languages and cultures, throwing in a host of details along the way: useful clays, mastodon fossils, sources of salt and marble, the raft of downed trees that blocked the Red River. When the *American Journal of Science* published Bringier's report in 1821, the editor and founder of the journal, the influential professor Benjamin Silliman of Yale University, was anxious over the apparent lack of system. Bringier's observations were "immethodical," he frowned, and his specimens inconsequential.[5]

Such concerns were characteristic of the insecurity that haunted intellectually ambitious Americans of the era: the *American Journal of Science*—widely known as "Silliman's Journal"—was financially precarious and always challenged for material, but it was becoming the premier American scientific publication, the one most likely to be read in European drawing rooms. Silliman clearly worried that Bringier's reports were not up to snuff. But however much an eastern establishment might want more method, more theory, more elevated prose, or simply more explanatory transitions, the staccato rush of Bringier's "Geology, Mineralogy, Topography, Productions, and Aboriginal Inhabitants of the Regions around the Mississippi and Its Confluent Waters" was consistent both with the intellectually omnivorous natural history of his time and with the curiosity of Americans hungry for more information about their new lands.[6]

Bringier discussed at length the recent "earthquakes and eruptions" of the middle Mississippi Valley. During the January hard shock, he had been in the midst of the event, dazed by ten- and fifteen-foot eruptions of water and carbonized wood "blowing up the earth with loud explosions" and by "the roaring and whistling produced by the impetuosity of the air escaping from its confinement." His horse "stood motionless, struck with a panic of terror," as "black liquid" rose up to its belly. For two long minutes, trees "kept falling here and there," and the "whole surface of the country" was riddled with holes like miniature volcanoes.[7]

But Bringier's account was not limited to his experiences during those overwhelming few minutes. He later returned to the area and did what he could to document and quantify what he had witnessed. Despite Silliman's evident disdain for his work, Bringier's efforts were within the best traditions of natural history at the time. He measured the cavities left by exploding sand blows, finding them as much as twenty feet deep, though already eroding, and he estimated that trees in a nearby earthquake lake had sunk about thirty feet. He did not put these exact measurements to an immediate or practical use, nor would his readers expect him to: in the scientific approaches of his time, a first step in coming to terms with a puzzling phenomenon was to measure it. Another was to talk with people who knew something about it. Bringier also wrote an account of how people in the region had responded to the quakes. His account has been as widely disdained in histories of the New Madrid quakes as it was by the small intellectual elite of his time. Yet Bringier's detailed discussion of the quakes' aftermath points toward an important reconsideration of the impact of the temblors.[8]

Bringier reported on the response to the earthquakes of a Cherokee town along the St. Francis River, a more than 300-mile-long waterway winding from southeast Missouri down through central Arkansas, paralleling and ultimately opening into the Mississippi. Bringier conveyed a vision recounted by Skaquaw, the Swan, to a gathering at the community of Crowtown in the first week of June 1812. Skaquaw began,

It was about one moon before the earth first shook, one night, when every thing was silent, and the sky as clear as spring water, that I was standing leaning on a stump, contemplating the blazing star, (comet,) [and] those everlasting lights which sparkle from one horizon to the other, when suddenly four lightnings departing from the four opposite points, came and alighted together at my feet, and there I perceived the blazing star; I first raised it on a chip, but perceiving it did not burn the chip, I tried with my finger, and found it was a tame fire. The moment it was in my hand, I saw two children come towards me, one from sun set and the other from sun rise. They were as bright as the sun of noon, and exhaled a perfume which laid my senses asleep for a few seconds; when I awoke I was in the hands of one of the children; my spirit had passed into the blaze of the star, and I perceived my body leaning on the stump where I had left it.

To this, Bringier reported, the assembled crowd responded "Atea," a sign of assent Bringier apparently regarded as parallel with his community's "Amen."[9]

The vision continued. The star children told Skaquaw that

the Ever-Great spirit, with your mouth, speaks to his beloved red children, that he has determined to put an end to . . . the mortal enemy of mankind, and save his children alone; the fire of war is burning already in all four corners of the earth. Watch for a sign, and the earth will soon shake, like a horse who shakes the dust from his back; but be sure to move away from St. Francis before the next sign manifests itself: go towards the sun set, and travel until you are stopped by a big river which runs towards sun rise; there stop, plant corn, and hunt in peace, until the last sign prepares you to hope for days of happiness.

In the community prophecy conveyed by Bringier, this Cherokee seer conveyed a profoundly spiritual interpretation that rendered the tremors as meaningful signs—signs directed against a "mortal enemy," quite likely understood by Cherokee listeners as the territorially hungry white Americans swallowing Indian lands. The quakes had made swamps rise, shaken forests, and boiled solid ground into quicksand. Good land for hunting had become marshy swamp. But these were not random or meaningless events: they were a message of significance to the children of the Ever-Great spirit. In the alarming destruction of their main waterway, in shakes that even in June still reverberated in smaller tremors, the communities of the St. Francis could find both direction and hope.[10]

The result of these prophecies, reported Bringier, "was the total evacuation of St. Francis river." Within two or three months, he noted, "all the Cherokees abandoned their farms, (and some were very good ones,) their cattle, and other property and removed, some to White river, and the greatest part to the Arkansas. Those that fixed on the White river have since removed to the Arkansas."[11]

The usual story about the New Madrid earthquakes is that they did not really change anything. Nothing much was going on in the New Madrid region before the 1811–12 quakes, and nothing much happened thereafter. The tremors are interesting as a scientific question, but not really as a cultural one: if a quake rocks a bottomland and almost no one is there to feel it, does it make a difference? Yet Bringier's account of prophecy, discussion, and emigration reveals that this long-standing way of framing the New Madrid earthquakes is mistaken: many people were there to feel the tremors. Dismissal of the earthquakes fails to recognize the patterns of life in the region south of New Madrid and reflects an anachronistic sense of geography. The quakes did not hit an unpeopled wasteland: they shook an area that had emerged over the previous decades as a culturally and politically significant area of refuge from American encroachment. Moreover, they transformed it.[12]

Before the New Madrid Hinterland: Environments of Conflict and Confluence

By 1811, the New Madrid hinterland was a region of newcomers. Many of the people there—Native American as well as European and American and African—had arrived in the previous generation or two. Some, such as Quapaws, French Creoles, and Osages, lived in communities that had been well established for decades or close to centuries. But all the competing Native and colonial powers in the region built upon a much longer history of overlapping settlement by communities making use of the same resources: rich agricultural land, abundant game, and accessible waterways.

Beginning after the year 900, and for much of the next four centuries, communities of people fueled by corn—a recent import—built a large urban center near the east bank of the Mississippi, opposite what is now St. Louis. At its heyday, roughly 1100–1200, the ceremonial and urban center of Cahokia (named for sixteenth-century people living on its ruins), had a population of perhaps 20,000 people, drawing in crafts, trade goods, and stores of crops from a wide area of the middle Mississippi. Cahokia was a dramatic marker of what is now known as Mississippian culture—its huge mounds still rise above the floodplain outside East Saint Louis. Yet Cahokia faced the problems of urban areas in many parts of the world. By around 1300, for reasons that likely included soil exhaustion, pollution, disease, political turmoil, and climate change, Cahokia declined and its populations dispersed.[13]

In the 1540s, Spanish explorer Hernando de Soto and his expedition of hundreds of people visited large and resource-rich Mississippian "corn chiefdoms" throughout what is now eastern Arkansas. The newcomers fought, bartered, spread pigs and germs, and ate heartily among these tightly packed and frequently bickering hierarchical city-states built on agriculture, abundant fish and wildlife, and networks of commerce and diplomacy. Yet by the time French colonists, traders, and settlers began to travel the Mississippi and its tributaries in the late seventeenth century, the middle Mississippi had been largely depopulated, the corn chiefdoms dispersed and collapsed. Soil exhaustion, epidemics, and warfare between towns likely all contributed to this widespread set of shifts—though the precise contours of such large-scale social change remain frustratingly blurry—and European violence helped destroy some of the last of these communities. Yet the area at the confluence of the midcontinent's major river systems, the Ohio, Missouri, and Mississippi, had long been a site of travel, trade, and interaction, and it would emerge as one again.[14]

Downstream of the confluence of those mighty rivers flowed much of the continent's rainfall, as well as the trade and communication of many historically powerful groups along the fertile but contested ground of the middle Mississippi. In the late seventeenth century, several divisions of the Osages ranged from home territories in what is now the state of Missouri, along the lower Missouri, upper Meramec, and Osage Rivers, out into hunting regions from the woodlands to the near plains. South of the Osages, the Quapaws held and traded the lower reaches of the Arkansas River, mounting raids on the Choctaws and Chickasaws while also trading with and aiding French garrisons as well as hunters in the White and St. Francis River basins. East of the Mississippi River, diverse groups of people, primarily Shawnees, Delawares, and other speakers of Algonquian languages, claimed traditional territory and sought new sites, as demographic shifts brought about by European settlers, their animals, and their microbes began to shift populations and their land claims across North America.[15]

Crucial to the resources and economies of the colonial-era middle Mississippi Valley was the now obscure St. Francis River, once considered the "principal river" in the middle Mississippi Valley aside from the Mississippi itself.* The St. Francis is a long winding river that flows roughly 460 miles south from southeast Missouri into northeast Arkansas. It begins in the low granite mountains of present-day Iron County, Missouri, gathers waterflow from tributaries in the St. Francois mountains of the Missouri mining district, flows through southeast Missouri and into northeast Arkansas in rough parallel with the Mississippi, and empties into the Mississippi River not far from the port of Helena, Arkansas. In the era of the New Madrid earthquakes, what is now the Little River near New Madrid was commonly termed the main branch of the St. Francis; now only the western fork is regarded as the St. Francis proper. Certain bends of the St. Francis draw

---

*It became the St. Francis under American settlement. Until the early nineteenth century the river was known to most Americans and Europeans by its French name, the Saint François, and to Native Americans by its Choctaw name, *Chobohōllay*, from *oca chobohōllay*, smoky water. The name is tantalizingly descriptive, but there is little in the present historical record to explain what looked smoky about it.

near the White River of north Arkansas, and before the drainage and damming projects of the early and mid-twentieth century it absorbed smaller tributaries from throughout southeast Missouri and northeast Arkansas, as well as the L'Anguille right near its outlet. At the time of the Civil War, the St. Francis River gathered waters from roughly 1240 square miles.[16]

Today, much of the reach of the upper St. Francis has been drained and farmed, with water from the Castor and Whitewater Rivers now diverted by drainage outlets directly to the Mississippi. What remains in free flow makes for excellent whitewater canoeing. Much of the middle and lower river, especially after it crosses the state line into Arkansas, is placid and silty, in places as much bayou as stream. The St. Francis is crisscrossed and braided with other streams but interrupted with several dams and small lakes, and frequently adjoined by wetlands and sloughs. Before 1812, the river itself had a clearer channel and more direct flow. Much of the river's basin was seasonal wetland, especially the territory between the St. Francis and the Mississippi, but the river was navigable and free-flowing. After 1812, the St. Francis became stagnant swampland, with channels obstructed both by downed forest and by changes in elevation of the land. Today, the lower St. Francis is important habitat for waterfowl and small animals; before 1812, it was home to an astonishing proliferation of animal, bird, reptile, amphibian, and—especially—insect life.[17]

European Powers and the Continental Interior

The early nineteenth-century New Madrid hinterland was shaped in part by European machinations, and in part by the movement and interaction of indigenous peoples. Under the colonial-era carving up of North America, the French laid claim to trading rights with the people of the middle Mississippi—and, secondarily, rights to settle among them and administer political authority. The first official French presence in the Mississippi Valley was the small, frequently relocated Quapaw trading site called Arkansas Post, or simply Arkansas, established in 1686 along the river that would soon take the same name. Soon thereafter, French immigrants, many of them from French Canada, began to settle villages along both banks of the Mississippi, as well as the late eighteenth-century trading site of St. Louis, near the confluence of the Missouri and the Mississippi. By the 1720s, the St. Francis basin, along with hunting regions of the White and Arkansas, supplied meat, oil, and fur to the burgeoning economies of the lower Mississippi.[18]

Early European settlement of the middle Mississippi Valley drew upon long-standing indigenous resource use: French newcomers farmed,

traded, hunted, and mined. By the time French expeditions of the 1710s and
'20s were exploring around the headwaters of the St. Francis, near-surface
lead districts were a well-known resource for Native miners who used shallow pit mines to dig the malleable metal. Soft, heavy, and easily shaped,
lead had long been used and traded through indigenous networks. Its mal-
leability and weight made it ideal for melting and recrafting into small
valuable items including decorations, jewelry, and ammunition. French,
Spanish, and American explorers continued to search unsuccessfully for
silver—Louis Bringier's travels were in part to seek out the site for a silver
mine—but lead continued to be a main environmental resource of the region through the period of American territorial takeover.[19]

The French and Indian War of 1754 to 1763 led to an imperial resorting.
France ceded its claim to the enormous, if barely French, territories of Louisiana. Britain claimed Florida and territory to the east of the Mississippi,
while Spain took lands to the west (it would later reclaim Florida after the
American Revolution). During the mid-eighteenth century, the northward
expansion of Spanish settlements into the Texas interior reshuffled people
and trade in the lower Mississippi Valley and limited the westward shift of
groups like the Osages. Simultaneous and related warfare roiled the North
American Southeast, where Britain and France sought proxy armies of Native Americans who knew the woods, while Native tribes sought to play off
European powers for what they could offer in goods or in leverage against
each other—or against the increasingly territorially avaricious British colonists claiming and clearing the seaboard.[20]

After 1763, Spain encouraged settlements west of the Mississippi River.†
However, the middle Mississippi Valley was even less Spanish than it had
been French: from language to officials to what people actually ate, bought,
and did, Native regimes from the Osage to the Illinois to the Quapaw were
generally far more locally influential than the abstract colonial authority
of sweaty provincial officials with little more to support their authority
than a title and a few wax stamps. Yet colonial policies shaped population
flow in significant ways. For the thinly stretched Spanish administrators,
it made sense to recruit buffer settlements that would discourage land-
hungry Americans who poured over the Mississippi eyeing "empty" lands.

---

† Formal cession took place in November 1762, but the Spanish governor did not arrive
in New Orleans until spring 1766. Potentially rapid travel up and down waterways of
the middle Mississippi region—rapid, that is, for small groups of healthy adults when
seasons and weather were favorable—stands in extreme contrast with the glacial pace of
European/American travel, which in this era was never speedy for anyone at any time.

One result was a liberal policy of land grants that encouraged eastern Native groups such as the Shawnees and Delawares as well as disaffected Americans, including Daniel Boone.[21]

In 1800, because of European financial and political pressure, Spain ceded Louisiana (much of the area between the Mississippi and the Rocky Mountains) to France under secret agreement. Finally abandoning dreams of a North American empire, France eventually then sold Louisiana to the Americans in 1803. American diplomats were interested in land, but they were perhaps even more interested in water. Securing control of the Mississippi meant reliable connection between the Ohio River and New Orleans, connection which would further the commercial and agricultural development of existing American settlements as well as allowing for new ones.[‡] Imperial re-sorting may have changed little for decades—traders often of mixed culture and blended families, not officials of any far-distant regime, were generally the most important local nodes of interaction throughout the midcontinent up through the early nineteenth century—but these European interactions would eventually lead to the torrent of Americans who reshaped the region along with much of the rest of the continent.[22]

Native American Newcomers to the Environments of the New Madrid Hinterland

At the same time as these European machinations, the politics of Indian resistance elsewhere in North America spurred emigration west of the Mississippi. Between the 1780s and the end of the War of 1812, emigrant Indian communities from the Southeast and the Ohio Valley dominated large swaths of present-day Missouri and Arkansas. In addition to settlements elsewhere in what is now Missouri, groups led by Cherokees, Delawares, and Shawnees, but comprised of people from many eastern groups, along with French Creoles and a scattering of traders, trappers, and people escaped from slavery, created a network of settlement and travel along the game-rich St. Francis River and nearby waterways of present-day northeastern Arkansas and southeastern Missouri. New Madrid was the northern nexus of a hinterland that stretched down the St. Francis into these settler communities. When the quakes shook the region, they remade the St. Francis:

‡ Robert R. Livingston, a chief negotiator of the Louisiana Purchase, immediately secured the right to develop a steam-power monopoly on the Mississippi: he was one of the financial backers of Nicholas Roosevelt and the initial voyage of the *New Orleans* in 1811–12.

no more a conduit for transit and communication, it became a swampy barrier. Understanding the dramatic change in these Indian communities thus provides new insight into the overall historical and environmental development of the region.[23]

From the 1770s through 1794, Shawnees in the upper Ohio Valley fought the American "Long Knives" moving into their home territory. Some, likely including the mother of the later Shawnee leaders Tecumseh and Tenskwatawa, moved to what is now southeast Missouri. By 1782, multitribal groups of Shawnees, Delawares, Chickasaws, and Cherokees worked with Spanish authorities to resettle in newly Spanish territory. In the decades after the American Revolution, more Shawnees, Delawares, and other Algonquian speakers from the southern Great Lakes moved across the Mississippi to what is now central and southeast Missouri. They developed networks of agriculture, hunting, and trade in the New Madrid hinterland, as trading houses established practices and patterns that would later be extended to the Rocky Mountain fur and pelt trade.[24]

Along one northward bend of the Mississippi, a multiethnic and polyglot settlement of Delaware, Shawnee, Creek, and Cherokee people established a trading site that became known as L'Anse a la Graise, or Greasy Cove, for the tallow boiling through which animal fat would be processed for use and sale—bear fat, for instance, provided lubrication, cooking oil, fuel for lighting, food seasoning, leather softening, and hair and body oil.[§] In the 1780s, white newcomers joined the settlement, and in 1789 it was recreated as Nuevo Madrid, a grandly planned (if rather shoddily executed) settlement organized by a American Revolutionary War colonel named George Morgan in cahoots with Spanish authorities in one of the complicated schemes that characterized the workings out of imperial power along these multiple and overlapping frontiers.[25]

New Madrid became in the 1790s a thriving trading port, where game and fish fattened coffers and households and the settlement contentedly "swam in grease." Like the polyglot settlement at Arkansas Post, New Madrid was a nucleus of both Native and white settlement and trade, as well as a node of colonial politics: Spanish officials required boats to dock at New

---

[§] Modern reaction to the name Greasy Cove might be summarized as "yuck." But for people before the age of supermarkets and cheap groceries, "greasy" had a wonderfully tasty, lip-smacking feel to it: being up to your elbows in grease meant that there were plenty of provisions for the winter and lots of goods to trade, children would have big bellies, and hunters could rest and tell stories. "Greasy Cove" is not a name of nastiness and disgust, but of satisfaction and plenty.

Madrid for inspection. Scots trader Robert Goah Watson, who arrived in 1805 as part of the initial American influx of the region, noted the community as "a considerable trading point principally with the Indians." The small community was thus a focal point in the growing traffic and population along this major riverway.[26]

Through the 1780s and '90s, settlements expanded the New Madrid area to the north, especially in Big Prairie, which was primarily American, and to the south, around Little Prairie, which like New Madrid was largely French and Creole. Some settlement began also to the west and further south, along the St. Francis River. Americans gradually formed a larger portion of new emigrants, pulled by generous grants from Spanish officials.[27]

The Osages, meanwhile, resented and resisted such Spanish grants, seeking to preserve their historic dominance of trade routes on the middle Mississippi and lower Missouri Rivers even as pressures of population and dwindling game pushed them south and west of their once-held territory. Further south, in the lower Arkansas River Valley, the Quapaws struggled to preserve their historic position as the brokers and diplomats of the region between the lower Mississippi and the *pays d'en haut*, the upper country of the Ohio Valley and the Great Lakes. Both Spanish and French negotiators went through the Quapaws in efforts to keep peaceful trade flowing, even as Quapaw communities undergoing epidemics and the pressures of cultural shifts changed, coalesced, and shifted location.[28]

In this context of territorial rearrangement and jostling for power, Cherokee parties settled—and helped create—the New Madrid hinterland, especially along the St. Francis River. In several waves, Cherokee-led groups of settlers began to make the New Madrid hinterland that the earthquakes would eventually disrupt. One early wave of Cherokee emigrants moved west after the American Revolution, when a faction was ousted from the southeastern Cherokee Nation in the political upheaval following the disastrous alliance of Cherokees with Great Britain. After 1792, Spanish administrators welcomed Indians as well as anti-American dissidents who might help fight the Osages. These early emigrants settled along the banks of the St. Francis and near New Madrid.[29]

A second substantial emigration began in 1795–96 and was led by Connetoo, also known as John Hill. In that period, Delaware and later Cherokee delegations petitioned the Quapaws for permission to settle upriver of them on the Arkansas. The Quapaws refused, insisting that no one, white or Indian, should settle upriver of them. Instead, the Spanish governor at Orleans instructed the Cherokees to settle and hunt on the St. Francis and the Delawares to take the White River. This neat division was one more

paper demarcation that did not in fact determine or describe the actual world: Delawares, Shawnees, and Cherokees settled on the White and the St. Francis, where they were soon joined by what one American official termed "Choctaws and Chickasaws from over the Mississippi."[30]

Other groups of Native eastern and southeastern emigrants settled on the St. Francis and adjoining White River in the 1790s, encouraged by Spanish officials eager for more anti-Osage buffer. In later histories of the region, early settlements are often framed as separate, but they were anything but: travelers, traders, and hunters (including hunting parties from the Shawnee and Delaware villages further north) moved frequently and freely among these Arkansas-area settlements. Such porosity indicates also the multiethnic character of many such settlements, as well as their connections with communities of Cherokees, Shawnees, Delawares, and others east of the Mississippi.[31]

Over the course of the very early nineteenth century, a steady stream of people left the Cherokee Nation for lands west of the Mississippi. As US Cherokee agent Return J. Meigs reported to the secretary of war, many went to "the River Arkansas," while others stayed nearer the Mississippi, in the game-rich areas of the St. Francis.[32]

The Indian settlements of the New Madrid hinterland grew significant and sizable. In 1801, Cherokees petitioned for land grants to support the increasing settlement along the St. Francis River, by then "one hundred gunmen," or about six hundred people. By around the turn of the century, a history reported that "the country between the mouth of the St. Francis and the town of Cape Girardeau was occupied by remnants of the Delawares,

FIG. 2.1. William Clark's map of the New Madrid hinterland. This excerpt is from an enormous map developed by the American explorer and administrator from about 1780 to 1810. It hung in Clark's St. Louis office, where he would frequently update it based on new reports from travelers. Clark's script is as hard to read here as it is in his journals of the Corps of Discovery he led with Meriwether Lewis, but the sheer density of this map serves as compelling graphical evidence of the many Indian settlements along the middle Mississippi before the 1811–12 earthquakes. Extending up into New Madrid, roughly parallel to the central, double-lined Mississippi River, is what Clark labeled the "River St. Francis." To its east are small ovals for ponds—this was floodplain lowland. To the west, along the river and the dotted post road, Clark labeled numerous small triangles "Cherokee settlements." This New Madrid hinterland would become an almost uncrossable "Great Swamp" after the earthquakes, but when Clark mapped it, the region was a center of Native American emigration and commercial activity. (Yale Collection of Western Americana, WA MSS 303, Beinecke Rare Book and Manuscript Library, Yale University)

Shawnees, Miamis, Cherokees and Chickasaws, in all about 500 families." In the fall of 1805, Louisiana governor James Wilkinson estimated that the St. Francis River settlements alone numbered over a thousand people. By 1811, at least two thousand Cherokees lived along the St. Francis—and their largest settlement was the largest in Arkansas at the time, although it does not even appear in most histories of the area. Precise numbers undoubtedly fluctuated with travel and hunting trips: perhaps the most accurate representation of the time is William Clark's map of the area, which shows villages dotted throughout the St. Francis.[33]

Meanwhile, the third major wave of Cherokee emigration to the St. Francis and Arkansas River settlements, the movement of the "Western Cherokees" in 1810, created a crisis within the Cherokee Nation. In 1805–6, in a struggle that was to presage later crises of forced removal, a faction of Cherokee leaders agreed to exchange land in their homeland for US territory across the Mississippi. This scheme was pushed by federal policy, implemented through Cherokee agent Return J. Meigs, but was bitterly contested by other leaders and many throughout the nation. In 1810, Tolluntuskee led West almost a tenth of the nation: perhaps 1,200 out of about 12,000 Cherokees. Their departure created deep political rifts, including the deposition of the principal chief who had supported removal (Black Fox, a widely respected figure, was later reinstated). Despite promises, no land was exchanged by American officials, who cited the resistance of the majority of Cherokees. Those who emigrated—known as the "Western Cherokees"—were regarded by the rest of the tribe as grievously separated, even traitorous.[34] Some settled on the St. Francis, others along the Arkansas, where Meigs pleaded with Washington officials for a fur-trading station, known as a factory, to serve as a "rally point" for Cherokee settlements in the region. Altogether, these waves of emigration had shaped significant population and culture in what was becoming eastern Arkansas.[35]

The New Madrid hinterland grew as an ellipse, with a northern focus at New Madrid and a southern focus at Arkansas Post. North of New Madrid were the Apple Creek Shawnee and Delaware communities of the Cape Girardeau district, halfway between New Madrid and St. Louis. A southern portion of this trading zone stretched down the St. Francis, with western tendrils reaching out to the White River and Arkansas River Cherokee settlements and the trading center at Polk Bayou (present-day Batesville)—whenever Osages left those trading routes undefended. The Arkansas Post factory provided the southern focus of this hinterland. Trading houses run by white and mixed-culture households knit together European, American, and Native communities in an economy of fluid movement and exchange among diverse communities of recent emigrants.[36]

United States Policy and Eastern Indian Emigration to the New Madrid Hinterland

The emerging Indian policies of the US government and the desire of eastern Indians for new land shaped land conflicts of the middle Mississippi Valley. Under the administration of Thomas Jefferson, tribes of the East were encouraged—though not yet forced—to move west. Jefferson's policies envisioned the western territories of the mid-Mississippi Valley as a reservoir that would absorb dissatisfied people and separate them from American settlers and settlements. While not a permanent arrangement, this would provide space for a gradual acculturation of resistant Native groups as well as orderly American settlement of regions east and—ultimately—west.

This was a simplistic and in many respects naïve view of geography, both physical and human. Still, Jefferson had already seen enacted an equally sweeping and artificial geometry of American environments, in the Land Act of 1785 that divided public lands into grids that took no notice of ridgelines, streams, geological features, stands of woods, or anything else on the face of the landscape. This plan for the orderly division and sale of land, in all its audacious abstraction, was peopling the eastern half of the continent (grids determined by this piece of legislation shape the city blocks down which many present-day Americans stroll). It was not perhaps unreasonable for those in Washington in the first decade of the following century to think that a top-down artificial division of Indian from white, with a boundary at a major river, might solve or at least ameliorate the tensions and conflicts of multicultural jostling throughout the eastern woodlands. This scheme of voluntary western removal and separation of peoples manifestly did not work, as subsequent conflicts in the woods of the St. Francis make clear, but federal policy was nonetheless one substantial force in encouraging Native American emigration in the first several decades of the nineteenth century, well before the forced removals of the 1830s.[37]

United States Cherokee agent Return J. Meigs and Creek agent Benjamin Hawkins thus encouraged emigration of the southeastern peoples with whom they worked. Hawkins negotiated with one group of Creeks, who scouted areas around the Arkansas River, but the interested chief died before the effort came to anything.[38] Meigs, by contrast, actively promoted several rounds of emigration, including the "Western Cherokees" whose emigration after 1810 came close to splitting the Cherokee Nation. Communities of the New Madrid hinterland received some level of federal support and recognition. American officials took over the responsibility for funding interpreters from the Spanish and French before them, and accepted matter-of-factly the extensive settlements of what one official in 1808 termed "the Cherokees on the River St Francis."[39]

At the same time, many Indian leaders themselves saw lands west of the Mississippi as an outlet from territorial encroachment. After the disastrous Treaty of Fort Wayne in 1809, in which Shawnees and other Algonquian-speaking related tribes gave up huge swaths of territory in what is now Illinois and Indiana, some Native leaders in Ohio called on Americans for new lands west of the Mississippi. For a time, leaders of several nations converged on the idea of emigration as a partial resolution to pressures of land, population, and natural resources.[40]

What might look desirable in Washington, or even by the Wabash, proved far more problematic in the rapidly settling woods and streams of the New Madrid hinterland. American settlers were quick to cry for help and intervention as Native settlements increased and friction grew along with trade. In 1804, the regional military commander endorsed the fears of local French and American residents of New Madrid about the "numberless Indian stragglers" pillaging river traffic and threatening settlements. He requested permission to deploy a company of militia against "the insults and depredations of lawless Bands of Indians." In particular, he argued that a "military establishment seems absolutely necessary on the river St Francis," in order to "keep the Indians in check, protect trade, afford Security and give confidence to our settlers in upper Louisiana."[41]

By 1803, an American official complained that "on the River St Francis ... are settled a number of Vagabonds, emigrants from the Delawares, Shawnese, Miamis, Chicasaws, Cherokees, Piorias and supposed to Consist in all of 500 families, they are at times troublesome to the Boats descending the River, and have even plundered some of them & Committed a few murders." As this official's unease reflected, the Native American settlements of the New Madrid hinterland were fluid sites, from which people traveled far in canoes, on foot, and on horseback, and through which diverse peoples passed. As during the Spanish and French regimes, many southeastern Indians crossed the Mississippi in the early nineteenth century to hunt or visit family for perhaps a season or two before returning to their homelands.[42]

The family relations and emigrations of Shawnee leaders Tecumseh and Tenskwatawa may be entirely representative of many families. By the turn of the century in what is now southeast Missouri, in addition to their mother, they had at least one and possibly several female relatives: possibly a half-Cherokee daughter of Tecumseh's, and most likely a niece in New Madrid who married a French Creole man. On several occasions Tecumseh traveled through the region from points east. In similar networks of travel and relationship, new settlements across the Mississippi were connected with eastern homelands, families, and debates.[43]

In 1809, officers of the local militia in New Madrid complained to William Clark of the situation. They resented not simply "the great number of Indians resident here interspersed amongst the Whites," but also "that all the Indian Tribes resident on the Ohio and Tennessee Rivers in their passage to and from the different tribes living in the South; pass through this place and the Village of Little Prairie." Such fluidity was unsettling to Americans like the New Madrid militia officers, who saw transient Native Americans as a threat.[44]

US federal policy toward these western settlements was riven by conflicts, partly because of political differences and partly because of the vast distance—physical and in effective governance—between federal edicts and local conditions. In the already well established tradition of American westering, many emigrant Americans set up homesteads on lands they did not own. These squatters from the United States pressed the Indian settlements along the St. Francis in the early 1800s. For a brief but tumultuous period in 1805, governmental policy dictated the eviction of white squatters and the complete cessation of private Indian trade. In September 1805 the US governor sent a detachment of army troops to clear out illegal white squatters in the St. Francis basin. That November, President Jefferson further instructed the newly appointed territorial governor, James Wilkinson, to accomplish "the depopulation of our loose settlements below [St. Louis] on the Mississippi & its branches—and the transfer of the Southern Indians to this Territory." American officials hurried to establish a more formal trading venture up the White River to pull the Native communities into closer commercial connection with agents of the US government.[45]

Indian agent John Treat complained in 1806 of the efforts of trading houses to secure the business of hunters through the extension of substantial credit. In addition to trying to capture the trade with many Indians directly, one private trading house extended "numerous Credits . . . as well to the Whites living on the White and St. Francis who collect the Peltry from the Indians at exorbitant prices . . . also the Indians readily obtain Credit one of whom (not a Chief) a short time since obtain to the amount of upward of \$3000 by which means more than one hundred Hunters are secured to them." This government factor was frustrated at the ways in which ordinary Indians, not even recognized leaders, were able to gain credit—loans leaking through the constraints of more regularized relations, goods dispersed and lost in the lowland forests and small streams of the middle Mississippi.[46]

Efforts to restrict trade and evict American settlers were never wholehearted and never very successful. Territorial governor Wilkinson used the intermingling of communities, and the usefulness to Native groups of American- and French-run trading houses among them, in objecting to the directive against illegal settlers. Complaining of the ungovernable state of the St. Francis, Wilkinson warned Secretary of State Madison that "I may find some difficulty (I fear) in dislodging eight families, which have taken refuge with a strong tribe of Cherokee Indians, high up on the same river." Wilkinson opposed the effort to clear out squatters, so his objection cannot be taken too simply at face value, but his concern may indeed accurately reflect the environmental dimensions of early settlement of the St. Francis and the New Madrid hinterland. A difficult terrain of lush and seasonally overflowed land could be challenging to governmental oversight and thus extremely useful to people seeking small plots to farm and wide areas to hunt, trap, fish, and travel.[47]

Ecological Bounty and the St. Francis Basin

The settlement of the New Madrid hinterland reflected how ecological conflict east of the Mississippi pushed communities to find new environmental accommodation in the West. By 1809, Indian agent Return Meigs noted of the southeastern Cherokee Nation that "the hunting life here is at an end." Game was so depleted in the eastern Cherokee, Choctaw, and Chickasaw homelands by the turn of the century that only by pushing far afield of traditional hunting lands could hunting parties return with furs for the booming peltry trade. Scouting parties sought the riches of western lands. Many found what they were looking for in the mixed ecosystems of the St. Francis basin, where hunters, trappers, and fishers could find ample stocks further from Americans.[48]

Both white American and Native American emigrants sought similar environments in the New Madrid hinterland—small-scale farming and husbandry, with an emphasis upon hunting, trapping, fishing, and trade. Such similar environmental use could lead to conflict. In 1808, a man named Elisha Evans died in New Madrid, almost insolvent, leaving seven orphans who were parceled out by the New Madrid Orphans' Court. The livestock of this resource-poor household suggests the mingling of American and Indian households. Among Evans's property his probate listed "a Bull about 4 years old running in the woods." But just before the bull appears a barely legible listing for "3 Cows with Calves." These had been apparently first listed as Evans's, but were then scratched out from the list, with an explanatory note that they were "the property of Kincheloe"—likely the William Kincheloe who signed a legal document related to a separate probate in 1814. The style of his naming in the Evans probate suggests that he might well have been Native American—few white people were ever called by just one name in these records. Elisha Evans's probate thus suggests that a bull left to run in the woods by a white New Madrid–area farmer might well herd with cattle owned by a nearby Indian livestock owner. This kind of mixing of livestock was the cause of conflict in the St. Francis River basin, as Cherokees and white Americans took or killed each others' livestock around the turn of the century and early 1800s.[49]

New Madrid hinterland terrain in which to hunt, trap, and fish also provided a refuge for people fleeing control. Indian trader Samuel Hopkins reported in 1825 that three men owned as slaves by a local Chickasaw were "forced from his savage cruelty" to flee to Indian villages of the sunk lands. Similarly, as a boy emigrating to Arkansas around 1810, American L. D. Lafferty remembered encountering below the mouth of Little Red River a settlement of thirty or forty African Americans. The oldest among them said they had fled Spanish New Madrid and set up this settlement on a large mound surrounded by "low swampy country." Lafferty's father called this settlement "Nigger Hill." The editor of Lafferty's memoir reported that eventually the people "were subsequently either killed or captured by the whites, and the colony broken up," but the very existence of an independent African American settlement in northeast Arkansas indicates how the terrain of the region offered possibilities not present elsewhere.[50]

African American use of the New Madrid hinterland never figures in formal histories of the region, but not all history is formal. In the very late nineteenth century, local white residents called an area of the Big Lake Swamp south and west of the town of New Madrid "Niggerwool Swamp" for its characteristic black, curly ground moss. The insult—a derogatory term for African American hair—is cutting in its casualness, but the name may possibly be a faint and distorted echo of actual African American embodiment, of people living by choice in an environment that for a time offered autonomy on the fringes of a slave regime.[51]

Interstitial land—land that was watery, seasonally overflowed, not steady prairie or roaring river—was useful for people whose primary modes of travel were by water, rather than overland, and who were not interested in large-scale monocrop farming. In colonial Louisiana, polyglot communities of Native Americans, escaped slaves, and runaway soldiers existed because they colonized watery land that was otherwise unwanted. Similarly, slaves working the rice fields of the Georgia and Carolina low countries in the eighteenth and early nineteenth centuries used knowledge of their tidal environments to their advantage: making small watercraft, trading, fishing, growing small plots of produce on pieces of higher ground. Small-scale, in-between territory offered a richness of animal and plant life that could sustain commerce and cultures.[52]

For hunters, the area between the St. Francis and the Mississippi was a bonanza. In his later years, Missouri memoirist Theodore Pease Russell reminisced about the hunting camps made by local farmers in the swamps of the Missouri Bootheel during the winters. "They would hunt bear and deer and turkeys, and what splendid times they did have and how fat they lived, roasted bear meat and wild turkey breast for bread," he remembered. "Oh! It makes me hungry just to think of it . . . and their cattle would come out of the swamps looking sleek and nice, after feasting on the cane all winter." Such bounty appealed both to American hunters and to Native Americans weary of depleted territory to the east. Yet as diverse groups moved to capitalize on these resources, conflicts over land and ecologies intensified.[53]

Trade in the New Madrid Hinterland

Indian and American (and European) emigrants created networks of commerce and communication throughout the New Madrid hinterland. From the perspective of early American officials, trade in animal skins, pelts, and fat in the St. Francis basin was frustratingly profitable, prolific, and hard to control. The lull in trade and settlement in the St. Francis basin in 1805 was short-lived. Americans were soon back in the region, aided by Native commercial and personal connections. By late 1808, illegal American traders and settlers were engaging in exchange and in bullying in the woods. Such "unlicensed Traders" were a constant problem for the early American administration of the Louisiana Territory. Early in 1809, Meriwether Lewis, recently a continental explorer but by then governor of the Territory of Louisiana, issued a proclamation dispersing "intruders" near the Shawnee villages along the Merrimac and "the Cherokee and Delaware Towns, on the river S^ت Francis, in the district of New Madrid." A government agent was sent out to detect illegal trading houses in the winter of 1808–9. Former co-captain William Clark (known to Native contemporaries as "Big Red Head") reported in early 1809 that Lewis planned to send the same agent back to "Cherokee Towns on the St. Francies River" to seize their substantial supplies of "merchendize and Whiskey." Whatever the intentions, such efforts did little to stifle what was by then a steady fur and peltry trade.[54]

Trade operated mainly at crossings of the St. Francis, including at a large-scale trading venture established by Connetoo and at the "usual crossing place" frequented by Delaware and Cherokee settlements. Several government agents proposed to set up official factories to pull the Indian settlements more into connection with the Americans. They followed the Native settlement patterns of the New Madrid hinterland: one of the recommended spots in 1808 was at Connetoo's trading post.[55]

For this brief but intense period in the late eighteenth and early nineteenth centuries, the New Madrid hinterland existed as an area of cross-cultural contact and exchange. In the complicated and interleaved history of the New Madrid hinterland, Connetoo played a significant role. Connetoo and William Webber (usually known as Red Headed Will Webber)** were two leaders among the Cherokee who had immigrated to the St. Francis in the mid-1790s. Connetoo's group settled a village on the slopes of Crowley's Ridge that became resettled as the Arkansas village of St. Francis. By the early 1800s, Connetoo was working profitably as a trader—though his business was illegal under the laws US officials attempted somewhat haplessly to maintain. Connetoo managed a wide-ranging and profitable trade in peltries and tallow: in the spring of 1807, Connetoo took in $4,800 in skins in one single set of illegal transactions witnessed (and lamented) by a government agent.[56]

Legal records from early Arkansas and Missouri reveal the recognized role of such subsequently forgotten early settlers of the region. When trading partners Peter Saffray and Francis Contelmy died along the St. Francis in 1806, most likely of the fevers endemic in the Mississippi Valley, their joint papers and other possessions were, according to legal documents, "Deposited ... in the hands of Wm Weber and Kanetou Both Chiefs of Cheroque Nation." A related affidavit in May 1806 by Anthony Campbell, identified as "lately resident at the river St. Francis," asserted that certain articles "said to belong to the estates of Safry & F. Contelmy were deposited in the possession of William Webber and Kanietaugh Cherokee indian chiefs on the River Saint Francis." Legal strictures recognized the local role of Indian leaders in this multicultural hinterland.[57]

Subsequent legal records involving Connetoo also reflect the conflicts over ethnicity and identity smoldering at the time. Under his American name of John Hill, Connetoo appeared on several juries in December 1811. Then, when involved in another case during the same court session (within days—the courts were highly episodic and thus highly compacted), he successfully claimed exemption from American prosecution as a Cherokee.

** Names and flaming red hair testify to the family and cultural blendedness of many Cherokees.

This ended his legal jeopardy but also his jury service, since only a "white male citizen" could serve on a jury.[58]

The complexity and volume of social and economic exchanges in the New Madrid hinterland appear in the historical record only in the early nineteenth-century probate records in which neighbors and the American courts sorted out the financial affairs of those who had died.[††] In the summer of 1807, William King, a merchant with ties to Pennsylvania, died suddenly "in a deranged state of mind." He left an estate of over $15,000—of which $13,455 was inventory from his store in New Madrid. Thirteen thousand dollars was an impressive inventory in 1807. That much merchantable goods indicates a significant trading zone. Similarly, if less spectacularly, when Jean Baptiste Barsaloux died in 1807, he left over $6,000 in assets: over $3,700 of inventory, over $1,500 of peltries, and a complicated pile of receipts, pledges, and notes—personal statements about monies owed or goods received that were one main form of exchange in a world largely devoid of banknote currency. Over $212 of these pledges were subsequently listed as "claimed by Indians," evidence that even though Native people might not be formally named in an American probate, their trade was a significant force in Barsaloux's community.[59]

In 1809, Pierre Dumay died at Little Prairie, a small village that would be completely destroyed by the 1811–12 quakes. Dumay left extensive debts, several children to be dealt with by the Orphans' Court, and complicated financial dealings that took at least until 1823 to straighten out (typically

†† These probate records are deeply exciting, in a geekily historical sort of way, because they were long held to be lost—burned in a courthouse fire or perhaps washed away in a Mississippi River flood. Luckily, I was encouraged by Lynn Morrow, the head of the Department of Local Records of the Secretary of State's office in Missouri, to go poke around. In June 2004, I visited the New Madrid County Courthouse to see what scraps might remain after fire, flood, or other calamity. I was very fortunate to be accompanied by Joan Feezor, a field archivist from Local Records whose sweet demeanor belies steel will and keen historical instinct. The clerk told us there were no old records around, but Joan insisted that we be able to look a little further. I started leafing through an enormous, dusty ledger that recorded old probate files, and amid the mid-twentieth-century listings I noticed a couple with odd case numbers that did not fit—numbers such as 1811 or 1807, that looked as though they could possibly be dates from the early nineteenth century. We requested that a few of these odd-numbered files be brought up from the dark basement—and then I am not embarrassed to say that Ms. Feezor and I did a little undignified victory dance right there in the courthouse conference room. Old probate files had apparently been scattered or misfiled, and then filed back into the probate book as they were recovered. Thanks to the diligence of the Department of Local Records, all of these probate files are now available through the Missouri Secretary of State's judicial records website, opening up an era's history that had long been lost.

for this period of fast-changing social and political structure, his legal records become less French and more English over those fifteen years). Dumay also left 1,076 pounds of lead shot. His probate is as difficult to sort out in the present day as it was in the early nineteenth century, but the interleaved debts and sheer poundage of the shot—very likely manufactured from local mines—suggest extensive trade with area hunters.[60]

Entrepreneurs held big plans for the New Madrid hinterland. John Merckel was a naturalized German-born merchant who died of a sudden illness in 1805. His property was only worth about $263 at the time of his death, but the range of goods indicates the interests that existed in his communities: his inventory included both an advertisement for an encyclopedia and a set of "Squaw's needles." (Indicating the cultural miscellany of the New Madrid region, the personal goods of the small-scale trader further included a dozen each of silver table and tea spoons, a spy glass, a pair of pocket pistols, and a silver cream jug.) When Benjamin Vanamburgh died in 1807, he left three orphan children who were apprenticed off to local farmers, a small yard of domestic animals, many farming tools, forty bushels of Indian corn, and 460 young apple trees. Those apples were probably meant to be consumed in liquid form: few apple varieties of the early nineteenth century were sweet enough to be eaten straight from the tree, but hard cider was a cherished refreshment. As American official Amos Stoddard noted, in 1803 settlements around New Madrid produced 740 gallons of whiskey (probably made from local corn) and bought 665 packs of peltries from Native trappers and hunters. Such inventories sketch the New Madrid hinterland as a lively zone of economic and social exchange.[61]

A New Understanding of New Madrid in Relation to the St. Francis River

Historical accounts of the middle Mississippi have passed over New Madrid as a small river port known as an early colonial settlement but otherwise remembered (if at all) because of the earthquakes that bear its name. They miss its importance because they represent the place as a dot on a map. "New Madrid" means the bounded village of houses, docks, and a few fields. Such a depiction of peoples and places is an anachronistic geography. That pointillist understanding stems from Anglo-American notions of landownership and the primacy of settled agriculture, neither of which held sway in the late eighteenth- and early nineteenth-century middle Mississippi Valley.

New Madrid was important not because all that many people lived there or farmed close by: instead, it was important as a nexus through which

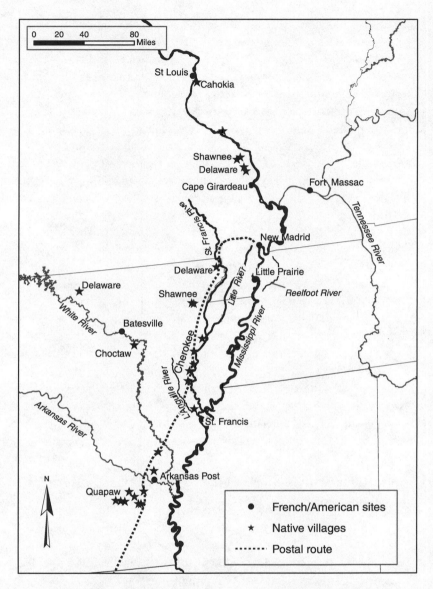

FIG. 2.2. The New Madrid hinterland, pre-earthquake. New Madrid, itself a small river port, was one anchor of a hinterland of settlement and trade that reached north to Delaware and Shawnee villages and south through Cherokee-led settlements to Arkansas Post. Before the earthquakes, the St. Francis River system—especially the Little River, which was known at the time simply as a branch of the St. Francis—was a well-traveled conduit for settlement, travel, and commerce in pelts, lead, and other valuables, including US mail. (Geographer: Bill Keegan)

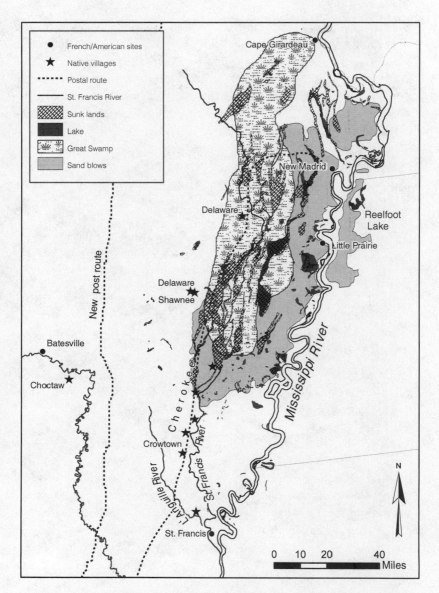

FIG. 2.3. The middle Mississippi Valley after the New Madrid earthquakes. The earthquakes uplifted portions of terrain, created the "sunk lands," covered wide areas with sand blows, and stopped up the free flow of the St. Francis River system. What had been a flowing conduit became an almost impassable swamp. Trade and travel could no longer go up and down the river—the former US postal route had to be remade, bypassing the "Great Swamp." Native settlers of the St. Francis fled west and south. Soon, their very existence would be forgotten. This map outlines only the clearly documented environmental changes, but they were likely even larger. (Geographer: Bill Keegan. Both maps based on William Clark 1810 map, maps by Jonathan Melish, Zebulon Pike, and Zadok Cramer, General Land office plats, 1807–1998, and *The Official Military Atlas of the Civil War*)

passed people, goods, and information. In 1806, New Madrid was the site of a "grand Council" of Delawares, Cherokees, Choctaws, and Chickasaws attempting to unite in coalition against the Osage. The area was, as the New Madrid militia complained in 1809, a zone of passage for those moving from the Ohio and Tennessee Rivers into the South, via the continent's major waterways. It was important in the same way and for the same reasons as Native American communities such as the upper Missouri Mandan and Hidatsa villages where Lewis and Clark and their Corps of Discovery wintered: relatively small settlements that channeled people, diplomacy, and trade through a much wider area. New Madrid itself was merely the mappable indicator of the much larger, more diffuse, but significant hinterland that stretched down the St. Francis River.[62]

The flow of rivers—and especially the flow of the St. Francis River—is crucial to an understanding of why New Madrid was before the earthquakes a vital and lively place, the anchor and opening for a hinterland of regional importance. People in the late 1700s and early 1800s relied upon rivers and waterways: as conduits and pathways, often navigated swiftly to connect communities over long distances. As William Clark complained of possibly hostile Sac Indians in 1809, "They pass from one side to the other of the Mississippi as they find it convenient." In the tumultuous period of the late eighteenth and early nineteenth centuries, rivers and streams provided connections since lost through the slicing of state boundary lines and the imposition of a different geometry of roads, drainage canals, railroad tracks, and highways.[63]

Separating the settlement of New Madrid from its hinterland thus reflects modern but not early nineteenth-century ways of defining land and people. In the probate for traders Francis Contelmy and Peter Saffray, the traders who died along the St. Francis in 1806, Contelmy was recognized as claiming "a tract of 750 arpents of land as a Settlement right situate[d] on the river St. francis about fifty miles from the town of New Madrid." (This would be, roughly speaking, somewhat below the south-west corner of the Missouri Bootheel.) He was also referred to in legal papers as a resident of New Madrid. Legal and commonplace understanding of lives was much more flexible in Contelmy's world than in subsequent history-writing about it.‡‡[64]

‡‡ It is possible that this flexibility may extend to other aspects of Francis Saffray and Peter Contelmy's legal record, as well. In a highly unusual process, the two men were probated together. This aspect of the historical record is a reminder to us that much that was important and obvious in the past was never written down—pregnancies visible to all for months, for instance, often enter family records only as a record of labor and

Based, apparently, from this settlement on the upper portion of the St. Francis River, Contelmy and Saffray had extensive enough dealings with Cherokees that their goods were deposited after their deaths with William Webber and Connetoo. This does not mean that these two smaller-scale Indian traders necessarily lived very close to the larger Cherokee settlement and trading house, as the crow flies or the river flows: on the contrary, all the people involved likely used the St. Francis and other waterways to connect places relatively distant by land.

The New Madrid Earthquakes and the Remaking of the St. Francis River

Social and economic relations were made possible by the connective flow of the St. Francis and its related waterways, especially what is now known as the Little River. When the rivers changed, so too did the communities that depended upon them. The creation of the sunk lands substantially altered the environment and settlement of this vital trading region.

Before the earthquakes, the St. Francis was no swampy backwater, but a riverine highway. Those who encountered the New Madrid hinterland in the twentieth century found a land remade by swamp reclamation, railroads, timbering, industrial-scale agriculture, and vastly altered social relations, as well as by earthquakes. Many concurred with the assessment of reclamation engineer Otto Kochtitzky, a crucial figure in the remaking of the Bootheel, that the St. Francis River east of Crowley's Ridge in Missouri "is no river at all," but a collection of promiscuous overflows and sloughs lacking a "proper individual channel." By the time he encountered the region, that assessment held true. For newcomers before 1812, however, the St. Francis was an important channel of communication.[65]

The St. Francis long formed a boundary marker on terrain and part of the New Madrid hinterland. Early settlers following George Morgan used the "St. Francois" as a westward boundary of the region in which they hoped to found a town, and the river formed a western boundary under the French and Spanish. Meanwhile, Quapaws used the St. Francis to explain the eastern boundaries of *their* claims to American explorers in 1805. Under American jurisdiction the St. Francis defined part of the boundary in the creation of District of Arkansas in 1806 and then the districts of New

---

birth. The joint probate for these two traders may indicate that Contelmy and Saffrey were regarded (grudgingly or implicitly) as life partners—or, equally plausibly, that they were simply work partners whose financial affairs were so mingled as to not allow for separate probate.

Madrid and Cape Girardeau. Early American settlers in 1804 petitioning the military commander of Upper Louisiana for protection from Indian marauders cast themselves as positioned on "a vast frontier from the mouth of the Ohio river to the river Arcansaw along the Mississippi, and then the whole length of the river St Francis on our back." Not an ill-defined morass, the St. Francis was a corridor of what American newcomers regarded as a perilous threat.[66]

Before the New Madrid quakes, the St. Francis had two main branches: by the late nineteenth century, the eastern fork was termed the Little River. This waterway was a navigable and accessible connector for a long stretch of the middle Mississippi Valley.[§§] At high water, a chain of bayous and lakes connected the Mississippi and what was then referred to as the St. Francis at New Madrid: this connection was a main reason New Madrid was such a good spot for a trading post. Waterways that were contingent and seasonal could be no less crucial for commerce: in the early nineteenth century, much travel was seasonal (muddy roads could be impassable; iced-over rivers might be easily forded; rivers at low water might be impossible to travel on but at seasons of high water be swift and smooth). One French official charged to develop the lead mining of what is now southeast Missouri recommended to court officials that "the economical and efficient means was to ship the lead on one of the smaller rivers, the St. Francis, which flowed into the Mississippi near the Arkansas River." He further explained that "since St. Francis was navigable during the rainy seasons, by placing the lead on pirogues near the mines, it would reach the Mississippi River in eight days." Such routes continued in use under American administrations in the early nineteenth century.[67]

Missionary and travel writer Timothy Flint noted in his widely read *History and Geography of the Mississippi Valley* that keelboats would travel the upper St. Francis before 1811, and explained why: "As its current was much gentler than that of the Mississippi, which, in this distance, is peculiarly swift and difficult of ascent, boats for New Madrid used to enter the mouth of the St. Francis, and work up that river to a portage, about twelve miles back of [New Madrid], and thence cart their goods to that town." That would all change in 1811. "Previous to the earthquakes," observed one trapper in the later nineteenth century, "keel-boats used to come up the St. Francis River, and pass into the Mississippi, at a point three miles below New Madrid.

___

[§§] As recent seismologists have discovered, in a much earlier era it had once been a braided stream, but became a more stabilized river with a clearer channel after an earthquake between five and three thousand years ago lifted adjoining terrain.

The bayou is now high ground." By 1819, lead was being transported overland by wagons for forty miles to the Mississippi.[68]

The creation of the sunk lands transformed the New Madrid hinterland
by transforming the environment of the St. Francis. The quakes themselves
were a sudden terror. During the quakes, waters rose so fast in some of the
swamps that a trader along the St. Francis was drowned. One newspaper
account reported widespread change to the St. Francis during the quakes:
"The St. Francis was at one time very low, at another overflowing the surrounding country." Decades later, the disruption from the quakes was still
amply evident. In 1858 geologist David Dale Owen reported that "the traces
of earth-cracks and sandblows are numerous, almost every where in the
St. Francis bottom. [S]ome of the earth-cracks are eight to ten feet wide and
six to eight deep." But this transitory if overwhelming drama led to larger-
scale changes to topography and water flow.[69]

The largest, most extensive, and most transformative topographical
depressions took place along the wide swath of the St. Francis, in what geologist Charles Lyell called the extensive "sunk country."[70] The earthquakes
dammed up portions of both the modern St. Francis and what is now the
Little River, to the east. The earthquake impoundment created a large area
termed the St. Francis sunk lands. Along the St. Francis (that is, what was
in nineteenth-century terms the western branch of the St. Francis), the
earthquakes created what became known as the Hatchie Coon Sunk Lands
and Lake St. Francis. To the east, the impoundment of the Little River (that
is, what Flint and other nineteenth-century writers called the main branch
of the St. Francis) formed the Big Lake sunk land, now a broad and shallow
fishing lake. The confusion of names perhaps accurately reflects the confusion of topography: the sunk land area became a swampy morass. Before the
creation of a modern dam that controlled water levels, Big Lake would expand and contract depending on local rainfall, creating a seasonally changing lake in place of a seasonally flooded forest. In 1856, the Little River was
declared "inaccessible swamp."[71] A longtime trapper reported in the 1860s
that in the St. Francis "sunken lands" stands of highland trees such as wal-
nut, white oak, and mulberry were at times—apparently times of high water—to be seen "submerged ten and twenty feet beneath the surface."[72]

The earthquake's tumult made dry land wet and former waterways dry.
Rivers filled with sand. A traveler named James Fletcher reported that "a
little river called Pemiscoe, that empties into the St. Francis ... is filled
also with sand." Here and in many other places after the tremors it was
reported that "on the sand that was thrown out of the lakes and rivers, lie
numerous quantities of fish." Local topography was further disrupted by

extensive slumping and cracking. After the quakes, a newspaper in Tennessee reported that "a gentleman in attempting to pass from Cape [Girardeau] to the pass of St. Francis, found the earth so much cracked and broke, that it was impossible to get along."[73]

Taken together, these changes rendered the St. Francis area into a broad, spreading swamp not usable for regular commerce and communication, rather than a more straightforward river with defined branches, reasonably deep channel, and seasonally useful flow. Downed timber clogged the St. Francis and the newly created Big Lake. Later surveyor Joseph C. Brown found the upper St. Francis (near the Missouri-Arkansas boundary line) to be so filled with old logs as to resemble an "old mill pond." Log "rafts" had formed, some functioning as floating islands. He noted that many of the cypress trees were still alive but concluded that "the land I presume has been sunk by the earthquake. The water flows about through the timber with great difficulty." Later seismologist Myron Fuller noted the hardiness of cypress in the sunk lands, explaining in his 1912 centenary report that in many places "the submergence was not sufficient to kill the cypress immediately, the trees lingering for years, some even to the present time, before succumbing to their changed environment." Contemporary tree-ring scientists agree, recording decades of stunted but continued growth in bald-cypress in the sunk lands because of the subsidence and movement of the earth in 1811–12.[74]

In the decades after the quakes, the damage caused to the St. Francis by the tremendous shakes was well recognized. In 1825, a trader complained that what he called the "east branch of the St. Francis" was "filled with timber" and thus "unnavigable for every species of boats except canoes." This local knowledge was long-lasting, even as it became little heeded outside the area. In the late 1890s, a number of deponents gave evidence in a long-running legal case about a drainage project in the sunk lands that was, after twenty-nine years of litigation, determined to have been a massive swindle of public money. Deponent Joseph Pelts gave testimony about "old logs taken out of the bottom of the river that have been bedded since 1811 and '12."[75]

The earthquakes blurred and spread the channel of the St. Francis. Joseph Pelts testified that "there is a great many places in all these sunk land rivers where the channel used to be before the shakes, it is choked up." "There is no straight channel to that river," one former reclamation worker similarly testified in the same trial, "so a man can't come at it." As witness G. T. Terrance explained of his young adulthood hunting, trapping, fishing, and farming in the sunk lands of the mid-nineteenth century, the St. Francis was "all water ... there ain't no bank." By the decades after the Civil War,

FIG. 2.4. Downed timber in the "Great Swamp." When engineers of the Little River Drainage District set out in the 1910s to clear and drain the earthquake swamps, they documented wide areas where timber blocked what had once been clear-flowing waterways in the New Madrid hinterland. That timber came in part from the 1811–12 earthquakes—but LRDD director Otto Kochtitzky actively denied the region's earthquake history, arguing that the swamp was simply a seasonal overflow and should be "reclaimed" for farming as completely as possible. (Image 6–06, from photograph album "The Ditches When They Were New," Little River Drainage District Records, courtesy Special Collections and Archives, Kent Library, Southeast Missouri State University)

the swamplands were only accessible by small dugout canoes like Terrance's. Even these had to move slowly, without any current to assist their travel. Operators had to pole or push their crafts or "cordelle" them, running a line out to a tree and then winching up even with it. One resident, a hunter and trapper, testified that a boat with a shallow foot-and-a-half draft could only travel down the Varner (now Varney) River to the St. Francis at high water. Even heavily laden canoes had trouble pushing through the thickets of clogging smartweed.[76]

In the 1890s, J. W. Estep described his experience in the mid-1870s in what he called the "Varner's river" tributary of the St. Francis where he had lived and spent time as a boy, fishing and frogging and tracking raccoons. In the river, he testified, he encountered "oak logs and other logs that fell down in the water . . . said to have been there when the 'shakes' was in this country." Under questioning, he explained, "I have been told by old people them logs fell there at the time of the earthquake.... They fell in the river and stuck up so the government boat [a craft whose work was at issue in the court case] couldn't get over them without sawing them off, and they pulled them out of the way." Of another cutoff where he had frequently caught fish, Estep explained that it was "timbered through there till a small cabin boat couldn't have went through there very well, unless it be a very high water time," until a "whisky boat" had gone through and "cleaned out" and "notched" the trees so it could pass (peddling whiskey remained one of the few historical constants of trade along the St. Francis, pre- and post-earthquake). Lumber cut for market got snagged and sunk in the whiskey boat cutoff, becoming a community resource when it was revealed at low water.[77]

The sunk lands resembled the swamplands common throughout the lower Mississippi Valley, but they had persistent, unique characteristics that identified them as having been created by earthquakes. Early twentieth-century geologist Edward M. Shepard argued that the small conical sand blows beneath the waters of the depressed sunk lands bubbled up with artisanal water that made the swamps of the sunk country more cool and pure than typical mucky slough land—and thus excellent habitat for bass and other fish that thrived in pure, cool water. Shepard further noted that the sunk lands themselves had sandy bottoms, not mucky as in most swamp-land areas. The waters of the sunk lands had been identified by an 1819 surveyor as containing clear (not murky) water, but only after nearly a century would that be identified as an earthquake feature. The sunk lands might seem to be ordinary, dismal swamp, but they were to a careful eye distinguishable as earthquake country.[78]

Environmental Upheaval and the Remaking of the New Madrid Hinterland

The environmental transformation caused by the New Madrid earthquakes had significant impact on communication, trade, and governance. In January 1813, Henry Cassidy, a judge of the District of Arkansas, wrote a letter of explanation and complaint to officials of Missouri Territory, which at that time had administrative control over the District of Arkansas. He had missed a deadline for registering a series of land claims by would-be Arkansas settlers, and he wanted officials in St. Louis to understand why. Americans in his district, he explained, had only learned very late of the relevant act. He had, nonetheless, set off from Arkansas Post with the necessary paperwork on 31 October. It was to be a daunting journey: "Since the earth quakes," he explained, the old water routes were not passable and "persons are compelled" to travel by land, a distance of "between four and five hundred miles." Cassidy reported that he and five others arrived on the St. Francis River about 120 miles north of the Cherokee village (likely Con-netoo's settlement), but there found usual crossings impossible: high water had been "occasioned as this deponent believes by the effects of the Earth-quakes which have choked the channel of the river and caused the waters thereof to overflow the low lands of the country." As he reported, "No per-son in company with this deponent would venture to cross the said waters." He himself, he implied, might have been willing—except that he was sick and out of provisions. Discouraged, the group turned back.[79]

Cassidy made a second attempt: he traveled back south, to the mouth of the St. Francis River, but because of his own sickness and foul weather he did not arrive there until 7 December—five weeks after setting out from nearby Arkansas Post and after the deadline for filing the land claims. Cassidy eventually made it to St. Louis by going up the Mississippi to the New Madrid District and overland from there. His trials, however, demon-strated how completely cut off the white American residents of the district now felt themselves because of changes to their environment: the earth-quakes had left them in an utterly "remote situation" from the seat of gov-ernment, unable even to file papers on time.[80]

This topographic challenge to communication and commerce remained for decades as a feature of the landscape. As regional booster Nathan Howe Parker sadly explained in his post–Civil War guide to Missouri, a strip of "beautiful and excellent lands" running for 350 miles along the western bank of the Mississippi was "entirely cut off and stands isolated from the interior of Missouri and Arkansas, by the great swamps lying west of it, and deprives and cuts off all communication from the interior southern part of Missouri and northern part of Arkansas." Private investors built a "pole road" cross-laid with poles and rails in the boggy passages immediately be-fore the Civil War to extend commerce from below New Madrid due west and across the Little River swamps (for a profitable toll, of course), and a privately maintained road provided a good-weather crossing of a particu-larly challenging portion. Another toll road was chartered and begun, but interrupted by the Civil War. Only the coming of timber railroads (and, later, drainage of swamps in the upper St. Francis) began to connect the sunk lands with the interior of the two states in which they lay.[81]

Places that were hard to travel proved hard to know, to chart, to assim-ilate. In the summer of 1812, official Silas Bent wrote to the US Surveyor

General about American settlers anxious to have their land surveyed, especially in the lower districts where all the deputy surveyors had beat a hasty retreat. Bent recommended some likely candidates, but warned that "the face of the country in the lower districts of New Madrid and Arcansas has been considerably changed by the Earth Quakes last winter." In particular, Bent emphasized that "the Saint Francois has intirely changed its bed for a long distance up." A year later, Bent was still discouraged by the prospects for finding a deputy surveyor to run necessary meridian lines in the earthquake zones. In the summer of 1813 he noted ruefully that "the country for some hundred miles below the latitude of the mouth of the Ohio has suffered much by Earthquakes for 18 months past—the proportion of claims is small there and many of them abandoned." Better, he advised, to start meridians elsewhere, where topography was less disturbed, expertise more available, and settlement more pressing.[82]

A dozen years later, surveyor Joseph C. Brown assessed the St. Francis basin with similar cadastral dejection. It was a place not worth knowing. He had not finished his surveying assignment in 1824—for if a line were to be run, he advised, it would take significant time and expense, "the whole being through swamps and cane breaks." Later Brown explained that "the vines and cane" along the St. Francis at the lower Missouri Bootheel state boundary made it impossible to accurately survey the actual riverbanks. The whole area was characterized, he wrote, by "wide swamp like appearance, dead timber and sunken land." His conclusion was clear. "I imagine it is not material whether the line that remains be run shortly, for the country through which it passes will perhaps never be inhabited." The surroundings resisted precision or exactitude but did not deserve them: useless, they would remain unplumbed.[83]

The state of the "Great Swamp" was endlessly frustrating to regional boosters of the early nineteenth century. It was hard to map, it was hard to navigate, it was manifestly unhealthy. Nathan Parker's 1867 guidebook noted apologetically that "a great portion of these swamps is not susceptible for the habitation of man." Only a few hunters and trappers made their way "during certain seasons of the year" through the "numberless group of islands" of the swamplands. Trader Samuel Hopkins agreed: the earthquakes left the St. Francis area in "ruins," he wrote to a government official in 1826, and "it never can be appropriated to any other than hunting and trapping purposes."[84]

The St. Francis was not in the end fully mapped until the 1840s, when formal survey plats were made for townships in the area. Even so, this cadastral capture did not in fact thoroughly grid the landscape. As a direct result, the surveying lines and state boundaries noted on maps today in the Bootheel region are in many places different from the lines as defined in legal documents.[85]

In 1836, senators L. F. Linn and Ambrose Sevier pleaded in Congress for more federal development of the public lands in the region, especially the "immense morass" covering much of the area from Cape Girardeau, Missouri, to Helena, Arkansas. Linn complained of the "rafts of various lengths" which clogged the St. Francis and restricted all river traffic. Such blocked navigation meant that "many high and fertile areas are entirely cut off from communication with the inhabited portion of the country." Yet this had not always been the case: Linn argued to a Congress ignorant of the region's tumultuous history that with the earth's upheaval in 1811 "the St. Francis, forced from its bed or ancient channel, was compelled to seek its devious way to the Mississippi through lakes, lagoons, and slimy quagmires." Currently the sunk lands were "in the form of basins, connected by sinuses [that is, channels]," which, once overflowed, become "stagnant pools" amid an "impenetrable jungle" whose rank "miasmata," or unhealthy airs, threatened the health of nearby areas.[86]

But!—the bold senators argued—what the earth had done, human labor could undo. Cutting out the rafts that blocked the St. Francis would enable salutary and salubrious free flow once again. The "noble stream" could in one season of clearing be "restored to its ancient channel." Dried-out swamps would then be shown in their full potential as "cotton lands of the finest quality, all of which are now entirely lost to the public, owing to the inundated state of the country." An investment in drainage could yield "a million or two acres of land surpassing in fertility the famed borders of the Nile."[87]

In his supporting letter, Arkansas congressional delegate Ambrose Sevier explicitly drew on the example of the removal of Red River Raft—the St. Francis rafts could, he argued, be removed "with equal ease." Sevier was perhaps more accurate than he intended to be. The Red River Raft, a centuries-old logjam that stretched as a 130-mile-long block to navigation and that was so thick hapless travelers occasionally mistook it for solid ground, was in 1836 undergoing massively difficult clearing efforts that had already dragged on for most of the preceding decade. The raft was cut out in 1838 ... only to reform slightly higher up the river. The Red River was not fully cleared of its raft until 1873.[88]

This impassioned advocacy did not accomplish any of the desired drainage in the nineteenth century—such efforts were a mammoth undertaking of the era of the First World War. By midcentury, waterways in northeast Arkansas were still "wretchedly bad" to cross, as the generally intrepid geologist David Dale Owen complained in 1858. By the time of the Civil War,

FIG. 2.5. "Sunken Lands of Arkansas." This 1869 illustration from a popular periodical shows the swampy morass of what had been a flourishing zone of Indian settlement before it was dramatically altered by the 1811–12 New Madrid earthquakes. Dead trees in the background were killed by tremors and subsequent flooding. This illustration accurately documents continuing Native American use of a landscape that could not be easily crossed, much less farmed. Yet by the turn of the century, both the earthquakes and the Indian land use documented in this image would be effaced from American knowledge of the middle Mississippi Valley. (*Frank Leslie's Illustrated Newspaper* 708 [New York, NY], Saturday, April 24, 1869, 85, courtesy of the Newberry Library, Chicago)

the St. Francis River remained a "belt of swamp lands and low ridges" running between Cape Girardeau, Missouri, and Helena, Arkansas. It was dejectedly known as the "Great Swamp" throughout the nineteenth-century in maps and travel accounts.[89]

The almost impassable swampland of what had been the New Madrid hinterland even entered American musical culture. In his 1958 song "Tennessee Stud," Arkansas schoolteacher and songwriter Jimmy Driftwood wrote of a young man forced to flee from Tennessee to Texas by the angry father of the young woman he had been courting. The young beau boasts of his horse that "I never would have made it through the Arkansas mud / If I hadn't been a-ridin' on a Tennessee stud." Driftwood knew his environmental history (not surprisingly—he wrote several of his songs, including "The Battle of New Orleans," to get his history students clear on the differences, say, between the American Revolution and the War of 1812). The song "Tennessee Stud" takes place in 1825, when riding on a direct route from

Tennessee to Texas would have required going over the area rendered impassably muddy in 1811–12.[90]***

Yet while bureaucrats and earth scientists found the Great Swamp a boggy challenge to field surveys or cleared fields, the areas remained useful in their own way. By 1826, trader Samuel Hopkins, who operated a trading site at Point Pleasant, wrote a fellow trader in Missouri that widespread environmental disruption to the St. Francis made it unsuitable for American settlement: the river was so hard to navigate that "the Indians are the only people who ever will inhabit it." Sometimes a briar patch is a good place to be. A local official complained in 1820 of Cherokee and Creek Indians on the lower White River who are "rather a nuisance in that place, and an Increasing one, they are becoming very obnoxious to the people and are forming a Dangerous resort for other Indians, to loiter about and steal from the Inhabitants." The areas of what had recently been a thriving zone of trade and travel were in the midst of demographic and environmental transformation, but they still provided at least a temporary and perhaps last "resort" for those increasingly dispossessed from other rapidly Americanizing parts of the Mississippi Valley.[91]

Indians of the Epicenters: The Flight from Earthquake Territory

For the people closest to the epicenters, the earthquakes were devastating and life-changing. Ample historical evidence has long documented the severe impact of the New Madrid earthquakes for the white and French Creole communities of New Madrid and Little Prairie, but gradually evidence has mounted for a Native American story that is equally significant. The earthquakes rent apart the networks of connection and trade of Cherokees, Delawares, Shawnees, French Creoles, and Americans in the middle Mississippi Valley, especially along the St. Francis River. The consequences of earthquake included spiritual upheaval and physical dislocation.†††[92]

*** In the song, the young man gets his young woman, and the Tennessee stud gets his Tennessee mare. Neither swampland nor irate fathers can stop true love. Prosaic commerce, however, proves somewhat harder to transport across deep, mucky swampland, and the Arkansas mud did block much of the ordinary travel and American settlement across the region for many decades.

††† And, as recent scientific investigation has demonstrated, these experiences of spiritual and physical upheaval may echo the reactions of earlier generations of middle Mississippi inhabitants who, when shaken with an earthquake around 1400 CE, burned their temple and threw the debris to the ground beside the ceremonial temple mound, on top of the nearby sand dikes that had riven the earth's surface.

Indians were as dismayed by the quakes as their white counterparts in the region. The US agent to the Cherokees, Return J. Meigs, recorded what one Indian told him about the effects of the quake on the St. Francis River to the south of New Madrid. The man—whose name Meigs did not note—told Meigs that "the surface of the earth on this side of the River St. Francis appeared to be compressed in such a manner as to force the water mixed with fine sand through small apertures as high as the tops of the trees." This was an early and in modern terms quite perceptive description of liquefaction. "Although he was at that time on a ridge," the man reported, "the water covered the surface several inches deep but soon disappeared." At this, the man and his Indian companion "were so astonished that they sat down and looked each other in the face without either attempting to speak for a considerable time."[93]

Indian communities closest to the epicenters apparently viewed the quakes as a divine call to reform and repentance, a spark to community-wide movement of spiritual revival. As with much of the Native history of that tumultuous era, the sourcing is slantwise and indirect. Accounts of this revival come mostly from a late nineteenth-century sourcebook on northeast Arkansas, one of a series of regional guides put out by the entrepreneurial Goodspeed Publishing Company, which made its money by collecting subscriptions from the many businessmen or local figures profiled in its beautifully bound volumes. The Goodspeed local history quotes a long account of how Native settlements of the St. Francis River area responded to the quakes—but, in typical late nineteenth-century fashion, does nothing to specifically locate the sources for its history-making.

This Goodspeed history offers tantalizing hints of people in a frightening and disrupted world seeking safety and solace in religious revival. These Cherokee-led communities—including Delawares, Shawnees, Miamis, and Chickasaws—engaged in a revived religious ritual "imploring the Great Spirit to avert his wrath." They built a "small hut" as a site for ritual. After a hunt, they cleansed their own bodies and faces and hung up deer by their forefeet in front of the hut, so that their heads would face the heavens. Then the community observed three days of fasting, prayer, and dream seeking. At that point, the Goodspeed history concludes, "believing themselves forgiven for every unwarrantable act of which they were sensible, and that the offering was accepted," the Indians engaged in "a mutual relation of their respective dreams" and enjoyed a feast of rejoicing and thanksgiving. Though the details are hard to confirm, the practice of spiritual revival was widely shared by many shaken by the New Madrid earthquakes.[94]

People throughout the New Madrid hinterland responded to this environmental disruption with flight. As Louis Bringier recorded, the prophet

Skaquaw among the St. Francis Cherokees offered a seamless interpretation of the disruption of comet and earthquake, urging his hearers to listen to the divine call to leave the area before an even greater sign. Indeed, they did exactly that. The St. Francis River Cherokees voted with their feet, fleeing the earthquake zones for territory further west. Similarly, Delaware and Shawnee groups occupying the northern regions of the New Madrid hinterland moved west after 1812, to northern Arkansas, northeastern Oklahoma, and to the southwestern corner of Missouri.[95]

The earthquakes helped accomplish environmentally what American officials failed to accomplish politically: moving Cherokees further west, to the Arkansas River and Ozark foothills. By April 1812, apprehensive American settlers reported to the secretary of war that "a large body of Cherokees are asscending the Arkansas River to settle." These were, the settlers implied, the same St. Francis River Cherokees who had recently murdered a white American and who, they feared, "will naturally when opportunity offers fall into their former Practices of Murder, Robbery and Theft." Connetoo's settlement may have remained longest: Bringier's account suggests these were some of the communities still along the St. Francis in the summer of 1812, and Skaquaw's prophecy was relevant to debates over whether they too should emigrate further west. In March 1813, Tahlonteskee wrote to Indian agent Return Meigs that "the Cherokees have now left the ponds and reach[d] the dry land and settled among the mountains. I now call on you hastily for a good place to build my house. For two years I stood in water with patience. I remained in that situation until my feet got cold. Kan'ne'too is still standing in the water." Eventually Connetoo also abandoned his village for Arkansas River settlements.[96]

Earthquake refugees crowded into existing Arkansas River Cherokee villages, stressing food supplies. Petitioning for a fur factory "high up the Arkansas" in 1816, Cherokee agent Return Meigs noted that not only were there at least 2,600 Cherokees in the area, but that they were joined by "remnants of tribes in that quarter, the Choctaws, Chicasaws, Delawares, the remains of the old Arkansas tribe, & Shawanoes."[97]

Disruption of early Indian settlement along the Mississippi by the quakes was common knowledge through the nineteenth century. In March 1812, the *Louisiana Gazette* in St. Louis reported that "the Cherokees who were exploring that tract of country between the Arkansas and White rivers have returned home, terrified by the repeated and violent shocks." An 1882 memoir held that "many of the Indians, who had been removed west of the Mississippi by our government, deserted their new country, and came back, supposing the Great Spirit was angry with them for having deserted the bones of their chiefs, their warriors, and their forefathers, and signified his displeasure by making the ground tremble under their feet." Only later did the significance of the earthquakes for Native peoples gradually cease to be part of the story of the earthquakes.[98]

The American squatters who had so pestered the Cherokees and Shawnees along the upper reaches of the St. Francis were apparently similarly discouraged or frightened away by the earthquakes. In 1814, correspondent William Russell reported to the Missouri land surveyor on prospects for land survey and sale in Missouri Territory, noting that the St. Francis basin "was, three years ago, tolerably thickly inhabited, and was principally covered with Private Claims; no part of the territory then exceeded this in fertility of Soil." That was then, however, and much had changed: "since which time," Russell explained, "it has been so materially injured by the Earthquakes, that almost the whole of the Inhabitants . . . have deserted that quarter of the Country." He concluded that "at this time, the Lands there would not bring any price; perhaps not even sufficient to pay for Surveying." An 1860 account reported that an early white settler family left the mouth of the St. Francis for a new home up the White River in "the year the yerth shook so." This white American family may have been resettling in the same area as some of the Indian families with whom they traded along the St. Francis.[99]

Like the depopulation of New Madrid itself, the emptying out of the St. Francis watershed was relative, not complete. At least three settlements, Creek, Delaware, and Shawnee, remained on the St. Francis by the mid-1820s. Point Pleasant, about twelve miles below New Madrid on the Mississippi, remained a meeting place: in 1820 a group of Cherokee leaders gathered at Point Pleasant to petition the Arkansas governor in their conflicts with whites and Osages, and by the mid-1820s trader Samuel Hopkins still operated a trading house there. And, as with the population decline of New Madrid itself, the emptying out of the region was rapid but ephemeral. A few years after the quakes, Connetoo's village on Crowley's Ridge was resettled by American pioneers flooding the region. Land speculators from east of the river quickly bought up likely plots: American settlement of what was becoming Arkansas briskly effaced prior Cherokee settlement.[100]

Natural disaster amplified pressures of population, ecology, and culture. Forces of American settlement—direct and indirect—had begun to disrupt the St. Francis basin communities well before the quakes. Already in 1804, the town led by Connetoo contemplated resettlement alongside the Arkansas River Cherokees to better resist the Osages, who were moving into new lands in response to pressures on game from American farms and commercial fur trading. Americans themselves were not far behind. In 1808, complained one agent sent to the St. Francis River towns by Governor

Lewis, "a certain Mʳ Hunt" was going among the "Cheraki town on the river St. Francis, to kill cattle that ... Hunt claimed as his property." Hunt was, the local Cherokees complained, "a very bad man insulting them and telling them that he will soon have them out of the Country that the land they live on is his and, others that have been removed from that place, those Gentlemen have marked a greate many trees as boundery lines."[101]

Such jeers were hard to counter: American land claimants faced remarkably few legal barriers when taking over Native households' farms, fields, and fences. Displacement from natural disaster after 1812 thus exacerbated existing pressures of population and American governmental intervention: the earthquakes were one in a series of transformative forces pressing against Native communities. Such displacement, moreover, carried its own pressures in the new homes made by the people of the shaken St. Francis.

War with the Osages

The New Madrid tremors had reverberations in many parts of Indian lands. The earthquakes emptied regions of the former New Madrid hinterland and intensified conflict in the areas further west where earthquake refugees fled. The movement of Cherokees and other Indian communities west of the Mississippi pushed those emigrants into political and environmental conflict with the Osages, a conflict dangerously amplified when earthquake refugees streamed out of the St. Francis basin into settlements claimed by the Osages along the White and Arkansas Rivers. Trouble with the Osages had been brewing for decades before the quakes. Earthquake flight made everything worse.

Over the latter decades of the 1700s, the Osages began to expand their central Missouri territory. In response to increasing settlement of the region around the Mississippi River, declining fur trade, and increasingly aggressive American land and trade policies, they moved west and south into lands held by the Quapaws. Recently emigrated Cherokees alternately jostled and allied with the Quapaws even as Osage hunting parties and fracturing bands increasingly claimed the same territory for fish, game, and farming. In 1808, William Clark forcibly maneuvered several bands of the Osages into signing away virtually all Osage land in Missouri in the Treaty of Fort Osage.‡‡‡ Rich lands in Arkansas were thus increasingly valuable to both Cherokee-led hunting parties and to Osage bands, as well as to the Quapaws

---

‡‡‡ Clark later sighed to an aide that "if he was damned hereafter it would be for making that treaty."

struggling to retain both their independent settlements and their historic role as go-betweens for trade and diplomacy along the Arkansas.[102]

Even before the earthquakes, conflicting land claims and resource use led to bloodshed. In the fall of 1805 territorial governor James Wilkinson warned the secretary of war that "the Cherokees of St. Francis River . . . vowed revenge for the loss of four Warriors Killed in a late attack of the Osages." Conflict between the Osages and the Cherokee raged in 1805–7, exacerbated by environmental pressures as game grew scarcer with more hunting and a long depression in the fur trade after 1808. Even before the earthquake, frustrations and violence simmered.[103]

The earthquakes pushed communities west, into the heart of these conflicts. As people from the settlements along the St. Francis moved into the White and Arkansas River Cherokee towns after the New Madrid earthquakes, they came into increasing conflict with the Osages. Larger forces continued to add pressures on already-stressed territory. After the War of 1812, and in the aftermath of a bloody and profitless war with the Creeks, Cherokees lost eastern territory to their erstwhile American allies. Emigrating west, they found themselves forced to fight Osages for hunting and trading rights. After 1817, large groups of Cherokees left the Nation for the Arkansas Territory in a land exchange that would presage the forcible removals of the 1830s. By 1825, there were about 3,500–4,000 Cherokees in Arkansas Territory. Ironically, the emigration of these "Old Settlers" was common knowledge in the late nineteenth century, but became forgotten over the twentieth and early twenty-first.[104]

These Cherokee emigrants might have resettled in the New Madrid hinterland of the St. Francis if they had emigrated before the earthquakes. Instead, they resettled in northwest Arkansas, where they came into conflict with the Osages, as well as other groups likewise forced west. The concentration of competing communities—many of them earthquake refugees, many of them fleeing overweening American cultural and territorial pressures further east—made northwest Arkansas a bloody ground.[105]

Over the next decade, and especially in violent outbreaks of 1816–18, Cherokees and allied Indians waged wars against the Osages in northwest Arkansas. In 1816, US government agent William Lovely engineered what he hoped would be a buffer: Lovely's Purchase was to function as boundary and neutral zone between Osage and Cherokee holdings. It served instead as a zone of conflict, though the fierce fighting did suppress settlement and help overhunted animal populations rebound. American agents arranged a Cherokee/Osage truce in 1818, but violent conflict in northwest Arkansas broke out again in 1819 and continued through 1826. This was a war of emigrants: Cherokees pulled together an alliance that included

Delawares, Shawnees, Weas, Peorias, Kickapoos, and Piankawshaws to resist the Osages.[106]

Earthquake refugees found themselves embroiled in these conflicts. A group of Cherokees wrote to Arkansas governor James Miller in April 1820, requesting Miller's help in bringing about "a reconciliation between us and our neighbors the Osages." One of those who signed the letter with his mark was Walter Webber—the son of William Webber, a leader of the earlier settlement along the St. Francis.[107]

This earthquake-exacerbated set of conflicts weakened all the factions involved. By the late 1820s, almost all of the Native residents of Arkansas, New Madrid refugees included, had been swept farther west, pushed by political pressure and military force into what would later be designated "Indian Territory." An avalanche of American settlement across the Mississippi after the War of 1812 vastly increased land pressures. The squatters decried by territorial officials became hordes of land grabbers who threw Indians off land fenced and plowed by Indian families. By the 1830s, as the tribes of the Southeast faced well-organized campaigns of forced removal from their traditional territories, the earlier toeholds in the Mississippi Valley had been lost.[108]

The Osages gave up much of their Missouri prairies in 1818, surrendered their Arkansas River hunting lands in an 1822 agreement, and relinquished hold on their last formal slivers of Missouri land in 1825. Cherokee hold on settlements in Arkansas likewise proved tenuous and temporary; war had decimated their political and economic strength. In an 1828 treaty, the Cherokees were granted a portion of Lovely's Purchase, in return for cession of the rest of their Arkansas lands. They were soon pushed further west, into Indian Territory.[109]

The New Madrid earthquakes exacerbated intense, localized conflict between Cherokees and Osages by concentrating more people in contested regions and destroying the attractive territory of the New Madrid hinterland. Without that rich area as a geographic base, emigrant Cherokees found themselves weakened. Environmental stress fed war with Osages, and both contributed to the fracturing of the Cherokee Nation, which was in an increasingly weak position with respect to the US government that would shortly claim almost all Indian land east of the Mississippi.

Meanwhile, former explorer William Clark and other American officials forced the Shawnees and Delawares of older Apple Creek settlements and newer Ozark emigrant communities into treaties and land cessions by 1825, with most pushed west by the early 1830s. The Quapaws reluctantly agreed to a treaty in 1818 that ceded almost all their traditional land. In the decade 1824–34, American officials pushed the Quapaws off the lower Arkansas

River and out of the new Arkansas Territory. Today little visible evidence of these many early nineteenth-century Indian settlements remains except for the small irony of the Quapaw Quarter, a historic district in downtown Little Rock, Arkansas, known not for Native progenitors but for gracious and well-restored Victorian-era homes.§§§[110]

The St. Francis Basin, Depopulated

The year 1812 marks the dividing line when Native American and American histories of the middle Mississippi diverge.[111] In December 1812, a year after the quakes and while many former Indian settlers of the St. Francis basin were struggling to remake homes given environmental challenge and attack by the Osages, Congress created Missouri Territory and the County of Arkansas.

The New Madrid earthquakes are not regarded as important in most histories of the region: the quakes may have been dramatic, and may have made for good stories, but they did not change much. But for many Indian communities of eastern North America, they came at a watershed moment. After the War of 1812, as Native groups witnessed the final failure of confederated resistance east of the Mississippi, they faced also a crisis of cultural adaptation or resistance. For many, including the Cherokees, this crisis would convulse and define them for much of the rest of the century.[112]

Missouri settler Justus Post, reassuring his family members in November 1815, wrote that "you need not be alarmed about the Indians—they will never venture upon this quarter of the world—the country is populating too rapidly to fear them." Post was one of many newcomers whose booster rhetoric often far overshot reality, but in this assessment he was wholly correct. His "country" was indeed "populating," but no longer primarily with Native Americans. Missouri Territory grew from an estimated 25,000 people in 1814 to more than 65,000 in 1820, almost all of them American.[113]

After wars, after earthquakes, after demographic and seismic upheaval, the middle Mississippi had a different geography. The resolution of territorial conflict, with most of the region's Native Americans of all tribes pushed west, marked a significant change in a midcontinent region where the terms of encounter, trade, and economy had for centuries been deter-

---

§§§ I grew up in one of them, wondering what the "Quapaw" in the historic district signposts meant. Arkansas schoolchildren today have a robust Arkansas studies curriculum that gives them a more thorough historical grounding, but the poignancy of such environmental traces endures.

mined by Native groups. After the 1820s, the environment of the middle Mississippi was primarily in the hands of white Americans.[114]

Of the early and substantial Indian communities of the New Madrid hinterland, little remained. White settlers in the 1840s in northwest Arkansas remembered at least fifteen Indian families remaining on Big Lake, and as late as 1836 about forty Indian families around Chickasawba—with some remaining until the Civil War. They continued long-standing lifeways of hunting, trapping, and trading with white neighbors until eventually forced out by larger landowners. Arkansas resident judge Charles Bowen reported that old mounds had crumbled into the Mississippi, that he had as a boy seen remains of an ancient fort, and that six miles from the Mississippi point called Barfield he recalled seeing a particular tree with "a hand carved in the wood, well executed, and pointing directly the way to Barfield." This hieroglyph on a tree was of the kind long used by mid-Mississippi hunters in the woods to mark their caches and note important crossings.[115]

Sam Hector, described in the late nineteenth-century Goodspeed history as "a truthful, upright citizen of Big Lake, who is proud of his Indian blood," recalled living in 1833 in a Native village called Chil-i-ta-caw, at the site of Kennett, Missouri, near Big Lake. Mr. Hector also described particularly poignant evidence of the former Indian occupation of the terrain: "Where these villages were said to have been located he has often seen apple and peach trees growing in the woods." Apples and peaches were European fruits much favored by southeastern Indians: relic orchards thus spoke a Native American history increasingly invisible in the American knowledge and history of the region.[116]

Only very recently have such environmental traces begun to come back into view. Efforts by local residents to secure National Historic Landmark designation for the remaining Mt. Hope Presbyterian Church Cemetery, much of which was bulldozed in the late twentieth century, have documented the site as the only remaining vestige of the town of St. Francis. This is likely the site of Connetoo's village, an important node on early trading routes and on the Little Rock–Memphis road. Peach trees in the woods, a mostly destroyed church graveyard: these are the pale, faded environmental traces of a substantial emigration unsettled by forces at once political and environmental.[117]

A dozen years after the New Madrid earthquakes, historical understanding of their demographic importance had already faded. Timothy Flint concluded in 1826 that after the excitement of the tremors, "New Madrid again dwindled to insignificance and decay; the people trembling in their miserable hovels at the distant and melancholy rumbling of the approaching shocks." Flint may have been right about the melancholy and misery, but

he was profoundly mistaken about the "again"—what he did not understand as a recent American emigrant is that the area *had* once been important, but after earthquakes, war, and a settlement boom became much less
so. Earthquakes exacerbated larger, national stresses to indeed force a once
significant area into "insignificance and decay."[118]

In an irony middle Mississippi swamps would share with late twentieth-
century weapons manufacturing sites and proving grounds, this depopulated area became rich in animal life.[119] After the earthquakes, the New
Madrid hinterland again became abundantly prolific—because it was a
landscape largely without people. As humans left, other creatures thrived.
By the late 1830s, the sunk lands were what one description of the region
approvingly termed a valuable source of "muskrat, otter, mink, rackoon,
some beaver, bear, deer, elk, and wild cattle," providing sustenance and
profit to seasonal hunters and trappers.[120]

Bison once roamed through much of the middle Mississippi Valley and
as far east as the Smoky Mountains: the Buffalo River in north Arkansas was
named for the wallows where they came to water. By the early nineteenth
century, the eastern herds (much smaller than those of the Great Plains, but
locally significant to both food supply and culture) had been hunted out
and displaced by habitat destruction. But in the canebrakes and sunk lands
of the postearthquake St. Francis basin, bison continued to find protec-
tive habitat. The sunk lands are rich in cane, a native bamboo whose leaves
provided nutritious fodder for overwintering bison and deer, as well as for
domestic cattle. A trapper who spoke with Charles Lyell reported that a
herd of three or four hundred buffalo had been discovered in 1844 on the
aptly named Buffalo Island. No record of them exists afterward, so these
remnants were likely soon hunted out—but bison roamed earthquake territory long after they had become a faint memory in most of the eastern
woodlands.[121]

Much of the sunk lands today are privately held, drained agricultural
land. Some areas are preserved: Big Lake National Wildlife Refuge, protected land rich in small mammals and birds, is centered on an earthquake lake close to the Arkansas/Missouri border.[122] Further south are the
hunting and fishing lands known as the St. Francis Sunken Lands Wildlife
Management Area, a patchwork of more than 26,000 acres across three
northeast Arkansas counties and administered by the Arkansas Game &
Fish Commission. (The lower reaches of the St. Francis are also drilled for
natural gas and timbered extensively in the St. Francis National Forest.)
As a large stretch of contiguous bottomland hardwoods, the sunk lands
are a significant habitat for migratory birds along the central continental
corridor of the Mississippi flyway. The St. Francis Sunken Lands Wildlife

Management Area is nationally famous as a place to hunt and fish. As in the early nineteenth century, hunters value the sunk lands for their waterfowl and animals—no longer bison, but deer and turkey, as well as smaller creatures including quail, squirrel, and rabbit.[123]

During Charles Lyell's visit to the shaken regions in 1846, an old trapper reflected this environmental truth by turning what was already conventional wisdom on its head. Lyell reported that the trapper "regarded the awful catastrophe of 1811–12 as a blessing to the country." The fur trade had exploded since much of country around New Madrid had "turned into lake and marsh." Beavers had been hunted out, but the year before had still seen the taking of 50,000 raccoon skins, 25,000 muskrat pelts to be turned into hats, 12,000 minks for ladies' dresses and coats; 1,000 bears and 1,000 otters, 2,400 smaller wild cats, 40 panthers, and 100 wolves. The earthquakes, concluded the old trapper, "had been the making of New Madrid."[124]

The hunter may well have been accurate about the animal riches of the region as he knew it, but no one who knew the Indian-led life of the pre-earthquake New Madrid hinterland could easily agree with his history. The earthquakes that changed the topography and nature of the St. Francis River were not the making but the unmaking of what had once been the New Madrid hinterland.

3 ✳ Revival and Resistance: Earthquakes on Native Ground

In April 1812, Chief Gomo of the Illinois River Potawatomis contemplated the tremors that had recently rocked his people's territory. The "Great Spirit is angry and wants us to return to ourselves and live in peace," he concluded. The reason for this anger was clear: "Many children have sold their lands. The Great Spirit did not give them the land to sell."[1]

Across many parts of North America, just as along the St. Francis River, Native American communities regarded the New Madrid earthquakes as powerful spiritual signs. From the upper Missouri River to the southeastern woodlands, many Indians interpreted the quakes as the calamitous anger of the Great Spirit at the loss of Indian culture and Indian lands. The quakes gave force to three movements: political and cultural confederation led by the Shawnee brothers Tecumseh and Tenskwatawa, religious revival among the Cherokees, and civil and military strife among the Creeks. Through each of these movements, Native peoples responded to disturbances in the earth by reasserting relationships with tradition and land strained by American territorial and cultural pressures.

In each of these three movements, Native American communities likewise created knowledge about the earthquakes. As Americans were doing at the same time in prayer meetings and revivals, Indian communities interpreted seismic disruption through spiritual conversations. Unlike Americans, Native tribes came to understand the earthquakes through political and cultural struggle against dislocation from their land.

While Tecumseh is not central to all these movements, his history is central to why they are all important. If most Americans know anything about the New Madrid quakes, it is likely some version of a story of how this powerful Shawnee political and cultural leader, speaking to tribes he hoped to recruit into an anti-American alliance, prophesied that after he returned home he would cause the earth to shake as a sign of the truth of his words and the power of his movement. It is a vague and lovely romantic story that folds neatly into narratives of noble Indians, lost causes, and mythic frontiers. Like Indians themselves, it is usually regarded as worth noting for narrative interest, but not the main business at hand.

This story of Tecumseh's prophecy treats him as a sole leader, his prophecy as a single event, and his spiritual call for Indian unity as unique. All are mistaken. Conversations about the spiritual meaning of the earthquakes and their affirmation of Indian resistance were commonplace in Indian communities by early 1812. Stories that involved Tecumseh and his brother Tenskwatawa were among the most widespread, but they were part of a larger whole. Naturalistic and apocalyptic prophecy was no mere romantic myth, but a central element in the two brothers' overall movement— just as in other, similar movements that had united Native groups across much of eastern North America in preceding decades. Limiting Tecumseh's prophecy to one sentence makes it easy to discount; coming to terms with years of prophecy reveals much more about how Indians built alliances in early America.

Further, what mattered tremendously in 1812 and '13 was what was said *about* Tecumseh and Tenskwatawa—often called the Shawnee Prophets—in white and Native towns and villages. The frequent discussion of the brothers' prophetic interpretations in local newspapers and passed-along news reinforces the meaning of earthquake prophecy in the terms of their era, in which spiritual speech about natural disaster was both meaningful and serious.[2]

Many among the Osages, Delawares, and eastern Cherokees held dramatic and sometimes violent rituals to repent of their reliance on white ways and customs. Farther south, unrest boiled over dramatically and violently in civil war and in the Red Stick War among the Creeks. The New Madrid earthquakes are an unacknowledged element of the set of conflicts Americans usually summarize as the War of 1812. More broadly, the earthquakes catalyzed reactions against American encroachment that helped shape the subsequent history of eastern North America and its many peoples.

Throughout these varied reactions, the role of earthquakes in Indian movements points to the changing status of knowledge of the New Madrid earthquakes over the more than two hundred years since they shook Indian and American communities. People in early nineteenth century North America put together an account of the natural world in part through storytelling, spirituality, and prophecy—just as they did through reports of natural phenomena in letters and newspapers. Yet all these forms of knowing would change status when regarded by the geologists and seismologists who began to research the New Madrid quakes in the late twentieth century. Stories of noble Indian resistance ensconced the New Madrid earthquakes more firmly in the realm of long-ago, even quaint, frontier culture, not seismic history. Grappling with how Native communities built a framework for understanding the disturbing seismic tremors of 1811–12 thus shows how events can be known—and suggests how they can be later forgotten, diminished, or ignored.

Tremors in Indian Country

The New Madrid earthquakes affected some of the chief groups of the eastern North American woodlands: Shawnees and other Algonquian speakers in the upper Ohio Valley; Cherokees who had over several generations drawn back from traditional homelands of the Smoky Mountains to occupy country at the north Georgia headwaters of the Tennessee River and its smaller branches; Creeks further south and west along the Gulf of Mexico coastal plains, and Seminoles gathering strength in the panhandle of Spanish Florida. On the night of the first hard shock, in mid-December, Christian missionaries living among the Cherokees in what is now northwest Georgia recorded that "houses shook and everything in them moved," while the chickens fell from their roosts and set up "a frightful crying." Continued tremors darkened wells, created sinkholes, and rocked buildings. Similar unsettling disruption was felt in other Native home territories. A trader writing from Natchez to his trading firm in Rhode Island reported on the "the Great convultion of the Earth" felt throughout Creek country, reporting that the "Jare or Shake" was "Similar to that of a Cradle to Such Degree" that people feared "their Houses might be upset."[3]

Like their American contemporaries, Indians shaken by the quakes were scared, fascinated, and curious. They sought to compare experiences and local reports, evaluate the extent of sensation and damage, and figure out just what had happened. Yet the New Madrid earthquakes held different meanings in Indian country than they did in the physically overlapping but culturally distinct American states and territories. Despite substantial parallels in the basic questions people were asking—what happened? how far away? how big was it? what does it mean?—Indian experience is best understood in the context of specifically Indian frustrations, hopes, and plans.[4]

At the same time, as quotations from missionary records indicate, the main window opening onto Native American communities in the early nineteenth century was constructed through American outsiders who recorded their impressions. Such sources are slantwise at best. From governmental agents to traders to missionaries to land shysters to military commanders judging strategy, Americans on Native lands all had reasons to be there that shaped their perceptions and records. Much is lost in this history of indirection. Missionary and US governmental records make clear, for instance, that among the Cherokees the New Madrid earthquakes fueled a convulsive spiritual movement that threatened to alter the course of the Cherokee Nation. Such records also make clear that the movement flared fast and then sputtered out, fading as rapidly as it had coalesced. Why? The sources we have leave the reasons for the movement's failure frustratingly unclear: conversations and negotiations in councils far from American eyes changed its course. Yet even in incompleteness, historical records make clear the widespread impact of the New Madrid earthquakes across the diverse Indian communities of eastern North America, and the high stakes—if not always every detail—of their interpretation.

Spirituality and Earthquake Reports from the Far West

In Native as in black and white communities, people across North America heard the quakes as a call to repentance. Jolted by God, they proclaimed repentance or sought revival. Like the Cherokee settlers along the St. Francis who gathered to hear Skaquaw's vision in 1812, many Indians shaken by the tremors witnessed the anger of the Great Spirit—the Mother of the Nation, the Divine Spirit, the Maker of Breath—and resolved to turn back to traditional ways and traditional lands.[5]

Something of this spiritual understanding of the New Madrid tremors is reflected in reports from the upper Missouri River. Most twenty-first-century understanding of the quakes is based around reports from people east of the epicenters. Most present-day maps of the effects of the quakes indicate very little reported effect to the west or north. This may, as most seismologists indicate, reflect the different nature of the earth's crust in western North America: the old, cold, more uniform crust of midcontinent carries and amplifies seismic waves more effectively than the more complex, fractured, and warmer rock of the West Coast, which tends to dampen them. Yet absence of information in 1812 does not mean absence of event. This reported seismic landscape brushes by the lack of textual record for much of the history north and west of St. Louis in this crucial early nineteenth-century period and ignores suggestions that other area earthquakes in the nineteenth century were felt in territories further west. What has been taken as a geological phenomenon, in other words, may be more of a social phenomenon, of who wrote down what reports in what form.[6]

Edwin James, the scientific officer on Major Stephen H. Long's expedition to the Rocky Mountains in 1819–20, reported that the New Madrid quakes rocked those living on the upper Missouri, causing widespread "superstitious dread." "The Missouri Indians," he explained, "believe earthquakes to be the effect of supernatural agency, connected, like the thunder, with the immediate operations of the Master of Life." James recounted an explanation for the tremors given him by an American trader: a large group of Otos had come to the American trader's village, furious at a report from a French trader that an Indian had been killed by a group of Americans who had seen him "riding on a white horse in a forest country." The American reported the Otos' angry conclusion: "It was certainly owing to this act that the earth was now trembling before the anger of the great Wahconda." Meaning and symbolism have clearly been lost across multiple cultural and linguistic translations: was this a story of specific murder, or a more general parable about American incursions, or both simultaneously? James was not equipped to detangle the strands of this account, so his reporting is all the more frustrating at several centuries' distance. What his account does make clear, however, is that the Native peoples of the upper Missouri felt and discussed the tremors. And, significantly, along the upper Missouri common explanation framed the quakes as divine reaction to American violence against Indians.[7]

Across much of Native North America, this same fundamental explanation held. The Great Spirit was angry at the crimes of Americans against Indians—and at Indians' failure to stop or react against them. John Dunn Hunter, an American kidnapped as a child and raised by the Missouri Osages, remembered of that winter that "the Indians were filled with great terror, on account of the repeated tremors and oscillations of the earth: the trees and wigwams shook exceedingly; the ice which skirted the margin of the Arkansas river was broken into pieces; and most of the Indians thought that the Great Spirit, angry with the human race, was about to destroy the world." In Osage territory as among emigrant Cherokees and along the upper Missouri, the Great Spirit was understood to be speaking to Native peoples through disturbances in the earth.[8]

Fears and Tensions of 1811

Reactions to the New Madrid earthquakes throw into stark relief the conflicts and anxieties surrounding American, British, and Native sovereignty in 1811 and 1812. Territory along the Mississippi, in southeastern Indian homelands, and in the *pays d'en haut* or upper country of the Great Lakes was contested by British, French, and Native American forces on their own and acting in concert. Alliances created during the English colonists' revolt still caught and strained like veterans' old scars. By the first decade of the nineteenth century, groups who had been allies or enemies during the

American Revolution and subsequent backcountry warfare suspiciously eyed movements for new alliances or revival of old campaigns.[9]

Beneath all tensions lay the quite literally overwhelming growth of American population. By 1812, in the territory between the Appalachians and the Mississippi, Americans outnumbered Indians by seven to one. The pressures they brought had ecological consequences, as game depletion stressed food supplies and trading networks. Overhunting and a depressed international peltry trade threatened formerly secure economies. By 1810, an Indian agent in Ohio reported that that "there is no part of this country now where 500 men can subsist ten days at a place on what the woods may furnish." In many communities, young people grumbled and grew restless as their hunting territories and garden plots were encroached upon by brash outsiders who built fences and killed game for profit. Land grabs by American settlers were followed by American missionaries' and officials' attempts at "civilizing" savage Indians. American missionaries pressed women's hoes into the hands of warriors, while mothers, grandmothers, and aunts who had long fed and clothed their families found themselves reliant on processed cloth and obliged to trade with white neighbors for corn meal. The illegal but widespread whiskey trade took a terrible toll on communities with few cultural customs to limit the use of powerful liquid intoxicants. In many places and among many peoples, prophetic spiritual leaders emerged in the early nineteenth century out of the collective dismay and resistance at these challenges.[10]

By the fall of 1811, many in the United States feared and anticipated a second war with Great Britain and with confederations of Indians in alliance with the British. Britain loomed over the northern border with the United States and patrolled the seas, eager to expand territory in Upper Canada and reclaim land lost when its colonies rebelled in the 1770s. War hawks in the American Congress eyed the farms and villages of Upper Canada as future US land, while on Atlantic swells British captains ostensibly on patrol for British deserters detained American ships and at times impressed their sailors. Many white Americans feared that the Shawnee Prophets' spiritual call, charismatic leadership, and grassroots rallying would succeed in uniting Native American peoples in a "general combination" to push back American settlement and reclaim traditional cultures and land tenure. As one American memoir recounted of the "eventful" year of 1811, "It was known . . . that nearly all the Indian tribes were more or less disaffected and hostile, and that in the event of a war with England an alliance would be formed among them." It was in this context of anxious suspicion that natural portents blazed in the sky and shook the earth.[11]

Tecumseh, Tenskwatawa, and Prophetic Omens

In the summer of 1812, writer Francis McHenry reported Indian prophecy about the preceding year's comet and earthquakes. From Georgia, McHenry wrote that "the Prophet, on his embassy to the Creek nation, in the month of August last, pronounced in the public square, that shortly a lamp would appear in the west to aid him in his hostile attack upon the whites, and if they would not be influenced by his persuasion, the earth would, ere long, tremble to its centre." McHenry warned his American readers that "This circumstance has had a powerful effect on the minds of these Indians."[12]

McHenry's is among the earliest written records of accounts of the New Madrid earthquakes that circulated widely in the oral traditions and printed media of many communities of North America. In both Native American and American accounts, a leader of the resistance movement—Tecumseh or Tenskwatawa or both conflated into one—was said to have predicted the earthquakes and the great comet. Such prediction was then taken as a sign of the rightness and spiritual truth of their multitribal effort to resist American settlement and reinstate Indian landownership and ethos.[13]

Earthquake prophecy is one of the few elements still faintly remembered in modern popular culture of a movement of advocacy with broad impact in the early nineteenth century. As he traveled through the Southeast and up through the middle Mississippi Valley beginning in the late summer of 1811, the Shawnee chief Tecumseh sought to unite tribes in a coalition that would reassert Native ways and resist American dominance. Along with his brother, Tecumseh organized followers who would eventually become a potent military force in the bloody conflicts surrounding the Great Lakes in the War of 1812. Had Prophetstown not burned in September 1811 and Tecumseh not fallen at the Battle of the Thames in the fall of 1813, we might today remember those conflicts as the "War of Indian Resistance" rather than as a mere sidebar to an essentially British/American war.* The earthquakes decisively ratified the brothers' efforts and contributed both to the military force of resistance in the northern borderlands and to the

---

* In chronicling Tecumseh and Tenskwatawa's movement, the lack of formal names is a recurring problem. We have no name for the Indian/American conflict around the Great Lakes that meshes with what we term "the War of 1812": we do not even have a handy English term for the powerfully important movement the two brothers spearheaded. "Movement of pan-Indian unity" is accurate, but depressingly pallid. Perhaps the best term, one that emerged from Americans who fought and feared Tecumseh, might be the "Indian League"—but in this and many chapters of Native American history there is no denying the perils of being named by one's enemies.

TECUMSEH.

FIG. 3.1. Tecumseh. Made more than fifty years after his death, this image does not represent Tecumseh's exact features or precise dress (he refused all European or American clothing). What it does show is the proud nobility of leadership ascribed to Tecumseh by admiring Americans after they killed him and his hopes of an Indian league during the War of 1812—and the way his role as military and political leader was neatly separated from the religious and prophetic leadership of his brother, who was cast as the sneaky, ignoble Prophet. In life, the brothers' roles were intertwined: both were often called "the Prophet," and they spoke powerfully of the comet of 1811 and the subsequent earthquakes as offering spiritual ratification of their cultural and military movement. (LC-USZ62–8255, Library of Congress Prints and Photographs Division)

Creek-on-Creek and Creek/American violence of the Red Stick conflicts of 1812–14. Their "Indian league" emerged out of communities under pressure, and the movement was grounded in spiritual as well as political resistance to the Americans.[14]

In most later histories, Tecumseh's New Madrid earthquake prophecy occurs as a surprising and exotic blip. In the early nineteenth century, earthquake prophecy was one strand of a movement with a long history and many facets. Tecumseh (or in Shawnee pronunciation, Tecumthé) and Tenskwatawa grew to adulthood in the late eighteenth century, when the Shawnees were responding to severe territorial pressure with both resistance and emigration. Tecumseh, the older brother, emerged early as a fighter and leader. He came of age resisting the newly independent American confederation and by 1805 was both civil and war chief among the Shawnees—an unusual joint honor, since those positions were usually separate. Tenskwatawa, born Lalawéthika, or the Rattle, was one of triplets, ungainly and without clear direction.[15]

Much changed for the brothers in the first decade of the new century. In the winter of 1805, terrible epidemics gave rise to prophecy, calls to repentance, and witch hunts among Delawares in Indiana, where Lalawéthika lived and worked as a healer. These spiritual purges were horribly violent and continued sporadically in related communities throughout the next few years. (In the spring of 1809, a US government agent reported with horror from the Cherokee towns along the St. Francis River that he had visited the nearby Delaware settlements "where, I found them burning their friends and relations, which they said were bad persons, such horrid seens, I never heard of before this.") Tending the sick during the initial 1805 purge and epidemic, Lalawéthika was struck unconscious in holy crisis. Awakening, he became a charismatic spiritual leader. Tecumseh returned to traditional attire and began to preach a return to traditional Native ways. Together, the brothers recruited followers and fellow prophets for a movement of cultural purification.[16]

By 1807, the two brothers were the center of a growing multiethnic cultural and political movement. Spiritual disciplines and missionary outreach recruited followers; community experiences of physical catharsis gained the movement immediacy and supporters. Early in 1808, Lalawéthika changed his name to Tenskwatawa, the Open Door. Americans soon came to know him (and often his brother) as the Prophet. In 1808, they established a new community near the junction of the Wabash and a tributary, the Tippecanoe. This village, intended to set a boundary between white and Indian lands, became known as Prophetstown. Resentment against the Treaty of Fort Wayne, which further dispossessed Shawnees in the summer of 1809, strengthened the brothers' movement. Through 1809, 1810, and 1811, Tens-kwatawa and Tecumseh made Prophetstown the spiritual base for an Indian league of vastly broad ambition.[17]

The brothers' movement and their use of the earthquakes as spiritual signs drew on Indian accounts of prophecy and prophetic leaders who mediated between the lived human world and the spiritual realm, most recently the movement led by the Oneida Iroquois prophet Handsome Lake in the late 1790s. The movement led by Tecumseh and Tenskwatawa also took place alongside the American Christian movement of emotionally affecting, physically demonstrative collective spirituality known at the time as the Great Revival—a movement which did much to shape their American contemporaries' spiritual reactions to the earthquakes. Rituals of cleaning, rebirth, and purified community and individual commitment had wide currency among diverse and often mutually suspicious groups of people in the tumultuous and anxious borderlands.[18]

Later history has sought hero and sidekick: the noble, dignified warrior Tecumseh—the "Napoleon of the west"—and the shadowy mystic Prophet certainly fit that bill.[†] Yet this subsequent separation of the two brothers' roles misreads traditions of Native leadership, in which spiritual power was important alongside tactical insight and physical prowess. For many people at the time—adherents and enemies—the two brothers had much more intertwined roles. As one early history reported, "There were two Tecumsehs"—the prophet and the warrior. Later Cherokee oral histories represented the Prophet as a main leader. Writing in 1809 from St. Louis, Indian agent William Clark (a few years back from his voyage of exploration with Meriwether Lewis and their Corps of Discovery) warned the secretary of war of "This Warlike Spirit or Indian *Prophet* who has his deputes in different parts of this Country destroying the quiet and tranquility of the diferent Bands and Tribes, and induceing them to War." Perhaps most accurate to the language and perceptions of the early nineteenth century was Creek agent Benjamin Hawkins's 1814 warning of "the Prophets" whose movement was creating such trouble. This blended and shared identity reveals both the

---

[†] In 1841, Benjamin Drake published a biography of Tecumseh that has shaped perceptions to this day. Drake publicized a much-retold story of a hunting contest when Tecumseh was a young man: after three days, Tecumseh returned to the village with thirty deerskins, while none of his competitors brought in more than a dozen. Similar stories told how he protected captives from slaughter by his own angry fighters and impressed even his enemies with his bearing and dignity. For Americans of the nineteenth century and, perhaps of the present day, nobly dead Indians are easy to admire.

opacity of internal Indian politics to many American and European observers and the central role of spiritual leadership to this Indian league.[19]

Later earthquake prophecies would build on the brothers' teaching from early in their advocacy, when they used naturalistic prophecies to reinforce their teaching and authority. Like prophets among the Ottawa, Chippewa, and Delaware, Tenskwatawa preached of an apocalypse that would engulf and overthrow Americans or those who did not return to traditional ways. During his initial period of spiritual rejuvenation, the Prophet saw a vision of a crab, a common mediator between human experience and the spirit realm, which would "turn over the earth so that the white people are covered" if Indians abided by the newly revealed teachings. This and similar prophecies would continue to emerge in the American press as the two leaders gained importance.[20]

American skepticism only reinforced the power of Indian naturalistic prophecy. Indiana governor William Henry Harrison wrote the Delawares in 1806 to admonish them against following Tenskwatawa. "But who is this pretended prophet who dares to speak in the name of the Great Creator?" he asked them. "Examine him." In particular, Harrison challenged the Delawares, "Demand of him some proof." Harrison called on examples from his own Bible. Could Tenskwatawa "cause the sun to stand still—the moon to alter its course—the rivers to cease to flow—or the dead to rise from their graves"?[21]

Yes, it turned out, he could: the Prophet called villagers together to witness him make the sun stop in the sky in what his American neighbors termed a solar eclipse on 16 June 1806.‡ Intertwining of natural and spiritual power ran through the brothers' entire movement and gave political force to the heavings of the earth in 1811 and '12.[22]

Pan-Indian Resistance and the Debacle at Prophetstown

When the New Madrid earthquakes struck, they were felt across much of North America. Such breadth reinforced the broadly unifying themes of

---

‡ Several astronomers had visited the Ohio Valley that spring to prepare for observations of this celestial event—a fact which Harrison evidently did not consider in issuing his challenge. Careful listeners can perhaps even now hear, echoing across several centuries, the sound of Harrison slapping his own forehead. Careful readers may also wonder whether Mark Twain had in mind the story of Tenskwatawa's adroit use of this powerful omen when writing the 1889 story of *A Connecticut Yankee in King Arthur's Court*, in which a time-traveling American uses a solar eclipse to very similar rhetorical and political advantage in sixth-century England.

Tecumseh and Tenskwatawa's advocacy. They called their followers to return to traditional ways, but their critiques were not a call to return to a Shawnee past, or a Creek or Chickasaw past. They were about an *Indian* past, and an Indian future. Tecumseh and Tenskwatawa envisioned a true "Indian league," in which, as Tecumseh explained in one speech to the Creeks, "all the Indian tribes on the continent [would] hold their lands in one common stock," with none able to sell territory to the Americans without the common consent of all. This was a spiritual as well as political imperative: as Tecumseh emphasized, the Great Spirit gave land to Indians in sacred dispensation. For all who heard it—ally or enemy—this was a potent message. American Henry Rowe Schoolcraft recorded that "in common with the western tribes the Creeks believed they were on the eve of a great revolution, through which the Indians would once more regain their ascendency in America."[23]

In the fall of 1811, Tecumseh spoke to many groups to rally support for this broad cultural movement. Later earthquake prophecy stories center on this trip, in which he expressed the key elements of the Indian league. In a speech to the Creeks, he urged that they should return to traditional customs, leaving aside the tools that both symbolized and enacted white acculturation. No more the loom that produced untraditional clothing and plow that bound men to farming. As one contemporary leader contemptuously explained, for men to labor in fields meant that Americans were reducing them to domestic beasts. Instead, Indian men should behave and dress as traditional warriors, proud of their animal-skin clothing, war club, and bow. Implicitly, too, Native women should work at traditional farming and clothing preparation, rather than learning to weave cloth, as generations of missionaries and American agents had urged. The domestic animals kept by American settlers were a particularly odious part of this cultural hegemony, and the two brothers—like other Indian patriots—bitterly resisted them. The rebirth that would follow the throwing off of white encroachment would be total: "Kill the cattle, the hogs, and fowls," urged the prophets who accompanied Tecumseh on his visits to other tribes; "throw away your ploughs, and everything used by Americans."[24]

Just before the New Madrid earthquakes, as Tecumseh delivered his stirring speeches among the Creeks and other southwest tribes, the military implications of this movement came to a head. In the fall of 1811, tensions smoked along the borderlands, between American settlers and Native Americans, and between the US and European powers. Adherents who gathered at Prophetstown chafed at delay and lack of action, while American officials fretted about when an attack might come. Before the Indian league could further coalesce, the Americans took initiative. In late

September 1811, as Tecumseh traveled among the southern tribes seeking adherents and common cause, American forces under William Henry Harrison marched on Prophetstown. In his brother's absence, Tenskwatawa attacked the Americans in a predawn raid that failed miserably. Early in November, after the ill-fated skirmish that Americans crowed as the Battle of Tippecanoe, Harrison burned Prophetstown.[25]§

Tecumseh's Journey and Signs from the Great Spirit

Far from the crushing defeat at Tippecanoe, Tecumseh journeyed to recruit potential allies that fall. His journey, one of a series of similar travels he made from 1809 through 1812, was important both to his Indian league and to the subsequent history making of the New Madrid earthquakes. "You see him today on the Wabash," his foe William Henry Harrison wrote in 1811, "and in a short time you hear of him on the shores of Lake Erie or Michigan, or on the banks of the Mississippi, and wherever he goes he makes an impression favorable to his purposes." From August 1811 to January 1812, Tecumseh traveled eastern Indian country to build his confederacy. He visited many peoples, including the Chickasaws, Choctaws, Creeks, Osages, western Shawnees and Delawares, Iowas, Sacs, Foxes, Sioux, Kickapoos, and Potawatomis and the multitribal communities along the middle Mississippi where the two brothers had family.[26]

Many accounts of Tecumseh's visit to the southeastern peoples are confusing or opaque—in part because Tecumseh and his traveling party made great efforts to keep their movements and goals secret from Americans and other enemies. He told his listeners to tell others that he was giving them advice to "cultivate the ground, abstain from ardent spirits, and live in peace with the white people." Traditions also suggest that Tecumseh cultivated an air of powerful mystery: one man in the 1880s recalled that when Tecumseh visited the Choctaws with only a few followers, wearing very little clothing, all of it traditional—a coarse-woven breechclout and a blanket over his shoulders—"the way he came, or how he left, was a secret."[27]

Yet however shrouded from American eyes, Tecumseh's visit was a major event in Native American communities: later discussions of his earthquake

§Harrison later successfully ran for president with a chipper slogan that touted his vice-presidential candidate as well as the Prophetstown victory over the Indian league: "Tippecanoe and Tyler, too!"

prophecy had the resonance of common currency. Many of those who did not hear him directly doubtless heard much about his visit. Tecumseh and his party conveyed the movement through many forms of experience, including his "Dance of the Lakes." Learning new rituals of movement and song, Tecumseh's hearers enacted preparations for unity and for war. "Sing the song of the Indians of the northern lakes and dance their dance," interpreter Alexander Cornells reported that Creek prophets taught their community after Tecumseh's visit. Many sources report that he distributed bundles of red sticks—red is the color of war in the Indian Southeast—as a calendrical device to count the days before coordinated action: those with the bundles would remove one stick every day, so that they would be ready to act together on the day of the last stick. This traditional coordinating practice was one reason why some parts of the pan-Indian movement were called "red sticks."[28]

Tecumseh relied upon prophetic power in his journeys of recruitment. He traveled with other seers, people of spiritual power who spoke of the will of the divine creator. One of his interpreters, a mixed-culture Creek man named Seekaboo (or Seekapoo), was a skilled linguist and apparently a charismatic prophet and leader in his own right. He remained among the Creeks and later encouraged uprisings among the Creeks and Seminoles against Americans: the movement tied to the Shawnee brothers spread through many nodes.[29]

Tecumseh's mission was dramatic enough in itself. It was made more so by natural signs that seemed to foretell powerful change, signs that would culminate in earthquake tremors. Tradition among some Creeks later held that he had accurately predicted "the fall of meteors" as a proof of his ability to rally Indians to "exterminate the whites."** Many who listened to Tecumseh and danced his dances in late 1811 also watched the brightening of the comet that year and understood it as highlighting his message. A meteor or comet was called a "fire panther" among the Cherokees and Shawnees. Tecumseh's own name meant Shooting Star or Blazing Comet, calling to mind for Shawnees the celestial leaping panther, both spiritual force and astronomical constellation, that was patron and protector for Tecumseh and Tenskwatawa's matrilineal clan. For many in Indian country, the comet

---

** This might refer to the brilliant comet, but is more likely a memory of a smaller star-fall. Before electric lighting the night sky was much brighter, more detailed, and more visible than most of us have ever beheld, and most people were much more aware of changes in the night sky.

blazed a shining omen of confederation and cultural resistance. The subsequent earthquake only further added to such prophetic resonance.[30]

Earthquake Prophecy

During Tecumseh's journey to the southern tribes, reports of earthquake prophecy helped build the brothers' Indian league. A narrative by George Stiggins, the American agent to the Creeks in the 1830s, tells much of this story. The son of a Natchez Indian mother and a Scots Virginian father, he was a farmer and later government agent who could communicate in English, Spanish, Latin, and some Natchez and Creek. His is an outsiders' perspective but a deeply informed one. Over the period 1831 to 1844, Stiggins wrote a long account of Creek history and culture which made clear the impact of Tecumseh's visit.[31]

Speaking before a Creek gathering, Stiggins reported, Tecumseh said that to show the success of his initial efforts, he would "assend to the top of a high mountain *in about four moons* from that time. And there he would whoop three unbounded loud whoops slap his hands together three times and raise up his foot and stamp it on the earth three times and by these actions call forth his power and thereby make *the whole earth tremble*." This prediction, Stiggins emphasized, was not perfect: observers at the time remembered the first shocks of the quakes occurring three months after Tecumseh's speech, not four. Yet the repeated nature of the actions—the three whoops, three slaps, and three stamps—seemed uncannily to echo the series of "hard shocks" which interrupted that following winter. As Stiggins wrote, "The earthquake happening so near the time that Tecumseh was to convince them of his power and truth ... they were certain that the shaking was done by him," and their "conviction of the event left no room to doubt anything he had said of the successful irruption of the Indians against the white people."[32]

Historian Thomas McKenney recorded in the 1830s that during Tecumseh's mission to the Creeks on the Tallapoosa, he spoke before the lodge of a leader, Big Warrior. He shared wampum and a war hatchet and distributed a bundle of time-keeping sticks, but then looked Big Warrior in the eye and said, "Your blood is white. You have taken my talk, and the sticks, and the wampum, and the hatchet, but you do not mean to fight. I know the reason. You do not believe the Great Spirit has sent me. You shall know." Many later historians dismiss accounts of Tecumseh's diplomatic efforts because they were based on stories collected after his death and the failure of his movement—they are martyr's tales, not history. Yet McKenney emphasized that he collected this account of Tecumseh from "the lips of the Indians, when we were at Tuckhabatchee in 1827, and near the residence of the Big Warrior." It is not difficult, hearing in mind's ear the eloquence of this quietly devastating statement, to imagine a listener (especially one reared in the traditions of public oratory and oral history), remembering these words clearly fifteen years later. What follows is factually disputed but rhetorically inarguable:

> "I leave Tuckhabatchee directly—and shall go straight to Detroit. When I arrive there, I will stamp on the ground with my foot, and shake down every house in Tuckhabatchee." So saying, he turned, and left the Big Warrior in utter amazement, at both his manner and his threat, and pursued his journey. The Indians ... began to dread the arrival of the day when the threatened calamity would befall them. They met often, and talked over this matter—and counted the days carefully, to know the day when Tecumthe would reach Detroit. The morning they fixed upon as the day of his arrival at last came. A mighty rumbling was heard—the Indians all ran out of their houses—the earth began to shake; when, at last, sure enough, every house in Tuckhabatchee was shaken down! The exclamation was in every mouth, "Tecumthe has got to Detroit." The effect was electric. The message he had delivered to the Big Warrior was believed, and many of the Indians took their rifles and prepared for war.[33]

The connections between political tumult and seismic tumult were inescapable for all those who heard this oratory or this story.[34]

At the same time, the quakes are in some ways not the main point of this narrative. McKenney emphasized that Tecumseh was perceptive about who was with him and who was not, that he made clear the stakes of not joining with his movement, that he used traditional symbols of diplomacy, exchange, and memorial (wampum, pipe), and that he was adroit enough about using contemporary concerns that people afterward could cast him as having done so about the earthquakes. All of those aspects of Tecumseh's quiet, determined exhortation of the southern tribes are crucial portions of this narrative; all of them are *true* in important senses, even if one particular element is not.

Versions of this earthquake prediction story proliferated, but the essential core remains. One Cherokee tradition said it was the Prophet who would signal his and Tecumseh's safe return by stamping to "cause the land of the Creeks to tremble." A Creek story reported a twofold signal: when Tecumseh reached his home, he would "strike the ground with his feet, at which the earth would quake, and the chickens begin crowing." This snippet of prophecy echoes the remembered alarm of birds and animals as well as the symbolism of the Shawnee leader's preaching—he was in a larger sense out to disturb the much-despised American chickens.[35]

Tecumseh's power and the force of his prophecy seemed to be borne out by the most powerful possible vindication by the Great Spirit in the form of the New Madrid earthquakes. Oral tradition throughout Indian country clearly left the image of Tecumseh as, in the words of one aging Choctaw in 1884, "a man to make the world shake."[36]

The first of the New Madrid quakes struck while Tecumseh was near the earthquake area, in Illinois or Missouri. He continued his work of recruitment and rallying in visits to the Shawnee and Delaware villages of southeast Missouri and to at least some of the Osage bands of Missouri.[37] A stirring account comes from the 1823 memoir of John Dunn Hunter, the American raised by the Missouri Osages. Hunter recounted a speech to the Osages in late 1812 in which Tecumseh used the recent quakes—felt in Osage lands—to press the urgency of the Indian cause.[38]

"Brothers," began Tecumseh, "the Great Spirit is angry with our enemies; he speaks in thunder, and the earth swallows up villages, and drinks up the Mississippi." To fully hear this requires mentally erasing Hollywood Indian talk: Tecumseh was speaking of real and pressing political anxieties, to listeners for whom the natural world existed in close relationship with both the human and the divine. And the events he described were only too literally true. When Tecumseh stood in recently shaken Missouri forest and described the swallowing up of villages, the drinking up of the Mississippi, and the thunder of an angry god, who of his listeners, Osage or white, could think of the last few months and not be moved?[39]

The dramatic and widely reported adventure of a group of Native American travelers near the epicenters served to broadcast the prophetic strength of the brothers' teachings. As one American correspondent reported, huge cracks opened in the earth along the Mississippi during the earthquakes, in one of which "seven Indians were swallowed up; one of them escaped; he says he was taken into the ground the debth of two trees in length; that the water came under him and threw him out again—he had to wade and swim four miles before he reached dry land." The conclusion of this rescued man was clear: "The Indian says, the Shawanoe Prophet has caused the Earthquake to destroy the whites."[40]

Many other American newspapers repeated versions of this story of travelers swallowed by cracks in the earth, emphasizing the miraculous survivor's conviction that the Prophet had called forth the quakes as calamitous judgment. This widely reported story may have its roots in Lalawéthika's visions during his period of prophecy as he became the Prophet, or in the spiritually and politically resonant speeches of Tecumseh at the gathering places of the peoples he visited that previous fall. This story may be rooted not only in their words but in the thematic resonance of the years of militant and naturalistic spiritual prophecy expressed by Tecumseh and Tenskwatawa and other local seers in widely spread Indian communities. Such understanding of the quakes as foretold, divine retribution for wrong-doings against Native peoples was widely reported in the American press and gave immediate and lasting shape to the Shawnee leaders' connections to the quakes.[41]

This metaphorical territory was contested. After the quakes, some who had become disillusioned with the Shawnee brothers' movement claimed that the Master of Life had shaken the ground because the Prophet himself was an imposter. Kickapoos and Winnebagos rode in delegation to Harrison and other officials in the spring of 1812 to appease the American authorities, claiming to have broken with Tecumseh and Tenskwatawa and arguing that the earthquakes were supernal indication that Indians should live in peace with the Americans. (On the other hand, this assertion may actually have been counterintelligence suggested by Tecumseh to allay American fears.) Whether sincere or strategic, these arguments demonstrate how thoroughly interpretation of the quakes became bound up with the two brothers' movement.[42]

For many people—Indian and American—movement of the earth ratified the strength and rightness of the coalition building and resistance led by Tecumseh and Tenskwatawa. Even for Americans who may not have given credence to Tecumseh and Tenskwatawa's spiritual power, the simultaneous disturbances in earth and sky and between peoples created apprehension and fear. As one American soldier who had marched on Tippecanoe recorded in his journal in April 1812, "We kept in fearful apprehension of an attack being made upon us by the Indians, whenever we should retire to rest; add to this the frequent shocks of earthquakes and the reader may imagine the unhappy situation in which we were placed."[43]

For many Native peoples throughout the Southeast, the earthquakes were not simply events, but signs. Signs, however, can be interpreted in many ways. The New Madrid quakes contributed to seemingly divergent movements in two communities engaged with Tecumseh and Tenskwatawa's resistance. In the Cherokee Nation, the earthquakes amplified a wave of religious and cultural convulsion that paralleled the Great Revival among their white contemporaries in its spiritual fervor and force. Among the Creeks, the earthquakes stoked violent conflict, contributing to inter-tribal war and to war with the United States.

Tensions in the Cherokee Nation and Beginnings of Religious Revivalism

The story of Tecumseh and his prophecy of the New Madrid quakes is widely familiar in American history, but poorly understood. The spiritual revival and reaction among the Cherokees in 1811 and 1812 is barely even known. But it is also historically meaningful, demonstrating how natural events amplified and expressed reactions to cultural change and political loss in early nineteenth-century Indian North America. As did many followers of Tecumseh, many Cherokees understood the earthquakes as a signal to reject ongoing Americanization and assert traditional Cherokee ways. Yet unlike in Tecumseh's movement for an Indian league, no one clear political message or political leader emerged from earthquake interpretations among the Cherokees. Their movement of cultural resistance and spiritual renewal was conflicted and ultimately short-lived.

For the Cherokees who occupied a vast nation of the southeastern North American woodlands, 1811 marked a period of political consolidation in the face of intense cultural, political, and military pressure, a period in which the earthquakes of late 1811–12 would serve as a destabilizing force. In the decades since the American Revolution, the Cherokees had been forced south from extensive territory once centered on the Smoky Mountains to a reduced area in what is currently northeastern Georgia, with portions of Alabama, Tennessee, and North Carolina. Most recently, American efforts to force a treaty in 1808–10 had resulted in the contentious departure of a sizable faction of Cherokee treaty signers to St. Francis and Arkansas River settlements west of the Mississippi. Those who remained in the Cherokee homeland sought to knit together rifts within the nation. To head off further factional treaty making, Cherokees reunited their long-divided Upper and Lower Towns, centralized decision making in a Cherokee National Council, and changed long-standing traditions such as matrilineal inheritance to accommodate American cultural influence. Such political adjustments came at a cost, but consolidated Cherokee nationhood.[44]

At the same time, a multitude of difficulties strained Cherokee society. Early in 1811, the Cherokee Council asked the American agent Return J. Meigs to help them remove unwanted white settlement, while allowing a small number of schoolteachers, smiths, and some builders of mills to remain until they could pass along their knowledge. In explaining this to the Moravian missionaries within the Cherokee Nation, Keychzaetel, or the Warrior's Nephew, emphasized that "we do not consider *you* as *white people* at all, but as *Indians*. You are here to love us, not to desire our land." Meigs wrote with satisfaction to his superiors that the Cherokees had reached such a level of civilization that they no longer needed American governmental assistance. US officials promptly reduced the price paid for furs and pelts and closed the Cherokee Nation trading post, stressing the Cherokee economy. Black Fox, the old principal chief, died. The Cherokee Nation experienced a severe famine and an outbreak of disease among horses. In the fall, the ominous comet streaked across the sky and rumors of war burned across the countryside. Loss of traditional lands and ways, newcomers flooding the territory and marrying into the tribe, aggressive mayhem by white settlers, an increase in public drunkenness and failures of public order generally—all these swirled about the log homes and community gatherings of the Cherokees. The nation was at a spiritual and political turning point.[45]

The result was a convulsive wave of prophetic visions, spiritual fervor, and community revival among the Cherokees. This movement began early in 1811, with "communications from the Great Spirit," but became much more intense after the New Madrid earthquakes. In community ritual and revival, many Cherokees voiced fears of disruption in the natural world and disruption of spiritual and community life threatening the core identity of the Cherokee people.[46]

The movement was first expressed as a spiritual reaction against white ways, customs, and manners. Traditional rituals, dances, and festivals had fallen out of practice—what good was a buffalo dance when no bison roamed Cherokee lands?—and this movement offered new ones. Some later remembered that it began with a "great medicine dance," in which women danced, a respected older man chanted, and a seer named Charley urged a rejection of white ways. Referencing potent symbols of white acculturation—clothing and domesticated animals—Charley urged the Cherokees to "kill their cats, cut short their frocks, and dress as became Indians and warriors."[47]

Prophetic experience spread through the Cherokee people. As Chief Keychzaetel told missionaries in February 1811, a short time before, three Cherokee travelers—a man and two women—had experienced a vision when they stopped for the night at an unoccupied house. In the vision, a crowd of Indians riding small black horses identified themselves as messengers from God and warned that recent difficulties had been brought about by acceptance of white ways. Whites had overrun their land, the once holy Beloved Towns were in American territory, even American stone milling had usurped Cherokee traditions. "The Mother of the Nation," warned the messengers, "has forsaken you because all her bones are being broken through the grinding." The very means of meal processing—the grinding of corn at mills, often mills run by Americans, rather than by stone pestles worked by individual women—spoke to the environmental and spiritual crises of the early nineteenth-century Cherokees. Yet the Mother of the Nation offered hope: she would return if whites were expelled and Cherokees returned to their former ways. The speakers amplified this with racial emphasis more commonly heard from white Americans in this era: "You yourselves can see that the white people are entirely different beings from us; we are made from red clay; they, out of white sand." The three travelers then saw "an indescribably beautiful light" showing four white houses that the Cherokees were to build in the reclaimed Beloved Towns to house "useful" white people.[48]

Responding to this call, spiritual leaders warned of cultural renunciation and called for revival of older ways. Agent Return Meigs warned secretary of war William Eustis in March 1812 that Cherokees were heeding "fanatics who tell them that the great Spirit is angry with them for adopting the manners, customs, and habits of the white people, who they think are very wicked." Repentant Cherokees were returning to the older ritual dances, and "I am told they discontinue their dancing reels & country dances which had been very common amongst the young people." As would be true in the slightly later Creek movement of prophetic revivalism and traditional resistance, the keeping of cats, cattle, bees, and domestic fowl were regarded as cultural dependence and degradation. Later Cherokee James Wafford recounted that revivalist Cherokees "abandoned their bees, their orchards, their slaves, and everything else that might have come to them through the white man."[49]

The 1811–12 revival movement was dramatic and potentially violent. Meigs informed Eustis that "in some few instances some have thrown off their clothing into the fire & *burned* them up." Many who enacted or heard about these rituals might think of the burning to death of accused witches in Delaware towns in 1805–6 and in the multiethnic religious community of Prophetstown in 1809, episodes that helped bring to the fore the prophetic leader Tenskwatawa. Among Cherokees, the rituals were more symbolic— fabric, not people, was put to flame. Some of the women, wrote Meigs, "are mutilating their fine muslin after being told by those fanatics that they are amongst the causes of the displeasure of the great Spirit."

Articles of clothing were a symbol of American acculturation and a frequent site of cultural clash. Hats marked a person as accepting American influence: many Shawnee and Cherokee traditionalists wore turbans. Yet there were deep divisions about which cultural forms to throw off and which to keep. At one meeting, wrote Meigs, a man "burned his hat as a sacrifice" and then "called on a young chief present to follow his example."

The young leader resisted this and, "putting his hand to his Breast said, 'It is not matter what cloaths I wear while my heart is straight.'" In the same way, Meigs recorded, "a young Cherokee woman told me that she was told that the Cherokees ought to throw away the habits of the white people & return to the ancient manners & that she told them that was nothing that they ought to become good people & leave off stealing horses & drinking whiskey instead of destroying their cloathing."[50]

The spiritual movement of 1811–12 thus gave voice to a complicated set of conversations about which elements of American culture to absorb. Could a strong Cherokee man keep his hat? With game depleted, was it not an expression of daring to steal horses instead of track deer? Tensions over mills and bees represented tensions within Cherokee communities: some American and European technologies were desirable, but which ones?

The Spiritual Crisis of Earthquakes among the Cherokee People

Reactions to the earthquakes among Cherokee people expressed intense spiritual searching and complicated spiritual conversation about how to negotiate between traditional ways and forms of knowledge and those of Christianity and the American nation. Ultimately, the earthquakes also fed into the movement of revivalism.

Non-Cherokee missionaries recorded much of this spiritual tumult. In their search for "information about the frequent earthquakes," several groups of Cherokees called on Moravian missionaries living among them at a site called Springplace Mission, and the earthquakes and their meaning were frequently discussed at the mission over that winter and spring. Since the missionaries kept a careful daily record of their life and work, their perceptions are a central source for the 1811–12 Cherokee reaction to the New Madrid quakes.[51]

A German-speaking Protestant sect emphasizing direct and personal experience of God, simple community and worship modeled after early Christian life, and missionary outreach, the Moravians established a mission in 1801 in the northeastern corner of present-day Georgia, near the plantation of the Cherokee Vann family. Springplace Mission soon encompassed an orchard, farm, diary, and school for Cherokee children, all of which operated through the labor of the missionaries, their slaves, and Cherokee students and associates. The mission was located along what soon became a new federal road from Nashville, Tennessee, to Augusta, Georgia, near the Conasauga River, which many Cherokee visitors used when stopping by the mission, and not far from ritually important limestone springs and a main field for Cherokee ball play. The European and Cherokee Moravians at Springplace thus hosted a busy stream of Cherokee and white visitors. Some stopped by to exchange news, others to have letters written, ask for assistance with blacksmithing or tailoring, recover from illness while traveling, bring children for instruction or check on family children in the missionaries' care, exchange neighborly presents, or trade goods in barter: a nice piece of moccasin leather for some corn, or candle tallow for some new-pulled white turnips. For much of its existence, from 1805 to 1821, Springplace Mission was led by the married couple John Gambold and Anna Rosina Gambold, who was the main writer of the mission journal.[52]

The Moravians aimed to change Cherokee culture, yet they were tolerated by some Cherokees, who sought education for themselves and their children. ("I have never heard anything bad about you," commented the Cherokee leader Shoeboot in the spring of 1812, "but there are also very bad white people." To this, the missionaries said, "We agreed with him.") The Moravian missionaries spoke little Cherokee and had to communicate through interpreters—often the children in their school. Yet even in its linguistic translations of Cherokee to German to English, and its cultural translations of missionary work, the mission journal provides the most abundant and direct information about Cherokee response to many forms of upheaval in the winter and spring of 1811–12.[53]

In the Cherokee Nation, the New Madrid earthquakes were yet another new and unprecedented shock. "Even old people," recorded Anna Gambold, "claimed that they had never felt such a thing in this country before." On the morning immediately after the first earthquake, Chief Bead Eye, the Trunk, and two of their friends came to Springplace Mission "very upset" to talk over the night's tumult. According to the missionaries, they said that "the earth is already very old, would it soon fall apart?" After all, the Cherokee creation story held that the earth is a great island, held suspended over a sea of water by four cords from the cardinal directions. Eventually, as the world grows old and worn, the cords will break, letting the earth sink once again into the primeval ocean.[54]

The Moravians rejected this form of apocalyptic fear. Instead, they "explained to them what causes earthquakes," emphasizing that the Christian God who made all the world "also had the power and force to punish in various ways humans who live in sin, and that He often had used such earthquakes for just such a purpose." The end days would indeed come, the missionaries agreed, but in the form of God's judgment. Earthquakes were thus a "warning to stop serving sin." Many of their gambling, horse-stealing neighbors, those who spent hours drinking and betting valuable family possessions on the neighborhood ball plays, would do well to heed such a call.[55]

In the Springplace journal, the missionaries have the last word. Yet that morning's conversation may have been much less one-sided than the missionaries realized. The Moravians regarded Bead Eye to be a layperson consulting religious authority. What they may not have known is that he was himself a spiritual leader. He was, as a Cherokee Moravian explained in the summer of 1813, "one of the *Rainmakers*," the "conjurers" who could arrange community rites so as to create beneficial weather. In his visit in the winter of 1811, Bead Eye may have regarded himself as a fellow spiritual expert consulting his peers, trying to learn more and weigh the wisdom of these "people of the book" against his own knowledge.[56]

Later the same day as Bead Eye's visit, another group arrived, "deeply concerned about and horrified by the earthquake." The Moravians spoke with the delegation about their conviction that even alarming natural events had meaning for those heeding the wisdom of a loving God. The leader Chuleoa, a long-time friend of the mission, brought a letter from fellow Cherokee leader Charlie Hicks. Hicks reported his experience of the earthquake, told his missionary friends of the "great consternation" of many around him, and urged the missionaries to provide the visiting group with "more exact instruction" about how to understand the events.[57]

"Our living quarters were moving so much that they seemed close to falling down," he wrote, and at the same time, "all the trees were also moving without the slightest wind." As many of his American contemporaries did, Hicks described times and sense-impressions with care: just before the houses began to sway, "a loud noise was heard from the north-northwest and some people saw a flash of light from right where the noise in the air had begun." Like many others, Hicks tried to add as much precision as possible to an experience that was disorienting in the extreme. He noted that several aftershocks occurred "this morning, between seven and eight o'clock," but that they were "not as strong as the earlier ones and without the slightest noise." Such observational care was buttressed for Hicks, as for many of his contemporaries, Cherokee and American, by a strong sense of religious meaning in such disrupting events: "Oh!" he added, "may we honor merciful God for His protection from day to day and call to Him for help in improving our lives in the future."

Charlie Hicks's account indicates the complex spiritual frameworks in which he and other Cherokees came to term with the earthquakes. At the time of the New Madrid quakes, Charles Hicks was a bilingual businessman and political leader. In 1817 he was chosen by the Cherokee National Council as second principal chief, and in 1827, principal chief. In the winter

of 1811–12 he was also in the process of coming to a Christian spiritual understanding—he was baptized by the Moravians in the spring of 1813—and
his explicit interpretation of the quakes was consistent with the Moravians'
theology. Like the Gambolds and their fellow Christians, Hicks understood
the earthquakes as a sign of humanity's weakness and dependence and the
need to rest faith in God. At the same time, Hicks's detailed description of
the loud noises and flashing lights that were close precursors of the shakes
may have carried a very different spiritual freight for him than for his
American contemporaries making similar observations.[58]

Traditional Cherokee spiritual frameworks established profound division between the order and predictability of the "upper world" and the disorder and change associated with the "under world." Human beings inhabited "this world," which held an uneasy place on this continuum of chaos
and harmony.[††] In Cherokee cosmology, it was important to keep separate
that which was associated with each section of the cosmic order. Cherokees
used dirt to put out fires—dirt was from this world, fire was from the upper
world, and water was from the under world. To mix upper and under worlds
by throwing water on a fire would bring pollution and chaos. Instead, this
world needed to mediate between the two, providing dirt to extinguish fires.
That a disturbance under the earth also involved, as Hicks reported, a loud
"noise in the air" and a "flash of light" thus threatened dire cosmological
consequences. In many respects, Cherokee spiritual fears after the earthquakes paralleled those of black and white American contemporaries—all
feared judgment and further destruction. Yet hints of how traditional beliefs
shaped specifically Cherokee experience of the widely felt tremors emerge
even through correspondence with Christian missionaries.[59]

As people in the Cherokee Nation discussed the earthquakes, many engaged with traditional beliefs and some asserted Christian interpretations.
Peggy Vann, a prominent Cherokee and recent convert (and niece of Charlie
Hicks), was running errands at her neighbors' on the day after the shakes.
She found "the poor people" in "consternation," worried by the midnight
disruption. "Some of them attributed the event to conjurors," reported
missionary Anna Gambold, "and some of them to a great snake who must

---

[††] The under world was a literal place, reachable through streams or the heads of springs,
but only for those who fasted to prepare themselves and had one of the underground
people as a guide. Many Cherokee stories warn of the deceptive dangers of this world beneath the waters, where chairs are made of living tortoises and soft-looking but threatening thorns, and where seemingly beautiful women might suddenly take off their hair
and offer huge snakes as riding horses.

have crawled under the house, and some of them to the weakness of the earth which, because of its age, would soon fall in." Snakes were powerful and respected figures in Cherokee spirituality and cosmology, capable of causing death and disease and—like crabs in Shawnee cosmology—often agents connecting the lived human world with the underworld of watery powers. Vann rebuked such traditional interpretations, insisting that the earthquakes showed "the love and the seriousness of God," urging the Cherokees "to take the salvation of their souls seriously."[60]

As the tremors continued, so did distress and discussion about them. In the following weeks, the Moravian community encountered "the change that the earthquake brought about in the character of many Indians." One old woman "apparently goes from house to house, wrings her hands, and cries bitterly." The missionaries heaved "many fervent sighs for these poor people" and hoped that God might "shake their hearts restoratively and permanently." Further groups came to Springplace Mission to discuss the disturbing events or send notes or emissaries asking for "news about the earthquake."[61]

By early February, the Moravian leaders had established their theological response: we "spoke with them in the usual way," recorded Anna Gambold of one Cherokee delegation, "and they seemed to approve of our words." She recorded in the mission diary that both of the visitors "looked very serious." Perhaps the earnest intensity of the missionaries' conversations struck the Cherokee interlocutors as appropriate to the importance of the event and its interpretation. The following day the Trunk (a neighbor whose frequent drunken beatings of his wife caused the Moravian community much distress), a previous visitor, Uniluchfty, and two others arrived, "driven" by "dismay about the frequent earthquakes." These visitors requested to hear "a lot about God," to which the Moravians responded with an encapsulated version of Christian theology and story of the Christian God's dealings with humanity. The response was surprising and highly gratifying to the Moravian leaders: "Never before had we noticed such intense attention by Indians to such talk. They sat for hours with folded hands and looked at us fixedly. Every time we were silent for a little while, they asked that we tell them *everything*, yes, *everything* we knew about God." After sitting for a while in the respectful silence that in the cultures of the eastern woodlands indicated careful listening, "one of them cried out, 'I cannot forget this! I will always think about this!'" All involved in the conversation seem to have been profoundly moved: as Gambold recorded, the missionaries "felt very unusual with these people." Such conversations reveal the profound spiritual searching that the New Madrid earthquakes inspired among

Cherokees and communitarian Christian immigrants, both groups of
people for whom the workings of the natural world held deep and ongoing
spiritual significance.[62]

Seismic tremors did not just mean spiritual trouble; for many people they
could bring spiritual uplift. In the Moravian household, the intensity of the
proselytizing conversations brought on by the earthquakes created a rich
sense of spiritual possibility, which Anna Gambold termed "a special feeling
of grace" during the weeks and months surrounding the hard shocks. For
them, the tremors represented a chance to bear spiritual witness, to reach
out to people who had not before been interested in the missionaries' interpretation of spiritual matters. Mother Vann, for instance, was troubled by
the quakes in mid-February and expressed a desire to be more like another
woman who was "indifferent" to them. For the Moravians, "This gave us
an opportunity to speak more intimately with her." Indeed, by Holy Week
in late March 1812, when the Moravians along with other Christian communities commemorated the events leading up to Jesus's crucifixion and
resurrection, Mother Vann and Charles Hicks had both grown closer to the
missionaries, and Hicks had expressed his wish to be baptized. Easter for the
Moravian missionaries meant profound and joyful thanks for the successful
ministry of "these past beautiful days."[63]

The tremors continued to rumble, and Cherokee leaders continued to
visit the missionaries to hear their perspective and discuss the troubling
omens. For the missionaries, the quakes were a potent reminder of the
power of God's forgiveness and the need for human redemption. For their
Cherokee traditionalist neighbors, they were also meaningful signs, but
in a different way: as in the simultaneous interpretations of Tecumseh and
Tenskwatawa's teachings, the tremors were a call to renounce American influence and return to Cherokee ways.[64]

Earthquakes Amplify Cherokee Spiritual Revivalism

The tremors that rocked the Cherokee homeland amplified the growing
movement of Cherokee cultural and spiritual rebirth. As Return Meigs
recounted, the Cherokees viewed the shocks as "the Anger of the great
Spirit," which they "are at this time in a remarkable manner indeavoring to
appease." As in Native communities closest to the epicenters, the tremors
led to the revival of old devotional rites. "They have resumed the religious
dances of ancient origin," recorded Meigs, and then "repair to the water,
go in, and wash: these ablutions are intended to show that their sins are
washed away & that they are cleansed from all defilement." Incorporating
traditional elements such as "going to water" that had long been a central

element of Cherokee spiritual devotions, the new rituals expressed repentance for cultural pollution.[65]

The earthquakes made the existing spiritual movement even more apocalyptic. In a letter to his Moravian friends in late February, Charlie Hicks wrote that because of the earthquakes, "fear and horror has spread throughout the entire nation." And indeed, "How can it be otherwise? Do not these belong to the signs that are supposed to appear before the [last] great day?" Many others expressed similar fears: one local man who came to Springplace Mission seeking meat for his hungry children in February 1812 asked the Moravians "*when* the end of the world would be."[66]

The teachings and dances of Tecumseh and Tenskwatawa affected Cherokee reactions to the earthquakes. In March, John Gambold recorded that Cherokee seers had predicted that white people would vanish when the sun went into eclipse. Gambold wished to speak to the Cherokee Council about the "real Light," lamenting that "the lies of the Shawnee prophets are circulated this far!"[67]

In the Cherokee Nation, this interpretation of the quakes as a sign to resist white influence gained particular Cherokee inflection. Big Bear, a leader who traveled in company with Shoeboot to Springplace Mission in February 1812, told the missionaries of a further vision that followed the quakes. As he recounted, a man was sitting before the fire in his house one night soon after the first December quake, deep in thought and worry over his two sick children—in an attitude of domestic contemplation like that which had inspired fellow visionaries Handsome Lake and Lalawéthika before him. Suddenly a tall man stood before him, "clothed completely in tree leaves with a wreath of the same foliage on his head," holding one child by the hand and carrying another. The small child, he explained, was God. "I cannot tell you now," said the man, "if God will soon destroy the earth or not." Certainly, though, God was angry because the Cherokees had ceded so much land, especially formerly sacred sites and towns. Cherokees should reclaim their blessed grounds and return to their proper rituals, dancing in thanksgiving before they ate the first crops of the season.[68]

The tall stranger then indicated that the father's two children were not truly ill, but had only "gotten a little dust inside of them." Like the Cherokees themselves, the vision implied, the children suffered only because they did not understand their real situation. Just as he had outlined what the people needed to do to resume their right relationship with their land and their god, he made clear how to heal the children. The tall stranger gave the father pieces of bark—a treatment typical of traditional Cherokee healing practices—and instructed him to give the children the bark along with a particular drink. They were immediately healed! After instructing their father in other remedies, the tall man said he would now return the child-god back home.[69]

Such visions had parallels with other prophecies by Native Americans of eastern North America, and certainly the child-like, tenderly carried god-figure might seem familiar to those who had heard Bible stories of baby Jesus. Yet as they were discussed, the visions may have emphasized differences in spiritual understanding rather than commonalities. The Moravians who recorded Big Bear's retelling of this vision "told him that we neither understand such dreams nor get into conversations about them." Theirs, they insisted, was simply the mission of telling of God's great love for all: "We very truly only wish," they told Big Bear and Shoeboot, that "Indians really get to know Him in His great love." In this, all present seemed to find common ground: "That is well spoken!" responded Big Bear. Shoeboot reflected on these different forms of knowledge about divine will and divine caring: "The white people know God from the *book*," he reflected, "and we, from other things."[70]

Continued tremors fueled further prophecies. By late February, the Moravians recorded with impatience that "we heard a great deal about dreamers and lying prophets." Cherokees in some districts dug holes or rushed to caves to hide themselves from prophesied hailstones "as big as a half-bushel measure" or fled to mountaintops to escape massive floods. By March, people were distraught at prophecies that the moon would become dark, huge hail would rain down, all the cattle would be killed, and the earth destroyed. Cherokee prophets foretold "a great darkness" during which "all the white people and also those Indians who had clothing or household items in the style of the white people would be carried away along with their livestock." Cherokee visionaries warned their neighbors to "put aside everything that is similar to the white people and what they had learned from them, so that in the darkness God might not mistake them and carry them away."[71]

In the postearthquake spring and summer of 1812, as war talk filled councils across North America, the Cherokee revivalist movement rose to a fever pitch. Perhaps the earthquakes might mean the end of the world completely? Johnston, a Cherokee student at the Moravians' mission school talked of new prophecies foretelling that "a *new* earth would arise in the Spring."[72]

For many who heard the Shawnee prophets, earthquake prophecy led to alignment with Tecumseh—at least for a time. Among the Cherokees, earthquake prophecy led to no similar clear commitment. Just as the revival movement seemed to be drawing to an apocalyptic crescendo ... it burned out. The spiritual movement represented a threat to senior Cherokee leadership, and many older leaders responded by allying themselves with naturalistic interpretations of the earthquakes, even with the Moravians' Christian theology, against the apocalyptic warnings of the new prophets. In March 1812, respected matriarch Mother Vann ridiculed warnings of end-times and pointedly offered to buy household goods being sold by frantic neighbors, in order to demonstrate that "she did not pay any attention to the lies." That same month, Cherokee Agent Meigs, who had a warm relationship with the Moravians, asked John Gambold to accompany him to the council at Oostanaula to speak to "the many lies that are confusing the nation so," and instead "proclaim the truth to the gathering." Though a newcomer to this gathering and unknown to many of the Cherokee leaders, John Gambold was welcomed by leaders seeking ways to "calm the spirits" of a increasingly volatile community.[73]

Gambold recorded that Chief Sower Mush upbraided the gathered people at the Cherokee Council. "Recently the earth has sometimes moved a little," he acknowledged. "This brought you great fear, and you were afraid that you would sink into it." Nonetheless, he continued, "when you go among the white people to break into their stalls and steal horses, you are not afraid." Yet such theft was by far the greater threat: "If they catch you in such a deed, they would certainly shoot you down, and then you, indeed, would have to be lowered into the earth." Moravian mission diaries are nothing if not earnest, but the sardonic humor characteristic of Cherokee public debate is still evident here. Sinkholes were not a main threat; stealing horses was. Through the traditional means of community meeting and acerbic commentary, leaders like Sower Mush worked unaccustomed environmental upheaval back into normal structures of Cherokee life.[74]

An older leadership sought to prevent chaos and panic; younger leaders likewise asserted themselves, channeling Cherokee fears by the end of 1812 into volunteer fighting forces allied with the Americans against the Creeks. This might repair relations and create a valuable alliance with the United States—and it certainly gave an outlet for the upwelling of collective disturbance expressed and shaped by earthquake-laden spiritual prophecy. By 1813, war, not visions, animated communities of the Cherokee Nation.[75]

For most Cherokees, knowing the Great Spirit through earthquakes and visions meant resisting Cherokee inability to keep Cherokee land and Cherokee ways. Many in the Nation understood the quakes as a sign of the Great Spirit's anger with them for their failure to adhere to and defend traditional culture and traditional lands. But at least one Cherokee leader interpreted the tremors more pointedly as a message of anger not for Cherokees, but for the whites who took their land. While he sat with Big Bear discussing Cherokee and Moravian views of the earthquakes in February 1812, Shoeboot was clear about who was to blame. As Anna Gambold recorded in the mission journal, he "said with a very meaningful expression, 'Many Indians believe that white people are responsible for this because they already possess so much of the Indian land and want even more. God is angry about this and wants to scare them through earthquakes to put an end to this.'" He concluded that "*All* the Indians believe very much that *God* allowed the earthquakes to happen."[76]

Spiritual and Prophetic Upheaval among the Creeks

Many in the Creek communities to the south and east of the Cherokees likewise saw divine anger in the earth's tumult. There, the prophecies of Tecumseh and Tenskwatawa and the upheaval of 1811–12 contributed to the civil conflict and later all-out war that people at the time termed "the Red Stick War."[77]

By 1811, in Creek country all nature seemed to be convulsing into crisis. Several decades of high tension over American influx and influence had added to several centuries of tension over acculturation and cultural shift among the Creek people: tensions were at the boiling point. As in the Cherokee Nation, events intensified an already existing spiritual politics into a period of prophetic upheaval. The two movements were united by river flow: Cherokee centers of prophecy lay along tributaries of the Coosa River that ran through Red Stick territory. Among the Creeks, the result was threefold: a prophetic movement, civil war, and international war between the Creeks and the United States.[78]

The conjunction of earthquake, comet, increasing unrest, and earthquake prophecy by the Shawnee leaders catalyzed a time of prophecy and panic among the Creeks. Many who heard Tecumseh and the holy people who traveled with him were inspired to take up forms of prophecy. Many hoped for what the prophets foretold: that "instead of beef and bacon they would have venison, and instead of chickens they would have turkeys." Prophets among the Creeks, especially Paddy Walch, High-Headed Jim, Captain Isaacs, and Josiah Francis, claimed to have special power and knowledge—Isaacs, for instance, had dived to a river bottom and received wisdom from an enormous supernatural serpent. They claimed to be able to command the forces of nature—to turn solid earth to swampland, make the earth shake, protect their towns, and make followers bulletproof.[79]

These recently inspired prophets began to identify particular people as workers of harmful witchcraft. As among the Delawares a few years earlier, frenzied mob violence overtook those identified as malevolent evilworkers. US agent George Stiggins later reported that "whenever such a person was found ... he was seized by a mob tied to a tree with ropes and lightwood piled around him set on fire and he burned to death." In this period of volatile fear and violence, according to Stiggins, "flying tales daily multiplyed and were exaggerated in all parts of the nation." Rumors "had no Father for they were said to be told by first one and then another and no body could ascertain who." This "ferment and agitation" had political repercussions: certain factions within the Creek nation used the witchcraft frenzy to move against political opponents. Prophetic frenzy grew through the spring and into the fall of 1813, ultimately involving perhaps half the Creeks. Civil unrest led to siege and battle, with those loyal to one leader, Big Warrior, leaving the Creek nation.[80]

For Tecumseh, Tenskwatawa, and their close confederates, the spirituality of their movement was intertwined with the goals of creating unity among Indians to resist and push back American cultural and territorial advance. For those who were influenced by the brothers' movement after Tecumseh's trip to the southern tribes, though, these tightly woven strands became unraveled. In tragic irony, the prophetic movement that emerged as the legacy of Tecumseh's visit served not to unite, but to divide the Creeks.

Creek communities splintered into Red Stick and American accommodationist factions, each fearful of losing more land. Hostilities soon erupted into a civil war in which "prophetic warriors" such as Paddy Walsh claimed their rituals could protect their forces and insisted on Creek fighters using only traditional weapons. Despite Tecumseh's admonition to wait for his leadership, to do the dances but wait until the entire southern confederacy could unite in a multilayered movement of cultural resistance, angry factions among the Creeks engaged in vicious, small-scale attacks on Americans and those allied to them in the spring of 1813. Following their prophets, Red Sticks tried to "destroy every thing received from the Americans," explained Creek agent Hawkins in the summer of 1813; adherents killed the hogs, cattle, and fowl of opponents and destroyed their looms and cloth, as in the Cherokee rituals of prophecy and cultural redemption. Disheartened by the destruction at Prophetstown and unable to leave the war up north, Tecumseh could not embark on another of his long journeys to shape and lead his planned southern confederacy. Deaths led to the clamor for revenge killings, and civil war burned on Creek lands.[81]

Earthquake prophecy ran throughout the Red Stick movement. As they fought Americans and other Creeks, prophetic leaders asserted that they could call forth environmental upheaval against their enemies: as one witness later reported, they "had power to destroy them by an earthquake, or rendering the ground soft and miry, and thunder." Prophets warned they would sink their opponents' towns. Tremors and sand blows functioned as weapons of threat and war, as the very earth promised protection to those obeying the visions of the Maker of Breath. Seers assured their followers that they could draw sacred circles around their towns to protect the people from attack and "render the earth quaggy and impassable." Such use of prophetic rituals to protect one's own forces and act against enemies had precedent in Creek society: community prophets were indeed known to turn back enemy armies (at least, when such rituals had been carefully and successfully enacted).[82]

But in this set of conflicts, both sides used the language of heaving earth. Red Stick opponents called for American reinforcements: "By showing them they are both feeble and ignorant, they will be crushed," exhorted leaders, "as neither thunder, quagmire, or the sun, will come to their aid." Similarly, "You may frighten one another with the power of your prophets to make thunder, earthquakes, and to sink the earth," Creek agent Benjamin Hawkins warned Red Stick leaders, but "these things cannot frighten the American soldiers.... The thunder of their cannon, their rifles and their swords will be more terrible than the works of your prophets." Well after the tremors had ceased to be felt that far from the epicenters, authority over seismic upheaval provided the framework for cultural resistance and military conflict.[83]

Attacks on Americans in and near the Creek homeland led to open war with the United States in 1813–14. Long-simmering bloodshed expanded into outright conflict in July 1813 when forces from thirteen Upper Creek communities attacked Fort Mims and killed some five hundred Americans and allied Indians. From July 1813 through March 1814 Red Stick forces fought an exceedingly bloody campaign of resistance to the Americans. During this period, almost half of all Red Stick men were killed in combat. Despite the ritual protections drawn around warriors by the prophet Francis, the new prophetic town of Ecanachaca was burned after the Battle of Holy Ground in December 1813.[84]

The conflict reflected none of the coordination that Tecumseh and his lieutenants had so long sought. Red Stick delegations appealed to Cherokees for help in an intertribal effort. Yet Cherokees who had heard similar prophecies and danced the same dances instead allied with the Americans to fight the Red Stick faction—betting on accommodation to help them keep their own land. Only a few dissident bands of Chickasaws and other southeasterners, as well as Seminoles from Florida—both Natives and escaped slaves—fought alongside the Red Stick Creeks. The anti-American Red Sticks sought Spanish and British support and engagement, but in the end fought a largely unsupported and unallied war.[85]

Earthquakes had helped start this war, but did not help the Indians win it. In the autumn of 1813, US forces under Andrew Jackson began a series of campaigns, pulling back on the first two but then ultimately forcing the Creeks into retreat. The Creek War ended in August 1814 with Creek capitulation and surrender of over 14 million acres of land. This conflict, as aged Creek Tustenuckochee sorrowfully observed from his people's exile in Indian Territory, "came near exterminating the Creeks."[86]

The consequences of this armed conflict swept even Native communities in which earthquake prophecy had not led to military force against the United States. The Cherokees, too, lost the war, as the United States consolidated holdings at the expense of enemies and allies alike. American forces seized Cherokee slaves along with those belonging to the Creeks; plundered Cherokee livestock, crops, and homes; and demanded that Cherokees along with Creeks cede 2.2 million acres of land in north Alabama as part of the peace settlement. As the shocked and saddened Moravians recorded, in December 1813 mission students returned from the local gristmill with "unpleasant stories of the Tennessee wagoners' wild and indecent behavior, on their return from the army, toward the Indians, mostly women and children whose husbands and fathers were away in the Creek Nation in the service of the United States." One former mission student whose father was off fighting the Creeks alongside the Americans had three of his pigs killed before his eyes and much of his corn taken; "they threatened him with the most horrible curse words and said that now revenge had been taken on the Creeks and soon the destruction of the Cherokees would be carried out."[87]

However different their reactions to earthquake tremors, and however limited their commitment to Tecumseh and Tenskwatawa's pan-Indianism, Cherokee and Creek communities were linked by the ferocity of American response to Creek warfare. American agents and even the War Department were shocked by the callous disregard for Cherokee allies, but General Jackson denied everything. Complicated negotiations in Washington were subverted by the hasty drawing of treaty lines on the ground. Ultimate agreement that would treat the Cherokees as allies was negated by overwhelming American political reaction against ceding hard-won land back to Indians—any Indians. American victory in the Creek War and the war against Britain and the Native American confederacy would spur a second crisis of removal from Cherokee Nation and lead slowly but directly to Indian removal and the Trail of Tears of the 1830s.[88]

War and the End of the Indian League

The crumpling of Creek campaigns to push Americans off their land expressed the fateful undercutting of the Indian confederacy. Even as Tecumseh and Tenskwatawa's movement gathered spiritual strength from seismic upheaval and military force from alliances with the British, it was weakened by being uprooted and stretched further from main bases of support in the upper Ohio regions. Defeat, lack of faith in British allies, and low morale ate away at Tecumseh's Native forces. Yet Tecumseh's plans continued: he continued his delicate dance of negotiation and deception with American officials even while sending belts of red-stained wampum to signal coming war to his allies.[89]

As British/American tensions heated up over the spring and summer of 1812, British encouragement of Indian allies' hostilities stirred the advocacy of the war hawks in the American Congress. For years the two brothers had been regarded by American officials as mere pawns of the British. Tecumseh was named as an ally of conniving Britain in President Madison's message to Congress on 1 June listing grievances against England and pushing for outright war. On 18 June 1812, the United States declared war with Britain, even as Tecumseh and his followers readied for a final, pantribal effort to drive Americans and their customs from the lands of the children of the Great Spirit.[90]

The initial stages of this tripartite conflict were promising for the military forces of Indian confederacy. In June 1812, American forces pushed tentatively into Upper Canada, announcing that no quarter would be given to any "white man found fighting by the side of an Indian." Meanwhile, Tecumseh's oratory pulled in wavering former American allies like the Wyandots, whose territory encompassed crucial American supply routes. In a move that stunned the American public and elated the northern allies, Detroit surrendered without a fight: Tecumseh and his allies had stopped the American advance. In the course of six weeks, Tecumseh's forces were able to take or force the abandonment of every American post on the upper Great Lakes west of Cleveland. In the heated atmosphere of fear and rumor, stealth and deception were critical: one story held that in a critical assault upon the American fort holding Detroit, Tecumseh led his men two or three times through a gap in woods visible to the nervous defenders, to give them the impression that several thousand Indian fighters, rather than a few hundred, were advancing alongside the British.[91]

In the fall of 1812, Tenskwatawa gathered allies around the brothers' old Prophetstown site, and support for their movement suddenly flared from an unexpected quarter. In northern Florida, the prophetic leader Seekaboo, left behind to recruit allies among the far-southern Indians, succeeded in rallying Seminoles and free blacks to attack American filibustering expeditions attempting to wrest away weak Spanish control of Florida. The Indian league that had been supported by dances, by prophecy, by comets, and by the tumult of the earth itself seemed to be weaving together.[92]

It was a brief moment. Through the long northern campaign of what historians now term the War of 1812, the Americans and the British/Indian allies fought a series of battles, skirmishes, and messy retreats in which rough conditions were often determinative: poor communication, miscommunicated orders, misunderstood bugle calls, and the difficulty of terrain led to "friendly fire" casualties between the uneasy British/Indian confederacy. Different factions were often unsure about who had won the last battle. At one crucial moment around Lake Erie, all sides knew that a naval engagement had taken place but could not see who won. Tecumseh and his military leaders had to draw grim inference from the fact that the British navy did not return to its berth: the decisive Battle of Lake Erie, in which British naval control of the lakes was destroyed, was later to be marked as a turning point in American control of the war.‡‡[93]

The Native-led repeal of the American advance into Canada was in the end much more determinative for Canadian history than for the Indian confederation: Tecumseh was (and is still) hailed as a national hero in Canada, but the cornfields of Prophetstown had by then been trampled and burned. After the Battle of Lake Erie, Indian forces were mostly fighting defensively, far from their own home territory, unable to launch offensive attacks. By the latter part of 1813 several allied attempts on American forts had failed; Tecumseh himself had been wounded. With the Detroit area no longer defensible against American naval attack, Tecumseh's federation was forced to fall back. Ill-served by British allies, abandoned by some of his followers, Tecumseh was killed at the Battle of the Thames in early October 1813.§§

---

‡‡ This was also the site of the stirring, if tactically questionable, injunction "Don't give up the ship!" This phrase had originally been the dying command of Captain James Lawrence of the USS *Chesapeake*, whose ship was in fact overrun by a boarding party as he expired. It was taken up—literally, as a vivid blue stitched banner—by his friend Oliver Hazard Perry in his command in the Great Lakes. At the climax of the Battle of Lake Erie, as his flagship, the *Lawrence*, was injured and listing, British officers called for Perry to surrender. He bellowed, "Don't give up the ship!" and blasted them with cannon fire. He had himself and his banner rowed to the *Niagara* as the *Lawrence* struck its colors—and then sailed back to break the British line in the *Niagara*, firing from both sides, and eventually forcing a British surrender. The phrase has been associated with the battle, the war, and the coming of age of the US Navy ever since.

§§ The exact circumstances of his death became highly politicized in the United States. Boosters for Kentucky politician Richard Mentor Johnson claimed loud and long (and possibly correctly) that he had killed the Shawnee leader. In 1819 Johnson successfully ran for a senate seat in his native Kentucky with the horridly cheerful campaign song "Tum ti iddy and a Rumsey, Dumsey! Colonel Johnson Killed Tecumseh!" In 1830, when Johnson was up for the presidential nomination (he was later to serve as vice president

With him died the Shawnee-led movement for Indian unity against cultural and territorial aggressors advancing under the American flag.[94]

By flickering firelight in countless Indian settlements, Tecumseh had inspired visions of unity across tribes and against American influence. Now, the forces Tecumseh had led were left hungry and decimated, with only paltry recognition or assistance from their British allies. Tecumseh was lionized in Canada and ultimately by many Americans, but his people failed to carve out the homeland he and they sought. Tenskwatawa continued to lead forces after his brother fell, but it was in service of a losing war. The Creeks lost many of their people and half their already reduced territory, while Andrew Jackson became the American nation's proud "Indian fighter." He was later to lead American forces in the two Seminole Wars that crushed the Shawnee-inspired Indian/African American resistance and wrested Florida from the Spanish in 1818. Creeks on both sides of the Red Stick War were forced to leave their lands in 1835–36. The Prophet Francis, whose advocacy was essential for the coming together of Seminole support for what would prove to be the final stage of the Red Stick War, was later tricked, captured, and hanged from the foreyard of a schooner by General Jackson in 1818. For most Americans, victory over Indians—any Indians, all Indians—was a necessary step toward expansion and empire.[95]

Just as the earthquakes marked and accelerated the changes in waterway territory that steam travel would soon effect on a grand scale, the tremors intensified the dramatic results of international war on the geography of midcontinent. Through the war, the United States tightened its control on the Tennessee, Ohio, and Mississippi Rivers. Waterways that had once provided smooth conduits between widely spaced communities were cut off by American military and commercial presence (and soon by mill dams and channel rerouting). As American trading vessels and steamboats began to connect river ports with increasing ease, Native Americans found crucial north/south communication checked. Movements of resistance were not ended by the war, but they were irrevocably splintered. Tecumseh's delegation could not easily have made the journey in 1821 that they did in 1811, certainly not as a secretive force. Similarly, a group of Indians would have had much more difficulty casually making their way along the Mississippi, as did the unnamed and unfortunate travelers swallowed by an earthquake crack. In military force and through sheer numbers, Americans came between Native American communities. The War of 1812 and associated In-

in 1837), political partisans put forward many wildly conflicting and highly politicized accounts of Tecumseh's death and possible burial site.

dian wars marked the end of coordinated trans-Ohio resistance and the opening of the Mississippi Valley to American settlement. The conflicts and the environmental tumult that accompanied and intensified them were a watershed for the United States, which moved into increased military, economic, and political power in world affairs, and also for Native peoples, who in the 1820s and '30s faced campaigns of cultural assimilation and forced removal. To many near the epicenters, the New Madrid earthquakes seemed cataclysmic. To many across the United States, the earthquakes paled in comparison to the territorial upheaval that earthquake-linked prophecy had attempted to resist.[96]

Loss and Memory

People in the eastern Cherokee and Creek homelands, Osages in the new Missouri Territory, Algonquian speakers along the Ohio, upper Missouri tribes uneasily greeting American exploring expeditions, and settlers in multiethnic towns near the quakes' epicenters all interpreted the seismic disturbances of 1811–12 as signs from the Great Spirit that Indian people should leave white ways (muslin, whiskey, livestock, fur trade, beehives) and return to older values, practices, and rituals. For these people, spiritual and cultural interpretations combined to make sense of a dramatic environmental crisis.

The earthquakes powerfully amplified existing spiritual and cultural movements of revival or sparked new ones. Just as Native models of leadership spanned prophetic teachings as well as military strategizing, Native revival movements spanned the spiritual, the cultural, and the political. From Cherokee to Osage to Oto lands, community and individual repentance were based around the failure to protect one's own ways in the face of overweening American cultural pressure, and community and individual responses were based around resisting that pressure, in ways symbolic, economic, or military.

Despite such resistance, the larger narrative of the fighting, prophecy, and population movements of 1811–12 in Indian country has to do with losses, large and small: loss of traditional lands, loss of traditional ways, loss of political leverage against a growing and increasingly aggressive neighbor. Yet sometimes smaller narratives can be important underneath larger ones. "Tecumseh" became a common name in late nineteenth-century Indian Territory among Cherokees and Creeks (as well as Americans, including the Union general William Tecumseh Sherman). For Native Americans still very much present, alive, and Native in the late nineteenth century, the legacies of that period of failed resistance were not simply ignored as failures, but were developed through story and tale, brought from eastern homelands in much the same way as holy bundles and sacred talismans.[97]

The long and traveled life of Tenskwatawa speaks to some of this complexity. After the bitter end of the Indian league he had constructed with his brother, he lived in exile with followers in Fort Malden, Ontario, before returning to Shawnee homelands in Ohio in 1824 and becoming an advocate of US governmental removal policies. When Shawnees of southeastern Missouri, descendants of the region's last sixty years of pioneering, scouted sites along the Kansas River as a destination for their forced emigration from Missouri, Tenskwatawa rode with them. In the Shawnee reservation, he was a leader in resisting Christian missionaries and new forms of tribal government. He died an old man in 1836, in a much reduced and newer iteration of Prophetstown, where his grave now lies somewhere beneath the site of present-day Kansas City.[98]

Stories of earthquake prophecy, of divine anger against land seizure and cultural loss, and of how the natural world would express righteous wrath expressed a form of truth about both natural and political events for Indian communities. For many Americans, the quakes heightened fears of general and dangerous uprising—of Native peoples and of their earth itself. Seismic upheaval also heightened fears of general and dangerous wrongdoing on the part of individuals and communities, a set of fears that was expressed through the language of the physical self and—as in Indian communities—through intense and somatic religious upwelling.

4 ✳ *The Quaking Body: Sensation, Electricity, and Religious Revival*

"Yesterday morning," reported one resident of what is now Queens, New York, in January 1812, "a shock of an earthquake was sensibly felt in this village." In ways that spanned the individual and the collective, the physical and the spiritual, the New Madrid earthquakes were indeed "sensibly felt" across early American society. Bodily knowledge shaped how individuals and communities came to understand the New Madrid earthquakes. In a twenty-first-century world, where earthquakes are registered on seismographs, blood tests and biopsies tell us how ill we are, and few mainline American churches—much less science departments—link spirit to body and mind, it can be hard to grasp just how embodied nineteenth-century responses to earthquakes were, and how profoundly those somatic experiences registered. But in the nineteenth century, bodily sensation was central to experiencing and understanding the quakes, on intellectual and spiritual planes alike.[1]

People across much of North America felt the earthquakes—and felt them with physical profundity. The tremors made their bodies feel unwell, disturbed, and strange. People were seized with shocks and jolts. Earthquakes were experienced as personal manifestations of ill health and imbalance. People also worried that earthquakes might induce environmental ills, such as dangerous miasmas and epidemics. But just as bodily symptoms of disease or injury could reveal information as much as induce fear, somatic experience provided a way to communicate about the quakes and a means to come to terms with them. To understand the earth's seismicity, early nineteenth-century witnesses of the New Madrid earthquakes used the sensations of their bodies. People compared alarming seismic jolts to other bodily experiences and attempted to quantify and tabulate their physical sensitivity to them.

The physical shocks of the quakes also gave the tremors spiritual power for many in early America. As their bodies were moved, so too were their souls. For many people, the tremors felt like electricity and like the over-whelming power of God. Reactions to the New Madrid earthquakes reveal the close relationship between truths of the body and truths of the soul in early American life.

In many Indian communities jarred by the shocks, the earthquakes sparked or amplified movements of spiritual and cultural revival based around renewal of traditional ways and resistance to American pressures. Despite important differences in the ways these movements played out among diverse Indian peoples, all were simultaneously political and cultural reactions to American influence. In contrast, in American communities of white and black people, earthquake shocks jolted people into awareness of their own individual sins or relationship with God and prompted them to repent and commit themselves to a Christian life. The New Madrid earthquakes sparked many Americans' participation in the movement of Christian spiritual renewal they called the Great Revival. Yet powerful community gatherings and shared rituals supported people undergoing revival in American frontier settlements just as in Cherokee towns. Important parallels unite the spiritual experience of the tremors across North America. Thrown off balance by seismic tumult, people across many communities feared for their relationships with the Almighty and each other and sought to set them right.

Through the shock of earthquake and the shock of the Holy Spirit, early Americans made sense of the earthquakes. They reached for metaphors as old as ancient Greek medical theories about the balance of bodily humors, and as new as the human-produced electricity that was cutting-edge technology at the time. Knowledge of the world around them was gained alongside knowledge of the sensations of the body and the truths of the very soul.

Earthly Disruption and Bodily Disturbance

In the winter of 1811–12, reports of strange bodily sensations appeared in newspapers, private correspondence, and public periodicals across eastern North America. Today's tremors in far-off places register as quantified reports of estimated magnitude or as visual displays of jagged seismo-graph readings. In the early nineteenth century, the New Madrid earthquakes emerged into common discussion as narratives of disturbing bodily experience.

Early Americans saw profound connections between their bodies and the New Madrid earthquakes. Many registered the quakes in terms of bodily disturbance. They described, quantified, and compared kinds and levels of somatic sensation to create communal knowledge of the quakes. People worried, moreover, about the impact of the quakes on their bodies, as the tremors were a sudden and dramatic new aspect of their environment. Finally, people used bodily metaphors to think about the changes in the earth, and even took the same approach to monitoring and recording earthly disturbances as they did for bodily disturbances.

In account after account, the repeated jarring dislocation of the earth from December through February registered chiefly as a set of weird and disturbing bodily symptoms. One woman surprised by tremors at night had "a strange sensation as if suffocating"—she leapt out of bed fearing the house was on fire. A group of women in an upper story of a building in Canada (most likely in Quebec) found themselves much more affected than those in lower structures. Several "complained . . . some days afterwards that they still felt an uneasy sensation in their heads"; one "kept her bed . . . for some days afterwards from indisposition arising from affright."* People overcome by the quakes might report "a sudden and deadly sickness, accompanied with a giddiness in the head." In Wheeling, then in the state of Virginia, several shocks of 6 February made many people feel "giddiness in their heads and a nausea on their stomachs."† Nausea was a common symptom of earthquake, in 1811–12 and through the later nineteenth century. Many who felt the New Madrid shakes had trouble keeping their feet and felt "dizziness and vertigo." People felt drunk, or seasick, or as if "on shipboard in a heavy swell of sea." In Hackensack, New Jersey, "Several ladies who were sitting, complained that they felt as if sitting on a poise [a balance for weighing objects], and were afraid of falling from their seats."[2]

Earthquakes shook up emotions as well as bodies. A report from Maryland after the January New Madrid shock noted that in addition to giddiness and nausea, "much wonder, and a little terror were among the consequences." Philadelphia physician Benjamin Rush observed that "the earthquake which took place on the shores of the Mississippi, in December 1811, produced silence, or great talkativeness, and moping stillness, or constant motion, in different people." Emotions flowed even later in the century: the 1886 earthquake in Charleston made some young men cry, but

---

* Modern people who experience earthquakes often have a prolonged sense of physical unease and disequilibrium: after a Southern California shock, a colleague told me she did not feel herself "for weeks and weeks."

† Virginia, like the Union itself, was then one entity. Virginia seceded during the Civil War, declaring itself part of the Confederate States. The area including Wheeling then separated itself from Virginia, rejoining the Union as West Virginia.

moved others to sudden laughter. Not just the physical body, but the whole sensing self was moved by tremors in the earth.[3]

Many who felt the New Madrid quakes thought what was happening was not a crisis in the surrounding world but within their own being. In January, "early in the morning about sunrise, as sitting at breakfast," a man in Arkport, New York, "had a strange feeling, and supposed at first that he was going into a fit, and removed his chair back from the table." That capacious term "fit" encompassed what might in modern medical terms be called stroke, heart attack, or epilepsy (as well as other conditions), and it indicated not only being suddenly overcome, possibly collapsing, but also a strangeness to oneself, the experience of no longer being in control of one's own body. People having a "fit" could no longer trust their own physical selves to hold them up or render accurate sense perceptions. For this man at breakfast, even observations seemed skewed:

> He then had a sensation as though the house was swinging, and observed that clothes hanging on lines in the room were swinging, as also a kettle hanging over the fire. He observed that his wife and family appeared to be greatly alarmed, and still supposed that it was in consequence of his apparently falling into a fit, but on enquiry found that all felt the same sensation.

Even his sense impressions of other objects going awry—seeing things sway in the air, feeling a swinging motion, witnessing the alarm of his family—seemed to indicate a problem within his body. Only talking with those around him made him realize that he was in fact experiencing a phenomenon outside his own body but with ramifications within it. Blurring sense and sensation, inducing sickness and ill-feeling, earthquakes affected sense of self just as they did arrangements of the outer world.[4]

In many similar cases, interpreting varied bodily sensations helped identify strange phenomena as part of an earthquake. Often people reported feeling sick, then reattributed the cause when the evidence of the outer world helped them organize their own inner sensations. A merchant visiting Charleston was just sitting down to breakfast with his companion when he felt "a giddiness in the head, and sickness of stomach" that he first attributed to "long fasting" since their previous meal. Only when "eggs rolled about on the table" and the house he was in "rocked from east to west, to and fro" was he convinced that the problem did not originate within his own body.[5]

One's body was not a constant register. "Scarcely a night passes that we are not sensible of some slight shock," wrote physician Alexander Montgomery from Kentucky to leading natural scientist Benjamin Smith Barton in Philadelphia. He then mused further, "The reason why we are more sensible of them in night than day I conceive to be that the noise of business and our different occupations so fully engage our attentions that we perceive then, none but the more violent shocks." Different bodies might experience the quakes differently, and one's own form might perceive them more or less keenly depending on mental and physical states.[6]

Reports of bodily sensation were grounded in a background of historical understanding that earthquakes could cause sickness and create epidemics. Americans and Europeans knew, for example, that the 1692 earthquake that devastated the British port at Jamaica had been followed by sickness that killed many survivors. Throughout the 1700s and into the early nineteenth century, reports of minor earthquakes in England and the American colonies included mention of their effects on health.[7‡]

In 1801, Noah Webster sought to chronicle earthquakes for a national history of the young United States. This history was a part of Webster's patriotic project to establish respectable American scholarship, alongside his reforms of spelling and his famous dictionaries of American English. Webster noted that earthquakes across the globe throughout the past century had led to epidemics—plague, distemper, "malignant ulcerated sore throat." Like many of his contemporaries, Webster ascribed the illnesses to the effects of "the noxious vapour ejected by the convulsions of the earth and eruptions."[8]

As in Webster's conclusions, fears of earthquake-related sickness drew on environmental knowledge as well as on historical precedent. Early nineteenth-century Americans shared an understanding that the human body was affected by the environmental matrixes through which it moved. This was a nineteenth-century inflection of what was essentially ancient humoral theory, marked by keen attention to the fluids and flows within the human form. Leading physician and congressional representative Samuel Mitchill observed several years after the New Madrid tremors that they "appeared to affect very sensibly both the body and mind of human beings." He noted that some people "had been deprived of their usual sleep, through fear of being ingulfed in the earth, their stomachs were troubled with nausea, and sometimes even with vomiting. Others complained of

---

‡On the other hand, in 1829 the *Cherokee Phoenix*, the newspaper of the Cherokee Nation, reported that recent earthquakes in Spain had cured "a great many persons with rheumatism." Seismic events had powerful—and not always predictable—effects on human health.

debility, tremor, and pain in the knees and legs." Mitchill noted that sickness followed in the seasons after the quakes, and ascribed it to a "loaded atmosphere," most likely "impregnated with sulphureous particles" from the quake. Atop a chemical explanation for how earthquakes change atmosphere, Mitchill—like Webster—layered a much older environmental suspicion of how bad airs might infuse and sicken human bodies.[9]

At the same time, bodily knowledge granted people resources as well as fears. People experiencing the New Madrid quakes dealt with their bodily consequences in the same ways they dealt with sickness. One observer in Beaufort, South Carolina, reported in February 1812 that "I was lying awake in my bed, when I began to feel my bed shake, but being witness to so many, I was not at first much alarmed." People became physically accustomed to temblors, just as they became acclimated or "seasoned" to the diseases of the country. A New Madrid–area resident who identified himself only as Major Burlison grew accustomed to the earthquakes in precisely the same terms in which he and his neighbors became acclimated to the endemic area disease of malaria, often called "the shakes" after its bone-rattling fever. "By degrees," he concluded about the continuing smaller earthquakes in New Madrid, "I grew familiar to the *shakes* myself."[10]

Many who felt the "hard shocks" wanted to convey their dramatic experiences and understand these baffling and overwhelming seismic phenomena. Yet they had no instruments to quantify the magnitude of what they had witnessed, no devices to record the earth's movement, and no technologies of rapid communication. Researchers began to implement seismometers and networks of seismographs only in the last decades of the nineteenth century. Before then, people who wanted to communicate across distance or across time had to come up with ways of describing and narrating their experiences.

One's own body was an instrument of seismic disruption. A visitor to Edenton, North Carolina, observed on 24 January 1812: "Yesterday at a quarter after nine we were alarmed with another earthquake; no mischief has been done, but many people were sensible of an indescribable motion, accompanied with a sensation of faintness or falling into an apoplexy." An earthquake was still an earthquake even if it did no "mischief": it registered in the changed sensations and sick or weakened feelings of those it disturbed. The sensitivities of the human form showed parallels with machines. In Suffolk, Virginia, during the 23 January shock, "persons who were standing or walking, felt a sensation similar to that produced by the rocking of a ship in a boisterous sea; many were thrown off their equipoise, and could with difficulty keep from falling. Every clock in the place was stopped." Bodies, like stopped clocks, registered disruption to the natural world.[11]

The language and concerns with which nineteenth-century Americans recorded the physical well-being of their families parallels the ways they recorded disturbances in the physical earth. Connecticut farmer Stephen F. Hempstead began his diary of his new home in Missouri after the main shocks of the New Madrid earthquakes, though he did mention substantial aftershocks late in 1813. In the summer of 1813, Hempstead recorded common but troubling ailments in the diary he kept for almost twenty more years. On Sunday, 1 August, he visited St. Louis and "found Son Thos. Relapsed again with the fever, have him bleed." Drawing off a plethora of bodily fluid was meant to calm the body, giving ease and drawing off the excess heat and inflammation signaled by a fever. Changing sensations were always of concern, and careful observers noted small changes in loved ones' states of being—much as those who felt the New Madrid shocks recorded their changing or gradually dampening physical sensations of the tremors.[12]

The notion of powerful forces seeking release united the understanding of body and of land. Both body and physical world were marked by attention to the buildup and release of long-gathering forces. By the late summer of 1813, Thomas Hempstead was no longer at the center of his father's diary or ministrations. Instead, Stephen Hempstead recorded that "the boil on my thigh broke last night having ben 12 days in gethering and relieved of much pain." Under the skin, in the interior of his body, foul forces had been amassing: his diary recorded the salutary release as putrid matter found a way out of his body. For people of Hempstead's time, this was a matter not just of physical comfort but of profound bodily balance. Health was a balance of fluxes and flows, usually interpreted as the literal fluids of the body. When ill forces brewed in the body's interior, getting them out brought relief from pain and also a return to health: a stoppage in the body had been unblocked, bad material released, and the body's normal balance could resume.[13]

Earthquakes were often described in similar ways. Frightened out of bed by the terrible roaring, cracking, rumbling sounds of the December temblor, a St. Louis man looked around for "the earth to be relieved by a volcanic eruption." One newspaper article that quoted river travelers from New Madrid on the alarming state of the area and its river immediately after the February shocks concluded after a litany of ills that as of 21 February "there had not been yet any volcanic eruption sufficient to draw off the cause." Much as a human body would be relieved by the lancing of a tense, full boil, the earth itself sought release and relief through powerful and noisy volcanic discharge.[14]

In a more subtle way, narratives of how people experienced earthquake shocks drew from widely shared conventions of observing and recording family health, particularly the keen observation brought to recording a beloved family member's final illness and death. Stephen Hempstead's diary is typical in the rhythms and language with which it describes the kinds of illnesses and health issues which faced most nineteenth-century families. All through the summer of 1820, as Stephen and his wife Mary Hempstead cared for children and grandchildren through an epidemic of measles, a "Bowel Complaint," and a "Bilious fever" that weakened and killed both young and old, Mary was also sickening. In early September, Stephen recorded that "my wife continues unwell." Two days later, "no Better take medicin dont have the desired affect." On 5 September, "no Better distressed for Breath." The next day she took to her bed. On 7 September he convinced her to allow him to send for a doctor. The practitioner applied a blister which "did not have the desired effect at all": she was still "distrest for Breath." Strong medicines characteristic of the heroic therapy of the early nineteenth century "opperated well but did not relieve her." That is, Mary Hempstead threw up or had diarrhea or sweats—all "operations" that might draw off an internal illness, restoring the body's healthful balance—yet even these interventions did not succeed in letting out the noxious forces that sickened her.[15]

On Saturday, 9 September, Mary Hempstead's illness came to a crisis. Stephen Hempstead and the family's doctor administered further blistering, strong doses of laudanum, and treatments of horseradish, mustard, and hot vinegar on her extremities, but these "did not help her breath or return the Natureal heat of her body." While her grown children came to her bedside, Mary Hempstead "had her Reason and Senses Clear and the Evidences of her Religion Bright without a cloud or mist of Darkness or dispeair." For a while, she "revived again" and sang a hymn. As she sang, "her breath failed her Several times," but she sang clearly at the last verse, which commends all earthly hopes to Jesus. At this, her husband recorded, "her Voice failed her and she fell a Sleep in Jesus in a Short time about 10 oClock in the Evening." The couple had been married forty-three years.[16]

Narratives of health and death such as Stephen Hempstead's careful diary recordings possess the same form and rhythm as narratives of earthquake, a similar attention to when forces are stronger or weaker, to the ebb and flow of fever or agitation in a restless patient. Moreover, they serve the same purpose: in writing a narrative of the earth's experience from the point of view of bodily experience, and then comparing it with other accounts by other people at different locations, observers hoped to make sense of that experience just as they might make sense of a new epidemic or an unknown fever. A writer in Wheeling, Virginia, observed of the December quake:

> It appears from those persons who happened to be awake at its commence-
> ment, that it came on gradually, until it advanced in strength to make the
> houses and bedsteads crack loudly. It then gradually went off. About eight
> A.M. it was again felt, but not so violent in its undulation as before. At a
> quarter before one P.M. it was again experienced, but much weaker.

Such language and concerns were much like those of people waiting by a bedside for the consequential shift that would tell whether a grave illness would lift or intensify, or a final push bring a child into the world.[17]

In accounts of a sick family member, concerned narrators trace a process with many elements, not all well understood. Those giving an accounting would include the whole sense of what the person was undergoing: When was the fever strongest? When were neighbors called in to help, or children called in to say farewell? In the same way, those who experienced earthquakes noted the preceding weather, the behavior of animals, unusual breezes, the sounds of the tremors, the way in which air seemed to change or stay still during them, and the ways in which movement or noise or smell or lights would come on, grow in strength, and then fall off.

Understandings of the flows and disturbances of earth and body show profound similarity. When Mary Hempstead was ill, her practitioners put strong salves on her arms and feet to draw off force and heat. So too might earthquakes on one part of the globe draw off the forces or flows of another. An article in August 1811 in a newspaper back in the Hempsteads' home state of Vermont reprinted news from London, reporting "dreadful forms and convulsions of nature in different parts of the world"—floods of lakes and oceans and "strange fluxes and refluxes of the sea." These were possible clues to a larger disturbance: "Those who recollect that similar events took place before the great earthquake at Lisbon are prepared to expect, that a similar convulsion has taken place in some distant part of the globe." Tremors and noises functioned as a barometer to the internal state of a dangerously disrupted earth. The language of these accounts—fluxes, convulsions—indicates the powerful bodily metaphors at work. Earthquakes were a symptom of the earth that could indicate far-off danger, just as inflammation or disruption in one particular part of the body could, in the nineteenth-century view of bodily systems, indicate an ailment or imbalance elsewhere. Frameworks of health and well-being shaped conversation about terrestrial processes.[18]

This understanding of earth as body held a long geological background. Many natural historians of the eighteenth century and earlier described the earth in organic terms, connected by subterranean channels and laced with veins of metals. In the early nineteenth century, as the New Madrid earthquakes shook the middle Mississippi Valley, geological frameworks were beginning to shift away from such bodily metaphors and toward more historical understanding in which layers of strata and mineral formations would yield a history of the deep past. The New Madrid earthquakes reveal the tail end of a geological understanding based on the contours and experience of the human form.[19]

※

As they came to terms with the New Madrid earthquakes, observers pieced together accounts not just of individual bodies, but comparison across many. From such comparisons across space and time, early nineteenth-century people sought to draw out more general conclusions.

Observers compared readings from distant places. In Charleston, during the January hard shock: "Many persons in different parts of the city were sensible of a shock at 8 o'clock in the morning." At Suffolk, Virginia, during the January shock: "The motion which the undulation gave was by wavering his body on the chair from the west to the east." In Augusta, Georgia, during the December quake: "Its course could not be determined we believe—from our own sensations we thought the house rocked from north to south, and that the east and west ends only experienced a tremulous motion." Individual bodily response might be doubtful (or doubted by the person feeling it), but the evidence of many bodies together was indisputable. Of the December shock in Augusta, Georgia, one correspondent noted, "Similar sensations and effects were produced in many other families in town." The gentleman in Arkport, New York, who first thought himself falling into a fit before realizing he was feeling an earthquake might have been reassured to find later, as the local correspondent reported, that "one of his neighbours felt the same, and on the opposite side of the river at the farm house and dwelling of Philip Church, the same motions and sensations were felt."[20]

Many observers distinguished various quakes based on how they felt. In Charleston, the 7 February quake had an "undulatory motion . . . much shorter and quicker than any we have before experienced." Such corporeal evidence created an accounting of the various shocks. In Savannah, Georgia, on 23 January, "the nausea of the stomach, giddiness of the head, and movement of unfixed bodies, were as sensible in this, as on similar occasions lately noticed." Moreover, different experience reflected the varying sensitivities of each individual. In the January quake in Suffolk, Virginia,

"Many persons were inclined to doubt the evidence of sensation, and were willing to ascribe the concussion to any but the true cause; but the same sensations being experienced in every part of the town at the same instant, left no doubt of its being the shock of an earthquake."[21]

Multiple reports of the same bodily sensation could help people sort out what earthquake events had occurred simultaneously. A correspondent to a New Haven paper wrote that a tremor was reported in New York and "the same earthquake was felt by myself and family, about the same time." The tremors produced a sensation of "dizziness of the head." The *Connecticut Herald*'s editor then noted that "a similar sensation was felt at the same time by one of the Editors of this paper, while sitting in his chair in the act of reading." Bodily immediacy underscored the shared nature—as well as the immediacy and ongoing occurrence—of the quakes.[22]

Earthquake accounts are filled with somatic nuance. Different bodies, observers insisted, registered earthquakes differently. The 23 January shock that was felt in Annapolis, Maryland, as a "deadly sickness" and "giddiness" was regarded by some observers as "singular," and not like others: it made no noise, rather showing itself in "a continued roll similar to that of a vessel in a heavy sea." Further, the shock was "sensibly felt by some, while others, although in the same room, and perhaps within a few feet of them, were not in the least affected by its operation." Those outside, "in the streets or open air, were insensible as to any extraordinary motion of the earth." Close bodily contact with the earth might allow for a greater sensitivity: a newspaper reported from North Carolina that an earthquake was "perceptible to the negroes that were sitting on the ground, at work." In Frankfort, Kentucky, a shock "was very distinctly felt in this city on *Friday evening* and occasioned considerable jarring of kitchen utensils, vibration of suspended articles, sensible motion, dizziness and alarm, in many intelligent and respectable families: while nothing of it was perceived by others or by watchmen, &c."[23]

A report on the January shock compiled in Washington, DC, noted that many different accounts from various locations had streamed in: "We continue to receive accounts from various parts of the country, of an earthquake felt at the same time as that which we have already noticed, as experienced in this city on Thursday last. They all agree in comparing the sensation caused by it, to that of fainting, sickness at the stomach, vertigo, or approaching apoplexy." Such a conclusion contains important but embedded natural scientific reasoning. Because the shocks were felt at the same time and produced the same effects, they must have had the same fundamental cause—they were not a local but a broadly dispersed geographic phenomenon with consistent effects.[24]

Sensations of the human body were recognized as different in different environments. Contrasting those living in substantial brick buildings with those in humbler homes, former Mississippi Territory governor Winthrop Sargent noted of the main December shock that "some persons living in wooden houses believe that they felt many small shocks between the hours of 7 and 8." Reports noted the difference between shocks that, while strong, would not wake sleepers, and those that would wake people from a sound sleep. In this way, early nineteenth-century observers began to articulate distinctions that would be central to twentieth-century standardizations of earthquakes.[25]

One important aspect of the use of sensation to evaluate the earthquakes was an effort to quantify and explicitly compare the feelings of the various shocks. Daniel Drake was a young physician in Cincinnati during the New Madrid earthquakes. In pages of careful listing, he recorded and tabulated the duration and time of each shock, along with the weather at the time and how large each shock felt to him. His careful tables of sense-impressions were intended and read neither as parochial nor as eccentric. They were serious evidence about the general workings of the natural world that contributed to his own standing as a scientific observer and medical authority.

Drake published his observations in a natural history of his home region, his 1815 *Natural and Statistical View, or Picture of Cincinnati and the Miami Country*. This book would help establish Drake not simply as a practitioner, but as a leading authority on American medicine, and in particular on the close connections understood to operate between human beings and the lands they worked and settled.[26]

In *Natural and Statistical View*, Drake presented bodily evidence in page after page of careful descriptions of the felt intensity of the myriad tremors. The first shock, on 16 December, was "a quick oscillation or rocking, by most persons believed to be west or east; by some south and north."[27] Certain shocks were perceived both "by the aid of delicate plumb-lines" suspended by curious residents and by "those persons who were at rest"[28] Drake used the varying perceptions of differently moving bodies—people at rest versus those in motion—to classify earthquakes. While the most violent two classes of earthquakes were felt by all and a third class of tremors were of "intermediate violence," a fourth class was "felt only by persons *not* in action," and a "fifth and lowest order of these multiplied agitations" comprised those "detected by pendulums, and the delicate sensations of a few nice observers, when at perfect rest."[29] Differentiating "nice" or keenly precise observers from those whose moving bodies would be less sensitive to smaller movement, Drake created a hierarchy of seismic bodily sensibility.

FIG. 4.1. Bodily evidence of past earthquakes. In 1912, USGS geologist Myron Fuller photographed a man (likely local guide C. B. Baily) and a horse, next to a tree and a small rise. In other terrain, this would be unremarkable. In a floodplain, it is dramatic seismic evidence: the change in terrain is a fault scarp. Like many early nineteenth-century people, Fuller used the human body as a useful form of comparison and evidence. He also demonstrated that the 1811–12 New Madrid earthquakes were only the most recent series of substantial earthquakes to rock the region: the tree growing out of the fault scarp is far more than a hundred years old. Fuller's report would languish, mostly ignored, for decades. Widespread logging in the region eliminated forests, and subtle changes made his evidence less convincing: most urbanized Americans lost any intuitive sense of how old a big tree might be. Recent researchers have returned to Fuller's report to ratify many of his claims—including very large earthquakes in the deep past. (Myron Fuller, *The New Madrid Earthquake*, 1912, Courtesy US Geological Survey Photographic Library)

This quantification of bodily sensation stands as testament to the translation of bodily experience into the raw materials of scientific study. Drake's work was important in early nineteenth-century conversations about scientific causes for the quakes, and it is cited still as seismologists attempt to sort out the patterns of intensity of the New Madrid shocks.[30]

In his somatic accounting of the Mississippi Valley quakes, Drake brought attention honed in the work of healing to the exploration of tumult in the surrounding environment. His reports of how an earthquake made people feel helped make his professional reputation. As a result of his attentive recounting and synthesis of regional natural history, Drake soon became widely known. He was invited to become a member of the nation's two most prestigious learned societies, the American Philosophical Society of Philadelphia and the American Antiquarian Society of Worcester, Massachusetts, in 1818. Drake was an energetic institution builder, responsible for medical societies, a medical school, a scientific association, debating societies, and other cultural institutions in Cincinnati. Starting with his native city, he worked for many years to assess the prospects and health conditions of the American borderland territories, publishing a mammoth two-volume *History and Geography of Mississippi Valley* in the early 1850s.[31]

While Drake's organizing initiatives would be remarkable in any era, his interest in what bodies could reveal about the workings of the natural world were completely typical of his time. The broad attention paid Drake's work shows the close connection between medical and natural scientific work in this professionally diffuse and intellectually aspiring world, as well as the close connection between realms of science and of medicine. These were thoroughly separated only in the later nineteenth century: in 1815, Drake's attention to the physical signs and bodily implications of local earthquakes indicate the deep connections between the work of healing and that of understanding the surrounding environment.

Revival

Just as the logic and experiences of the body made sense of community experience of earthquakes by allowing conversation about and comparison of differing sensations, bodily experience also shaped spiritual reactions to the earthquakes. Earthquakes shook up spiritual faith for many in American communities—just as they did for Native American contemporaries, such as the Chickasaws who learned Tecumseh's dances and the Cherokees along the St. Francis River who responded to seismic shocks with rituals of spiritual renewal. The jolts of earthquakes acted as a jolt to the soul, prompting reflection, repentance, and revival.

Experience of earthquakes was multisensory, so it drew on religious commitments enacted through many senses. Tremors and sensations, having fits and being strange to oneself, feeling jolts and shocks—these experiences connected moments of spiritual truth with the physical surprise of being overtaken by earthquake. Fascination with the flow and sensations of electricity helped give a somatic vocabulary to the twinned bodily crisis of earthquake shock and contemporaneous revivalism. The New Madrid earthquakes fed fears for individual and global crisis, indeed for the end of the world: they also strengthened the movement of individual spiritual re-birth and collective faith gatherings and physical experience of God's truth, a set of experiences that contemporaries called the Great Revival.§[32]

Earthquake narratives were a regular part of individual and community spiritual practice, and registered the profound spiritual significance of the quakes. Such narratives appear not only in writing, but in the songs that expressed truths of the heart. Community singing was a bedrock of early nineteenth-century spiritual life. Widely sung hymns, along with stories from the Bible, provided the basis for spiritual understanding of what was happening when the earth shook. Several hymns (such as "A Call to the People of Louisiana" and "Earth Quake 1812") were written about the New Madrid earthquakes. Songs of faith tell a great deal about fears of earth-quake in this era, and about the hopes of revival and salvation prompted by crises of bodily well-being.[33]

The hymns about the New Madrid earthquakes reflect the fear and dis-ruption of the quakes, and in particular the spiritual meaning of the roar-ing noises of seismic commotion. The deep underground roar and boom-ing frequently heard by earthquake-rocked people—what many called "the thunder of hell"—meant that their souls were in danger. "I could but stand and wonder," ran the New Madrid hymn "A Call to the People of Louisi-ana," "Expecting ev'ry moment to hear / Some louder claps of thunder." To people of the early nineteenth century, there was no ambiguity about what such "louder claps" might mean: take heed, their hymns warned, of earthly omens, lest the signs of apocalypse and damnation take your soul unawares. A traditional Southern gospel song expresses the spiritual threat of such underground noise:

§ Historians often use the term "the Second Great Awakening" to refer to this early nineteenth-century movement of revivalism and reform, and they often frame this movement through the sermons of New England pastors. In doing so, they emphasize links with New England's Great Awakening of the early and middle eighteenth century. The people involved with frontier revivalism, though, did not regard their movement as an outgrowth of anything from New England. They drew inspiration from the original Pentecost, the day on which God's Holy Spirit descended in tongues of fire upon the fearful and tremulous apostles hiding away after their leader's crucifixion, giving them the ability to preach the Gospel boldly and in many languages (as narrated in the Bible at Acts 2). Christian believers celebrate Pentecost as the beginning of Christian preaching and the Christian church. Just so, many in the borderlands hoped, their spirit of revival and renewal would bring equal blessings in their own territories and even around the globe.

> I heard a mighty drum
> It was way down in the ground.
> It must have been the Devil, Lord
> Turnin' round and round.[34]

Amplifying this loud and dreadful noise, "sulphureous smell" filled the air around the earthquakes' epicenters. The stench of sulfur—especially in combination with burning lights and crevasses in the earth—had both pragmatic and intensely symbolic meaning. Sulfur smell was important in the world of the early nineteenth century: sulfur was a common reason why springs were not palatable water sources. Several witnesses reported formerly sweet springs of drinking water that became noxiously fouled with sulfur after the New Madrid earthquakes. Yet sulfur was powerful: water from sulfur springs was no longer good for drinking, but it was a potent medicine, "dosed out" to vivify or heal. Mary Hempstead visited such a healing spring in 1818 as part of her unsuccessful efforts to cure her long-running illness.[35]

The characteristic rotten-egg stench of sulfurous emanations makes sense in modern physical explanations of seismic upheaval: anaerobic decomposition of organic matter smells like rotten eggs, and when a lot of it is suddenly brought to the surface whole areas can reek. But sulfur stink, like the terrifying roar of underground thunder, also represented the threat of damnation to fearful frontier Christians. The devil smells of burning sulfur. Evidence from the New Madrid quakes seemed to speak of a world in which forces of evil, represented by foul and threatening sensory experience, could suddenly burst forth from underground. Evidence from the body had meaning for the soul.[36]

The earthquake cracks and chasms around New Madrid represented an undeniable material danger, but being swallowed by earth also held a set of symbolic meanings. As evangelist and writer Timothy Flint reported of the cracks around the destroyed village of Little Prairie, "The chasms in the earth . . . were of an extent to swallow up not only men, but houses, 'down quick into the pit.'" His readers, widely familiar to an extent almost unimaginable today with the stories and cadences of the King James Bible, would have readily understood his reference. In his passage, Flint quotes the book of Revelation, in which an angel dispatched from heaven "laid hold on the dragon, that old serpent, which is the Devil, and Satan, and bound him a thousand years, And cast him into the bottomless pit, and shut him up, and set a seal upon him, that he should deceive the nations no more."[37]

Such interpretation was widely and vividly present. In "Earth Quake 1812," one of the hymns written about the New Madrid quakes, the yawning chasms opened by the shocks are a call to repent before having to endure the same suffering on an eternal scale: "The pit shall receive you, while mercy shall leave you, / To mourn your sad, awful and direful case." The upheaval of repeated earthquakes is frightening in any cultural context but carried particular resonance in a culture in which earthquakes are a familiar form of God's judgment through history, where sinners are threatened with being swallowed into an abyss, and where the devil lives underground.[38]

Christian North Americans in 1811 might not have felt earthquakes before, but they knew from their Bibles that earthquakes meant God's judgment.** They might well be familiar with the long-standing if complicated view of earthquakes in Christian theology. Eighteenth-century theology had long interpreted earthquakes as retribution for human sin. Christian tradition emphasized the tumult of the natural world surrounding Christ's death: many medieval European sites memorialized earthquakes ascribed to the moment Jesus died, when according to Matthew's Gospel he cried out, the earth shook, the curtain of the temple tore in two, rocks split open, and bodies of the dead were raised out of their graves.[39]

In the largely Protestant early nineteenth-century United States, earthquakes remained a clear sign to unbelievers to turn to God. "The Stony Heart," a popular camp-meeting hymn, cast upheaval in the earth as a sign of how creation might respond to God's will—in contrast to the hardened hearts of unbelievers: "The rocks can rent, the earth can quake," ran one verse; "The seas can roar, the mountains shake; / Of feeling all things show some sign, / But this unfeeling heart of mine." As in the New Madrid hymn "A Call to the People of Louisiana," the quakes were a "mighty call" to reflect, repent, and return to the Lord.[40]

Those who felt the earth shake in 1811 could draw on a long tradition of preaching that used environmental calamity as a call to repentance. When a tornado struck in the midst of a camp meeting in Chillicothe, Ohio, in 1809, the preacher tore on, urging repentance even as the surrounding forest tore apart. Just so with the New Madrid quakes. Preachers and their listeners—and in this era many listeners *became* preachers—recognized the tumult about them as a sign to deepen or return to spiritual devotion.[41]

** As, for instance, in Psalm 104:32: "He looketh on the earth, and it trembleth," or Psalm 60:2: "Thou hast made the earth to tremble; thou hast broken it: heal the breaches thereof; for it shaketh." This tradition continued into the New Testament (as in Hebrews 12:25–27): end-times are heralded by a series of earthquakes in the book of Revelation, beginning with chapter 6, verse 12: "And I beheld when he had opened the sixth seal, and, lo, there was a great earthquake; and the sun became black as sackcloth of hair, and the moon became as blood."

Many early nineteenth-century Americans who felt the quakes thought themselves to be in the midst of the end-times. One Virginian awoken by the 16 December tremors immediately "called all the children up, and prepared them for worse events." A man named Jacob Bower, from near Beards Town, Kentucky, later wrote of being afraid to go to sleep lest he awaken in hell—a fear amplified by surrounding cracks in the earth. Disruption and terror of the moment were only the most immediate aspect of what they might presage.[42]

Such calls were heeded. A "gentleman" writing from Beaufort, South Carolina, wrote a Philadelphia friend that "the people here are very much alarmed, and the civil power have appointed a day of fasting, humiliation, and prayer that the evil which seems to await us may be averted." Such rituals drew on a long tradition in which governments responded to calamity with collective repentance and self-denial, a tradition that would continue in nineteenth-century reactions to epidemics and other disasters.[43]

At the same time, many of the tales that circulated about the earthquakes emphasized that those with solid faith could persevere even when not on solid ground. Preacher Charles Finley reported the spiritual triumph of a particularly devout woman named Gardiner. She was often rebuked and ridiculed for her "shouting"—emphatic, energetic, vocal affirmations of God's love and truth—not only at services but during daily chores. Yet on a day of hard shocks, "while the houses were rocking and the chimneys falling, as though the dissolution of all things was at hand, sister Gardiner ran out shouting and clapping her hands, exclaiming, 'Glory, glory, glory to God! My Savior is coming! I am my Lord's and he is mine!'" Tested by peril, sister Gardiner held fast in rejoicing.[44]

For the strong in faith, earthquakes brought not eschatological fear, but eschatological rejoicing. Peter Cartwright, who as one 1882 memoir described him, was a "famous old circuit-rider, who could preach sermons, scare up sinners, whip rowdies, and tell good stories with any one," hinged several tales on the willingness of frontier folk to welcome the end-times and the coming of the savior. As he told (often), he was preaching in Nashville in mid-January 1812. Early one morning he was walking on the hill near where he had preached, when he saw a slave woman of the house coming down to the nearby spring, carrying her bucket on her head. Just then the tremors hit, causing loud commotion and panic: chimneys collapsed, scaffolding fell from nearby buildings under construction, sleepers awoke, people screamed in the streets. The girls of the family who owned the slave came running outside, crying out for her to pray for them. The woman turned from the girls and began to shout in jubilation: "My Jesus is coming in the clouds of heaven, and I can't wait to pray for you now; I must go and meet him. I told you so, that he would come, and you would not believe me. Farewell. Halleluiah! Jesus is coming, and I am ready. Halleluiah! Amen." Off she went, shouting and clapping, still carrying her empty pail on her head.[45]

Similarly, up in Logan County, Kentucky, a preacher named Valentine Cook sprang from bed during the January shake and ran out toward the east[††] rejoicing, wearing nothing but his nightclothes. As Cartwright told it, Valentine's wife, Tabietha, ran after him and cried out, "O Mr. Cook, don't leave me." "O Tabby," he called out, shouting and jumping on, "my Jesus is coming, and I cannot wait for you!" These tales shine with the polish of long telling. They are part of a culture of storytelling, of larger-than-life tales that tell larger truths.[‡‡] Yet it would be a mistake to dismiss these or other good tales out of hand. Just as Davy Crockett did in fact end up in close company with an improbably large number of bears, many of the western settlers Peter Cartwright exhorted did hold a very real sense of a personal savior whose coming in glory they needed to be ready at any moment to welcome.[46]

These immediate and insistently unpolished responses—running off to meet Jesus in your underclothes, shouting with joy with a water bucket on your head—are entirely indicative of the spiritual upwellings of the early nineteenth-century United States. In the borderlands regions where the quakes were most felt, Christianity—like Christ—was intensely personal and physical. In American settlements, as in the Psalms, earthquakes are an earthly threat that can serve to emphasize the faith and security of God's people.[§§] For a believer like sister Gardiner, the demonstrable instability of the physical world highlighted the enduring steadfastness of a loving creator. In a hymn sister Gardiner may well have sung, God's mercies are "broader than earth! And firmer than mountains!"[47]

Frightened or confused by tremors, Americans throughout the most affected regions intensified a religious response shaped by bodily experience. The camp meetings of the Great Revival, which rushed across the

---

[††] East is the traditional direction of the Resurrection and eternal life, as well as literal sunrise: older North American country cemeteries usually lay out coffins with feet toward the east, so that the dead can see Christ at his coming.

[‡‡] In his autobiography, Cartwright took pains to deny the tall tale that he had won a fight with the legendary Mississippi keelboatman Mike Fink.

[§§] As Psalm 46 promises (verses 2–7), "though the earth be removed, and though the mountains be carried into the midst of the sea; Though the waters thereof roar and be troubled, though the mountains shake . . . the Lord of hosts is with us; the God of Jacob is our refuge."

midcontinent borderlands even as the New Madrid earthquakes shook them, taught a bodily knowledge that had spiritual truths.

Just as earthquakes did not create but did powerfully intensify movements of spiritual renewal among Indian communities, the New Madrid earthquakes did not create the American revivals of the early nineteenth century, but they certainly intensified this spiritual movement. The frontier areas where the New Madrid tremors were most dramatically felt were exactly where this evangelical movement was taking spark, and the New Madrid earthquakes hit just as the revivals began. The spring of 1800 witnessed what one nineteenth-century church history called "the most extraordinary revival of religion that ever happened on this continent, or perhaps in the history of the church since the Day of Pentacost." This was a revival in Gasper Ridge, Tennessee, at which striking manifestations of God's spirit filled and vivified the crowd. The Gasper Ridge meeting was followed by an even larger revival at Cane Ridge, Kentucky, in 1801. As revivalist Peter Cartwright later wrote, after Cane Ridge, "The heavenly fire spread in almost every direction"—especially west and south. Earthquake and spiritual movement met at the confluence of North America's mighty rivers.[48]

The physical experience of the New Madrid earthquakes fueled the intensely physical experience of the meetings of the Great Revival. In a community rush of religious feeling and Christian "conviction," people learned faith through bodily experience. Physical changes bore testament to God's power: unbelievers were thrown to the ground, convulsed, and overtaken by muscular contractions, tears, or a sometimes alarming variety of sacral sounds and motions. The spasmodic movements, barks, and grunts of some of these physical forms of religion characterized what soon became known as "religious exercises." People shouted and hollered, falling to the floor and slumping to their seats in dramatic physical reflection of spiritual crisis. Congregations wept, fell, went rigid, and fell into trances. The "jerking exercise" of sudden intense movements could spread to dozens, even hundreds of people at a time. In a practice termed "treeing the devil," believers would go down on all fours, barking and snarling. Some congregants were led by their rush of spirit to a kind of holy leap-frog in which women and men would crouch and jump over each other. More commonly, believers engaged in the "holy laugh," a solemn laughter in response to vigorous sermons and moving hymns.[49]***

***The "Shakers" famously engaged in similar strikingly spasmodic ritual dance. They were more organized, extreme, and feminist than most camp-meeting Christians, but the group (formally the United Society of Believers in Christ's Second Appearing), took

The whole setting of revivals fostered the physical intensity that led to spiritual regeneration—and soon to parallels with the overwhelming physical experience of earthquakes. Outdoor, multiday sessions of preaching, praying, and hymn singing pulled in people from many miles all around. They were led by preachers who rode a formal circuit for organized denominations, as well as by others—some of them women, children, slaves—who might be "called" on the spot and followed no set doctrine except the revelation of the spirit. Warm-up sessions might occur during the day, but the main sessions of camp meeting usually happened at night. One 1819 preacher described the "awfully sublime" scene, late after dark, with the flickering light of torches and candles casting shadows from masses of people and surrounding tree branches. Such scenes were particularly moving to people accustomed to going to bed soon after natural light faded and getting up with the rising sun. Sessions were organized around the moment of crisis and self-reflection known as "conviction," a moment that was often manifested through physically dramatic outer signs. Those convicted of their own sins, in full recognition of their own helplessness before the Almighty, were prepared for the change of heart and spiritual commitment that was spiritual conversion.[50]

Meetings were organized around physical support for those in the throes of conviction, people an 1818 hymnbook described as "slain of the Lord" and "brought into the glorious liberty of his children." A "mourning bench" provided a place for those who needed prayer and help as they recognized their own sinfulness. Nearby might be what skeptics called the "glory box," where people drew near to conversion, often with weeping, convulsive movement and falling down. Thoughtful attendees at one 1804 meeting cut down scores of saplings to about chest height, to give affected mourners something to hold onto and "jerk by."[51]

What one preacher called these "bodily exercises of the most wonderful character" were central to what worked about camp meetings. They were also central to what critics distrusted about them. Skeptics made fun of the barking, jumping, jerking, and collapse of congregants and nudged elbows over the consequences of such unrestrained physicality. One rival minister noted in his diary after Cane Ridge that a large number of local women who had been "careless" at the revival were now "big"—or dead in childbirth.[52]

People of the early nineteenth century felt God's truth in their bodies. For ministers, bodily movement was one sign of preaching success. To

part in many conventional camp meetings and held to the same theology of bodily expression.

believers and doubters alike, this was roughly physical religion, in which frontier preachers might tear off their coats and challenge a heckler to a fistfight.††† Preachers would forcibly embrace the necks of would-be rowdies who came to disrupt their meetings and then hang on while exhorting them: the struggle for souls was alike physical and spiritual. In one often-told story, Peter Cartwright forced an obstreperous ferryman to pray by fighting him midstream and then forcibly baptizing him in the name of the Devil, until the half-drowned man gave in and recited the Lord's Prayer.[53]

Those who resisted the spirit of holiness were often the ones most physically affected: only when sinners gave over to prayer would the fierce seizures abate. As a later memoirist recounted, one "large, drinking man" was seized by the jerks and found himself unable to run away (nonbelievers often took to their heels in the face of such intense religiosity). He took out his bottle and cried to the crowd that he would "drink the jerks to death," but was so convulsed that he could not take a drop. He kept struggling, fell and snapped his neck, and died in the midst of the praying crowd. Like Jacob wrestling the angel, both men and women "wrestled and prayed" with the power of God through the most intense of the great revivals as they made their way through conviction to conversion: salvation was only won through fierce struggle.[54]

In this era of tumult and conflict—within the earth and upon it—physically demonstrative spirituality united Americans with their Indian contemporaries. The broad-chested fighting preachers of frontier American evangelical religion may offer a rough parallel to common Native American ideals of leadership in which the spiritual and the physical were alike potent. Movement was similarly important to both American and Indian revivalism. During the Creek prophetic movement that followed Tecumseh's visit and the earthquakes, spasmodic bodily movement stood for spiritual allegiance. "Shake your own war clubs, shake yourselves," urged the Shawnee prophets who visited Creek towns in 1811: movement and dance spoke for commitment to Indian resistance and against American encroachment. A man named Samuel Manac recalled encountering the Creek prophet High-Headed Jim in 1813. When High-Headed Jim came as part of a delegation from a nearby town, "he shook hands with me, and immediately began to

___

††† In the 1997 movie *The Apostle*, the evangelical preacher played by Robert Duvall does exactly that, winning over a younger skeptic by thrashing him. Duvall recalled that he wrote the scene after reading about a successful episode in a noted evangelist's preaching career. This could well have been Cartwright, but need not have been—there are many similar stories. The physical and spiritual power of evangelical preachers has been an enduring theme of American folk Protestantism since those days of backwoods revival.

tremble and jerk in every part of his frame, and the very calves of his legs would be convulsed, and he would get entirely out of breath with the agitation. This practice was introduced in May or June last by the Prophet Francis, who says he was instructed by the Prophet." For Native Americans as well as for Christian communities, movements of body could demonstrate allegiance of the spirit.[55]

The New Madrid earthquakes fed the waves of revivalism in American communities just as they did in Indian lands. When the earthquakes hit, many felt themselves in need of revival. Camp meetings took place throughout the earthquake zones. In July 1814, Stephen Hempstead rode out to one in his neighborhood. Landslidings reminded people of their own backsliding. The Reverend James Finley was once preaching to a group of "sin-hardened men" in a cabin when a tremor shook them. As everyone stumbled and grasped for solid ground, Finley adroitly jumped on top of a table and cried, "For the great day of His wrath is come, and who shall be able to stand?" Few preachers could ever find a more dramatic illustration.[56]

In the most-shaken regions, the earthquakes "brought many to grace." As evangelist Cartwright later recounted, "The earthquake struck terror to thousands of people, and under the mighty panic hundreds and thousands crowded to, and joined the different Churches." Methodist membership in the town of New Madrid grew more than sixfold in the year following the quakes, even as many residents fled the area. After the earthquakes, remembered Arkansas Methodist evangelist William Stevenson, "it was easy to preach to the people; all seemed to be humbled and came together in crowds to hear what they must do to be saved." Baptists likewise saw numbers roll in. In January, a visiting Baptist minister from Tennessee told the Moravian missionaries to the Cherokees that the recent earthquake, along with a severe windstorm, had damaged many buildings in his area and "had a great impact on the characters of many residents in Tennessee."‡‡‡ [57]

For many in the American borderlands, earthquakes and "conviction" had parallel manifestations. The experiences of being overtaken by earthquake and of getting the Spirit were remarkably consonant. These processes happened in and through the body, striking the body's registers with the same language, descriptions, and accompanying emotion. When struck with a quake or struck with God's truth, people of the Ohio and Mississippi Valleys shook, jerked, and fell to the ground in anguish. What one earthquake hymn called "this shaking, this direful quaking" referred equally to the once-solid earth and to the bodies of those upon it.[58]

---

‡‡‡ At the same time, many of their neighbors poked fun at the short-lived faith of so-called "earthquake Christians."

Electricity

Evangelist Peter Cartwright wrote about his experiences preaching in the Ohio Valley right around the time of the New Madrid quakes in ways that make clear the parallels of body and spirit. §§§ He noted with triumph that in one session in an otherwise solidly Baptist district, "suddenly the power of God fell on the congregation like a flash of lightning." And, as he described it, the effect was very much that of lightning—or of its close terrestrial af-filiate, earthquake: "The people fell right and left; some screamed aloud for mercy, others fell on their knees and prayed out aloud; several Baptist members fell to the floor under the power of God." Preachers who "struck fire"—the term for successful exhortation—acted on the body just as mas-sive tremors did. Fleeing or fearing further earthquakes, frontier Christians felt God's truth as a mighty shock.[59]

As Cartwright's story implies, powerful preaching and powerful earth-quakes were alike electrifying. In the late eighteenth and early nineteenth centuries, language and experience—and sometimes, preaching—drew together the bodily effects of electricity and spiritual rebirth. A former slave described his conversion: "Like a flash, the power of God struck me. It seemed like something struck in the top of my head and then went on out through the toes of my feet. I jumped, or rather, fell back against the back of the seat." Those under "powerful or pungent convictions" were de-scribed at the time as being "struck"—and just like people hit by a blow or by a lightning bolt, those struck by spiritual power sometimes fell senseless. The numbers of people "struck down" was to believers a powerful witness to the revivifying power of God in the new territories. One Connecticut pub-lisher wrote of "sudden and irregular storms of fervor" at Methodist reviv-als, where "the very air . . . seemed impregnated with electric fluid."[60]

Background knowledge about electricity provided the basis for many Americans' understanding of earthquake sensations and spiritual jolts. Many early Americans had, indeed, an easy familiarity with the sights and

---

§§§ Nailing down the precise movements of an itinerant preacher is no easy task, so it is not exactly clear where this story took place. It was in about 1812, in the northern end of the Wabash Circuit of the Tennessee Conference, which encompassed some of Indiana, but this was a four-week's circuit (that is, it would take Cartwright four weeks to ride round it), and an energetic preacher on a good horse could cover some serious ground—preacher Lorenzo Dow logged 10,000 travel miles in 1805 alone. Wherever this exact meeting was, Cartwright's story and much of his preaching experience was cer-tainly taking place in and among the people who felt the New Madrid quakes and whose movements and settlements were affected by them.

sounds of domesticated electricity, despite its very recent birth as a scientific field. Alessandro Volta had only created the galvanic battery—also known as a voltaic pile—in 1800. Volta had alternated disks of copper, zinc, and brine-soaked cardboard to produce electrical jolts (each set of disks was termed a "galvanic cell," named for Luigi Galvani's previous experiments with the nerves in frog's legs). Volta's galvanic batteries allowed researchers and showmen to generate and manipulate electricity more easily than they had been able to using the Leyden jars created in the 1740s for storing static electricity.

Electrical equipment brought the physical experience of electricity into early American culture. Arrays of Leyden jars, or as they would now be termed, capacitors, powered the famous electrical experiments of theorist, politician, and tinkerer Benjamin Franklin in the 1740s and early 1750s.**** Franklin's work was part of an explosion of interest in electricity, particularly in its therapeutic or healing potential, in Europe and in North America. Leyden jars were certainly capable of packing a painful (even lethal) wallop—animals perished in droves in electrical demonstrations—but they were challenging to create, maintain, and charge. Globe machines sometimes shattered, spraying bystanders with fast-moving shards. Volta's creation of the galvanic battery sparked new designs, new devices, and new insights into the complicated relationship of electricity and magnetism. Backwoods reports of the feeling of electricity in 1811 and 1812 thus reflected very recent scientific insight.[61]

Electricity was a surprisingly accessible phenomenon, even in the very early nineteenth-century borderlands. People across North America were eager and enthusiastic consumers of this new experimental science, flocking to electrical demonstrations where they could see, feel, and be touched by the exciting and mysterious electrical fluid. Traveling electrical lectures were part of an early social economy of American science, in which entrepreneurs with a flair for showmanship might make a living—or at least a temporary living—from scientific lectures, often featuring dramatic and engaging spectacle (see coins glow! see a dead ox's eyelids open and close!).[62]

**** A Leyden jar is essentially a device for storing the charge of static electricity. On a day of low humidity, a Leyden jar constructed out of a small lidded container, a nail, a little aluminum foil, and some water can provide almost endless amusement for household children. Properly charged by a piece of PVC pipe rubbed vigorously with wool socks, even a fist-sized Leyden jar delivers a biting zap: substantial and linked arrays of Leyden jars like those Franklin manipulated are—as he experienced—powerful indeed.

Science and medicine were close companions in electrical fascination. Traveling lecturers charging fees for colorful exhibitions worked from demonstration kits with small wooden compartments worked into saddlebags, almost identical to those carried by medical practitioners of the same era. Itinerant lecturers like the energetic T. Gale—a traveling entrepreneur who left so few records that we do not now even know his first name—instructed fellow Americans how to build their own devices to produce and administer therapeutic electricity. He and other early experts urged a do-it-yourself medical electricity smoothly in line with the do-it-yourself democratizing impulse within medicine of his time. The careers of traveling showmen like T. Gale can be frustratingly enigmatic. Itinerant boosters, like circuit-riding preachers, left few written records and never stayed in one place long. But Gale, along with exhorters and travelers like Peter Cartwright or Tecumseh, was one of many to bring new forms of knowledge and enlightenment deep into the backwoods of early nineteenth-century North America.[63]

Through public and bodily demonstration, electricity was simultaneously party trick and object of investigation. Crowds clapped as coins sparked in the darkness, luminescent metallic leaf outlined witty phrases, electrically charged handshakes made neighbors jump, and the "electrified kiss" of smiling young women repelled all suitors.†††† Rubbing the elastic gum of the caoutchouc tree, commented Louis Bringier in his 1821 article in which he also described the effects of the New Madrid earthquakes, could make a small object adhere to a wall "until all the fluid is dissipated." The fluid he meant was electrical fluid, and it was both familiar and fascinating. As he commented, "these experiments made in a cold winter's day, afford some amusement." When those who described the New Madrid quakes wrote of how earthquakes looked or felt like the discharge of an electrical battery, their descriptions or conclusions might not mesh easily with modern science or perceptions, but they were drawn from ample experience.[64]

Many people agreed that the quakes must be connected with electricity because of how they *felt*. In his household laboratory, Franklin had once attempted to tenderize a turkey through the discharge of an array of Leyden jars. Instead, he tenderized himself. When he recovered, he noted that the resulting "universal Blow" caused a "violent quick Shaking of my body."

---

†††† These demonstrations were more or less chaste, but the subgenre of electrical pornography gave full rein to the erotic implications of electricity. Benjamin Franklin's reputation as both genius and sexual adventurer were the subject of particular wit in late eighteenth-century France. In one French watercolor, he is portrayed as having enthusiastic sex with an Indian figure representing America, with electrical machinery as backdrop.

Many in the Mississippi Valley felt much the same sensation over fifty years later, and they too ascribed their strange, sudden, unnerving feelings to electricity. "The shocks of the earthquake," observed an 1823 travel guide to the new state of Missouri, "produced emotions and sensations much resembling those of a strong galvanic battery." Instrumentation and objectivity had not yet made nineteenth-century theorists distrustful of bodily experience. "Emotions and sensations" were a reasonable form of evidence for the nature of scientific phenomena.[65]

Many people felt the electricity of the earth in the New Madrid quakes. Medical authority Samuel Mitchill quoted at length a letter from a "gentleman" from Russellville, Kentucky. After listing the symptoms brought on by the shakes—nausea, vomiting, tremors, debility, pain in the lower legs and joints—this Kentucky correspondent concluded that "the shocks seemed to produce effects resembling those of electricity." In the early nineteenth-century world in which the New Madrid quakes were so overwhelmingly felt, this statement of bodily effects was an important claim about the nature of the quakes. Bodily experience and environmental truth were usually held to be closely related. What felt like electricity, might be. The roaring, flashes, prickling sensations, and even smells associated with thunder and lightning were valuable clues to the fundamental nature of earthquakes.[66]

In 1812, some in Charleston felt the New Madrid earthquakes as a series of "strong and sudden jirks"—the form of sensation that many associated with electricity. Another generation of Charlestonians, those who lived through a substantial 1886 quake centered underneath them, likewise reported the feeling of being pricked with pins and needles. One woman who felt the Charleston quakes experienced them as a tingling "beginning in her big toe" and then climbing upward to the rest of her body. Based on these experiential reports, Dr. Peyre Porcher of Charleston argued in an 1886 article in a Philadelphia medical journal that the recent earthquakes in his city were expressed in part as electrical phenomena. The tremors felt like electricity—in particular, argued Porcher, like discharge from a galvanic battery. Such feelings indicated the physical behavior of quakes. "The commotion, attrition, and tremblings of vast masses of earth might not only squeeze out water, sand, and gases," argued Porcher, "but very naturally generate electricity on a tremendous scale which would affect those who were specially sensitive."[67]

Response to Porcher's arguments about physically felt seismic electricity, however, also reveals how scientific explanation had changed since the New Madrid earthquakes. John Guitéras, likewise a Charleston physician, argued in 1887 against Porcher and his evidence of the body that "seismologists . . . have shown that electricity has nothing to do with earthquakes, either as cause or effect." Bodily responses and nervous afflictions could instead be ascribed to the confusion of memory or "hysteroidal impress." Significantly, Guitéras *contrasted* bodily sensation with the behavior of electromagnetical devices. Guitéras argued that complete calm had been observed in "magneto-electric apparatus during the earthquakes." "Barring the mechanical disturbances," he noted, "it will be found that no effect whatever was produced upon these instruments, nor was the magnetic needle in the least affected."[68]

By the 1880s, a few medical authorities like Porcher could, like their colleagues of very early nineteenth century, draw conclusions from the human body to the larger physical world; but others like Guitéras could contrast the evidence of felt body and cold machine, separating those forms of evidence and giving credence to apparatuses unsusceptible to hysteria. Right at the beginning of the development and deployment of seismograph networks, just in the decade when researchers in Europe and Japan were struggling to refine and deploy self-registering seismometers, those struggling to understand an 1886 Charleston quake were already in a vastly different instrumental and scientific world, in which assertions about jolts, shocks, and pins and needles were starting to sound ignorant and old-fashioned. Electricity was not the explanation for seismic upheaval, no matter what people's tingling toes said. Tracing the history of human sensation and of how physical sensation is experienced reveals the history of what constitutes credible evidence about an unnerving natural event.

Tracing the history of earthquake sensation also reveals the many sources of metaphor, of inspiration, and of reasoning in early America. Feeling the earthquakes, people made connections. In describing the unusual sensations of earthquakes, people reached for metaphors as old as the Bible and as traditional as sea travel, but also as up-to-the-minute as artificially produced electricity.

Bodies and Earthquakes: Conclusions

Reports of the New Madrid earthquakes reveal how the body served in early nineteenth-century science simultaneously as instrument and metaphor: a way of recording, communicating, and comparing an unusual, frightening, overwhelming experience, and simultaneously a way of thinking through that experience.

Through bodily sensation, people across North America pieced together knowledge of large-scale movement in the earth. Seismic tremors in the early nineteenth century brought people together across large distances in shared sensations that were regarded as an accurate gauge of change in the natural world. Putting together sensations across realms of experience, people who felt the New Madrid earthquakes connected environment and spirit through the feel of electrical shock. But this understanding of somatic truth was not limited to the analytic. The feelings of earthquakes also conveyed spiritual knowledge. Tremors told the truth of the soul's need for rebirth.

Disturbed in their bodies, many early North Americans feared for their souls. They heeded the evidence of the body and understood the shock, jolts, and jars of earthquake as a shock of the Holy Spirit. They listened, they moved: just as they felt warnings in their physical selves, they acted out their renewal and faith through physical movement and demonstrative noise. Singing out their fear for end-times in the unreal illumination of late-night gatherings of faith, they sang, too, of their bodily and spiritual allegiance to a power greater even than shaking earth.

Such insistent physicality stands as an important reminder that human bodies exist in a context. They inhabit a physical context of sometimes unsettling or dramatic environmental changes. And—perhaps equally as important—human bodies exist in the context of spirituality, intellect, and imagination. In the context of the New Madrid quakes, one's own self provided information about surrounding phenomena, and identical sensations indicated the webs of connection linking self to soil.

Paying attention to this use of bodily sensation can help mark the moment when learned people began to distrust the body's evidence. That is an important shift. For earthquake science, that shift happened with the increasingly sophisticated development and networking of seismometers in the early twentieth century. Before that instrumentalization, accounting for one's physical sensation could be a valid and nuanced way of conveying information about an event. After those developments (as in earth science today), evidence of the body is regarded as suspect and even embarrassing: a seismologist who feels woozy because of an earthquake will not reference that nausea in her field reports or publications. But people who wanted to contribute to scientific discussion in the early nineteenth century would do so straightforwardly and with keen attention.

This shift to suspicion of bodily sensation and even excision of bodily sensation from scientific work has taken place at different times in different scientific disciplines and in different national contexts. Narratives of the New Madrid earthquakes capture the role of the body in a crucial era of medico-scientific understanding, in which people strained to reach toward exactitude, measurement, instrumentation, but did not yet have the tools to fully grasp them. In the absence of instruments, they themselves were the instruments.

Recognizing the prior history of bodily experience and bodily sensation reveals the larger sweep of the history of how we construct knowledge not only about health and disease, but about other aspects of human experience. Unrestrained physicality is no longer a hallmark of most American Christian experience. People who bark and jump at a sermon can still find supportive communities in a few faith traditions, but in most places they are more likely to be removed from church halls than called up front to preach. Plenty of churches, especially churches in Southern African American denominations, still preserve vocal call and response and visible physical reactions to moving exhortation, but even those tend to be restrained in comparison to the wildly noisy and energetic responses of early nineteenth-century camp meetings. Many Americans might shout "Amen," but shouting of the shock of the spirit would not make sense as a theological statement for most Americans. The world of American Christianity has far less place for strong, sweaty, noisy, active bodies than it once did.

People of the nineteenth century felt earthquakes down to their bones and deep in their souls. For those who experienced the New Madrid quakes, attention to bodily state was not just a concern of a particular afflicted individual, but a widely shared interest in reporting, understanding, and interpreting the surrounding physical environment. Their spiritual and scientific understanding of what was happening all around them was shaped by the sensations of their own bodies. A history of medicine in this era and place meant also a history of geological events. As people in seismic zones throughout the world understand, earthquakes certainly disturb modern bodies too, but the importance of communicating, understanding, and interpreting those sensations has attenuated steeply and almost completely. In that changing history of bodily sensation rests a changing history of earthquake knowledge.

5 ❋ Vernacular Science: Knowing Earthquakes in the Early United States

In the winter of 1812 a physician named Alexander Montgomery wrote from his small town of Frankfort, Kentucky, to the well-known Philadelphia naturalist Benjamin Smith Barton about the "frequent shocks of an Earthquake" that had alarmed people up and down the Mississippi and Ohio Valleys. Montgomery knew that his former teacher, based in a big city at the center of many kinds of scientific conversation at the time, would be interested in a thorough firsthand account of the effects of the quakes.

During the severe February tremor, Montgomery reported, chickens were thrown from their roosts and "persons stood on their feet with difficulty." In places, the damage was intense: "Parts of some Houses of Brick have been Shaken down; But in Louisville at the distance of about eighty miles below us and on the banks of the Ohio, the Damage done Houses has been considerable. There is scarcely a brick House in the town that has not lost Chimnies—the walls of others have been very much *Cracked*." The quakes changed the environment in peculiar ways, fouling a spring that now stank of sulfur, caving in Mississippi River banks, and destroying one island entirely. All nature seemed in upheaval: "Dogs howled—Cattle bellowed—& Horses intimated their dread by running from place to place."[1]

After the February tremor, Montgomery also wrote to another colleague in a letter that was then published in a local newspaper. In this second letter, he reported that "large quantities of a substance, which, if not pumice stone, possessed many similar qualities, and evidently the products of volcanic fires, were picked up from the Kentucky river opposite this town."[2] Dr. Montgomery observed and apparently tested the "substance": "It is extremely light, has not suffered any alteration by any degree of heat we have been able to apply, and will not *retain* heat at all." He proposed to share this scientific curiosity with like-minded investigators, sending one piece to a chief scientific institution of the very early nineteenth century, the museum run by Charles Willson Peale in Philadelphia, and another sample directly to Barton. Based on the unusual, pumice-like floating rocks, as well as reports from local Indians and the mountainous region which he thought was the source of the "violent concussions," Montgomery concluded in his letter to Barton that the quakes must be of "volcanic origin."[3]

Alexander Montgomery's records of the New Madrid earthquakes now consist of one handwritten letter and one several-times-republished fragment of a letter. Little more apparently exists of his life or of the rocks he sent.* Yet even this brief, incomplete evidence shows what a vital world of scientific imagination he inhabited. In important and unrecognized ways, Dr. Montgomery stood in conversation with people like William Leigh Pierce and Louis Bringier. Their letters and reports about the "hard shocks" of that tumultuous winter reveal a previously ignored realm of scientific communication, theory, and practice in the early United States. These earthquake accounts open a window on a busy scientific world amid the small farms, canebrakes, and hunting grounds of the American borderlands. In the early American republic, many different people, many of them off in the backwoods and boondocks, shared a lively interest in contributing to scientific explanations.[4]

This world of rough-and-ready experimentation and practical tinkering, of scientific theories that appear alongside crop yields and commercial advertisements, rarely appears in our histories of the early United States. We modern people expect earthquakes to be the subject of science, but this vernacular science is not what we expect. It is too messy, too informal, too imbued with the prosaic concerns and everyday enthusiasms of country-store discussion. As rough-cut as butternut home-dyed cloth, such exchanges are just as easily overlooked. Yet people clothed in butternut claimed territory, fought wars, and changed cultures. To overlook the substance of similarly homespun scientific exchanges like Alexander Montgomery's is to ignore

---

*I have not located either of these samples—if they ever even reached Philadelphia to begin with. Some of the Peale museum collections burned in fires in the 1850s, and most of the rest were sold to New England museum entrepreneur Moses Kimball or to his close collaborator, showman P. T. Barnum. Much of Barnum's entertaining display disappeared in a spectacular fire in 1865. Some remaining specimens (presumably the inflammable ones) made their way into other collections, including the Smithsonian, and most of the Kimball collection eventually ended up at Harvard's Museum of Comparative Zoology, but only after some years spent jumbled in a barn, with labels removed and everything askew, the stuffed birds largely disassembled from their bases. Energetic searching by archival staff could not locate Montgomery's "substance," but it is certainly possible that somewhere on a dusty shelf in Philadelphia, Washington DC, or Cambridge a piece of Kentucky River rock still sits quietly unlabeled and unregarded, now a mere relic but once viewed as offering insight into fundamental questions of the earth's movement.

a fundamental aspect of early American life—and of American scientific thinking. The vernacular science revealed by such exchanges was a vital element of American society in this early nineteenth-century period of change and expansion, and it forms the unrecognized bedrock for the explosion of invention, innovation, engineering, and institution-building that would ultimately transform the United States into an industrial, economic, and educational powerhouse after the Civil War.[5]

Vernacular science was ambitious. It was at once local, embodied, and specific, and at the same time highly theoretical and interconnected across vast distances. The earthquake thinking of people like Alexander Montgomery relied on all their senses and knit together the human body, the weather, and the earth beneath their feet.[†] Their models of the earth were dependent on extremely local and limited observations, and yet connected events across the globe. Early Americans debated volcanoes and the forces of electricity as possible causes for earthquakes. They hazarded conjecture about comets and their relationship to what lay below the earth. In all their thinking, surmising, and trading of information, they at once thought through the world and how it worked and sought to comprehend territories that were new and largely unknown to them. Seized by sudden, alarming, highly interesting phenomena, many people in the early American republic recorded how they wondered and thought about the way the world works—and about what constituted the nation in the process of expanding across those very earthquake zones.[6]

Letters, Artifacts, Experiments:
The Materiality of Early American Science

In his attempt to use phenomena of the unknown wilds to puzzle out geological questions, Alexander Montgomery represents many who felt the New Madrid quakes and used them to push forward their own scientific questioning. In networks of correspondence, in the coarse materiality of artifacts and objects, and in often idiosyncratic efforts to use devices to

---

[†] At the same time, scientific approaches in this period certainly also encompassed spiritual concerns. One series of articles in a Georgia paper reflecting on the causes of earthquakes closed with a quotation from the Psalms: "Lord, what is man that thou art mindful of him and the son of man that thou visitest him?" This chapter focuses on natural scientific explanations, while other chapters focus on interpretations of the spirit, but that division is tactical rather than substantive, an artifact of the need to understand and sort through themes in the voluminous writing about the quakes, rather than a distinction essential to the sources.

measure or register earth movement, early Americans sought to harness the evidence around them to make larger claims.

In this period, letters were a crucial form of scientific conversation. Interpersonal correspondence was, after all, the only way of gathering information in an emerging scientific context that lacked a centralized system. Like many others who experienced the quakes, Montgomery wrote his letters to colleagues, and—like many others—he clearly expected that if his missives were found interesting they would be reread and excerpted or published. One of Montgomery's accounts appeared first in a local paper and then in one of the earliest books written about the New Madrid quakes, Robert Smith's 1812 compilation. From private letter to newspaper to book: a country doctor's observations were read over much of eastern North America.[7]

Montgomery's discussions of the pumice-y floating substance also indicate who participated in scientific conversations. He described the substance to noted Philadelphia scientific authority Benjamin Smith Barton in a February letter, and to a colleague he named simply as "A.F.," who then forwarded the letter on to a newspaper, from which Robert Smith picked it up and reprinted it in his book-format compilation. Still another unnamed colleague agreed to deliver a specimen to Barton in Philadelphia.[8]

Participating in the sciences of description and evaluation could help those colleagues, like Montgomery, gain connection. For much of the nineteenth century, science in North America was an open field. Formal training or certification of expertise was much less important than demonstrated skill in observation—the same kinds of observational skill used by hunters and trackers or animal breeders, as well as physicians. Being acknowledged as a colleague in observation made someone a participant in science.[9]

This was a science built up from observations and material objects. Montgomery was typical of his time in focusing discussion of large earthquakes onto a small piece of rock—the pumice stone that he discussed in his correspondence, a sample of which he forwarded to Peale's Philadelphia museum.‡ Through correspondence, Montgomery attempted to engage in

‡ The museum's energetic founder, the patriotic painter and impresario Charles Willson Peale, envisioned a popular and accessible workplace and demonstration site for modern science. He displayed mineral and other collections, stuffed birds and preserved reptiles, often obstreperous live animals, his own *trompe l'oeil* paintings, and, famously, a complete mastodon skeleton. Despite many creative efforts, he was never able to make his museum the publicly funded institution he wanted. In 1809, Peale retired and handed over the museum to his son Rubens Peale (all the many Peale sons were named after famous painters). Rubens Peale was even more interested in commercial promotion than his father, and many felt that the intellectual caliber of the museum fell swiftly after '09.

shared scrutiny of his collected floating substance. "You will observe," he wrote to Benjamin Smith Barton, "the substance has been exposed to the action of a very intense degree of Heat; & I have no doubt myself, of its being of a Volcanic origin." Like Montgomery, other observers noted similar floating pumice-like rocks in major rivers after the earthquakes and came to the same conclusion, interpreting these artifacts as evidence that volcanic eruption had caused the tremors.[10]§

In his willingness to try the properties of this intriguing substance through heating it, Montgomery was like many other early Americans who used small-scale, often informal experimentation to determine properties of the surrounding environment. In the years surrounding the earthquakes, George Hunter and William Dunbar dragged heavy equipment up boulder-strewn river courses to try the chemical properties of the Arkansas Hot Springs, and Stephen Austin and Henry Rowe Schoolcraft traveled and assayed in the lead districts of Missouri in the course of exploration and commercial development. As Benjamin Franklin knew only too well, experimentation was not for the faint of heart. Only a few years before the New Madrid quakes, explorer Meriwether Lewis had licked a rocky bluff to see what it was made of—almost poisoning himself in the process. Further, when Sacagawea, the young Indian translator for the Corps of Discovery, fell gravely ill, Lewis similarly made use of a sulfur spring, "the virtues of which I now resolved to try on the Indian woman." (He and Captain Clark agreed that the water did indeed help her.) Alexander Montgomery's examination of intriguing rocks drew on traditions of experimentation—trying and tinkering—central to the practice of medicine and science in his era.[11]

Ironically, Montgomery was apparently trying to signal his participation in serious scholarly exchange by building links with an institution whose own reputation was in the midst of a serious downhill slide.

§ Contemporary geologists do not share Montgomery's conviction that his floating rocks were in any way related to the earthquakes. They suggest that perhaps a piece of half-burned coal fell off a steamboat, or that he had found pieces of lignite—the rough coal ejected from near the earth's surface by sand blows and other "liquefaction features." To trained geologists, lignite looks nothing like pumice, but to an untrained observer, it might: the first time I saw a sample of gnarled, rough-surfaced lignite from the New Madrid earthquakes, rent with holes and uneven spaces, it certainly struck me as looking pumice-y. Current researchers do not know how lignite could have ended up near Frankfort, and more tellingly that does not strike them as a critical question. Modern science is less tied to the limits of a local perspective. A few floating pieces of mineral are unlikely to affect the larger questions of what causes quakes like the New Madrid events. The present epistemological insignificance of Montgomery's carefully gathered samples speaks volumes about transformations in earth science since that cold and shaken winter.

Montgomery tried to use his letter to transcend distance, writing as if he and Barton were together looking at the rock on a table before them. Montgomery went on to estimate the source of the rock from the speed of the river current and the time elapsed between the last hard shock and the appearance of the stone in the Kentucky River across from Frankfort (based on a four-mile-an-hour current, he thought "the eruption must have taken place between three & four Hundred miles above us.")** Faced with limited and highly local observations, he examined a piece of floating stone to deduce the source and nature of the shaking of the earth beneath him.[12]

Montgomery was not alone in collecting and interpreting material artifacts. Early in March 1812, William Leigh Pierce published an open letter in a Savannah, Georgia, newspaper defending his own veracity and motives against some who were skeptical of his account of the recent earthquakes. The evidence, he argued, was present in their own town: "Let those who please to examine for themselves," he advised, "visit the Library Room of this city, where specimens of the matter thrown up in the recent convulsions may be seen." Artifacts of the quakes, brought to local display, would amply attest to the powerful effects of the earthquakes Pierce described.[13]

Pierce and Montgomery—like Louis Bringier—were writing in an era marked by the collection of artifacts whose exchange was a crucial currency in scientific legitimation. Ornithologists, naturalists, geologists, collectors of Indian curiosities, and sundry investigators of the late eighteenth and early nineteenth centuries worked in a tradition of natural history as conveyed by and represented in samples and specimens. After returning from his journey of exploration and beginning his work as a settled government official, William Clark would often permit visitors to St. Louis to look through his "Indian Museum," a collection of Indian relics, unusual animal skins, preserved pelicans, elephant teeth, and other artifacts. Mineral cabinets were a source of individual and civic pride. Late in the nineteenth century, remembering his days piloting the great steamboats of the Mississippi, Mark Twain described one of the standard curios of earlier planter households: a "convention of desiccated bugs and butterflies pinned to a card," collected and recorded by the household's daughters. Twain skewered the young women's scientific curiosity just as he did their artistic skill—the insects, like badly rendered copies of *Washington Crossing the Delaware* or highly decorated but badly played guitars, were all evidence to him of the small intellects and small horizons of the country life he himself had fled.

** Everyone I've shown this account to is puzzled by how he thinks that the rock could have floated to him, given that the rivers flow *from* Kentucky *to* Missouri: clearly some aspect of Montgomery's explanation is not well conveyed in his writing!

Yet however caustic, Twain's memory testifies to the central place of such collections and such collecting. Whiskey-pickled reptiles, medical oddities, and glittering minerals might all be found on the shelves of those interested in natural philosophy and the workings of the world.[††] The New Madrid earthquakes were centered on the fringes of the American frontier, but curious objects and samples associated with them connected people in a thick matrix of shared community and curiosity.[14]

One artifact gathered from the New Madrid earthquakes did become—quite literally—the stuff of natural history. Daniel Drake and other writers reported with fascination that along with lignite, warm water, and foul stench, long-buried fossils were blown out of the sand blows near the earthquake's epicenters. A French traveler of the late 1810s named Edouard de Montulé was particularly interested in these "curious objects." A local German resident showed Montulé "the femur of a mammoth" spit out by a sand blow, a bone that matched the skeleton Montulé had seen displayed in Philadelphia, where it was a famed centerpiece of Charles Willson Peale's museum.[15]

But Montulé also witnessed an even more significant item: the German also showed Montulé "a head which might be compared to that of a monstrous ram, with the horns reversed." This was a momentous find: as a later geography of the Mississippi Valley recorded, "From one of the fissures formed during these convulsions, was ejected the cranium of an extinct musk-ox (*Bootherium bombifrons*), now in the possession of the Lycaeum of Natural History of New York." This skull provided the first example of a woodland musk ox skull in scientific literature. That initial 1828 description is still referenced in our contemporary scientific research. Material apparently spat out from the rendings of the earth's surface during the New Madrid quakes became part of the natural historical record of the Mississippi Valley and its past inhabitants.[‡‡][16]

[††] Reptiles picked in whiskey could be transported and stored with success. Not so for mammal parts: a large batch from the Lewis and Clark expedition, forwarded by President Jefferson to Peale's museum, proved verminous and unsalvageable: "The uncleaned bones bred the Insects which afterwards fed on the Skins."

[‡‡] I say "apparently" because we cannot check: like Alexander Montgomery's carefully gathered rocks, the original skull has since been lost. Somewhere in the lurching, uneven process in which nineteenth-century gentlemanly societies gave way to professionalizing universities and sleek modern research institutes, a single bovine skull was broken or mislabeled or mislaid—buried through bureaucratic accretion or carelessness just as it had been buried under the earth before being suddenly surfaced in the early nineteenth century. Further, present-day east Arkansas archeologists Claudine Payne and Marion Haynes told me that based on a more recent bison skull that they recovered

Geology and Science in the Nineteenth Century

This musk ox skull was important not just as a curiosity, but as a piece of evidence in ongoing, international scientific discussion fueled by economic interests. The New Madrid earthquakes were particularly interesting to many people because they took place at a time of inquiry and debate over geological phenomena and principles. In part, this inquiry was prompted by raw commercial interest. In many parts of the globe, mineralogy and mining science tantalized rulers and republicans alike with the promise of subterranean wealth—shining metals, portable fuels, and building materials. The sciences of the earth played an important role as part of the European-led exploratory voyages and expeditions of the late eighteenth and early nineteenth centuries. At the same time, European and American developers sought to capitalize on valuable lead deposits in Missouri just north and west of New Madrid—soon, they would seek iron in nearby veins.[17]

In the decades surrounding the New Madrid earthquakes, debates over the character and processes of the earth raged in European journals and scientific societies. Was the earth's surface shaped by sudden catastrophic events? Or, as Scottish naturalist James Hutton argued in his 1785 *Theory of the Earth*, was it shaped over thousands and thousands of years by uniform, gradual forces? How were rocks formed? Through deposition by water—the position of the German geologist Abraham Gottlob Werner and the so-called Neptunists? Or through the forces of "caloric," or heat, as advocated by the "Vulcanists"?[§§] In the first few decades of the United States, Werner's arguments for the role of deposition by water gained acceptance, but the very process by which solid masses were created remained open to continu-

---

from a construction site, a skull that had been fractured by the processes of burial and erosion, it seemed to them unlikely that a similarly large skull could actually have survived intact after being blown out of a sand blow. Most likely the now-lost musk ox skull did not actually come flying out of an earthquake fissure, but was revealed through some much more ordinary process of flooding or erosion and discovered after the quakes. Yet I share what I imagine to be nineteenth-century readers' fascination with this explosion onto the earth's surface of an early relic of extinct North America—and I take from the accounts of this now-lost skull a respect for the ways in which objects from an earthquake were seen to indicate quite literally the history under the earth.

§§ These terms are, of course, far too simplistic representations of what were nuanced and often overlapping forms of argument, and they have fallen out of favor in the history of the earth sciences—yet for those not already familiar with some of the arguments animating early geology, they can give a helpful foothold in sometimes confusing debates, and perhaps convey to us the challenge of finding fundamental causal explanations for some of the most basic processes of the geological world.

ing research and debate. Similarly basic questions about the composition of the earth bedeviled and fascinated scientific observers. Was the earth's interior solid or liquid? Rent by gaseous chambers or subterranean fires? Lakes, perhaps? How thick was the crust? Curious readers found very little scientific consensus on which to draw until roughly the 1830s, when the general conception among scientifically informed Europeans and North Americans was of a slowly cooling molten earth covered by a thin crust.[18]

Over the last decades of the eighteenth century and the first decades of the nineteenth, a profoundly new and historically based geology emerged from the fieldwork, correspondence, and publications of European writers and field investigators. This geological historicism was buttressed by new ways of understanding physical evidence, especially fossil remains. Beginning in 1808, leading vertebrate anatomist Georges Cuvier collaborated with taxonomist Alexandre Brongniart to argue that fossils were not incidental but geologically indicative: different kinds of fossils characterized formations—and thus constituted a layer-cake history of stages of organisms. In subsequent publications, Cuvier pushed his colleagues to think more broadly and creatively about almost impossibly deep time, arguing that present processes, features, and phenomena could provide the basis for inferences about events so long ago as to be unobservable. This approach represented a geohistorical view of physical evidence as well as an expansive view of time and geological processes.[19]

In bustling European cities, understanding the history below the earth became a central theme and fascination to an extent perhaps surprising today. The Geological Society of London, founded in 1807, was a site of animated discussions. There investigators debated the identification of strata and shared exciting new finds. In 1830–33 British geologist Charles Lyell published his ambitious and self-consciously monumental *Principles of Geology*, a work in which he argued, like Cuvier, both for the vastness of deep time and the vitality of present causes as explanations for enormous changes in the deep past. When he traveled to the United States, and to New Madrid, in the following decade, he was a minor celebrity.***[20]

*** Lyell's work became even better known as a result of Charles Darwin's publication of *On the Origin of Species* in 1859, a book in which Darwin, himself a respected figure in geology, famously drew upon Lyell's and other researchers' recognition of the power of small changes summed over vast time. Applying that insight to living beings, Darwin explained how the forces of natural selection acting on heritable traits over an exceedingly long period could produce species differentiation through a process we term *evolution*. Geological arguments held powerful consequences in the nineteenth-century world.

Especially in North America, geological investigation (and farmers' plows) yielded massive and intriguing bones of mammals and what seemed to be giant lizards, skeletons that natural philosophers studied, entrepreneurs sold, and showmen hawked with equal enthusiasm. Fossils like the musk ox skull became a feature of American popular culture in the middle and late nineteenth century. Like electricity, geology was not only an exciting and lively scientific field but a popular sideshow.[21]

Thinkers in Europe sought knowledge and analysis of what they still regarded as a "New World." French savants eagerly read about developments and findings of early American geology—Edouard de Montulé was one of many. The wild and unknown places of the "far West" were endlessly fascinating to researchers who sought in fossil bones, fallen meteors, huge waterfalls, and indigenous peoples to work out theories of natural history. After 1812, leading European scientists such as Lyell and Alexander von Humboldt incorporated the New Madrid quakes into their theories, observing possible connections with far-distant quakes and other geological phenomena.[22]

Knowing the Earth, Knowing the Nation

For Americans, the search for knowledge about the earth held potential patriotic as well as commercial meaning. Geological knowledge and geological interest were widespread and valuable, if spotty: a Mississippi Delta salt trader in 1813 noted that a particular site might from its "marly appearance" look "calcareous," but instead "have rather the Silicacous cast." Careful planters strove to assay and improve the characteristics and potential of their soils. Geology was a main form of scientific knowledge in the early United States.[23]

In similar fashion, economic and intellectual imperatives pointed early American scientific observers toward explicitly nationalistic goals. The search for natural resources shaped private and public scientific endeavors, from the ongoing search for coal deposits to the survey of live oak trees conducted by a US Navy desperate to build more ships after the War of 1812. In 1809, William Maclure completed the first geological map of the United States, thus introducing North American geology in a formal way both to those who lived there and to European scholars themselves in the midst of a flurry of such maps and surveys.[24]

Beginning in the 1810s, formal institutions gave shape to such interest. The American Geological Society was founded in 1818. Though small and short-lived, it presaged an explosion of organizations devoted to the earth sciences in the middle and later nineteenth century, culminating in the 1879 founding of the United States Geological Survey, or USGS. In the three decades before the Civil War, a majority of American states established geological surveys. The surveys sought—and seek, in the present day—to help private entities capitalize on mineral, oil, and other deposits. At the same time, the surveys represent a main way state governments (and, eventually, the federal government) supported and engaged in science.[25]

Yet the desire for truly American sciences went beyond even these pragmatic and commercial goals. Many patriotic American observers of the late eighteenth and early nineteenth centuries resented the dependence and condescension that marked their relations with the European intellectual elite. "Too long has it been fashionable," wrote the editors of the *Medical Repository* in 1804, "for our people to seek scientific news from transatlantic regions, while they neglected the manifold novelties by which they were surrounded at home." Instead, the editors urged that readers "turn their backs to the east, and direct their views to the inviting and productive regions of the interior of America." In this context, to offer local observations about seismic movement was at once intellectually productive and patriotically fulfilling.[26]

Doing Science in Early America

Like Montgomery's letters, certain of the accounts of the New Madrid earthquakes are easily identifiable as scientific, in a sense recognizable over the distance of centuries as well as to colleagues at the time. Interested Americans might read earthquake investigations published in explicitly intellectual venues or discuss them in meetings of the self-improvement societies that were a hallmark of early American cultural life. Some did both: members of the School of Literature and the Arts, a blossoming literary and scientific association created by Daniel Drake in Cincinnati,[†††] spent a session in their first year of 1813 discussing a presentation on the recent earthquakes, and several years later some of those same listeners likely read the

---

[†††] The terms "science" and "arts" were bigger, more comfortable, and more relaxed terms in early nineteenth-century usage, companionably bumping shoulders on a wooden bench at a local debating society or on a velvet couch in a port-city drawing room. In the era of the New Madrid quakes, "arts" could easily include mathematics alongside language and geography, as well as history and chemistry. An organization like Drake's for the literary and philosophical arts might easily encompass discussion of the causes of natural events.

observations on the quakes Drake published in his 1815 *Natural and Statistical View, or Picture of Cincinnati*. Reports on the New Madrid earthquakes appeared in the few explicitly scientifically oriented journals—scientific in a very broad sense, ranging across fields now separated today—that did exist in the early United States. Samuel Mitchill, professor of botany and materia medica at Columbia College, published his compiled observations in the *Transactions of the Literary and Philosophical Society of New-York* in 1815 after presenting them at a meeting the year before.[27]

But there was no "professional" community of science, and no hard line dividing experts from dabblers, the well-informed from the curious. Many earthquake accounts may strike us as prosaic and homey—a paragraph or two in a small-town paper, a breathless torrent of description in a letter to family back East. Yet their many repeated compilations insist upon the contemporary coherence of what can seem scattered or informal, dashed-off observations.

Throughout accounts of the New Madrid earthquakes runs the theme of intentional, self-conscious observation and theorizing about how the world works—in other words, elements that can be productively and usefully understood as scientific thinking. No one piece of this mass of writings about the earthquakes tells a very clear picture about scientific thinking in the early United States. But pulled together, the hundreds upon hundreds of earthquake accounts reveal much about how science was done—through networks of correspondence and readership, sustained with great difficulty across great distances—and about nineteenth-century models of the earth and its history.[28]

The New Madrid earthquakes made borderland regions important. The earthquakes were felt most strongly in the middle Mississippi Valley and Ohio River Valley communities of what were then the American borderlands. The underlying reason for this is geological: soft sediment shakes. Living in Frankfort, Kentucky, on the Kentucky River, a tributary that flows into the Ohio, Dr. Alexander Montgomery experienced strong shaking from the main New Madrid tremors. Yet the consequences of this difference in how earthquakes affect various kinds of environments were not simply geological, but social and intellectual. They gave people living along the arteries of the Ohio and the Mississippi firsthand information and perspective that other people wanted. Firsthand experience of the quakes gave their observations credibility and importance in networks of scientific communication.[29]

On the outskirts of the young United States, a virtual unknown like Montgomery could use networks of correspondence to learn more about the quakes and then send his observations, information he gathered, and the strange rock he collected out to the larger world, weaving himself into the fabric of scientific communication. The rivers connecting him to the New Madrid Bootheel region gave him the mysterious mineral substance he investigated as well as news from travelers about the earthquakes' effects. On the borderlands of the young United States, Montgomery was nonetheless central to the environmental disturbance that fascinated his contemporaries.

At the time Alexander Montgomery wrote him, Benjamin Smith Barton was a leading American naturalist. He had published a major botany of North America and written on racial theory (at the time, a significant aspect of natural history) as well as ornithology and medicine.[30] Barton and Montgomery's communication indicates the networked aspect of early American science—not yet in many formal ways, but in overlapping, informal networks that would soon begin to coalesce into institutions like the stately American Philosophical Society in Philadelphia, where Barton was a leading figure, or Daniel Drake's fledgling, ambitious Cincinnati association.

Early American science was an intensely personal and interactive venture. Benjamin Franklin was central to the founding of the Philadelphia Junto in 1727, which organized the Library Company of Philadelphia and led to the founding of an early version of the American Philosophical Society in 1744 (the APS was formally founded in 1768). Decades later, in the 1800s and '10s, Daniel Drake was similarly busy founding not only his School of Literature and the Arts, but also a circulating library, the Cincinnati Lyceum, a debating society, a theatrical club, a medical society, and a major medical school. Franklin and Drake were unusually energetic and successful as well as polymathic, but the many friends and colleagues they convinced, cajoled, corralled, or were encouraged by in all this association making were entirely typical of the thirst for information and connection in the late eighteenth and early nineteenth centuries, as well as the impulse to form associations and networks in growing communities both large and small.[31]

Discussion of what the quakes were and why they happened took place in a matrix of communication with many holes. Poor mail service and lags in reporting time created endless frustrations for the scientifically curious—correspondents complained that the few scientific journals published in North America might take literally years to be available in other cities or the countryside. Yet this matrix still served to sustain connection and community. The science of the New Madrid earthquakes was diffused, a network with many nodes. It was a science of newspapers and correspondence, not a science centralized in a few journals or at a few leading institutions. The term "scientist" falls far short of capturing this widespread interest (the term itself didn't even exist until the 1830s). Early nineteenth-century people interested in science might call themselves "naturalists" or "natural historians" (old-fashioned ones would call themselves "natural philosophers"). Mostly, though, those interested in explanations of the natural world, like Alexander Montgomery, would not necessarily have a formal name for that interest, or a formal affiliation—much less formal salary—to give it shape. What they shared was curiosity about why the world worked as it did and a conviction that careful observation and measurement, diligent recording, and assiduous comparison could reveal mechanisms, correlations, and ultimately causes in the world around them.[32]

Compilations of firsthand reports were at the time at the forefront of scientific understanding of earthquakes. From the mid-eighteenth century to the mid-nineteenth, earthquake study—like work in the earth sciences broadly—emphasized the collection and compilation of reports and information. Although by the 1820s European researchers were beginning to collect historical earthquake accounts to look for seasonal, meteorological, or geographical patterns, little parallel effort had yet been mounted in North America, so New Madrid compilations represented an important contribution. Collations of many individuals' accounts served as the beginnings of such seismic catalogues. And collations and collections were a crucial way in which knowledge of the quakes dispersed and ramified.[33]

In the intellectual world of the New Madrid earthquakes, direct observation and careful attention pointed the way toward truth. One compilation on the quakes published in 1812 was, as the title emphasized, "Collected from Facts." Particular, personal observations could be important for a scientific understanding of these tremors. After reports of myriad puzzling physical sensations felt by numerous of his neighbors who only gradually realized they were feeling an earthquake, one correspondent wrote that "if the same or similar motions have been felt at other places, doubtless it will be communicated. I should like to hear it accounted for on rational principles."[34]

Through such informal, widespread exchanges, early Americans built a culture of scientific curiosity and entrepreneurship, a culture that nurtured the steam engines, industrial mills, agricultural improvements, railroads, and innovations in communication and transportation that reshaped not only North America but much of the world. Such developments did not spring merely from a few sites or a few innovators: rather, the revolutions of technology, science, and engineering that made the United States modern between the Civil War and the world's first Great War were all built upon the widely engaged-in tinkering, curiosity, surmise, and scientific investigation of early American culture.[35]

Capturing Time and Movement:
Simultaneity, Measurement, and Instrumentation

Accounts of the New Madrid quakes show the overwhelming scientific imperative of the nineteenth century: Measure! Quantify! River trader William Leigh Pierce was unusual in his proximity to the epicenters and the keenness and depth of his language and observations, but he was like many observers in his emphasis on the quantifiable and verifiable in his descriptions.[36] His specificity was typical. A traveler stopping by the Moravian mission to the Cherokees reported that in the Cherokee town of Taloni "in a field 13 sink holes appeared as a result of the earthquake, the largest of which is 20 feet deep and 120 feet in circumference and is supposed to be full of greenish water." Observers estimated the sway of various tall objects affected by the tremors (the steeple of the historic Maryland statehouse was seen to vibrate six or eight feet at the top of its 181-foot height) and measured the temperature of warm water issuing from a North Carolina mountainside after the quake (142° Fahrenheit).[37]

Many earthquake observers sought to measure and quantify *things*. They also sought whenever possible to quantify *time*. Efforts to measure the effects of the shocks were tied to a deep interest in simultaneity and timing. Were the quakes felt everywhere at the same time? Or did they ripple across North America? Were they related to other earthquakes of early 1812? The quest to bring exactitude to the knowledge about these quakes was also a quest to capture time. Standard time zones were not introduced into the United States until 1883, a period when hurtling railroad trains made standardizing time practices across large distances a national imperative (one accomplished by private railroad companies). Sorting out when events occurred was a crucial task in an era before standardized timekeeping and one that held hope of revealing relationships between far-distant events.[38]

Timing as exact as possible was an important element of New Madrid accounts. Former Mississippi territorial governor Winthrop Sargent noted that the first main shock "occurred at my house within one or two minutes of 2 o'clock, correct solar time, upon the morning of the 16th of December 1811." People who felt the quakes often tried to give an account for when they happened and how long they lasted. Especially in further-away areas where the shocks were perceptible but not terrifying, observers tried to be precise about their notations of time. "I find by T. Parker's regulator," wrote correspondent W.V. to *Poulson's Advertiser*, 8 February 1812, "that my watch was slow 3 minutes 30 seconds; this will give the correct time 4 hours, 27 minutes, 30 seconds A.M." W.V. closed the account "with the assurance that you may depend upon the correctness of the time," and a plea that these observations be put together with "similar observations in different places, by comparing which together, an idea may be formed of the centre from which the numerous late shocks have proceeded."[39]

Paradoxically, concern for the exact timing of events shows most clearly in widespread frustration with the difficulty of preserving an accurate time sense in times of crisis. A "severe shock of an earthquake was felt in this place," noted one observer from Augusta, Georgia, in December 1811, "which continued by the best account we have been able to obtain, between one and two minutes—some persons think much longer—but considering the awful nature of the occurrence, and the hour of it, we are aware that its duration could not be accurately determined, without noticing its commencement and termination by a correct time piece."[40]

Some writers collated and tabulated their measurement and qualitative observations. Several reports on the quakes—like William Leigh Pierce's—included tables of shocks, with notes on their timing, the number of shocks, their duration, and their comparative intensity. Pierce, indeed, was fastidious, noting that the first shock alarmed his party "precisely at two o'clock on Monday morning the sixteenth." In Cincinnati, Daniel Drake recorded the duration and time of each shock, along with detailed information on the weather and precipitation at the time and how large each tremor felt. Such charts, tables, and detailed lists tie the observations of earthquakes to those of epidemics, city populations, and other details of western lands that appeared throughout the literature of American exploration and settlement: in observing disaster, as in evaluating farmland, people of the antebellum United States frequently combined narrative description with quantified and tabulated detail.[41]

Such interest in timing stems not only from an overall valuation of precise measurement and quantitative description, but also from widespread interest in relating local disruption to experiences elsewhere. Similar sensations could indicate similar (or the same) cause. A report from Chickasaw Bluffs (near Memphis) reported that the December 16 quake "occurred at 30 minutes past 2 o'clock, in the morning . . . the same time that it seems to have been felt in the Atlantic states, and in this country." Close timing connected phenomena across distances.[42]

Many people were relieved when their personal experiences connected with larger phenomena. One subscriber wrote to *Poulson's Daily Advertiser* to report feeling shocks while lying in bed awake in Philadelphia. He awoke his wife, and together they watched the curtain of their bed and windows shake and "very sensibly felt the undulation." Two other women in a room above theirs also saw the shakes: "The leaves of some flower-pots which were in their room, were so much in motion, that one of the ladies thought it arose from the draft of an open window." Yet until reading Poulson's paper the household cast what they had experienced as local and specific: "I have met with no confirmation till I saw the account from Alexandria in your paper."[43]

As with the Philadelphia's household's shaking flowerpots, objects as well as sensations told the movement of the earth. In one of the startling effects of the quakes, an Ohio surveyor reported after the first December sequence of tremors that his compass needle refused to settle. During the 7 February quake, wrote one eyewitness in Alexandria, Virginia, "a looking glass suspended from the wall, continued to oscillate for half a minute after the earthquake had ceased." People along the Eastern Seaboard noted that the January shocks were hard enough to stop their clocks. Sometimes objects could register tremors too small for human sensibilities, as when one observer wrote from New Haven, Connecticut, to insist that "the concussion of December sixteenth, was perceived in this town. An apothecary saw vials suspended to the ceiling, oscillating; but the motion was too small to be perceived, except by very sensible objects." Everyday objects became instruments of observation.[44]

In the era of the New Madrid quakes, such ad hoc instrumentation was the only possibility. Self-registering seismometers were only commercially available beginning in the 1880s and would only become widespread in the decades after 1900. Despite the plethora of increasingly sensitive, exact, and diverse instruments available in European centers by the late eighteenth and early nineteenth centuries, people living in North American cities, much less North American borderlands, had access to very few devices for measurement or registration, especially relative to seismic motion. When the earth moved and they wanted to figure out how much or in which direction, early Americans therefore responded the same way they did when their guns jammed or their kitchen pots broke: they tinkered.[45]

Cobbling together from what was at hand, many people used ordinary devices and household objects as makeshift instruments in an attempt to understand the bewildering world around them.‡‡‡ In March 1812, an American soldier being mustered to fight the Indian league and the

‡‡‡ Such homespun instrumentation proved useful well into the nineteenth century. After an extremely damaging earthquake hit Charleston, South Carolina, in 1886, the US Geological Survey distributed instructions and seismoscopes to interested residents who volunteered to record numerous aftershocks. In some cases, these instruments consisted simply of a bowl filled to the brim with molasses that would slosh out when a tremor hit.

British wrote in his journal that the most recent shocks were reported to have caused more than eight inches of oscillation further south, but had only moved the earth about three inches where he was stationed, at Vincennes: he and his comrades had measured the shocks by suspending a lead ball with a thread from the ceiling. One woman wrote from Charleston, whose loose soil made the New Madrid quakes particularly momentous, that she and her friends awoke horrified at the midnight convulsions of their lodging house: on the third story, their rooms "rocked like a basket." When the worst of the shocks had passed, no one could go back to sleep, so the ladies "suspended a pair of scissors by a string to the mantle piece," where "the most trifling motion would occasion them to vibrate." Fascinated, the group "sat watching them until daylight," seeking in trembling shears some hint or reassurance about more devastating shocks to come. With a certain flair, a household in Annapolis suspended an ostrich egg a foot from the first-floor ceiling after the huge January tremor, then watched in amazement as the enormous egg oscillated four inches from side to side during subsequent smaller shocks.[46]

An engineer in Louisville named Jared Brookes went further: he gave a room of his house over to an array of "pendulums and vibrators" whose "smart motion" revealed movement too subtle for human perception and showed the character of motion, whether up-and-down or side-to-side. Brookes set up different-sized instruments to catch different-sized shocks: "delicate vibrators" alongside pendulums of varying lengths and a spring six inches tall. In journal entries almost every day over the four months of the greatest shocks, Brookes told the story of earthquakes through the changing movements of his array: one "considerable shock" on 29 February "acts first on the spring vibrators, then on all, pendulums act largely, trembled off in about three minutes; then . . . more placid motions at short periods." His instruments "perceived" motion too fine for humans to catch.[47]

Yet Jared Brookes's record incorporated far more than mere instrumental observations. He narrated earthquakes through the many senses and fields of experience of the sciences of his time. Tremors that triggered his pendulums might be associated with "a circle round the moon" or "atmosphere without the least elasticity, one might say dead and brittle." All nature was involved in the experience of earthquake. Brookes recorded on 27 January that one "violent shock, as sudden as the arrival of a cannon shot" was associated with peculiar weather in which "smoke rises in erect columns to an uncommon height; the animal system disposed to relaxation, much complaint on that account; at 10 at night vapor thickens; some light wind from south." Those attempting to puzzle through the mysteries of earth action sought evidence in their own forms and in the familiar objects of everyday life. Seeking explanations, they examined connections with a lived universe, connections that often crossed boundaries of what might today exist as separate categories.[48]

Earthquake Weather: Connecting Earth and Sky

In this deeply agricultural world, weather was one of the first correlations that many observers—like Jared Brookes—sought in trying to make sense of the quakes. Some, such as the careful Dr. Daniel Drake, devoted pages to a narrative ordering of time of day, temperature, winds, and precipitation during distinct tremors. Compiler Robert Smith included in his book extracts from a meteorological table, specifying measurements of exact timing, wind direction, temperature, barometric readings. Even short narrative accounts frequently included mention of the weather before and after specific tremors.[49]

A common theme was the unusual qualities of the weather surrounding strong shocks—"warm as in the Indian summer," for instance, in the middle of December. Many people noted that the earthquakes were accompanied by fog, haziness, or a darker than usual atmosphere. An engineer forty miles below New Madrid on a flatboat recorded that "the atmosphere was filled with a thick vapor or gas to which the light imparted a purple tinge resembling but different from Indian summer or smoke." Such darkness conveyed foreboding and strangeness.[50]

The strongest shakes seemed capable of changing the weather, and bad weather was thought to foretell a strong tremor. John James Audubon "remarked a sudden and strange darkness rising from the western horizon" before the shock he experienced. Afterward, "the heavens brightened as quickly as they had become obscured." Audubon's comment is telling: many people who experienced the hardest shocks observed that the weather changed immediately afterward. Jared Brookes noted that after the second huge shock of 23 January, "it began to rain transparent ice in drops of the size of pigeon-shot, for two hours." Others looked to changes in weather to forecast a quake—William Leigh Pierce noted the "lowering appearances of the weather" that led him to fear a further shock. Another letter writer from South Carolina noted a few days after the 7 February hard shock that "from the looks of the weather today, I am afraid we shall have another."[51]

And yet even in the decades after the New Madrid quakes, earthquake weather remained frustratingly hard to figure out.§§§ As was the case for

---

§§§ Interest in "earthquake weather" is not simply part of early modern ignorance: reported in California in the 1860s by Mark Twain, it continued well into the twentieth

many scrutinized relationships between natural forces and earthquakes, in the early nineteenth century and today, despite observers' most careful attention, connections often proved elusive. Winthrop Sargent reported carefully on the weather throughout his experience on the Mississippi River during the December temblors. He, like many observers, experienced cloudy weather and rain before many of the larger shocks, but then also recorded early on the morning of Tuesday, 20 December, "a shock of greater severity and duration than any of the preceding—the sun rose brilliant with a clear sky." "Serene" weather could reign over the most horrible of quakes; weather did not necessarily indicate the trouble brewing in the earth.[52]

Sargent's apparent surprise at the calm clarity of morning weather in the face of a great quake indicates the way in which seemingly flat observations carried implicit arguments—here, about the failure of expected correlations in disruptions below the earth and in the sky. Though most nineteenth-century accounts contain little in the way of explicit theoretical declarations, that does not mean that theorizing was absent. While often seemingly atheoretical, earthquake accounts carried musings or arguments about causes and effects, evidence and truth, in their structure and emphasis.[53]

Despite the lack of clear relationship, these early investigations reveal much about the science of the nineteenth century. Observers regarded a world without clear boundaries between the heavens and the earth. What happened in the skies could indeed be closely linked to what happened in the ground under their feet. This was a holistic universe—and at the same time a scientific one—in which the kinds of categorical distinctions made by later sciences did not yet apply. Both science and environmental perception were profoundly and in some ways usefully diffuse. Weather was an accessible venue for considerations of what causes what, and of how such statements could be constructed. Conversations about earthquake weather capture a moment when the very relationship of correlation and causation was in the process of being worked out.**** [54]

---

century. Yet when I heard Californians after the Loma Prieta quake of 1989 refer nervously to "earthquake weather," it was always as something that *someone else* believed in. Over the past two centuries, the concept moved from scientific inquiry to folk knowledge to vaguely embarrassing folk superstition.

**** When Scottish naturalist John Bradbury and his crew encountered a group of families gathered to pray together after the December 1811 hard shock at a log house downstream from New Madrid, one man blamed the recent comet for the disruption: the comet had "two horns," he explained, "over one of which the earth had rolled, and was now lodged betwixt them: that the shocks were occasioned by the attempts made by the earth to surmount the other horn. If this should be accomplished, all would be well, if

Noise and Causation

The search to relate weather and earthquake further points to another aspect of nineteenth-century earth science: for observers of the early 1800s, the realm of possible causative powers was extremely broad. The forces that overwhelmed the senses of people who experienced earthquakes—noise, light, smell—might move the earth, too. The sheer sound of the earthquakes represented not only a descriptive feature and a spiritual threat, but a potential causal agent.[55]

Careful observers struggled to express what they had heard with precision and clarity. As Alexander Montgomery wrote to Benjamin Barton, "The shock was preceded by a hoarse rumbling noise, somewhat like distant thunder." Many similarly strained to hear differences in kinds of earthquakes. Timothy Flint reported a few years after the quakes that people remembered different kinds of shocks: those in which motion was up-and-down were much louder and more explosive than horizontal shakes, but at the same time less destructive. The earthquakes could even confuse sound. Jared Brookes recorded during a period of "considerable shove and vibration" that "sound (as often of late) seems, as it were, to have lost its rotundity, and matter its sonorous properties." He continued uneasily: "The peal of the bell, the beat of the drum, the crowing of the cock, the human call, although near at hand seem to be at a distance, and the different reports seem to steal, in a manner silently, separately, and distinctly upon the ear."[56]

Many reports included extensive discussion of the apparent directionality of the quakes: they usually seemed to come from a certain direction and travel to and sometimes past the observer. Such interest was consistent with the more professionally developed seismology that began to be articulated in midcentury. British engineer Robert Mallet, a pioneer researcher into seismic waves, was the first to systematically study the directionality of earthquake damage (and hence, he inferred, earthquake waves) in order to determine the quake's focal point in his investigations of a destructive temblor in Naples in 1857. A central puzzle—for people who directly experienced the New Madrid quakes as well as for scientists who studied them later—is what caused these tremors. In describing the directionality of the earthquake's sounds and appearances, many people of the time began hesitantly to theorize about where the quakes came from, how far away they started, and thus what might have caused them.[57]

---

otherwise, inevitable destruction to the world would follow." Bradbury drew only gentle humor from this incident, but it reveals the widespread impetus toward explaining these alarming events using the information or observations at hand.

Through sound, observers—listeners—might gather useful information about regions far distant. In this, experience of earthquake might reveal more about the sound world of the early United States. Far-off sound could carry through nonindustrial environments in ways quite foreign to most present-day experience. One 1848 article noted, for instance, that the characteristic cypress trees of Mississippi River swamps and bayous produced a resonating effect that transmitted sound for extremely long distances. Hearing noises could be one way to begin to shape an opinion of places too far to see.[58]

Much of the nineteenth-century interest in the sounds of earthquakes has to do with the potential relationship between noise and weather. Common wisdom from the early modern period up through the twentieth century granted tremendous power to loud noises. Battles were long held to change the weather and bring on rain. In the era of modern armament, many people believed that it was the noise from cannon and gunfire that caused the clouds to weep. When rains were scarce on the High Plains, Civil War veterans knew that artillery fire at Gettysburg had caused several days of heavy downpour, so they suggested similar noise might help. An old soldier from Denver wrote in to the federal government during the Dust Bowl to recommend that 40,000 Civilian Conservation Corps boys stage sham battles with $20 million worth of ammunition: the noise would stir up the rain. "Try it," he urged; "if it works send me a check for $5000 for services rendered." In the reports of earlier American earthquake echoes a similar conviction that being rocked by sound could transform the very environment.[††††][59]

In accounts of great noise, there is a sense that attributes from one realm of experience might transfer to another. Eliza Bryan, resident of New Madrid, noted in 1816 that "this country was formerly subject to very hard thunder; but for more than twelve month before the commencement of the earthquake there was none at all, and but very little since, a great part of which resembles subterraneous thunder." Bryan did not explain the mechanism that would connect earthquakes with the disappearance of above-ground thunder, but she clearly drew some relationship between these phenomena. Her description implies that the thunder has in some sense gone below ground: the same force and power that used to character-

---

††††Currently in New Zealand, "hail cannons" are used extensively to protect crops. They fire an audible "shockwave" that is meant to disrupt the "embryo" of a forming hailstone. New Zealanders complain that their noise is a major feature of adjustment for those moving from city to supposedly quiet countryside.

ize weather above the earth now characterizes its disruption beneath. Loud rumbling—formerly of thunder, now of quakes—was to her a significant indicator that this movement of forces had taken place. Accounts of the New Madrid quakes thus reveal a world of interconnection between aspects of environment that a modern view would regard as separate.[60]

A Science of Global Connection

Many who wrote about the quakes posited causal connection between events in the Mississippi Valley and those elsewhere in the world. In so doing, they joined local, even parochial, experience of the backwoods with theories of earthquake causation current in the most refined and well-read circles.

Many newspapers that carried news of the New Madrid quakes also carried snippets of reports about earthquakes elsewhere. Certainly, the New Madrid quakes continued for many months, so they overlapped with a number of earthquakes elsewhere. But such coverage also reflects a facet of early nineteenth-century scientific conceptions: a sense of global interconnection, in which experiences in one part of the world may indicate a more profound disturbance elsewhere.[61]

Earthquake accounts frequently linked the New Madrid tremors to recent reports of mysterious earthquake-like rumblings reported by travelers at sea and on land throughout the globe. One merchant ship, for instance, had heard loud explosive noises in the night, and had sailed through a raft of floating pumice. Meanwhile, that summer volcanic explosions rocked the island of St. Vincent, and in October 1811 Mount Etna was reported to have erupted from six apertures at once. Not mere curiosities, these were terrestrial symptoms of deep and powerful disturbance. Though these events did not necessarily involve the shaking of earth, they were regarded by scientific authorities like the German geographer Alexander von Humboldt as part of a larger expression of seismic phenomena.[62]

The sheer scale of environmental upheaval made clear to many people who felt the New Madrid quakes that they must have repercussions on a massive scale. On 4 January 1812, *Niles Register*, a national newspaper, noted that "some dreadful calamity may have been experienced in a distant part of the world—probably South America; judging from the violence of the sensations felt in different parts of the union." Similarly, the preface to an 1812 collection of New Madrid earthquake accounts noted that the "many and repeated shocks of Earthquakes" felt so strongly "in our southern and southwestern States," signal problems not only there but elsewhere: they

"indicate that there has been some terrible, and perhaps destructive eruption of the Earth, somewhere to the south-west of us, perhaps Mexico, New-
Spain, or Quito, of which we are hereafter to have tidings.—As the great
Earthquake which sunk a part of Lisbon, was felt in Scotland, 1100 miles
distant."[63]

A good deal of theorizing focused on the curious sequence of many large
earthquakes.#### In the fall of 1755, within less than a month, large earthquakes leveled Lisbon, Portugal, and shook New England. Then in the spring
of 1812, bare weeks after the New Madrid tremors, another enormous quake
in Venezuela devastated Caracas and helped end the short-lived Venezu-
elan revolutionary government. As Samuel Mitchill had done, a number
of thinkers surmised connections between these seemingly paired catastrophes. Missionary, travel writer, and man-about-the-Mississippi-Valley
Timothy Flint wrote of the season of New Madrid quakes that on one night
in particular people observed "continued glare of vivid flashes of lightning,
and of repeated peals of subterranean thunder, seeming to proceed . . . from
below the horizon." These signs were significant not simply for that particular local environment, but also for telling the truth of events far distant:
"The night, so conspicuous for subterranean thunder, was the same period
in which the fatal earthquakes at Carraccas occurred, and they seem to sup-
pose these flashes and that event parts of the same scene." Such accounts
look within the earth for explanation but are also concerned with knitting
together events across the globe into one coherent story of environmental
disruption, with perhaps a central set of causes. They bear witness to the
search for a global science, even in the small towns of New England and
along the American frontier.[64]

Volcanoes in the Mississippi Valley

Volcanic theorizing served for many New Madrid observers to connect
North American upheaval with disturbances in far-off nations, while also

Intense interest still today surrounds the seemingly sequential nature of earth-
quakes—major quakes often seem to be followed by one or two others in quick succession elsewhere in the world. Tracing such connections was long dismissed by expert
commentators as understandable but deeply naïve folk science; seismologists emphasize
that the earth is a seismically active place and that we simply tend to notice earthquake
activity in the immediate aftermath of a large event. But as geologists Susan Hough
and Roger Bilham point out, recent work on remotely triggered earthquakes raises the
strong possibility of real, if still little understood, connections between many of these
historic quakes.

providing an explanation for the alarming phenomena that so distressed them. Many people who experienced the earthquakes thought they had found themselves in the middle of a volcano. In the same language in which he might discuss the "operation" of an emetic on a sick family member, a man on a Mississippi River boat at the time of the December shock wrote to his father that he felt "a distant convulsion, which we conjectured to be an earthquake, caused by the eruption of some volcanic operation far to the west of the valley of the Mississippi. We hope in God that its seat was far from human habitation." Volcanoes were not regarded as a strictly local phenomenon—those far away might be felt, and might be causing the alarming rearrangement of the local environment.[65]

Just as domestic healers might scrutinize one part of the body to understand the health of another, eager reports sought news of possible eruption far distant. One St. Louis newspaper reported that "within 30 miles of the great Osage village, on the head waters of their river, and 180 miles from this town, it is said, that a volcano had ceased to burn for the last three years, and it is thought to have broken out in some quarter of our country." Like a bodily rash that might erupt after a period of quiescence to reveal a true inner disorder, the breaking out of volcanic forces might indeed explain much of the tumult along the Mississippi.[66]

American accounts passed along several such Native reports. Compiler Robert Smith reported the experience of riverboat captain John Vettner, who with his crew was five miles below New Madrid during the February shock. Vettner reported to audiences further east that "some Indians, who had been out in search of some other Indians that were lost, had returned, and stated that they had discovered a volcano at the head of the Arkansaw, by the light of which they travelled three days and nights." Such seemingly exact reports of a glowing volcano supported an overall if somewhat diffuse sense that volcanic activity must be responsible for the chaotic experiences produced by the quakes.[67]

Moreover, reports of the quakes demonstrate a vision of western North America only just beginning to come into focus. As Robert Smith wonderingly surmised in 1812, in introducing additional material on the earthquakes, "we may yet receive authentic information of effects still more tremendous, and possibly of a volcanic eruption at no very remote situation." Many people looked for explanation to come from some as yet undiscovered far-off event, likely a volcano. Such speculation built on earlier reports of volcanoes in the Southwest, part of the geographic haziness that inspired Jeffersonian-era exploration of both the upper and lower tributaries of the Mississippi. No one could assume that all was known: both the surface and the below-ground nature of western lands were still very much up for determination.§§§§ The search for volcanic explanation not only underscores the lack of European and American knowledge about the early nineteenth-century American West, but also reveals an alternate imaginative geography: like early modern maps with sea monsters in the untraveled far waters near the margin, early nineteenth-century Americans carried with them a picture of North America in which potential fields of smoking volcanic mountains lurked over the unknown horizon from newly settled regions.[68]

Reports of western volcanoes gathered strength from widely reported and puzzling underground noise and activity in many parts of the West— smoking mountains, pumice stone, and rumbling noises in the territories as yet little explored by Americans or Europeans. William Clark was very interested by a noise like distant artillery that he heard while exploring above the falls of the Missouri on the Corps of Discovery in 1804. He concluded that the nearby "Medicine River" must have been named because of "this unaccountable rumbling Sound, which like all unacountable things with the Indians of the Missouri is Called Medicine." Timothy Flint remarked on the loud noises and volcanic appearance of the Ouachita Mountains in south-central Arkansas (site of what is now the Arkansas Hot Springs), concluding that they were "probably volcanic," though they had not been subject to recent eruption.***** In the sensory world of early America, noises strongly suggested the kinds of subterranean upheaval and movement long associated with volcanic eruption.[69]

Earthquakes and volcanoes were—and are, in the present day—often experienced together. Contemporary science recognizes that volcanoes can trigger localized earthquakes, just as local oil or water drilling can, and that this is a different causal mechanism from tectonic earthquakes.††††† In the

§§§§ In 1824, a guide to Kentucky concluded that "there are not remains of land or burning volcanoes in Kentucky, nor of any considerable fresh water lake"—the existence of volcanoes in Kentucky was still viewed as a distinct possibility in 1824.

***** Native communities, on the other hand, apparently interpreted these noises in spiritual and cultural, rather than seismic, terms: another early nineteenth-century writer noted that "loud explosions are frequently heard among the hills, somewhat resembling the blowing of rocks with gun powder. These noises, the Indians say, are made by the spirits of white people, working in the hills, in search of silver and gold mines."

††††† The relationship between drilling and seismic activity remains incompletely understood and is under intense debate because of the practice of hydraulic fracturing, or fracking, in current mineral extraction, which people in certain areas blame for increased earthquake activity.

nineteenth century, many people experienced earthquakes and volcanoes as related phenomena and sought to explain that relation.[70]

Many people who felt the New Madrid earthquakes drew upon a notion of large, far-reaching subterranean chambers and connections—spaces that would link and explain volcanoes and earthquakes—which had deep roots in classical and early modern theories. The notion of disruption to subterranean chambers is essentially an Aristotelian view of earthquakes which was further developed by seventeenth-century theories of the earth. German writer Athanasius Kircher, for instance, viewed the subterranean globe as a system of fiery canals and chambers explicitly parallel with the circulation of living beings. Volcanic eruption, Kircher argued, represented the explosion of underground, fire-filled chambers. This fundamental notion influenced many later models, including those which explained the New Madrid upheavals as a form of volcanic action.[71]

In the decades after the enormous quake that devastated Lisbon in 1755, many European theorists continued to link earthquakes with volcanoes. English parson and scientific observer John Michell offered several new observations about what would now be termed the epicenters and precursors of the massively destructive Lisbon quake. Michell was one of many to hold that earthquakes were likely connected with volcanoes, perhaps through the action of subterranean fires. In 1763, Rudolf Erich Raspe, best known as the flamboyant author of *The Surprising Adventures of Baron Munchausen*, republicized Robert Hooke's late seventeenth-century theory of dynamic vertical displacement of the earth's crust, which treated earthquakes and volcanoes as connected manifestations of the underground fermentation of sulfurous minerals. Writers disagreed about what exactly was combusting, or how, but through the era of the New Madrid earthquakes many researchers held a general sense that something—physical or, later, chemical—must be burning or exploding.[72]

In the early 1810s, Leopold von Buch and Alexander von Humboldt took up the connection between earthquakes and volcanoes in a different way, postulating that earthquakes were caused by "craters of elevation," inflations or elevations of the earth over a potentially broad area, caused by the buildup of pressure in subterranean vapors. Separating volcanic activity into two kinds, they argued that true volcanoes like Vesuvius (so-called "craters of eruption") were rare. Instead, most volcanic activity was similar to the "craters of elevation" Humboldt had witnessed in South America, in which the slow spread of basalt under the earth's surface caused a bulge through which veins of magma would sometimes erupt. For Humboldt, this model of vast subsurface disturbance, in which specific eruptions were only the most visible and sensible aspects, helped explain why volcanoes could be so much more extensive than their surface manifestations might indicate—and could therefore be connected with earthquakes at large distances. Other geological researchers including Charles Lyell and Charles Darwin soon rejected Humboldt's notion of "craters of elevation," but the apparent connections between earthquakes and volcanoes informed many geologists' writing.[73]‡‡‡‡

Many of the North American witnesses trying to figure out the New Madrid quakes were familiar with this welter of theories of earthquake causation. They drew liberally from the many available European scientific explanations, especially the notion that earthquakes were caused by explosive combustion (often of an unclear source or cause) in subterranean chambers. "We regard Earthquakes as the consequence of the expansive force of classic vapors," argued one pseudonymous author in a series of articles in a small-town Georgia newspaper, who then discussed the mechanics of how such a subterranean explosion would be felt and experienced given North America's geology. Similarly, Isaac Lea, writing in the *American Journal of Science and Arts* in 1825, argued that earthquakes were caused by explosion or combustion in the subterranean channels that lead down from the mouths of volcanoes. Like many, Lea argued that these channels could extend for long distances, even connecting volcanoes to the sea—thus, he argued, salt water and ocean fish were, he claimed, sometimes seen to be thrown out of volcanoes such as Vesuvius. "The existence of those extensive cavities," he argued, "satisfactorily accounts to us for the fact of earthquakes being severely felt at great distances from volcanoes," as for instance in the case of the New Madrid quakes (which had woken him in Pittsburgh): after all, there were no volcanoes within a thousand miles of New Madrid. Perhaps not surprisingly, modern and technologically up-to-date Americans such as New Madrid compiler Robert Smith thought steam must play a key role in both volcanoes and earthquakes.[74]

New Madrid accounts that asserted or implied a causal relationship between volcanoes and earthquakes were consistent with European scientific theories at the time, which posited subterranean chambers, combustion, and explosions of various kinds as connecting dramatic geological phenomena. Scientific theories linking earthquakes and volcanoes also demonstrate how far-off ultimate causes were thought to be. Proximity was not thought necessary for events in the globe to be related: volcanoes thousands of miles distant might be causally connected with upheaval in the Ohio Valley.

---

‡‡‡‡ In modern terms, Humboldt's major contribution to the study of earthquakes was his wave theory of earthquake transmission, which was generally adopted.

Earthquakes were also consistent with *physical experience* of volcanic eruption, in ways perhaps foreign to modern understanding and unacknowledged within our histories of seismology. In the frameworks of the early nineteenth century, earthquakes and volcanoes manifested many of the same terrestrial symptoms: loud noises, smoke, flashing light, white ash, foul and acrid smells. The phenomenon of "thundering noise" was common to both earthquakes and volcanoes. Earthquakes in the early nineteenth century were characterized not simply by the shaking of the ground, but by noise, commotion, and "the emission of flames, water, vapours, &c." (The "etc." here is perhaps particularly disturbing—what else might that encompass?) In the era of the New Madrid tremors, earthquakes looked and felt like volcanoes.[75]

Exactly what created this experienced relationship was not necessarily clear. Many North American writers conjoined keen observation with a kind of theoretical agnosticism. In his compilation on the New Madrid events, Samuel Mitchill added extensive appendixes giving observations and theories about other earthquakes worldwide. He cited a "Mr. Drouet" on the possible global connections of the Caracas earthquakes, and in particular on their possible relationship to volcanic activity:

> On the theory of earthquakes, this gentleman, after examining the several hypotheses of those who ascribe them to fire acting upon subterranean fuel and forming very elastic vapour; to fire acting upon water and converting it to most powerful steam; to fire producing both rarified air and rarified water; and, lastly, to the subtile, pervading, and irresistible force of electricity; leaves the subject, in that respect, like his predecessors, exactly where he found it.

Most observers seemed relatively satisfied with this frustrating but suggestive mix of causal arguments: it was not precisely clear what the volcano/ earthquake connection was, and it was not clear what caused volcanoes themselves, but the various posited theories could nonetheless suggest questions to ask and observations to make.[76]

Some writers plainly dismissed any link between the New Madrid events and volcanic eruption. "I shall not pretend to hazard a conjecture as to the sources of those mighty concussions," wrote Winthrop Sargent in 1812, "nor am I perfectly satisfied of the existence of volcanoes in the west and northwest, notwithstanding the report of lava, scoria, and pumice, in those quarters." The evidence of the earth was simply too equivocal. Stephen Austin was similarly struck by the devastation and distress of the people around New Madrid, but noted, "There is not in any of these places the smallest appearance of Volcanic Matter tho. there are numbers of marks of the most violent operation of heat—as well in the coal which was discharged as in a kind of Sinder which has been found in small quantities." The earth's upheaval was clearly related to fire, heat, and discharge of coal and mineral matter from beneath the earth, but Austin and Sargent were unwilling to encapsulate that evidence as being volcanic.[77]

Yet many of Sargent's or Austin's contemporaries thought that the earthquakes must be related to volcanoes *somehow*. Philadelphia printer Robert Smith noted that the information he had continued to receive since the original compilation of material—his was, after all, the first collection of reports about the quake, and was made apparently within months of the 1812 shocks—had demonstrated that the tremors had been even more damaging that he had first allowed. He expressed his own openness to the breadth of this phenomenon: "There appears indeed considerable ground for expectation, that we may yet receive authentic information of effects still more tremendous, and possibly of a volcanic eruption at no very remote situation." In this, as in much of the discussion about the quakes, lay a clear sense that further information may yet lead to quite different conclusions. Writers readily acknowledged the lack of completeness of their own information—their knowledge was not ossified, but constantly growing to encompass new evidence and new events.[78]

The connections formerly understood or posited to exist between volcanoes and earthquakes appear now only in very small typeface in very minor footnotes in histories of earth science. Yet this was an important and substantial element of conversations about earthquakes in the early nineteenth century. The theoretical agility and flexibility of earthquake-volcano theorizing signal one reason why theories now understood to be mistaken belong in our history writing in respectable font size. The character, if not the substance, of such conversations remained a powerful current of later scientific thinking in the United States. Far from being a kind of ill-informed scientific dead end, volcanic theorizing in early New Madrid earthquake writings reveals the dynamism and willingness to revise and rethink in early American knowledge making about the nature of western terrain.

Constantin Volney, Native North Americans, and Knowledge of the West

The relationship of earthquakes and volcanoes was particularly of interest in the American West, since the influential French writer Constantin Volney had argued in the 1790s that earthquakes and volcanoes—conjoined expressions of the earth's "subterranean fires"—were absent from North America west of the Alleghenies (they had, he argued, shaped much of the topography of the Eastern Seaboard and eastern Great Lakes). For Volney and many of his readers, human culture could demonstrate geological truth: his evidence had to do with supposed lack of words for "earthquake" and "volcano" in languages of Native Americans.[79]

Volney's reports were widely read in Europe. In the United States, they were both widely read and widely lambasted—the 1804 Philadelphia edition of his work is liberally sprinkled with acerbic commentary and indignant objection by his patriotic American translator. Yet despite many Americans' impatience with the shallowness of Volney's research, he was a well-known European authority and his arguments had to be reckoned with. Well into the nineteenth century, people reporting on western territories referenced his arguments.[80]

Writers familiar with areas shaken by the quakes argued that Volney was simply not reading the geological or human evidence. Many Americans argued that both experience and geological testimony clearly showed prior seismic tumult of many regions in North America which this European tourist was simply not prepared to read. Gottfried Duden, an optimistic German traveler and writer whose reports brought thousands of German emigrants streaming into Missouri in the 1830s and '40s, wrote in disagreement with Count Volney in about 1830: "In Volney's work we read also that earthquakes are completely unknown in the western states, that in the Indian languages the words earthquake and volcano are completely lacking, and that south of the Great Lakes there is no trace of them." But, he argued, the Missouri region had a long history of earthquakes, and "everywhere in my neighborhood there are traces of volcanoes."§§§§ [81]

Such assertions of volcanic and earthquake history may reflect subtle truth both about who could witness natural events in the western American borderlands, and what those events were. The New Madrid quakes were felt most strongly in the Mississippi and Ohio Valleys, the borderlands of American and Indian territory. In those areas, free white Americans and their black slaves encountered and negotiated with many Native peoples. Indian knowledge occupied an uneasy space in American understanding of terrain: American newcomers were eager to erase Native knowledge and claims, to deny deep connection with the "new" lands they themselves coveted in the western territories, and yet—as in their eager forwarding of accounts of recently broken-out volcanoes—they were often dependent upon Native American contemporaries for the most basic information and aid.[82]

§§§§ Elephant Rocks State Park in Missouri is indeed a striking—and beautiful—magma formation.

White and Indian knowledge of terrain thus existed in uncomfortable and sometimes conflicting relationship. As William Leigh Pierce reported, "Some Indians, from the country adjacent to the Washita," held that "the Burning Mountain, up the Washita River, had been rent to its base." At the same time, for Pierce's audience to give such Native observation credence, it had to be endorsed with more cultural reliability. A slave's report on earthquake sounds, for instance, might be credible if endorsed by a slaveowner's "confidence." "This information," Pierce continued, "I received from a Settler at the Hills [Walnut Hills, just above Natchez], and his appearance was such as to attach credit to his information."[83]

In his letter to his former teacher Benjamin Smith Barton, Alexander Montgomery noted that "we have not as yet been able to acquire any certain information as to the *seat* of those violent concussions although, general opinion has placed it in the country of mountains which lie between the Head of the Arkansas & Red Rivers." In that wild western geography, only a few people could give an accurate report, yet knowledge from those people was accepted only provisionally by the American reading public. Montgomery explained that "report *derived* from the Indians has stated that immense Volcanic-eruptions have occurred in these regions." As he emphasized, however, "This however, wants Confirmation."[84]

Montgomery's comments reveal common themes of earthquake conversation: widespread interest in the source region for the quakes, and the common supposition that they originated in the West—sometimes the very far West. His remarks, moreover, signify that what could be apprehended locally might not indicate the full truth—regional manifestations might be different from the ultimate cause. His letter also implied a kind of potential flexibility to the cause. Who knew what might be taking place in the unknown regions past the areas of American settlement? Montgomery's rhetorical shrug is a way of indicating how wild, dangerous, and unknown those places were.

His framing of Indian knowledge is also consistent with the attention in New Madrid accounts of the status of the observer as credible witness. Many reports of the most dramatic effects of the quakes were second- or thirdhand. One early report—of volcanoes in Buncombe County, North Carolina, and of the fall of a noted landmark, the Painted Rock—was debunked early in the process of conversation and exchange, and later commentators wished perhaps to avoid being seeing as similarly duplicitous and sensationalist. To emphasize their legitimacy, most reports included some mention of who exactly had seen, heard, smelled, or felt the phenomena being discussed. Whenever possible, editors and readers looked to verify seemingly outlandish claims through common experience—a newspaper account of the impact of the December shock in Richmond, for instance, emphasized that "an Earthquake was witnessed by many people in the city." Yet the location of the New Madrid quakes in borderland areas meant that only relatively small groups of people were able to testify to their most dramatic manifestations.[85]

Native American knowledge was frequently held at arm's length, regarded with skepticism along with the reports of drunken rivermen—a profound irony, given that these might well be the people most familiar with the areas of the quakes' most intense impact, and that as local inhabitants many of them would be familiar with subtle signs of environmental change. In contrast, those deemed "respectable gentlemen" were viewed as credible observers. One traveler through the Creek nation, for instance, could provide a valuable report because he was "well known in this city." Many seeking to plumb the environmental truths of the recent "commotion" were anxious to gain corroboration from multiple accounts and from those "personally familiar with the incidents they related." Those whose reports had earlier been corroborated were particularly trustworthy: Robert Smith endorsed one quoted account of the February tremor in his 1812 compilation because it was "from the same respectable individual" who wrote an account of a previous quake. Another correspondent from Wheeling, Virginia, was "formerly a resident of Philadelphia, and from his well known character for veracity and intelligence, his statement, it is believed, may be implicitly relied upon."[86]

But Alexander Montgomery's report of volcanic explosion also quietly indicates the role of Native American knowledge in reports of quakes. Distrusted—after all, their report "wants Confirmation"—reports of "Indians" were nonetheless a key part of how his fellow Americans were coming to terms with the environment of North America. Further, such tales of volcanoes in the heart of the continent likely reflected physical reality more closely than a superficial reading might suggest. After all, what Montgomery cites as volcanoes could well be reports from Native people fleeing the widespread, alarming, and dangerous phenomena of sand blows in the epicentral regions.

Very likely, the tales of volcanoes from Indians—easy to dismiss at the time, because they were coming from people who were not given much credibility, and now, because present science makes no room for volcanoes in the middle Mississippi Valley—may in fact reveal something useful and accurate about what was actually going on. Though there is no geological evidence today of volcanoes in a current sense along the middle Mississippi, there is ample and growing evidence of widespread and sometimes enormous sand blows that would indeed look quite impressively volcanic.

Indeed, sand blows are still sometimes colloquially referred to as "sand volcanoes." The New Madrid earthquakes were experientially "volcanic" events—just as those traveling Indians said.[87]

Naming the experience of the New Madrid quakes "volcanic" recognized the importance of eruptions throughout the most affected regions—eruptions later science would term sand blows, of liquefied sand rather than magma, but eruptions nonetheless. In the descriptions, metaphors, hints, and even exaggerations about the quakes, and particularly in surmised subterranean connection between earthquakes and volcanoes, is an abiding concern for the breaking open and breaking down of the stable surface of the earth. Fear of volcanoes was also a fear that all that was underground would not safely stay there but would instead move around, come to the surface, or open up to swallow people or structures. In much the same way, writers tried to bury Indian knowledge, make it conditional or secondary, yet the stubborn realities of the erupting earth show the truth of "Report *derived* from the Indians." Kept submerged, Indian knowledge erupts unexpectedly to the surface.

Lightning and Electricity

In the era of the New Madrid earthquakes, electricity provided a way to think through truths of the body and the spirit—and the natural world. Electrical thinking provided links with the spirit and the body, and with exciting scientific advances. In the new science of electrical fluid, many observers not only sought explanation for earthquakes, but worked out how American science would be done.

"Ever and anon," reported senator L. F. Linn in the 1830s, calling for federal funding to help mediate the quakes' environmental disruption to the St. Francis River in eastern Arkansas, "flashes of electricity gleamed through the troubled clouds of night, rendering the darkness doubly horrible." The senator's prose was perhaps overwrought, but his observations were typical. Many people who felt the "hard shocks" of earthquakes associated them with lightning and thus with electricity. This was in part because of the overwhelming nature of both kinds of experience. In the middle Mississippi Valley, thunderclouds mass together into a roiling purple-black wall that stretches from one side of the horizon to the other and speeds across the sky in a fast-moving phalanx. Midwestern thunderstorms can bring sharp hail in the middle of summer. When they get right overhead booms of thunder shake the very air. Even in thick modern buildings they are awesome in a grand and breathtaking way. In small shelters or out in the open they are hard to imagine and endure. "The thunder and lightning which prevail on this river" wrote one 1818 Mississippi traveler, "are truly grand." They were certainly the loudest and brightest things most premodern people had ever experienced.[88]

Visible flashes of light during the biggest of the New Madrid shakes reinforced the connection between earthquakes and electrical storms. In Knoxville, Tennessee, on 16 December, one person saw "in a direction due north, two flashes of light . . . much resembling distant lightning." During the tremors, many observers experienced lightning and thunder as phenomena of the earth itself, not simply of the sky above. Erupting from sand blows, people near the epicenters saw "flashes such as result from the explosion of gas, or from the passage of the electrical fluid from one cloud to another." People in many places heard rumbling noises, often coming from below the horizon and preceding the flashes of light.[89]

One such account is suggestive of the range of people involved in making record of their experiences. In Maryland, visitors to a house

> mentioned that they were much alarmed at about 11 o'clock last night, by a tumbling, as they thought, in the earth, attended with several flashes of lightning, which so lighted their house, that they could have picked up the smallest pin—one mentioned that the rumbling and light was accompanied by a noise like that produced by throwing a hot iron into snow, only very loud and terrific, so much so, that he was fearful to go out to look what it was, for he never once thought of an earthquake.

This passage offers a hint of how women's everyday household knowledge might have entered into accounts which on their face reflect only male authorship—women, after all, were the main people to use small pins. Changes to everyday experience means that some comparisons are all but lost. Was the "loud and terrific" noise of hot iron in snow a hissing and roaring of sudden steam? Or the popping and banging of ice crystals sharply changing state? Was the visitor who made that original comparison a woman, accustomed to cooling her heavy flatirons in available snow? Or a man, thinking of hot stable tools or branding irons? What is still clear is that the sounds and visual experiences of the quakes were linked, and that they together felt and sounded and seemed like thunder and lightning.[90]

"Subterranean thunder" signified to New Madrid observers that the same process was going on underground as usually happened in the sky. Many people who had experienced earthquakes in earlier centuries and in other places had witnessed lightning and heard thunder. Similarly, those who felt the New Madrid quakes often saw and heard evidence of electricity in the earth itself. A "gentleman from Knoxville," who was traveling near warm springs in North Carolina during the December shock, described an awe-inspiring scene: "The fulminating of the mountains was accompanied with flashes of fire, seen issuing from their sides—each flash ended with a snap, or crack, like that which is heard on discharging an electrick battery, but infinitely louder." This "gentleman" was part of a world so new to electricity that the spelling was still not standardized, but his conclusion was clear: "the Earthquake was caused by the electric fluid."[91]

In the years and decades after 1811–12, many Americans saw the electrical nature of earthquakes as self-evident not only because of physical experience—what they felt, saw, and heard—but because of theory. Following Benjamin Franklin's lead, other natural philosophers of the mid- and late eighteenth century had been fascinated by the role of "electrical fluid" in the economy of the earth. Beginning in the 1720s, Puritan divines wrote on the connections between earthquakes and lightning. A particularly popular performance in Britain and later in the United States was Joseph Priestley's demonstration of the effects of an earthquake on a model town destroyed in spectacularly dramatic fashion by electrical discharge. Electric forces were especially promising for helping explain some of the puzzling but dramatic evidence of geological formations. Investigating unusual angular, fragmentary deposits of galena in the mining districts about seventy miles northeast of New Madrid, English traveler and scientific observer George W. Featherstonhaugh surmised that recent earthquake activity was the cause, and that the geological formation including the Iron Mountain may have been "thrust up" and "shattered" "when an electric power of great intensity passed along these lodes."[92]

For many European scholars, electricity offered a plausible explanation for earthquakes. Earthquakes must be due to some form of electrical discharge. One British theorist, William Stukeley, reasoned in the mid-eighteenth century by explicit analogy between processes of earth and sky. If lightning (and thunder) are created when electricity passes from a charged cloud to one that does not possess a charge, what happens when a charged cloud instead passed its electricity to the ground in a lightning strike? Stukeley's answer: seismicity is the earthly equivalent of discharge of electricity, the terrestrial analogue of a thunderstorm.[93]

Theories of lightning and earthquake flashed into public awareness as a consequence of a fierce "lightning-rod debate" in Boston in the mid-1750s. The exchange followed a set of frightening and unusual seismic events in November 1755. As with the connections between the New Madrid and Caracas quakes, contemporary theorizing focused on earthquakes that happened close together. On 1 November, a catastrophic earthquake leveled Lisbon in a major world event that became the referent for European discussions of earthquakes for several generations. Then, before the news of that quake had even had time to reach North America, New England was shaken on 18 November by a significant earthquake of its own. Less important on the global scale, the Cape Ann earthquake was nonetheless a major event for scientifically minded people in North America.[94]

The 1755 Cape Ann quake was the occasion for a public dispute between two intellectual leaders—Harvard's Hollis Professor of Natural Philosophy, John Winthrop (a descendent of the Massachusetts Bay Colony founder), and the Reverend Thomas Prince, pastor of the Old South Church—about the cause of the earthquakes. For four months in 1755–56, the two exchanged increasingly irate volleys in the press. Drawing on recent scientific inquiries, Prince surmised that the recent erection of lightning rods according to the demonstrations of Dr. Franklin had in fact made the city vulnerable to earthquakes. For the Boston cleric, such lightning rods were a theological issue—the temerity of humankind attempting to avert God's thunderbolts—as well as a scientific one, having to do with electricity being drawn to the iron rods and then stirring up unrest in the earth as a consequence. Winthrop attacked him with vigor and flourish on both theological and scientific grounds, and he decisively won the debate.[95]

Dr. Franklin's "iron points" proliferated throughout American cities as both a scientific and a political intervention: builders of the Maryland State House in 1772 constructed an imposing lightning rod both to protect its stately dome and to emphasize the colony's allegiance to Franklin's theories over those of the scientists connected to the court of King George III. Even so, fears for how lightning might be related to earthquakes were not so easily dispelled.[96]

Though in European science electricity was losing ground as an explanation for earthquakes by the 1810s, in North America the relationship of electricity and earthquakes remained an active question. William Stukeley's theories of earthquakes were a touchstone for decades after he wrote, providing the basis for musing and debate about the events of New Madrid. Many people writing about the New Madrid quakes assumed something electrical in their causes.[97]

In 1816, a remarkable piece of public science pulled together electricity, lightning, and earthquake weather. This anonymous article on "The Cold of the Present Season" attempted to explain the abnormally frigid temperatures of 1816 that created concern and scarcity around the globe. Later scientists would ascribe the extreme climatic conditions of that year to the April 1815 eruption of Mount Tambora in Indonesia. The result was an international disaster—"eighteen hundred froze-to-death"—in which frigid growing seasons stunted crops in many parts of the world. On 8 June, 1816, one Maine farmer recorded a three-hour snowstorm during this "year without a summer." Some environmental historians see the extremely harsh conditions of 1816 as one reason for the flood of emigration out of the New England states in the late 1810s. In addition to the financial strains of the Panic of 1819 and the lure of newly opened western regions, many long-suffering Yankees may simply have found that miserable year to be the last straw.****** 98

In the pages of a leading paper in the nation's capital, "On the Cold of the Present Season" connected the current cold temperatures with the New Madrid earthquakes of four years earlier. The argument drew explicitly on Stukeley's work as well as on earlier scientific assertions about electricity, and it expressed ideas that were in common currency. Though this article brought those threads together in a particularly pointed way, earlier publications had made many of the same points. The anonymous author argued that one major force producing heat is "the circulation of the electrical fluid, through the atmosphere, and over the surface of the earth." This circulation is caused in part by disequilibrium between the surface of the globe and its enveloping atmosphere. Earthquakes "take place after the electrical fluid is very unequally diffused and when by some cause the equilibrium is restored; so that an earthquake may be compared to an electrical shock of great magnitude." An earthquake, in other words, represents the shock of restored electrical equilibrium—exactly along the lines of Stukeley's arguments, which made earthquakes the terrestrial parallel of lightning strike.99

The article pointed out that the world had recently witnessed an unusually large number of disturbances in the earth: the events in the Mississippi Valley as well as the devastating Caracas tremors which had helped bring down Simón Bolívar's revolutionary government in Venezuela early in 1812. Around the same time, shocks hit New Granada and other West India islands and California; on 27 April, a volcano at St. Vincent Soufrière erupted and the island Sabrina in the Azores was built by ocean eruption. All of these dramatic events had equilibrated the electrical fluid between the atmosphere and the ground. Without a differential, there could be no earthquakes—and also, as the article argued had been the case in recent seasons, there had been many fewer lightning strikes than usual. Without an electrical differential, the earth was subject to very little circulation of electricity. All was quiescent and still—and therefore there was little heat to warm the globe.100

****** That peculiar year has continued to inspire. It is the subject of the alt-classical trio Rasputina's "1816, The Year without a Summer," the first track on their 2007 album *Oh Perilous World*.

This one short article sews together many pieces of the fabric of early American science. The article's arguments are syncretic and broad-reaching. Pulling out strands of scientific argument over the past half century, the author stitches together a far-reaching set of theories about earth systems. The article's context is even more revealing: drawing on a letter published by the Royal Society of London over fifty years before, working out complicated theories about climate and landforms, it was nonetheless published in an ordinary East Coast American newspaper, in the humble context of everyday commercial and agricultural news. Just below this article appears a brief snippet about a "Productive Cow" in Connecticut that supplied the astonishing average of fourteen quarts of milk a day. Searching for American science in the early nineteenth century means looking *in the vernacular*, among the productive cows and the pragmatic theorizing of a deeply scientifically curious nation.[101]

Conclusions

Shaken by the Mississippi Valley quakes of the 1811–12 winter, people in many parts of North America sought to report on their experiences, compare them with what other people experienced, and draw conclusions that would make sense out of events that baffled and scared them. Just as in medicine, meteorology, and other sciences of the time, American thinkers placed their emphasis on local observation and physical experience. This emphasis made their science vernacular not simply in the sense that observations were made with unaided eye, ear, and nose, and that they were made public in everyday publications, but also in the sense that they were widespread and drew on widely shared notions of causality. Just as in religious revivals, the scientific endeavors of nineteenth-century Americans reflected a faith in the capacities of ordinary people, in ordinary situations, to potentially come to grips with complex phenomena.[102] Trying to figure out the New Madrid earthquakes was something that many different people all over the United States were doing in the early nineteenth century.

Despite the terrifying and bewildering aspects of these quakes, and despite not having any one clear answer for why they took place, early nineteenth-century people thinking about natural phenomena sustained a conviction that the earth gave its evidence straightforwardly. That evidence might be hard to understand, and it might be frightening and overwhelming, but if enough observers could only put together enough of their observations, the evidence would ultimately be there to read. And, perhaps most importantly, not only there for an elite few to read, but for anyone who had access to the phenomena.

While often seemingly atheoretical, earthquake accounts carried musings or arguments about causes and effects, evidence and truth, in their structure and emphasis: it was interesting to note where noises or tremors come from, since that might reveal what caused them; smells of sulfur might be a useful clue to volcanic origin; careful observation of weather surrounding the tremors might provide the key to unlock the complex relationship between climate and earthquakes. Natural systems were dynamic—indeed, they were often extremely volatile—and they were interactive. What happened in one aspect of the natural world could affect what happened in another. Frightening noises at sea might indeed have some connection to upheavals in the earth thousands of miles away. Tremors in the earth could alter weather—in what ways, it was not always clear: such relationships proved elusive. Yet in the early nineteenth-century science of earthquake upheaval, the *search* for such relationships animated the intellectual world of a far-flung and scientifically curious nation.

Historians have failed to notice this vernacular science because we have been looking for science in the wrong places. Typically, historians of science read scientific publications or lab notes. Cultural historians or political historians might read newspapers. Social historians leaf through yellowed sheaves of family letters. Only rarely and in special circumstances do we all read material usually belonging in the other domain. Certainly historians of science, who have looked hard and earnestly and without result for the kinds of formal publications and well-demarcated venues that transmitted and developed science elsewhere, have by and large not checked the crop reports or looked through wild accounts of volcanic eruption in the midcontinent. We have therefore missed a great deal in the early culture of scientific conversation and observation in the United States.

Science in early America did not, in the main, happen in and through scientific journals or international gatherings or the other forms of communication that helped develop science in European states of the same era. Science in early America was something that many ordinary people *did*: by writing down what they saw or felt, by weaving networks of discussion, by tinkering around and measuring and comparing. Small-scale, diffuse, seemingly inconsequential scientific exchange, practice, imagination, and theorizing pervaded early nineteenth-century American life to a degree that is both astonishing and unrecognized in the ways we usually write our histories.

The vernacular science of early America provided the technical skill, the creativity, and the impetus for the steam craft, harvesting machines, conveyor belts, telephones, labs, efficiency studies, and universities that would over the following century and a half remake the American scientific and technical scene and thus much of the modern world. In the wide-eyed reports of a long-ago and long-forgotten earthquake are the foundations of a world neon and new, foundations easy to overlook because of their ordinariness and brevity, their breeziness and exclamation points. But those dissections of causality in the front pages of small-town newspapers, those small-scale exchanges of breathless observation and earnest measurement, are the foundational planking of our scientific and technological present, a future built by Alexander Montgomery and his correspondents in the scientifically curious and intellectually ambitious world shaken by the New Madrid earthquakes.

6 ✳ Sunk Lands and Submerged Knowledge: How War, Swamps, and Seismographs Hid Evidence of the New Madrid Earthquakes

Early in the nineteenth century, the New Madrid earthquakes came under intense discussion in many places and by many people. They were available for surmise, for metaphor, for rallying call; they provided evidence for theoretical assertion, the basis for spiritual insight, a reason to go someplace else, fast. By the late nineteenth century, they had almost disappeared from popular knowledge or scientific investigation. A few researchers might still mention the tremors from time to time in scientific literature, but almost no one in the earth sciences focused on them. The quakes were part of local lore, without much significance or even validity to any outsider. In the long period from the Civil War to the last decade of the twentieth century, changes in the social history of the middle Mississippi Valley, to the environment around the epicenters, and within seismology allowed the 1811–12 tremors to subside beneath the surface of attention and memory. Earthquakes that had once been the subject of eager conversation and exchange came to be almost completely forgotten.

The disappearance of common knowledge of the quakes reflected the sinking and submergence of the land itself. People of the eighteenth century and the first decade of the nineteenth used the rivers south of New Madrid, especially the St. Francis, to travel the middle Mississippi Valley. Travelers and emigrants settled on and near the St. Francis; they went up and down its channel to trade or trap or carry news; and when it was high enough they floated lead from southeast Missouri mines down its currents. Travelers preferred the St. Francis to the Mississippi as a navigable route. In early 1812, when news of the earthquakes began to reach a wider audience, local people reported that the tremors caused the hinterland south of the New Madrid, especially around the St. Francis River, to subside—in places significantly. This subsidence created the "sunk lands." The sunk lands were a transformed terrain: what had once been a network of streams that might seasonally overflow had become a swampy morass in which current and channel were hard to detect and the trunks of trees killed by the earthquakes blocked once-flowing channels. Earthquake subsidence created a lasting, spread-out barrier to easy travel or transit.

Not only are the sunk lands an apt metaphor for the muddied obscurity of the quakes, but by the late nineteenth century, they were themselves the object of doubt and denial, a terrain in which observers no longer saw evidence of earthquakes. By the late nineteenth century, some local observers and many commentators from elsewhere in the country had begun to question the once obvious environmental knowledge that earthquakes had made the land sink. Scientific researchers and settlement advocates argued that the sunk lands were instead caused by the collapse of subsurface strata, unrelated to any quake, or by ordinary floods—or that there had been little change in what had always been a swampy backwater. The sheer extent of the sunk lands suggested the implausibility of their having been formed by earthquakes.

The sunk lands became a topography of doubt. Terrain that had been a crossroads became land that no one really knew. Even the very extent and exact boundaries of the sunk lands are today difficult to trace with precision: features that have been contested for two hundred years are challenging to map.

In the subsidence of knowledge about the sunk lands lies also the subsidence of knowledge about the earthquakes themselves. From the American Civil War through the late twentieth century, knowledge of the sunk lands and the earthquakes that created them was both forgotten and denied. The devastating rift of the Civil War and subsequent changes to the racial order and land use of the former New Madrid hinterland turned attention away from any evidence of earthquakes. In repeated and powerful erasure, floods swept away telltale sand blows. Boosters and technological enthusiasts saw swampy, fertile land as ripe for drainage and reclamation, denying the forces of nature in order to assert the potential of human technology and advancement. During the same period, advances in the technology used to study the movement of the earth at once created the discipline of seismology and focused seismologists' attention away from earthquakes with so few technological traces. Later in the twentieth century, the new insights of plate tectonics directed researchers' efforts toward plate boundary quakes like those of the American West Coast, even as an earthquake prediction scare brought about by decidedly questionable science made New Madrid seismicity seem questionable by association. By contrast with the seismicity of California, in which technological and disciplinary change brought earthquake events into sharper focus, the same developments served to obscure the earthquake history of the Mississippi Valley. All of these processes together—social, environmental, and scientific—hid the New

Madrid earthquakes in plain sight, submerging them from common knowledge and from scientific study.

Much of the history of the earth sciences since the mid-eighteenth century is a story of the increasing ability to interpret the earth's past, first through the reading of strata and formations and more recently through tools such as bore holes, mineralogical analysis, and electronic pulses directed into the deep layers of the earth. Indeed, the ability to *think historically* was, as much as any one insight about place or past, the signal achievement of late Enlightenment earth science, one that shapes the field to this day. The story of earth science is a story in which geologists, seismologists, and other scientists have pushed further back in time, creating clearer and sharper views of many aspects of the earth's past.[1]

Yet the past is not only discovered and learned. It can also be covered up, forgotten, and ignored. The story of the New Madrid quakes from the Civil War to the late twentieth century is a story of the submergence of certain forms of evidence by inattention, by forceful denial, by the forces of environmental transformation, and by the simple turning away to see something else instead. These forces combined into a submergence of deeper historical perspective that hid from view dramatic environmental events of the relatively recent past.[2]

Memory, Forgetting, and Denial

The forgetting of the New Madrid earthquakes was slow and gradual. In the early decades of the nineteenth century, continued tremors brought home the reality of Mississippi seismicity. By midcentury, tremors continued to be felt even in Kentucky and Ohio. Rather than a jolting shock, such tremors became an annoyance. As an 1856 local history noted, people around the former epicenters "paid little or no regard to them, not even interrupting or checking their dances, frolics, and vices."[3]

By 1853, a collection of regional humor stories could use earthquakes in Kentucky as the backdrop for a story of comeuppance: colleagues goad a blowhard lawyer named Cave Burton into telling his well-worn tall tale of a trial in Kentucky in 1834, when a hung jury came thundering down the back stairs of a courthouse in the midst of a dramatic trial and was taken for an earthquake, causing panic in the courtroom and a near riot in the town square. The whole story, however, is a setup: merely a way to keep Burton talking while the other lawyers take turns sneaking out to a back kitchen to enjoy a new barrel of oysters—a rare treat—before he can hog them all. By midcentury, earthquakes had become a joke rather than a calamity or threat. The actual tremors were a quaint memory, the province of old-timers.[4]

Environmental signs ceased to be read as earthquake evidence. By the twentieth century, sand "kindo' throwed up" in the sunk lands was regarded not as a remnant of the earthquakes, but simply as a side note about the terrain in evidence given for a court case. A 1916 survey of northeast Arkansas soils described extensive sand blows and their effects on the fertility of fields, while maintaining absolute silence on what caused them. By the time of the Great Depression, "sand blow" was no longer a concept associated with earthquakes of the midcontinent. Instead, the term referred to the massive dust storms of the Dust Bowl.*[5]

Scientists began to deny that the earthquakes had ever occurred. At an 1883 meeting of the American Association for the Advancement of Science, speaker James MacFarlane argued that stories of a midcontinent earthquake were mere legend. The sunk lands of northeast Arkansas and southeast Missouri—the present wetlands that were once the heart of the thriving New Madrid hinterland—were formed when an underlying limestone strata dissolved. Disruption that lasted from 1811 to 1813 was a dramatic and prolonged episode of "mere subsidence." Earthquakes had nothing to do with the phenomena long associated with them. Discussion was apparently intense: "A question as to the truthfulness of the reports from that region," reported *Science* later, "brought out very contradictory opinions in the discussion." In the late nineteenth-century terms of increasingly professionalized science, that represents a near fistfight—in this case, over who had or had not seen trees upright in Reelfoot Lake.[6]

In response, at the 1892 meeting of the Geological Society of America, geologist W. J. McGee marshaled textual and environmental evidence—illustrated by "lantern views"—for the reality of the New Madrid earthquakes. Still, MacFarlane's skepticism represented the emerging view. Similar denial was well articulated in popular venues. In an 1872 history of Missouri, railroad magnate and regional historian Louis N. Houck noted that "the facts regarding this earthquake have almost faded from popular memory. Many otherwise intelligent persons now pretend to believe that it was no extraordinary occurrence, and that if it occurred at all, it has been grossly exaggerated."[7]

---

*In early twenty-first-century American life, the once-obscure term "sand blow" returned into common online use to describe the mammoth desert storms experienced and filmed by American troops in Afghanistan and Iraq.

Advocates of late nineteenth- and early twentieth-century environmental improvement—timber harvesting, swamp drainage, railroading, the creation of farms out of sunk land—minimized the earthquakes' effects, even their existence. Such arguments proved lastingly influential. When graduate student Leon Ogilvie began in the late 1960s to take stock of the Bootheel's land, in a thesis that became widely cited, he drew largely on reclamation advocates' arguments in asserting that the earthquake—carefully phrased in the singular—"had little effect on the local terrain." The sunk lands commonly attributed to seismic disruption "predated the earthquake." The notion that earthquakes had created extensive swamplands was part of the "distorted propaganda" that followed the earthquakes. Instead, the quakes merely intensified the long-standing environmental problem of low-lying and seasonally inundated land. By the 1970s, a Harvard seismologist would dismiss as "hysteria and superstition" the notion that the Mississippi River could possibly have been forced upstream by midcontinent seismicity. More measured analysis similarly concluded in 1970 that the sunk lands were an alluvial feature—part of ordinary change in the Mississippi basin because of seasonal flooding and changes to water flow—not the result of the 1811–12 quakes.[8]

Moreover, the perceived geography of the New Madrid quakes shifted and shrank. To the extent that they were discussed at all, the quakes become a phenomenon limited to a small stretch of the middle Mississippi Valley—no longer an event that happened on Long Island, in Virginia, in villages of Upper Canada. This gradual localizing and regionalization of the quakes linked them ever more strongly to the areas around the perceived epicenters. Tracing the history of that region—the former New Madrid hinterland spanning southeastern Missouri and northeastern Arkansas—thus helps explain how the quakes could fade from cultural as well as scientific visibility.

Social Change in the Postearthquake Middle Mississippi Valley

The War of 1812 and the Indian wars of the same years of crisis left Native American groups little able to resist the flood of American settlers interested in moving to the Mississippi Valley. Following these conflicts, an avalanche of Americans, European newcomers, and African-descended slaves poured in. This rush was the result of governmental policy as well as long-standing settlement patterns. The US government sought to reward soldiers of the War of 1812 and populate western regions through a system of land grants in the western territories across the Mississippi, including the New

Madrid hinterland. This substantial influx meant that new people with no direct knowledge of the region around the epicenters rushed to claim land.[9]

Difficulties in travel and communication—difficulties vastly exacerbated by the earthquakes' effects—long delayed any exact understanding of the environmental change to what had been the New Madrid hinterland. Land surveys predating the earthquakes showed maps of farmlands that were now lakes. Damage to the settlements of Little Prairie and New Madrid was well reported in early nineteenth-century media, but the transformation of the sunk lands was both wider spread and harder to capture—especially when many Americans did not have a clear picture of the pre-earthquake state of the area to begin with. The difference between a lush, sometimes inundated forest and a morass of downed trees and permanent swamp is important, but hard to appreciate when framed in brief descriptions. Would-be settlers who had heard or read about the bountiful hunting and fertile soil of the middle Mississippi Valley thus found themselves wading beside water moccasins in the midst of virtually impassable swampland. Some newcomers continued west past the sunk country, finding in Crowley's Ridge, the striking narrow uplift paralleling the Mississippi River, the high ground similarly valued by Cherokee settler Connetoo and his community a few years before. Those who stuck with their land claims in the sunk lands faced both legal and topographic challenge.[10]

As with the New Madrid certificate land claims, legal tenure of lands in the "sunk country" of the New Madrid hinterland caused extensive legal conflict. Estimates of the expanse of disputed sunk lands range from 60,000 to 300,000 acres—the very vagueness of which speaks to the lack of settled knowledge of such unsettled country. In what is now northeast Arkansas, large landholders who found themselves after the earthquakes in sudden possession of sprawling shallow lakes staked new claims to adjoining acreage under the doctrine of riparian rights—that is, rights to land on the banks of a body of water. The very newness of terrain flummoxed legal adjudication: what was to be done with dry land that turned into someone else's lakefront? Such claims generated enormous and long-lasting controversy, which ultimately reached the US Supreme Court.[11]

Land once used as community resource became concentrated into a few hands over the late nineteenth and early twentieth centuries. Powerful landed interests were creative: when lawsuits went against them, planters led by local landowner Robert E. Lee Wilson were able to get state law changed. Even new federal law worked for the large landholders—in 1921, federal statute gave planters "preferential rights" to buy disputed sunken land. Over the course of a century, large owners consolidated and expanded their rights and their holdings in what had been the New Madrid hinterland.[12]

In the late eighteenth and early nineteenth centuries, the New Madrid hinterland was a place of far-flung communication and connection, where visitors from the southeastern Cherokee homelands might stop by for a season or a long hunt; relatives from Shawnee villages along the Miami might come to call on sisters, nephews, aunts; rivermen would bring goods and news from the trading ports of the Ohio River; and hunters would distribute dried bison meat from summer expeditions out on the tallgrass prairie. The late colonial and early American New Madrid hinterland was sometimes a zone of discontent and trouble, but it was also a zone of movement, exchange, and interaction. Once the earthquakes shook up those rich hunting lands, making streams into swamps and forest into morass, both the environment and its people began swiftly to change. By the early twentieth century, the area that had once been the New Madrid hinterland was increasingly remote and insular, not well known or well visited by those outside, and divided by severe stratifications of money and skin color.

The Civil War

In the Civil War, the former earthquake country around New Madrid became the site of ferocious guerrilla war, largely because of seismic features that prevented the movement of large land forces and facilitated raids by small groups. Earthquakes created territory good for small-scale partisan conflict. But paradoxically, memory of that warfare erased memories of the earthquakes that made the skirmishes possible. The region of the New Madrid epicenters came to be associated with terrible battle, not terrible earthquakes. Before the war, the New Madrid hinterland could still be seen as earthquake territory. After, it was remade into a terrain of racial and social stratification.

Throughout much of the South, the Civil War was a breakpoint in memory.[†] Families, fortunes, political allegiances, and lives were remade by the convulsive and violent resuturing of the Union and the emancipation of four million Americans. In the areas around the New Madrid epicenters, the war broke and remade memory of the past in ways that were intensely local but had repercussions for how the larger seismic history of the United

---

[†] The war literally destroyed records of earlier history. CERI historian Nathan K. Moran notes that in plotting the locations in the Southeast where he would expect to find firsthand reports of the New Madrid earthquakes, but cannot, he can trace the path of Sherman's march to the sea.

States would later be understood—as a phenomenon of the expanding West Coast, not the benighted continental interior.[13]

The in-betweenness of the region shaken by the New Madrid earthquakes was exactly at issue in this war to split a nation, as in earlier conflicts between mutually suspicious peoples. In the war, Missouri stood for the Union—but just barely. Like Kentucky, it was a slave state, but one controlled by the United States, not the Confederacy. Illinois was a free, Union state, while Arkansas, Tennessee, Texas, and the entire lower Mississippi Valley were part of the rebelling Confederacy (along with the aligned regions of Indian Territory and New Mexico Territory, which made up the future states of Oklahoma, New Mexico, and Arizona). The areas most affected by the New Madrid earthquakes lay between the western and trans-Mississippi theaters of the war, with the Mississippi River the crucial and contested conduit between them. For all sides in this civil war, just as in the War of 1812, a key goal was control of communication and shipping along the Mississippi.[14]

A crucial battle for control of the Mississippi helped determine the guerrilla character of the rest of the war near the epicenters, and also demonstrates the rewriting of earthquake history into military history. The battle of Island Number 10 decisively opened the upper and middle Mississippi to federal control. In April 1862, a horrible two days at Shiloh ended in bloody victory for the Union and began the series of massive land battles that would consume lives both Northern and Southern for the following three years. Coming after victories on the Cumberland and Tennessee Rivers, the battle of Shiloh placed the Union in control of central arteries of shipping and communication—indeed, in control of much of the Mississippi Valley. Yet the middle reaches of the river itself were still held, if barely, by Confederate forces that had fallen back to claim New Madrid and Island Number 10, a small island just upstream from the town of New Madrid, in a sharp southward bend of the river that made the island outpost a key blocking point for commerce or Northern military advance.‡ New Madrid was soon abandoned under Union assault, but powerful Confederate artillery guns based on and near the island controlled movement down this central point of the river.[15]

---

‡ Despite the early nineteenth-century attempt to standardize Mississippi River islands by labeling them sequentially from the mouth of the Ohio, the islands themselves, formed as they were by depositions of water slowing as it left a channel turn and wiped away by subsequent floods and changes to the current, were far too dynamic to be held in place by mere vocabulary. Island Number 10 subsequently washed into the Mississippi; the battle that holds its name is now marked by a sign near Tiptonville, Tennessee.

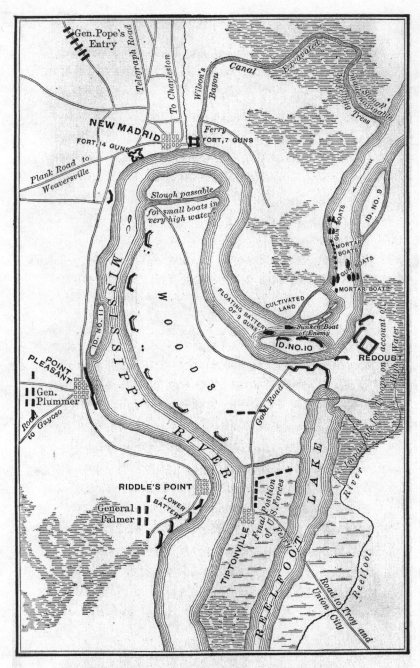

NEW MADRID AND ISLAND NUMBER TEN
MARCH 3-APRIL 7, 1862

Union forces took Island Number 10 early in April 1862 by a combination of panache and hard slogging. Troops aided by "contrabands"—former slaves who had freed themselves and fled to Union forces—dug a bypass canal above New Madrid to deliver much-needed supplies to Northern forces downriver. Meanwhile, the Union gunboat *Carondelet* daringly dropped past the heavy artillery of Island Number 10 in the dead of night as a raging thunderstorm provided cover. Two nights later, also under cover of a spring thunderstorm, a second gunboat successfully joined the *Carondelet*. From their position past the Confederate post, the two boats bombarded and took the artillery pieces, forcing the Confederates to abandon their outpost. This decisive silencing of a main blockade meant significant strengthening of Union communication and an extension of Union control over the confluence of the Missouri, Mississippi, and Ohio Rivers. Much of the West came under federal power, even as Union forces struggled to gain control of the rest of the river.

Lost in the battle itself, and lost in all subsequent retelling, is the environmental history of this crucial hairpin turn on the Mississippi. What made the Confederate outpost hard to reach were swamps—earthquake swamps spilling over from Reelfoot Lake. What made the hard-slogging Union able to cut through the narrow point above Island Number 10 (in addition to sweat and effort) was the shallowness of the swampy area: Union crews were cutting through trees that had been killed fifty years before, in a shallow swamp created by the New Madrid earthquakes. This was a battle fought in earthquake country.

Island Number 10 was one of the three key battles in which national forces took the Mississippi and forced the rebellion in the middle Mississippi West into guerrilla form. Over the course of that same month, a Union fleet under flag officer David G. Farragut forced their way past the forts defending the mouth of the Mississippi and ultimately seized control of the

FIG. 6.1. Battle of Island Number 10. The role of Island Number 10 in the struggle for control of the Mississippi rewrote the meaning of the New Madrid area for generations. This map illustrates the island as a point of blockade: swamplands surrounding this crucial hairpin of the Mississippi River protected Confederate artillery from attack. Federal gunboats eventually forced their way past the batteries at Island Number 10, and federal soldiers and escaped slaves ("contrabands" in the ironic Union term) won by cutting a shallow canal through the slough above New Madrid. The book makes no mention of the role of the New Madrid earthquakes in creating this terrain: those were earthquake swamps. (John Fiske, *The Mississippi Valley in the Civil War* [Boston: Houghton, Mifflin, 1900], map after p. 102.)

city of New Orleans. In the spring of 1862, Union commanders attempted to take the similar hairpin turn of the lower Mississippi by the bluffs of Vicksburg, Mississippi. That singularly defensible and critical outpost—"the key," as president Abraham Lincoln understood, to both upstream and downstream commerce—held out for over a year against repeated Union efforts. Over the spring and summer of 1863 a daring naval and overland campaign under Gen. Ulysses S. Grant enabled Union troops to besiege the city of Vicksburg. The suffering city surrendered on Independence Day, 1863. As the Confederate flag fell from Vicksburg, so too did Confederate hopes to control much of their heartland.§16

Soon after the fall of Island Number 10, Confederate rebellion in the contested states of the middle Mississippi became a war of backcountry partisans, fought in former earthquake epicenters on the seismic terrain of the swampy sunk lands. After several land battles in northwest Arkansas in 1862, Arkansas and Missouri became a backwater of organized military conflict, but a maelstrom of violent backwoods guerrilla war.** The Confederate Congress passed the Partisan Ranger Act in April 1862, giving official fighting status to those who deserted from formal forces to range through the backwoods, and this act was reinforced by local Confederate orders in Arkansas. Taking to the bush became a legitimate act of loyal Rebels. That suited many just fine: as one guerrilla fighter explained, "I wanted to be my own General." At the same time, loyal Unionist civilians in some areas found bushwhacking to be the most effective way of aiding federal troops. Even after the fall of Island Number 10, freebooting or loosely Confederate pirates raided Union supply ships and frequently sold the captured stores back to Union troops. Union forces, meanwhile, created specially equipped riverboats that could speedily offload men and horses in an effort to com-

§ Independence Day was not again celebrated in Vicksburg until after World War II.
** At the March 1862 battle of Pea Ridge, a large but ill-disciplined Confederate force under Earl van Dorn failed in an attempt to cut off the federal supply routes led by Brig. Gen. Samuel R. Curtis. Van Dorn's forces largely disbanded or deserted, many becoming partisan fighters in southern Missouri and northern Arkansas. Thus at the end of April 1862, no organized Confederate force was able to oppose Curtis's Federal Army of the Southwest in its march across southern Missouri to reenter northern Arkansas. A Confederate assault against Union forces at Prairie Grove in northwest Arkansas in December 1862 did not result in a clear victory for either side, but meant the end of large-scale Confederate efforts to reassert control over northwest Arkansas or reach into southern Missouri.

bat river piracy. Especially in the former earthquake zones, jayhawking and bushwhacking *were* the war.[††17]

The guerrilla conflict of the Civil War seared the collective memory of all who experienced it. The partisan warfare of the area around New Madrid and the Arkansas/Missouri Ozarks was exceptionally brutal, terrorizing families, noncombatants, and anyone of suspicious loyalties or with corn or hogs to raid. Thuggish raiders only sketchily affiliated with either side defined a war of long-nursed grudges and bitter reprisal.[‡‡] Near Warrensburg, Missouri, in 1862, a train car of unarmed Union recruits was ambushed and killed, many of them scalped and mutilated. In southeast Missouri, Confederate partisan Alfred Bolan boasted of killing forty Union. He was subsequently beaten to death by a Union spy who mangled him with a piece of plowshare. Bolan's broken and bloody body was then exhibited to great satisfaction by Union infantrymen relieved to be rid of his attacks. Leroy William, son of a murdered Unionist in Conway County, Arkansas, began a vendetta against Confederate partisans. He soon became known as "Wild Dick" for charging into groups of rebels holding his horses' reins in his teeth and firing pistols with both hands.[§§] The First Arkansas Cavalry

[††] *Jayhawkers* originally and proudly referred to anti-slavery activists from Kansas and was later used as an insulting term by Southern patriots against all Yankee partisans. *Bushwhackers* was first used in the late 1840s by writer Washington Irving to refer to skilled, tough frontier woodsmen and soon came to refer to any sort of slinking, guerrilla-type ambush fighters, but especially was used during the war by Northerners as a term for Confederate partisans. Both terms, though, soon became as confused as the loyalties of the various backwoods thugs about whom they were employed, referring to various forms of guerrilla warfare and backwoods brigandage that characterized the war in much of the middle Mississippi Valley and other border regions.

[‡‡] Outlaws Frank and Jesse James got their start as irregulars under guerrilla leader "Bloody Bill" Anderson—part of a group that entered Maj. Gen. Sterling Price's lines with human scalps hanging from their bridles.

[§§] Charles Portis's 1968 novel *True Grit*, and the two popular Hollywood movies made from it (the 1969 John Wayne version and to a much greater extent the version directed by Joel and Ethan Coen in 2010), drew with perhaps surprising accuracy from this history of guerrilla conflict in the borderlands. A fourteen-year-old girl avenging her father's killer might be historically improbable, but stray uniformed corpses forgotten deep in rocky mountain caves were certainly not. Nor was the climatic conflict in which two bickering Confederate veterans reach back into their war experience to win a shootout against a guerrilla band: one by an extremely accurate long-range shot with a breech-loading rifle—a key innovation of the Civil War—and the other by taking reins in his teeth and galloping forward while blasting pistol fire on both sides. More than one drunken, washed-up, wounded veteran of the late 1870s had been taught such skills by their nations' convulsive war.

Volunteers was formed as a federal response to this partisan warfare in 1862. They were generally short on horses (Confederate bushwhackers stole most of their horses, then creatively stole their mules, too), but these "Mountain Tories" used knowledge of the shared terrain to combat Rebel guerrillas. They were intensely hated by Confederate irregulars.[18]

The environments of earthquake territory were central to this guerrilla Confederate resistance through the latter two-thirds of the war. Swampy conditions in east Arkansas and the Missouri Bootheel—the overflowed regions created and furthered by the quakes—favored guerrilla warfare and suited Confederate bushwhackers. As one Arkansas Confederate partisan explained in a letter back home, "When we camp, we hunt for a swamp and then move before day." Much as escaped slaves had a half century or more before, defenders of slavery in the early 1860s used the inaccessibility of the swamplands of east Arkansas and southeast Missouri as a strategic resource. Missouri State Guard brigadier general Merriwether "Jeff" Thompson, the "Missouri Swamp Fox," was active in the Bootheel and eastern Arkansas. In spring 1865 Union messengers sent to deliver terms for the surrender of Thompson and his "Army of Northeastern Arkansas" had difficulty even locating them to negotiate with. An environment of overflow made quick and dirty action possible and frustrated the large infantry movements that commanders on both sides understood as vital to military success.[19]

St. Francis River environments that had once allowed community consolidation instead facilitated raids by those who best knew the local terrain. Stymied in his spring 1863 attempt to disrupt federal presence in southeast Missouri when he was unable to dislodge Union forces who had retreated to the fortifications of Cape Girardeau, north of New Madrid, Confederate general John S. Marmaduke ordered a retreat back to Arkansas, which his men slowly accomplished under heavy rear assault. Meanwhile, Marmaduke sent forward Jeff Thompson with a few hundred men to secure the Missouri bank of the St. Francis at Chalk Bluff, the crossing at the Missouri/Arkansas border where the St. Francis flows through a break in Crowley's Ridge. Working with little except axes and frontier skill at making do, Thompson's men constructed a rough crossing comprised of a log raft for artillery and a floating log bridge guyed into place with grapevines and a few ropes.[20] At night, Marmaduke's men crept over the rough bridge as their horses swam alongside in churning current, escaping from pursuing federal forces. Veterans later claimed a dramatic finish: just as the last of the Confederate column crossed on 1 May, the pursuing federals came into view on the river's north side. As the last Confederate crossed, cavalry officers cut the vine ropes with their swords. (Other commentators claim that the bridge simply collapsed, but all agree that it was swept downstream.)[21]

Yet the swampy environment that had enabled a Confederate survival would not enable Confederate victory. Though Marmaduke and his troops did ford the challenging river, their slow retreat meant that they took such heavy casualties from Northern fire that he was subsequently forced to abandon his second advance, leaving the area as he had found it, in Northern hands. Even after their success in crossing the St. Francis, Marmaduke and his forces still had a long, mucky, hungry, mosquito- and sickness-plagued crossing of the St. Francis swamps before reaching resupply camps. The raid and subsequent retreat marked the end of Confederate ambitions to retake southeastern Missouri: thereafter guerrilla skirmishing, not cavalry advance, would mark the war in the region.[22]

Battle, not seismic upheaval, would be associated with the looping curves of the middle Mississippi River in postwar American life. By 1866, when exhausted former foes and jubilant former slaves sought to put the war behind them, popular memory had significantly shifted. One northern river traveler passing New Madrid on a barge trip in 1866 noted that it was a "town of several hundred inhabitants. . . . It derived its importance during the war from being the point of supplies for Grant's army during the operations on island Number 10. Saw a couple of beautiful women there while the boat was stopping."[23] Twenty years before, New Madrid had been the destination for a geological tourism, a rustic, out-of-the-way, but vital place for British geologist Charles Lyell to visit and investigate. After the war, it was one of the less significant wartime points of strategic interest and a place for young gentlemen to enjoy the sight of lovely ladies.

Race and Land in Former Earthquake Territory

Through a series of local changes that tracked national tensions of race and land, the people and places of the New Madrid epicentral region underwent a series of transformations in the late nineteenth and early twentieth centuries, transformations in which earthquake history was rewritten by present circumstance. These changes were so wrenching that they erased past societies and past meaning and uses of land. The Native past of the region was pushed back to the vague shadowy realms of prehistory. The settlement of the New Madrid area by Native American pioneers of the late eighteenth and early nineteenth centuries, along with the role of the region's earthquakes in pan-Indian spiritual and cultural revival, was buried by harsh decades of black/white racial tension. Conflicts over land and race erased prior histories.

In the early twentieth century, environmental changes to the New Madrid hinterland were accompanied by a strict geography of race. Formerly ethnically mixed settlements gave way to all-white communities. A local resident named J. W. Estep testified in the 1890s about the African American swamp reclamation crew he had encountered growing up in the sunk lands in the mid-1870s: "Us boys had a good deal of fun out of the negroes. They were a new thing in our country, they were sorto' afraid of us. We deviled them a little just for our fun." At the same time, Estep described a boyhood in which he worked on the family farm during farm season, and then at other times "fished and hunted and wallowed around on the river, and done nothing, you might say." Estep described a seasonal pattern of resource use that would have been familiar to Indian settlers of the 1800s and 1810s, but as an older man, he dismissed his childhood of seasonal river life as a lot of "doing nothing." Estep's experience reflected profound social rifts that began after the Civil War and continued well after the Great War.[24]

Demographic change from Native and white to white and black began with the remaking of society and economies after the Civil War. African Americans emigrated into Reconstruction Arkansas, and to a lesser extent into Missouri. In reaction, many white residents organized themselves through the Ku Klux Klan, creating what one federal official termed "a reign of terror, intimidation, and murder" against African American residents, even as Jim Crow laws of segregation and inequality rolled back the modest gains for black residents in the 1890s. An early twentieth-century land boom in the Missouri Bootheel and northeast Arkansas exacerbated social and racial tensions. Carrying traditional Northern ideas of American racial hierarchy, new white settlers from Indiana, Ohio, and Illinois created "sundown towns," white settlements where African Americans were forbidden from living—or even lingering after sunset.[25]

This utterly altered and utterly polarized racial geography of the middle Mississippi Valley had grave consequences for scientific knowledge of the area and its past. In 1909 the Academy of Natural Sciences of Philadelphia sent off an expedition along the St. Francis and related rivers of northeast Arkansas to identify Native American artifacts before curio seekers and pottery sellers destroyed them all. The Little River, now a smaller tributary of the St. Francis but once known as its main branch, drains Big Lake near the northeastern notch of Arkansas, just below the Missouri Bootheel. It looked extremely promising—but the expedition was forced to a full stop at the Little River community of Lepanto. Clarence Moore, the archeologist in charge, explained in his official report that "our quest . . . came to an end owing to the hostility against negroes, entertained by the natives along the river above Lepanto, who maintain a negro dead-line, permitting no colored person to go among them." It was a dead-line indeed. Moore noted, "As this race prejudice has resulted in the murder of a number of negroes, we did not deem it fair to expose to slaughter men who had served us faithfully for years." When expedition workers are at risk of being murdered because of the color of their skin, it is difficult to do good science.***[26]

Changes to the Missouri Bootheel were recorded with a raconteur's slow-drawn wit by planter and social commentator Thad Snow. In 1910 he moved from Indiana to southeast Missouri, then an "agricultural frontier" of cotton lands. Starting in the 1930s, he began to publish in the *St. Louis Post-Dispatch* and eventually publications like *Harper's Magazine* (where he extolled the wisdom and diligence of a favorite mule, Kate). His memoir *From Missouri* was published after World War II by a national press; he spent time in the winter of 1949–50 attending economics seminars at Harvard, where he rubbed shoulders with people like John Kenneth Galbraith.††† In his home region, Snow came to be renowned for his physical strength, strong opinions, extensive library, fearlessness around snakes, and capacity for hard liquor. He also came to be reviled for what many of his white contemporaries viewed as dangerously socialistic, left-leaning views, especially about the necessity for change in the social and racial order of the Bootheel.[27]

Thad Snow's commentary captured social and environmental changes that would remake his region and its perceived past. Snow chronicled the boiling up of racial tensions as Southern cotton planters fled north in the early twentieth century to escape the boll weevil that was voraciously consuming the staple crop of Dixie. Many Southern planters sought to relocate safely past the "boll weevil line," north of which they thought a winter freeze would halt or at least contain the insect's spread.‡‡‡ By the early 1920s, cotton farming was booming in the Bootheel and northeast Arkansas.[28]

*** In the tones of distance and understatement in which so many early twentieth-century field accounts framed their hard-earned conclusions, Moore explained that "our sole motive for referring to this disagreeable episode is that when an amply equipped expedition abandons a most promising region, a valid reason for doing so should be forthcoming." Presumably the black workers on the expedition regarded the reason as valid enough. The legacy of enforced ignorance was long-lasting: even into the twenty-first century, these regions of northeastern Arkansas remain archeologically understudied—of over 10,000 identified known sites in the region, very few have been examined.

††† Indeed, the copy of *From Missouri* I read was donated to the Harvard Libraries by a fund established by Galbraith.

‡‡‡ Planters were mostly wrong about the boll weevil: in the insect version of hibernation, boll weevils can live through most Bootheel winters. By 1913, despite planters' optimism, weevils had crossed the state line into Missouri, continuing their progress of about sixty miles a year after their introduction from Mexico in 1892. In favorable conditions, the insects can go through a life cycle in under three weeks, destroying cotton fields as they

To farm this new crop, white landowners sought black agricultural labor. African Americans from further south, looking for better working conditions, arrived to cultivate cotton in the Bootheel and northeast Arkansas. The population of several Bootheel counties almost doubled in the century's first decade, but white "timber workers" did not welcome these "cotton Negroes." In a particularly vicious and widespread campaign of intimidation and violence during harvest season 1911, they made that unwelcome clear through nighttime mob terror, beatings, shootings, and the burning down of rooming houses. Local papers became adept at fashioning euphemisms for lynching. In the initial years, many of these newly arrived black laborers worked in what Thad Snow sardonically termed a "highly organized fashion" under foremen carrying whips and guns. Some landowners locked their workers in a stockade at night; some later served federal time for violation of peonage statutes. In contrast to these work conditions, for many poor families—especially black families only a generation or two removed from formal race slavery—the family-based system of tenant farming seemed to promise more future, more freedom, and more dignity.[29]

In the decades after the Civil War, the former New Madrid hinterland of northeast Arkansas and Bootheel Missouri became simultaneously a center of sharecropper agriculture and of white racism. Populations boomed, and land use was transformed. The system of sharecropping, formally if rarely known as tenant farming, emerged out of the period of Reconstruction and became widely employed through much of the middle Mississippi Valley. A landowner would essentially rent land to a "cropper," a head of household who would work alongside his family for a portion of the crop. On paper that does not seem entirely unreasonable, but unfortunately cotton is not grown on paper. Once sharecroppers borrowed the seed, perhaps the mules, certainly the groceries to get them through a season—usually from the landowner—they were often fortunate to break even. Sharecropping became essentially a form of debt peonage tying families to increasingly worn-out land.§§§[30]

---

thrive. "Boll Weevil," a blues song famously recorded by Lead Belly beginning in the 1930s, told the sad story throughout the Delta: "The first time I seen a boll weevil, he was sitting on the square. / The next time I seen a boll weevil, he had his whole family there." A national eradication program began in the 1970s, but boll weevils were not wiped out in Arkansas and Missouri until 2009, when American cotton fields were declared 98 percent clear of the pest. Still, cotton farmers in northeast Arkansas voted staunchly against the assessment that would pay for weevil eradication in the 1990s, holding to the notion that the boll weevil was not a problem in their area.

§§§ "I never picked cotton," sang Johnny Cash in one mighty piece, "like my mother did, and my brother did, and my sister did, and my daddy died young—workin' in a coal

The economic challenge and protests associated with sharecropping on former earthquake territory would remake the meaning of the area for many Americans. The terrible Mississippi flood of 1927 washed away the fields as well as hopes of many tenant farmers and other poor agricultural laborers. As the country's depression deepened after the 1929 stock market crash, drought in 1930–31 destroyed farm incomes in the middle Mississippi Valley. The policies of the early New Deal were meant to ameliorate but instead intensified inequity in the region: programs that paid planters to plow up "excess" crops left out sharecroppers entirely, as planters simply defined their tenants as day laborers not entitled to any share of subsidies. An entire rural population was downsized. The result was desperation for many. As Thad Snow observed, "Even Negroes lose color when they neglect to eat for too long a time."[31]

Dramatic and well-publicized protest by the Southern Tenant Farmers' Union brought the middle Mississippi Valley to national headlines as the epicenter of social tension, not seismic trouble. The STFU was a pioneering and racially integrated group begun in a small northeast Arkansas town in the summer of 1934. Led by gifted organizer, farmer, and preacher Owen Whitfield, the STFU brought the plight of sharecropping families to public awareness throughout the nation through the Missouri Sharecropper Roadside Demonstration of 1939. On 10 January 1939, roughly 1,500 recently homeless tenant farmer families set up camp along public roadsides—state-owned land where, leaders had determined, they could not be moved. In bitingly cold winds, the protesters sheltered themselves with quilts on stakes, heating with twig fires, with household effects about and children on the ground.**** "Let our camps do the talkin' for us," Whitfield urged the demonstrators, "They'll talk loud enough, especially if it rains and snows on you." By the second day of the strike, reporters from papers in Chicago, New York, Baltimore, Louisville, Des Moines, and elsewhere were present to gawk, question, and write excited stories for the national press. By day 10, the tenant farm families were legally cleared off. Many "roadsiders" were evacuated to camps where they were in Snow's acerbic terms "guarded and cared for," as local political leaders had promised, "except for

---

mine." The song was about growing up in Oklahoma, but Cash himself was born into an Arkansas sharecropping family; from early childhood until he was eighteen, he did "a full day man's job in the cotton fields," as he described it late in life. His music reflects a precisely articulated bitterness toward that grueling and dirt-poor labor.

****In an echo of concerns about a "dead line," organizers were careful not to situate any of the encampments too close to the Arkansas state border—Arkansas planters were regarded as quick to defend their racial and social prerogatives with shotguns and hunting rifles.

unessential items, such as food, fuel, shelter, and water." One group of "socially minded St. Louis people" bought a small Ozark plot for a group of roadsiders, where they resettled—a twentieth-century cognate, perhaps, of the independent free black settlement that white traders derisively termed "Nigger Hill" in the early nineteenth century.[32]

Race-based terror and racially integrated community movement were what most newspaper-reading Americans would know of the region after the 1930s. These protests were dramatic, visible, and only partly effective. Through the next several years after the STFU strike, federal agencies investigated and tried to implement reform. Yet in the end, almost nothing changed: only US entry into what was becoming the Second World War effectively jump-started the American economy and offered jobs elsewhere for sharecroppers throughout the South. Fundamental racial tensions in the areas of the New Madrid epicenters endured.[33] ˏ

In the postwar New Madrid hinterland, the types of crop changed and systems of sharecropping ended—no one needs children and mules out in the fields when John Deere can supply machines instead. But despite changes in crops and personnel, large farms remained the dominant form of environmental use in southeast Missouri and northeast Arkansas. Cotton reached its greatest extent after World War II, overtaken in the 1950s by soybeans; rice became a major crop in the 1970s.[34]

Monocrop agriculture and its racial fault lines focused attention away from the seismic history of the region. Farming also physically erased evidence of the quakes. Plowing and harrowing further spread out the white granular deposits of sand blows, destroying smaller sand blows or blurring larger features. "Where the land has been cleared and cultivated," drainage advocate Otto Kochtitzky noted in the late 1870s, "the sand piles have generally been worked down into the adjacent soil." Land leveled and drained by large-scale machines is land less easily read for subtle signs of environmental disturbance.[35]

Over the hundred years since the New Madrid quakes had shaken northeast Arkansas and southeast Missouri the human geography had been transformed, from mobile multiethnic communities traveling up and down rivers to hunt and trap, to trade, and to get news and stories, to a black and white landscape of large landowners and resentful poor whites, of even poorer black people and of a culture of suppression and silence.[††††][36]

––––––––––

†††† Into the early twenty-first century, a few people in Missouri and southern Illinois still spoke middle Mississippi Valley French and played music distinct from that of the Cajun lower Mississippi or the Québécois. But these few communities of "Creoles" of the "Illinois Country" often find themselves forgotten in North American cultural catego-

Obscured Landscapes: Industrial Agriculture, Floods, and Undergrowth

Evidence of the New Madrid earthquakes was hard to see even a few years after the quakes, and human changes to the epicentral environments made them even harder to perceive. The New Madrid earthquakes were powerful precisely because they shook a deep bowl of loosely consolidated sediment, the Mississippi Valley embayment. They therefore did not leave the same clear environmental traces as quakes which displace or cut across hard, resistant topography. Sand blows and crevasses soon erode, fault scarps soften into roundness, quake-induced lakes start to look much like the regions' other oxbows and sloughs. Earthquake deposits of sand on soil may look much like deposits left by regular floods; floodwaters in turn sweep away layers of prior geological evidence.[37]

In the New Madrid hinterland, human activity heightened these changes to seismic terrain. Drainage and industrial agriculture flattened, regularized, and imposed a new geometry on the regions around the epicenter, erasing evidence of the quakes. They receded from public as well as scientific awareness. The straight lines of land survey and drainage canals—especially the Little River Drainage District canals, which cut across the landscape in long, straight arrows—imposed a rigid geometric order on the rumpled surface of a seismic zone. In the late nineteenth and early twentieth centuries, a series of related interventions transformed the surface environment of the New Madrid hinterland, obscuring or simply drawing any attention away from the physical traces of long-ago earthquakes.

As an environmental parallel of the calamity of war, the frequent floods of the Mississippi helped wipe away awareness of past convulsions of the landscape. Over the late nineteenth and early twentieth centuries the main natural force to threaten the epicentral region was not shaking earth, but moving water. In the 1870s, future drainage engineer Otto Kochtitzky once barely saved his Bootheel corn harvest from a spring rise by storing it on top of a haystack. He engaged a steamboat which, with its shallow draft, was able to pull up across the flooded fields and anchor to an upper branch of a cottonwood tree, where he loaded the corn directly from the haystack. Area residents had similar stories in which enterprising boys hungry for supper

---

ries, their music, food, language, and customs dispersed by generations of geographic mobility and folded into the better-known French heritage of Louisiana and Quebec. As a local musician told me at a bluegrass festival, "The Québécois are the French who lost, the Cajuns are the French who left, and the Mississippi Valley French are the ones who got forgotten."

swam horses out onto flooded fields to harvest corn underwater. Even allowing for a certain narrative exaggeration, such stories told an essential truth of life near a mighty river.[38]

Flooding remade a seismic landscape. In addition to smaller and more local overflows, massive floods periodically overswept the New Madrid hinterland. A devastating sequence in 1912 and again in 1913 set the stage for the agricultural problems of the 1920s. The great Mississippi River flood of 1927 covered nearly 26,000 square miles and affected eleven states. Surveying the massive rise in April 1927, one army engineer commented, "The Mississippi is no longer a river, but a lake extending from St. Louis to New Orleans." New Madrid residents lived in upper stories or took refuge from the floodwaters atop Indian mounds that had not yet been plowed flat. Mississippi River currents reasserted their old right-of-way west of New Madrid, in the area that had been connected with the river during seasonal high water during the era of Native American and colonial lead production. The Mississippi once again flowed behind New Madrid down the St. Francis basin, rejoining its usual bed just north of Helena, Arkansas. Then again in 1937, another winter flood. To save the town of Cairo, Illinois, the Army Corps of Engineers dynamited the levees on the Mississippi Valley floodway, flooding much of the Bootheel. Such floods pushed long-ago earthquakes away from public memory. Moving waters commanded the forefront of attention.[39]

Raging floodwaters swept away environmental evidence of prior histories on the land. Other aspects of the environment similarly challenged investigation. Muggy mid-Mississippi swamplands were hard to reconnoiter or plumb. Extensive vegetation quickly covered evidence of seismic or human activity. Railroad workers in the Bootheel had difficulty laying rail in the spring of 1901: much of the lumber cut and left lying along the trail the previous fall and winter was completely hidden by the tall grasses. This was, moreover, no easy landscape to study: sluggish rivers and their swarms of mosquitoes, stands of thick cane, tangles of vines and poison ivy, even wild cats and wolves all made the zones of earthquake damage hard to access or assess. Physical labor in the sunk lands was dogged by illness, especially the malaria or "ague" that was a constant feature of the terrain. A crew working in the sunk lands at one of the channel-clearing projects of the late nineteenth century might have two hundred hands employed, but only a score capable of work at any time. The Missouri Botanical Garden, based in St. Louis, sent out an expedition along the St. Francis and southeastern Missouri in 1892 and 1893, a comparatively late period to be botanizing a few hundred miles from the garden's doorstep. But the sunk lands were exotic and little-known territory a hundred years after the quakes that created them.[40]

The environments of the sunk lands thus posed profound and mucky challenges. Well into the 1870s, the best passage across them was still on "corduroy roads," locally called "pole roads" because they consisted of poles laid crosswise across the top of a levee. Mississippi County, Arkansas, long had two county seats, Blytheville and Osceola, because the road across the interior low country was impassable much of the time. Even veteran railroad surveyors sometimes got lost in the "gumbo" of the west and east Bootheel area termed the Nacormy Swamp. The town of Pemiscot, near the swamp, was (and is) said to be a Native American word for "liquid mud."[41]

Well past the Civil War, the challenges of overland travel in the lowland areas affected by the earthquakes continued to shape settlement in the region. The towns of Kennett and Caruthersville developed on either side of the swamp and were connected only in dry years, by the Hickory Landing plank road. Railroad booster Louis Houck recalled that in the late years of the nineteenth century the two towns were as isolated from each other, and travel took about as long, as "between Africa and South America." "It took about as much time to go from Caruthersville to Kennett," he maintained, only partly joking, "as from Kennett to New York." The swamp, railroad engineers concluded, was an area "outside the world."[42]

Railroading and Timber Harvest in Earthquake Country

In and through this morass, railroads promised connection. Even as they did so, they created a landscape no longer recognizable as earthquake evidence. After the Civil War, developers constructed a maze of short-line private railroads throughout the former New Madrid hinterland. These rail lines at once obscured certain forms of knowledge and made possible others. Railroads reconnected areas of sunk lands with the interior of Missouri and Arkansas, remaking the connections undone by the earthquakes. They also dramatically reworked the local landscape by making possible large-scale lumbering of area forests.[43]

Louis N. Houck was a main figure in this regional enthusiasm for railroad building and thus a main figure in the rewriting of earthquake history. He laid track throughout southeast Missouri in the 1880s through the 1890s, constructing three railway systems and periodically seeing them collapse into receivership. By 1901, local residents boasted that all of southeast Missouri was within six miles of a railroad thanks to Louis Houck. (This claim could not quite sustain close scrutiny, but the interlaced network of Houck's

FIG. 6.2. Louis Houck, Houck railway workers, and one of the engines that helped bridge and transform formerly impassable earthquake swamps in Southeast Missouri in the 1880s and '90s. (Box 2368, Item 69, [Houck Louis 2368 069], University Photograph Collection, courtesy of Special Collections and Archives, Kent Library, Southeast Missouri State University)

and competitors' lines did cover a great deal of the region.) Local traditions of guerrilla warfare continued among railway magnates. Houck's long-raging battle with national financier Jay Gould became legendary: at one point in 1896 a contingent of Gould's lawyers in a sleeping car woke when a freight train loaded with railroad ties slammed "accidentally" into nearby switching, obliterating it, while Houck's crews escaped with several rail cars to a different jurisdiction outside the county, aided by telegraph line sabotage, destroyed timbers, and additional derailments by Houck workers.‡‡‡‡[44]

‡‡‡‡ Houck roads were of notoriously poor quality, "little more than two streaks of dust and the right of way," according to disgusted Missouri railroad commissioners in the

Throughout the New Madrid hinterland, railway development served to remake swampland. In a local version of efforts playing out in many locations in the Mississippi Valley, Houck commenced his railroading by dovetailing his work with the emerging movement for swamp reclamation. An 1875 Missouri state law prohibited county aid to railroading, so Houck defined his railroad construction as land reclamation. He convinced the county court that his railroad embankment would be constructed high enough to serve as a levee; a system of drainage ditches and culverts would then protect surrounding areas during St. Francis overflow. In return for this public beneficence, he would then be granted public land. In the early 1880s, he began this scheme of combined swamp reclamation and swamp railroading.[45]

Houck's and the other regional lines helped open the former New Madrid hinterland for scientific investigation. Railroad and timber personnel were the local sources of information for the few researchers to visit the region in the late nineteenth and early twentieth centuries. Timber company engineers were some of the first to promote a mid-1930s program of increased study of area seismicity pushed by Saint Louis University seismologist Father James B. Macelwane. Around the turn of the century—around the same time that the Missouri Botanical Garden sent out botanizing expeditions into southeast Missouri—Louis Houck made a practice of sending unusual botanical specimens discovered in the course of his crews' railroad work to the garden for cataloging. Houck was himself an early historian of southeast Missouri. Railroads and timber companies—often working in tandem—were a main way in which this region became known.[46]

At the same time, Houck's and other developers' "swamp railroading" destroyed earthquake evidence. In 1891, crews used local materials to solve an engineering problem in the western Bootheel. They scooped up the plentiful sand near what is now the town of Kennett to establish a proper railway roadbed, jacking up existing rails and ties to spread the sand beneath them. Such practical jury-rigging and use of resources locally to hand were entirely typical of early railroad building in many environments of the United States. What was less typical was the source of this particular building

---

1890s. The chief argument in their favor, argued detractors, was that they were "still better than walking, at the price"—though the edge was perhaps slight. Yet as a benefactor who helped found Southeast Missouri State University, Houck leveraged public opinion—especially during his conflict with the Goulds—to become a local hero, the "Father of Southeast Missouri."

FIG. 6.3. Logging camp. In the late nineteenth century, new families moved to the region south of New Madrid to harvest the area's vast resources, converting old-growth trees into board feet of timber. Their labor would destroy earthquake evidence: no longer could a researcher like Myron Fuller compare the size of old trees to draw conclusions about the timing of earthquakes. (Image 1487-5-02, Himmelberger-Harrison Lumber Company Records, courtesy of Special Collections and Archives, Kent Library, Southeast Missouri State University)

material: the local sand used by railroad crews was from the sand blows created throughout that area during the 1811–12 quakes.§§§§[47]

By the turn of the century, the extensive forests in the former New Madrid hinterland of southeast Missouri and northeast Arkansas attracted attention from railroad companies, who recognized the profit to be made from shipping mid-south timber to markets further north. Railroads ran lines to promising areas, establishing boom towns around timber mills in what had been areas of virtual inaccessibility. The hinterland forests were still wild territory: when panthers came prowling through, mules became unmanageable, and when timber mill whistles blew early in the morning,

§§§§ This simultaneous use and effacement of earthquake features continues today: it is common practice for builders to use sand from sand blows in the foundations of houses in the New Madrid area because the sand provides such good drainage.

they might be answered by the howl of a wolf. Harvesting the hardwoods and, later, the pines of the region helped domesticate it. The fur trapping that still characterized use of the sunk lands by many local residents in the 1870s was no longer possible or profitable by the mid-1890s.[48]

Large landholders consolidated their holdings and their ability to profit on the resources of former earthquake territory over the late nineteenth and early twentieth centuries. The Jonesboro, Lake City and Eastern (JLC&E) Railroad laid track to Manila, Arkansas, late in 1900, offering access to timber and the hunting and fishing tourism of the earthquake-created Big Lake. By 1890 entrepreneurs had developed a market in wild ducks shot on Big Lake. Soon thereafter the JLC&E was shipping large quantities of fish, fur, and game north and east. The trade in wild fowl and game lasted until Big Lake was named a federal game reserve in 1915 by President Woodrow Wilson, at the behest of large landowners such as plantation owner Robert E. Lee Wilson, who had hunting clubs on Big Lake. Just as he maneuvered through the courts and legislatures to diminish the rights of smaller property holders, Wilson worked political connections to exclude independent hunters from the area.[49]

Though influential in the reworking of former earthquake land, local railway lines were short-lived. Large landowners' restricted access to the sunk lands eventually helped shut down the extensive local railroad network in the middle Mississippi Valley, which operated for only a few decades. The "Good Roads" movement of the 1920s, in which Thad Snow was a local leader and Louis Houck a local opponent, helped bring an end to the era of privatized, piecemeal infrastructure development which Houck so typified. By the late 1920s, almost all the "Houck roads" were taken out of service and their rights-of-way abandoned. Passenger and mail service on the JLC&E was discontinued shortly after World War II. By the late twentieth century, almost all the railroad lines in the Bootheel closed. Remaining, railroads were used exclusively for freight shipments. People and most of their possessions rode on new-built highways while old rail lines were subsumed by the region's fast-growing vegetation.[50]

Meanwhile, the mechanization of agriculture throughout the country led to steep population declines in the former New Madrid hinterland. By the 1970s, not natural resources but poor understanding of probability helped drive the economy of the Bootheel: the Lady Luck casino was constructed in the 1970s in Caruthersville, a town that had once anchored rail traffic harvesting the region's timber, game, and watermelons. Older environmental histories became more challenging to see and were little heeded.[51]

Draining—and Denying—the Sunk Lands

The interconnected railroading and timbering of the former New Madrid hinterland obscured earthquake history in the region. During the same decades, large-scale drainage of the sunk lands was accomplished through outright denial of that earthquake history. The main engineer and advocate of sunk lands drainage and reclamation, Otto Kochtitzky, argued strenuously in the early twentieth century that the region's earthquake history was exaggerated and misleading. Through his advocacy, Kochtitzky not only promoted swamp drainage but also shaped historical and scientific perspectives on the region.[52]

Kochtitzky built on decades of advocacy for draining swamplands. In the years after the New Madrid earthquakes, senator L. F. Linn had argued that their effects could and should be remediated. Linn strongly urged the cutting and clearing of "rafts" blocking the St. Francis and other rivers. Commentators throughout the middle and late nineteenth century similarly called for swamp drainage and land reclamation in northeast Arkansas and southeast Missouri. In the Swamp Land Acts of 1849, 1850, and 1860, the US Congress gave away 64 million acres of publicly owned wetlands to fifteen state governments, in return for the efforts of draining and "reclaiming" them. Corruption blossomed. Throughout the middle Mississippi, shady corporations and sketchy investors claimed to drain vast areas of swampland in return for free land. One particularly egregious scheme in the Bootheel, begun in 1870 by Confederate veteran William S. Sugg, led to a twenty-nine-year court case but little actual drainage. Similar if less graft-ridden efforts converted wetlands to farmlands across a wide swath of middle America. From 1850 to 1930, Midwestern wetlands shrank by 70 percent. In the former New Madrid hinterland, this large-scale reworking of wetland was accomplished largely through the coordination of the Little River Drainage District.[53]

The Little River Drainage District effaced evidence of the quakes that had created the sunk lands. The region's earthquake history was also effaced by the central figure of the LRDD, Otto Kochtitzky. Over decades, he waged a long argument that the New Madrid quakes were in fact an exaggeration based largely on the kinds of inflammatory sources available about them, frontier "tall tales" that had been taken for objective science.[54]

Kochtitzky grew up in the Missouri capital and first encountered the New Madrid hinterland as a young land surveyor in the mid-1870s. Beginning in 1877, he helped engineer and manage a barely profitable railroad across the Little River swamps, on a route first chartered as a plank road in the 1850s before work was interrupted by the Civil War. Kochtitzky's fam-

FIG. 6.4. Floating dredge. Like enormous chomping monsters, steam-powered dredges and draglines drained and cleared the Great Swamp of the St. Francis with noisy, smoke-belching efficiency in the 1910s and '20s. Their crews utterly reworked the terrain, making it accessible to American agricultural development, and in the process hiding or destroying clues to a prior seismic history. Machine technology would be similarly crucial in seismologists' turning away from research on the New Madrid earthquakes in favor of more abundant instrumental data from the West Coast and other regions of more frequent earthquake activity. (Item 4-07, from album entitled "Types of Machines That Built the Levees and Excavated the Ditches," Little River Drainage District Records, courtesy Special Collections and Archives, Kent Library, Southeast Missouri State University)

ily enterprise was typical of the small, rickety private railroads that Houck began regionally to buy or replace, and that larger national magnates like Gould would render obsolete. Houck's simultaneous railroad enterprises fueled a popular enthusiasm for drainage, and from the mid-1880s to the late 1890s, Kochtitzky managed the first channelization of the Little River.***** Around the turn of the century, he moved into "land dealing and drainage engineering"—endeavors which were in that period completely entwined.

***** During a brief period away from the Mississippi Valley, Kochtitzky became a city clerk of Mount Airy, North Carolina, where he served on the commission to apportion the estate of the original American Siamese twins Chang and Eng Bunker, who had retired to farm after managing a touring career in Europe and the United States and who died in close succession in 1874.

FIG. 6.5. Pulling willows along a drainage ditch. The Little River Drainage District is at industrial scale; the largest such project in the world when it was built in the 1910s, today it employs almost a thousand miles of ditches and three hundred miles of levees to drain 1.2 million agriculturally productive acres. Yet much of the work was accomplished by hand, as in the hot, mucky work of clearing ditches of clogging willow, pictured here. ("Ditch No. 1, willow pulling job, water 35 feet," item 494, Little River Drainage District Records, courtesy Special Collections and Archives, Kent Library, Southeast Missouri State University)

Thus began Kochtitzky's main career: advocating and then building a vast system of drainage that marks the region to this day.[55]

In a mammoth engineering scheme overseen by Kochtitzky, planned early in the century and begun during the Great War, southeast Missouri's pervasive wetlands were drained by the LRDD. The effort was led by engineers who had recently helped construct the Panama Canal. From 1914 through 1928, crews dug a network of ditches, canals, and levees across the Bootheel, in a vast swath a hundred miles long and from ten to twenty miles wide, running from Cape Girardeau to the Arkansas line. The district had an original construction cost of over $11 million and was at the time the largest drainage system in the world. It comprised almost a thousand miles of ditches and over three hundred miles of levees. Steam and electric draglines, floating dredge boats, and sweating day laborers moved more than a million cubic yards of earth—on a scale of environmental reworking surpassed only by the Panama Canal.[56]

FIG. 6.6. Regularizing the Great Swamp. In this image of a finished set of drainage ditches, a Little River Drainage District photographer documents an utterly new geometry of order successfully imposed upon the wild, jumbled, swampy terrain of the Missouri Bootheel by this massive early twentieth-century environmental intervention. The Little River Drainage District drew on the same machines, staff, and techniques as the Panama Canal. ("Ditch No. 81 at Jct 81 & 84 Looking North," item 3094 V8 023, Little River Drainage District Records, courtesy Special Collections and Archives, Kent Library, Southeast Missouri State University)

The LRDD remade the visible topography and water flow of the region in implementing two main changes. It diverted Ozark highlands water that formerly flowed into the New Madrid hinterland into more direct routes to the Mississippi. Further, the district replaced the irregular creeping fingers of the area's myriad small streams with straight drainage ditches at right angles to each other, roughly one mile apart, all ultimately draining into Big Lake. The Little River Drainage District cuts sharp lines across southeastern Missouri, in echo of visible earthquake faults like the San Andreas. It imposes rigid geometric order on the rumpled surface of a seismic zone.[57]

Similar if smaller-scale drainage efforts altered the topography of the southern New Madrid hinterlands in Arkansas. In 1893, local planters created the St. Francis Levee District in Arkansas. Throughout the next few decades, other drainage districts were organized throughout the much-disputed sunk lands. Taken together, the environmental interventions were immense. In 1939, the US Army Corps of Engineers installed the

FIG. 6.7. Newly cleared land, newly straightened waterway: an earthquake zone re-made. Standing as ghostly sentinels, deadened trees shape the stark geometry of the New Madrid hinterland drained and regularized by the Little River Drainage District in the early twentieth century. Unrecognizable as a seismic zone, this new environment resembled another landscape of modernity soon to become all too familiar: the trenches of the Great War. (From album, "The Ditches When They Were New," item 5-07, Little River Drainage District Records, courtesy Special Collections and Archives, Kent Library, Southeast Missouri State University)

world's largest siphons on the St. Francis at Marked Tree, Arkansas, to control flooding and further drain the sunk lands.[58]

Overall, sunk lands drainage was, in backers' terms, a tremendous success. Before 1907, over 90 percent of the land administered by the Little River Drainage District was covered in swamp. Afterward, nearly 96 percent was cleared. Taxable property in the LRDD's counties more than doubled. Currently the Little River Drainage District drains 1.2 million acres and encompasses roughly 540,000 acres across seven counties.[59]

Yet environmental transformation added to the increased social stratification of the region and its people. Despite the acclaimed regional success of these many projects, they worked against many smaller landholders, many of whom were burdened by drainage fees which aided large timber companies eager to gain value for land they had deforested. By the 1920s, many smaller landowners had become sharecroppers.[60]

Even as the draining of the sunk lands consolidated the economic and social power of landowners, it consolidated a rewriting of history on the land. Through his engineering and his advocacy, Otto Kochtitzky denied the influence of the New Madrid earthquakes on the area's topography. "No part of this valley," he asserted, "was either lifted or sunk by the New Madrid earthquake jars occurring during the years 1811 and 1812." Reelfoot Lake was not formed by an earthquake, Kochtitzky argued: it was simply a "partially filled channel of an old water course, possibly the Ohio River, abandoned in the early period of the upbuilding of the level plain now filling the eroded valley." In the period of early settlement of the southeast Missouri region, Kochtitzky argued, it "was all flat and overflowed swamp." A few squatters hunted and traded on the low, flat, wet, ridges, but little else could be done with the terrain. Nothing of significance had changed in 1811–12.[61]

Drainage interventions effaced evidence of seismic shocks. Along the Little River, baldcypress that lived through the quakes were subsequently killed by the dredging and leveeing of the LRDD, resulting in the loss of that source of tree-ring data. Kochtitzky himself reinterpreted signs on the terrain. Trees in the sunk lands, he argued, had not completely died in 1812. This was to Kochtitzky clear evidence that they had not been killed by sudden tremors. (Later dendrochronologists would point out the seismic hardiness of baldcypress, whose extensive root systems make them resilient to forces like wind and earthquake: even when damaged, they can live for decades despite compromised growth.) The flowing water of the swampland had long been noted to be clear-running and shallow. This feature of the sunk lands was interpreted by geologist Edward Shepard in a 1905 article in the *Journal of Geology* as evidence that sunk land water bubbled up from artisanal sources through apertures created by sand blows. Kochtitzky, however, cited an 1819 land surveyor's description of clear water as evidence that the surveyor had seen no trace of "earthquake effect."[62]

In particular, Kochtitzky asserted that the sunk lands were caused not by earthquake subsidence, but by subsequent "unusual seasonal or overflow condition." Surveyors in the 1850s who had described such swampland as "sunk land" were merely accepting "local legend." Fluvial, not seismic, processes could explain the region's entire topography. Deposition of Mississippi River sediment by Pemiscot Bayou, for instance, impounded the water of Little River swamp to create Big Lake in Arkansas. The lake was created by normal processes of overflow and sedimentation, not by any sudden subsidence. Kochtitzky reinterpreted previous geological evidence of the earthquakes as "pure romance." The sunk lands were not earthquake features, but "naturally formed swamps." As such, they were available for the

FIG. 6.8. Cotton field near Kennett, Missouri. This is a photograph of agricultural plenty—and of environmental transformation. Cotton bolls grow in the late '20s in fields wrested out of swamp by the massive Little River Drainage District. Almost a century later, researchers would investigate sand blows in fields such as this. (Item 15–06, from photograph album page entitled 'General views of reclaimed land today,' Little River Drainage District Records, courtesy Special Collections and Archives, Kent Library, Southeast Missouri State University)

kinds of interventions "reclaiming" wetlands throughout North America. Kochtitzky's history making was one precondition for the environmental changes he helped bring about.[63]

The widespread misapprehension of the region's environmental history, Kochtitzky argued, was due to the unwarranted reliance on "exaggerated, unreliable, and misleading" accounts. Kochtitzky reviewed not only original accounts from the early nineteenth century, but subsequent histories by Louis Houck and geologists like Shepard. Ironically, he quoted Houck's lament that the earthquakes had been much forgotten in order to argue that they *should* be further forgotten—he agreed with Houck's analysis, just not his lament. Kochtitzky remade the environments of the Bootheel and rewrote their history. One recent history of the Bootheel notes that the earthquakes did not change the main barrier to population inflow in the nineteenth century, the swampily overflowed topography. Yet earlier nineteenth-century sources make clear that earthquakes did not *change* this barrier to settlement: they *created* it.[64]

After Kochtitzky, the swamps had simply always been there. Popular consciousness paralleled this forgetting of the earthquakes' effects. Even geologists working the region can assume small lakes in the epicenter region are just seasonal overflow—before checking a map to see that Myron Fuller had marked them as earthquake features. Environmental characteristics shaped culture. The Bootheel region of southeast Missouri is often ignored and sometimes ridiculed: occasionally other Missourians have suggested giving the Bootheel back to Arkansas. Thad Snow would later famously term his home district "Swampeast Missouri." The self-image of southeast Missouri (as well as northeast Arkansas, which has no such handy moniker) is tightly bound with the cypress sloughs and muggy summers of its remaining swamplands. In personals ads in contemporary media, some people self-identify that way: one young man introduced himself to the online world in 2010 with "Hi, I'm swampeast." Swamps characterize regional self-conception, but not regional history making.[65]

Seismology and Changing Knowledge

As the former New Madrid hinterland was being reworked and reinterpreted in the late nineteenth and early twentieth centuries, so too was the entire study of the earth's movement. Seismology came of age in the same turn-of-the-century decades that the New Madrid sunk lands were being cast as ordinary and remediable swamps. Up-to-date science of European and Asian research centers might seem to have little to do with steam shovels in the humid undergrowth of Missouri Bootheel swamps, but both reshaped understanding of the historic earthquakes of 1811 and 1812. Broad, international changes to the science of moving earth had as profound an effect on the understanding of the New Madrid earthquakes as did the highly particular interventions of local boosters and developers.

In part, changes in the understanding of the New Madrid quakes rested on the diminished credence given to individuals' accounts of their own earthquake experiences. In the early nineteenth century, compilations of such narrative accounts were understood to comprise scientific discussion. Such compilations continued to be in the mainstream of scientific investigation well into the nineteenth century: leading researchers Richard Oldham and Robert Mallet oversaw similar, more extensive catalogues of earthquake accounts in the 1850s and '60s. Similarly, in 1869 geologist Nathan Shaler employed historical accounts to discuss the New Madrid earthquakes. But though Shaler was in the mainstream in 1869, his textual methods and approach would soon be secondary to the main instrumental thrust of earthquake research. By the turn of the century, transformations in technology and scientific models turned attention within the earth sciences elsewhere.[66]

Between the Civil War and the late twentieth century, two profound changes shaped the discipline of modern seismology: the instrumentalization of seismic observation in the decades surrounding the turn of the century and the reconceptualization of the earth's composition and movement in a stunningly swift period from the early 1960s to early '70s. Both of these revolutions created new views of earthquakes and seismic action. Both offered new insights. Both transformed and pushed forward earth science. Both led to a near total erasure of the firsthand, localized, anecdotal knowledge of the New Madrid earthquakes, as evidence not accessible to instrumental observation and not well explained by the plate tectonic model.

Moreover, both these changes to scientific understanding created a regionally differentiated understanding of seismology in North America. Instrumental arrays and theories of global tectonics sharpened focus on the American West Coast even as they blurred understanding of earthquakes in the Mississippi Valley. In the early twentieth century, as crucial seismic infrastructure and staffing were being established in California, advocates were unable to rally popular or scientific interest in the less instrumentally apparent seismicity of the New Madrid region.

Seismology in the United States was largely built up through study of the West Coast by governmental and secular educational institutions. By contrast, in the Mississippi Valley seismological efforts were much more limited, and they remained strongly marked by the Jesuit renaissance within the earth sciences and tied to Jesuit universities and networks. California seismology illustrates much of the main developments of global seismology. New Madrid seismology, by contrast, illustrates a history of institutional frustration, an inverse history of that which illuminating revolutions fail to reveal, that which advances in science serve to hide.[67]

Denial of Earthquake Threat on the West Coast and Continental Interior

In the late nineteenth and early twentieth centuries, denial of New Madrid seismicity was of a piece with denial of California earthquake history and earthquake risk. After a Northern California earthquake of 1892, San Francisco boosters sought to minimize the damage to their state's reputation. Attitudes toward California earthquakes in the late nineteenth century reflected both fatalism and denial: with little hope of being able to mitigate or protect against sometimes damaging California quakes, leaders and the general public minimized the severity of the problem and rebuilt quickly without altering patterns of architecture or engineering.[68]

One way of minimizing the earthquake threat was to emphasize the environmental hazards elsewhere. For this, the undeniably intimidating tornadoes of the continental interior were made to order. A hundred years of California earthquakes, editorialized the *San Francisco Chronicle* in 1892, were far less damaging than "a single cyclone in the Mississippi Valley." No one would think to compare the West Coast tremors to Mississippi Valley *earthquakes*. "One Western cyclone," opined the Sacramento paper that same year, "will do more damage than all the earthquakes California has ever known." Coming just as railroading entrepreneurs like Louis Houck set out to create new infrastructure in the former New Madrid hinterland and drainage advocates led by Otto Kochtitzky were beginning to rework its history in the service of massive environmental remaking, such emphasis on threat from tornados rather than earthquakes helped remake the perceived dangers of midcontinent.[69]

Then, in April 1906, a huge earthquake and subsequent fire devastated San Francisco, leaving the city center a smoking ruin. Media outlets around the country recounted the horror of the desolated area, but even as communities counted their dead, city leaders began to manage information and perceptions as well as aid. A concerted public relations effort by community leaders emphasized the calamity as San Francisco's "Great Fire"—a long-understood and potentially more controllable risk—rather than earthquake. This PR effort was a master stroke, influencing not just public but historical memory of the 1906 events over much of the twentieth century. A similar sustained effort rolled into work after a tremor rocked and damaged Santa Barbara in 1925. "Forget earthquakes!" urged the Pasadena *Star-News* to understandably jittery residents and investors. These well-managed and successful efforts had effects far beyond the Golden State. Past experience of Mississippi Valley earthquakes was erased not only by active efforts to forget them, but by efforts to forget *California* quakes.[70]

Even after the 1906 San Francisco quake, California boosters used midcontinent tornados as an effective foil. If the Mississippi Valley was a zone of cyclone hazard, then the occasional West Coast tremor was not especially alarming by comparison.[71] A 1925 editorial cartoon made fun of Midwestern rubes clinging to church steeples as a tornado roared past to destroy the countryside: these are the people, small-minded and ignorant, who would fear *earthquakes*? When a successful 1939 film version of the popular book *The Wizard of Oz* had a young girl named Dorothy and her charming dog Toto carried bodily away by a Kansas tornado, the environmental differentiation was complete. Newcomers might have to cope with the occasional earth tremor in the booming, golden West Coast, but there was certainly no safety in staying stuck in the mud of the tornado-ravaged continental interior.[72]

Meanwhile, increasingly well-informed experts on California seismology looked on with dismay and even indignation at this rewriting of seismic hazard. For American seismology, the 1906 San Francisco quake was a foundational event. The Seismological Society of America was founded in 1906 after the great San Francisco quake, and began publishing its *Bulletin* regularly in 1911.[73] Public discussion of seismic events reflects a profound early twentieth-century irony: earthquakes were increasingly denied in the American public sphere even as they came into clearer focus among those with expertise. Whereas in earlier eras, expert and lay knowledge were in close communication, by the early twentieth century public and professional perspectives reflected sharp divisions in these ways of understanding earthquakes. This divergence would become even wider with respect to perceptions of Mississippi Valley seismicity.

Seismic Devices and the Birth of Modern Seismology

Modern seismology was birthed in the last decades of the nineteenth century and the first decades of the twentieth as researchers developed and refined seismic sensing instruments and created techniques to interpret those instruments' recorded data. Instrumental improvement brought about a truly global science of observation in a relatively short and exciting period of technological and institutional development. This instrumental genesis straightforwardly revealed much about the seismic activity of many parts of the globe, but would have ironic consequences for holding back understanding of the New Madrid earthquakes.

The late nineteenth-century triumphs of seismic instrumentation built upon slow changes in the study of earthquakes since the middle eighteenth century. Spurred by the massive earthquake that devastated Lisbon in 1755 and a series of small earthquakes felt throughout Britain in 1750, researchers in Europe and especially Great Britain sought to learn more about earthquakes in the late eighteenth century. At that time, knowledge of earthquakes still drew from the ancient notion that the earth's surface was being disrupted by activity in subterranean chambers tunneling the globe, but new sciences offered potential insights: perhaps earthquakes were the surface consequence of subterranean explosion according to the newly developing principles of chemistry, or perhaps the electrical fluid had something to do with the mysterious and powerful movements of the earth. Researchers became much more specific in their inquiries into exact experience and evidence, giving precise and elaborated accounts of the effects of earthquakes such as the one that devastated the Italian region of Calabria in 1783.[74]

Through the middle nineteenth century, researchers undertook ever more specific investigations of particular earthquakes and developed ever more extensive catalogues of historical earthquakes. Seismological research took on a new dimension with energetic experimentation led by Robert Mallet, a creative British engineer captivated by the dynamics of earthquakes and their effects on different materials. Beginning in the late 1840s and continuing into the early 1860s, he pioneered the use of artificial explosions to study wave propagation in materials ranging from wet sands to granite outcrops. The precise measurements obtained in this kind of experiment allowed for increasing mathematical analysis of earthquake wave propagation, and indeed an increasingly mathematical way of understanding earthquakes.[75]

As in the New Madrid earthquakes, dramatic seismicity in many times and places—from first-century China to enlightenment Italy—spurred innovators to create first seismometers (devices to sense seismic movement), and then seismographs (devices which recorded that movement).[†††††] By the 1840s a number of devices had proved both promising and frustrating in various European centers, and the British Association for the Advancement of Science established a committee to create some sort of reliable, self-recording instrument. A flurry of development and improvement resulted in functioning seismometers in Göttingen and Prague by the late 1840s.[76]

Seismology was changing during roughly the same period that the former New Madrid hinterland was being reworked. The breakthrough of

---

††††† These various and creative efforts would then spur exacting historians to extensive if not very productive debate about priority and bragging rights. As with many aspects of the history of science and technology, assertions of the *first* seismometer or seismograph tend to get fuzzy upon closer inspection. Was it indeed the Chinese instrument operated from the first through the sixth centuries, a ceramic vessel with dragon-headed spouts dropping weighted balls into the mouths of surrounding ceramic toads? Eighteenth-century European bowls full of mercury whose sloshing indicated slight tremors? What serves as a true recording device? And do mere proposals count? From this somewhat quarrelsome literature emerges a consensus that in many places and times, useful devices like sloshing bowls were independently and concurrently developed—the horizontal pendulum in particular appears to have been invented several times in the nineteenth century—and that an intense period of instrumental tinkering emerged from lively exchange among European academic centers in the late nineteenth century during a truly remarkable period of technological innovation.

reliable, reproducible, self-registering, commercially available technology came in the 1880s. In 1880, researchers secured the first demonstrated, lengthy seismograph record of earthquake motion as a function of time, and commercially available seismographs were available later in that decade. The transformative potential of these new instruments began to be realized in April 1889, when Ernst von Rebeur-Paschwitz demonstrated in a letter to *Nature* an apparent correlation between signals his devices had received in German stations and a simultaneous earthquake in Japan.##### Drawing on these precedents, main developments of modern seismometry took place in the foment of intense work in earthquake study between about 1880 and the Great War, among an international group of researchers based especially in Italy and Japan.[77]

Leading researchers rapidly constructed networks of reporting stations with increasing coverage and standardization. This emerging discipline of seismological observation was part of a movement for international cooperation in scientific observation. By the late 1890s, despite various national suspicions, researchers had established a number of venues for international cooperation and had to a large degree standardized seismological research. By the turn of the century, seismographs were set up on each inhabited continent, and in 1902, hardy observers even managed several months of measurement from Antarctica. In the first decade of the new century, the International Seismological Association could boast publications, newsletters, communication, and a growing earthquake catalogue generated by an increasing number of seismological observation stations and increasingly sophisticated instruments.[78]

By the end of the nineteenth century, most researchers moved decisively away from earlier forms of knowledge, in which the body served as instrument, personal narratives were scientific documents, and improvised devices helped individuals record tremors. Researchers in Europe, Asia, and to a lesser degree in North America studied particular earthquakes, often through structural damage to buildings as well as changes to landscape; they catalogued historical earthquakes over long periods in several parts of the world; and they began to record and analyze seismic waves through seismic sensing equipment. Engineers began to design buildings to withstand earthquake shaking. Rapid technological innovation and improvement linked observers across the globe and gave a quantified basis for mod-

---

The perhaps unsurprising irony of discovery operates here: it is now unclear whether the amplitudes Rebeur-Paschwitz recorded were caused by the Japanese quake. Yet his demonstration of teleseismic observation was convincing and galvanizing at the time, and his report spurred further intense development.

ern seismology. Together, these developments ushered in an era of modern seismology by the turn of the century.[79]

Consolidating Expert Knowledge of the Earth

In the United States, developments in the subfield of seismology paralleled the consolidation and institutionalization of American science more broadly. After the Civil War, scientific endeavor in the United States was marked by elements still familiar from the early nineteenth century. Local observers in many places and parts of society struggled to contribute to a larger set of conversations that often seemed far from them and at times unreachable. Yet much changed over the last decades of the nineteenth century. The cause of "pure science"—self-consciously noncommercial investigation—gained momentum, along with the use of the term "scientist." In a move of lasting institutional significance, a group of investors in 1876 founded the Johns Hopkins University in Baltimore explicitly on the model of European, especially German, institutions. Hopkins was the vanguard of halting but influential efforts to reform academic and scientific work, emphasizing the new laboratory sciences over the more local and commercially oriented scientific pursuits of an earlier era. Much of the research in the earth sciences still took place under the rubric of commercial prospecting and for-profit endeavors, but researchers were increasingly supported by professional associations and trained at schools such as the Massachusetts Institute of Technology and the Columbia School of Mines. In 1879, the US Geological Survey was created out of the consolidation of the four previous federal surveys of the trans-Mississippi West. By the early twentieth century, American seismology was one of a number of sciences coalescing into modern disciplinary form.[80]

The regional focus of American seismology was dramatically determined by the 1906 San Francisco earthquake. Institutions and research questions focused on the West Coast. Curiosity, money, and equipment all followed this devastating and famous quake. Researchers installed networks of seismographs in Northern and Southern California in the late 1920s. Before too long into the twentieth century, to be a seismologist in the United States meant to work in California.[81]

Moreover, to be a seismologist was to operate and interpret seismographs. Twentieth-century study of earthquakes was accomplished through instrument and through visual representation. The jagged peaks and valleys of seismograms—the long paper record onto which a seismograph needle scratched a record of the earth's movements—substituted for the word pictures of heaving earth or waving trees as researchers sought the origins of particular quakes and of earthquake phenomena generally. Such visual language represented an abstraction of information very different from the narrative accounting of horror, prayer, grief, and panic that had long characterized disaster reporting in many times and cultures.§§§§§ The fundamental evidentiary base of seismology shifted, in broad terms, from words to numbers and graphic traces in the early twentieth century.[82]******

Such instrumentation and institutional cooperation held powerful consequences not only for how earthquakes could be observed, but who observed them. By the first few decades of the twentieth century, many researchers had become skeptical about the role of public observation. In the North American context, seismology was similar to other sciences in an increasing sense that only expert observations could be useful. New Madrid drainage engineer Otto Kochtitzky expressed distrust about the ability of untrained observers accurately to report earthquake events. He traveled to Charleston, South Carolina, in 1887 to investigate the effects of that city's devastating earthquake as part of his emerging investigations about the earthquake history of the New Madrid region. Kochtitzky had taken particular care to investigate the fissures and crevices reported in Charleston. In several places where the largest fissures had been reported, fissures supposedly one to three feet wide, he found no crevice yawning more than an inch. For Kochtitzky, the conclusion was simple: "Those good people simply couldn't tell the truth about the earthquake which gave them so great a fright." His skepticism about local observers' accounts in Charleston reinforced his dismissal of past local observers' accounts of the creation of the New Madrid sunk lands through earthquake subsidence.††††††[83]

§§§§§ In September 1923, a Sydney newspaper reprinted a portion of a very early earthquake record from the Australian Jesuit observatory, along with the telling commentary, "That zig-zag line, recorded on Dr. Pigot's seismograph, means disaster to Japan. Reduced to lines, there is nothing about it, to show that it is the mark of a nation in mourning."

****** At the same time, visual images are no more intrinsically self-evident than any other kind of evidence: despite the images some of us may remember (dimly) from geology classes—the images of nice clear P and S waves caused by earthquakes—the actual raw evidence of seismographs is a *mess*: jaggedy waves all over the place in a seeming tangle of overlapping indicators. Only very careful and painstaking work can resolve these traces into clear, coherent patterns.

†††††† Contemporary photographs do not bear out Kochtitzky's dismissal. One image from after the Charleston earthquake, for instance, shows a crater easily as wide as and far longer than the man seated next to it on the ground.

To many in the newborn discipline of seismology, networks of instruments seemed to make networks of observers obsolete. After the 1906 San Francisco quake, leading American seismologist John C. Branner wanted to establish an observer corps of reliable locals who would record and send in their physical experiences. In the new *Bulletin of the Seismological Society of America*, he and fellow seismologist Harry O. Wood championed the notion of an observer corps. Yet when an earthquake struck central California on 1 July 1911, Branner was dismayed by the disinterest of his peers in utilizing just such a network of observers. Branner himself was in Brazil for the summer, and in his absence no one in California bothered to collect the impressions of people in the impacted zones.[84]

When he returned from South America, Branner and his graduate student E. C. Templeton did indeed use extensive local interviews, along with investigation of damage to environments and structures, to identify the epicenter of the earthquake. This signal success was marred by their frustration at the failure of informal, nongovernmental efforts to provide more prompt and widespread local perspective. Branner subsequently sought to get the US Weather Bureau to add earthquake reporting to duties of weather observers (which in 1914 it did), but his efforts to recruit local observers and to make prompt field observation central to seismological research met with sustained passive resistance from scientific peers as well as from a public interested in ignoring or minimizing earthquake threat. Only with great effort were Branner and the Seismological Society able to build a more institutional and ongoing commitment to local observation.[85]

Such shifting in who was a source was profound, and reflects an increasing divergence of expert perspectives from local ones on the New Madrid quakes. In a 1905 report, geologist Edward Shepard ratified the knowledge of "the countryman" of the region, noting that a fault scarp near Gayoso, Missouri, had indeed produced "the place where the land sunk." Yet by 1905, "countryman" knowledge was no longer consistent with most scientific assessment of the New Madrid hinterland regions. The evidence of the earth had begun to be insufficiently convincing, in part because of changes to the face of the countryside, but also because of larger shifts in who constituted a credible observer of scientific fact.[86]

Rather than observation by happenstance, dependent upon individual initiative, earthquake science required coordinated effort by large groups working under central direction. In several places, leading seismologists of the early twentieth century attempted to standardize, attune, and interpret the firsthand, experiential reports of people who felt earthquakes. Overall, the increasing instrumentalization of seismology changed what it meant to have a network of observers. Rather than those who happened to be attentive at the particular moment when an earthquake struck, observers had to be prepared, trained, and well coordinated, with equipment, spaces, and routines prepared ahead of time.[87]

In a parallel and ultimately more lasting development, seismological observation stations gained staff, routines, and whole cultures over the course of the twentieth century. To maintain, much less read and compare, often delicate instrumentation required teamwork over long periods, routinization of methods as well as machines, increasingly specialized spaces to anchor and operate seismological sensing equipment, and increasingly specialized staff to tend to and interpret it, from lab techs to PhD students to departmental secretaries.[88]

In California, such institutions were generally secular, associated either with universities or with research institutes. By contrast, Mississippi Valley seismology failed to garner the institutional support or technological infrastructure of the West Coast, and it was long marked by Jesuit history. For many decades, Jesuit observers and institutions were the main nodes of interest and research on midcontinent seismic movement. One key figure, the Reverend James Macelwane, attempted to establish in the New Madrid region the structures, techniques, and arrays that were shaping seismological knowledge in California. For the most part, he met with failure: the same global changes in seismology that promoted work on California seismicity flummoxed research into the midcontinent.

The Frustrations of Father Mac:
The Mississippi Valley Was Not California

As researchers strung together and staffed observational networks along the West Coast, James Macelwane struggled to do the same for the New Madrid region. His largely unsuccessful efforts reveal the different patterns of institutions and research that shaped the divergent course of the two regions over the twentieth century. His work also reveals a separate, often ignored thread in the global development of seismological science.

Much initial North American seismological instrumentation was created and implemented through Jesuit stations. Jesuit physics professor Frederick L. Odenbach at John Carroll University in Cleveland pioneered seismographic instrumentation in the United States and in 1909 developed the Jesuit Seismological Service, a network of nineteen observation stations across the United States and Canada. It was the first such network in North America and at the time the only continent-wide network employing uniform instrumentation. Cooperation, however, proved difficult to sustain, and the network was short-lived, ceasing in 1922. It would be revived three years later in response to efforts out of the midcontinent Saint Louis University.[89]

Founded in 1818 as the first university west of the Mississippi, Saint Louis University carried into the Mississippi Valley the emerging Jesuit focus on the observational sciences of earth and sky. The same emphasis on international and far-flung networks and a reliable, educated, disciplined observer corps that enabled the international missionary order of the Jesuits to lead in meteorological study also positioned them to shape early seismology. SLU###### was an early center of weather study in the continental interior: two Jesuit professors from the theology and philosophy program began sending meteorological observations to federal government observers in 1860. (Thus Saint Louis University is one of the main sources of the pronouncements of "first/wettest/hottest/longest west of the Mississippi" or "since meteorological record-keeping began" that are a constant theme of St. Louis civic life.)§§§§§§ Later, the school was proud to host the first department of geophysics in not just the trans-Mississippi West, but the Western hemisphere: created in 1925, the New Madrid region was a main program focus.[90]

The founding head of that program, the Reverend James B. Macelwane, SJ, was for much of the early- to mid-twentieth century both a successful leader in American seismology and a frustrated advocate for more in-depth regional seismic research on the midcontinent. Macelwane grew up in Ohio, joined the Jesuit Order at twenty, and earned undergraduate and two master's degrees at Saint Louis University in the early 1910s. He was assigned to help disassemble and repair the SLU seismograph, a Weichert inverted pendulum seismograph installed in 1909 in the basement of a university building.[91] That endeavor led to the first of his 133 publications. Macelwane taught physics at SLU for several years before moving to California for further graduate training. He earned a PhD at the University of California, Berkeley, in 1923 in physics with a seismological dissertation—the first such degree awarded in the United States.

Despite higher-ups' occasional efforts to make everyone instead sound out "S-L-U" in what is presumably a more dignified way, locals invariably and familiarly refer to the local landmark institution as "Slew."

§§§§§§ Want to know when the first formal Episcopal church service west of the Mississippi was held? Local historians will tell you! (October 1819.)

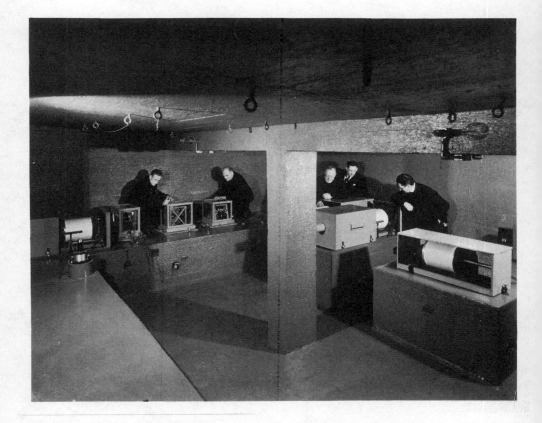

FIG. 6.9. Jesuit seismologists tending instruments, Saint Louis University, in the 1920s. This is a picture of the height of scientific cool: to be a seismologist in the early twentieth century was to work with seismometers. This is also a picture of intellectual aspirations: researchers led by the Reverend James B. Macelwane (second from left) attempted to establish seismic arrays in midcontinent parallel to those of the West Coast. They were largely unsuccessful: California, not Missouri, became American "earthquake country." Only in the late twentieth century would a new generation of researchers and a new generation of instruments again establish the seismic consequence of midcontinent. (Courtesy Saint Louis University Earthquake Center and J. B. Macelwane Archives, Saint Louis University)

Macelwane's California work was foundational for his efforts in the middle Mississippi Valley. In California, Macelwane and California seismologist Bailey Willis initiated a program of closely spaced seismic stations in an active earthquake zone, a program that eventually implemented a dozen stations. Macelwane was later to push for a similar monitoring array in the

New Madrid region, but was never able to muster the public and institutional support to fully implement it.[92]

In 1925 Macelwane returned to St. Louis as a SLU professor of geophysics and inaugural director of the new Department of Geophysics. He worked at the school until his death in 1956. As researcher, educator, and tinkerer, "Father Mac" introduced students to the work of making and reading seismograms and caring for seismic measuring equipment. Macelwane reorganized and revitalized the Jesuit Seismological Service in 1925 as the Jesuit Seismological Association, the eastern answer to the predominately western-US Seismological Society of America. The SLU department installed Wood-Anderson seismographs at their own array in 1927.[93]

Macelwane worked tirelessly to bring emerging California-style earthquake awareness into the Mississippi Valley. Almost alone in his era, he was a public advocate for study of New Madrid seismicity and for civic planning for risk reduction. In particular, he worked to expand instrumental networks. Macelwane advocated among seismic experts, Missouri-area industries, and the St. Louis public for a seismological network of "Ozark Earthquake Investigation." As Branner did in California, Macelwane and his colleagues sought to organize a "postal card survey of intensity distribution" to study continuing New Madrid area seismicity. He used his local prominence to push for greater funding for seismological instrumentation, staffing, and research; he also used it to try to increase public awareness and earthquake planning. Following the implementation of the Northern and Southern California seismograph networks of the late 1920s, SLU began operating Wood-Anderson seismographs in the 1930s in Little Rock, Arkansas, and Cape Girardeau, Missouri.[94]

What work there was on New Madrid existed largely because of Macelwane's advocacy. His work was recognized in his field, and the National Research Council gave him responsibility for the study of New Madrid area quakes. He gained the cooperation and support of the Carnegie Foundation and the US Coast and Geodetic Survey. He worked with colleagues on a catalogue of known earthquakes in the region. In the 1930s, Macelwane argued for seismically sound building practices in Missouri, Tennessee, and the middle Mississippi Valley, as well as augmented instrumentation. By 1934 Macelwane could optimistically speak of all recent progress in research on midcontinent seismology.[95]

Yet Macelwane's efforts never came to fruition. The "preliminary sketch" of Missouri's seismic history that he started with a collaborator apparently never progressed into a completed work; seismic monitoring in the region never approached the close monitoring that Macelwane had helped implement on the West Coast. Throughout his work, seismological study in the midcontinent was hamstrung by technical limits. Early and midcentury earthquake locating depended upon comparing reports from several recording stations, but the number of recording stations in the middle Mississippi Valley remained too small to effectively locate the source of most earthquakes.******* 96 Macelwane's advocacy was never enough to rouse much public response or success in funding. The most instrumentally promising regions were those of frequent large-scale seismicity. The West Coast was by midcentury clearly the place to invest more seismic monitoring and research, not the apparently quiet reaches of the middle Mississippi Valley.

Further, for most of the twentieth century the primary earth science investigations of the mid-American continent were taken up with the search for coal, oil, and gas reserves.97 In the context of rapidly developing insights yielded by fossil fuel exploration and by earthquake study centered on the West Coast, the midcontinent seemed neither seismically interesting nor economically rewarding. Researchers looked forward, to the increasing precision and abundance of instrumental data, and outward, to the West Coast's recent, demonstrably severe earthquakes. Almost no one looked back, at the outdated and seemingly outlandish reports of cataclysm in the midcontinent.

James Macelwane's contributions to teaching and research in seismology were honored by his professional peers—notably, in the 1962 creation by the American Geophysical Union of the Macelwane Medal to honor outstanding young researchers (several researchers in New Madrid seismicity have received this award). Yet Macelwane's work on New Madrid seismicity quickly faded to professional and regional obscurity, submerged beneath the larger currents of interest in his professional field. In 1973, SLU seismologist Otto Nuttli reintroduced the quakes as a subject worthy of scientific study. His landmark publication marked the slow resurgence of interest in New Madrid. But it took the better part of two decades for researchers to follow up with more widespread and detailed seismological study. Over the latter twentieth century, multiple developments within seismology shunted

------

******* In a particularly frustrating technical failure, the records in the Wood-Anderson seismograph were being changed just as an earthquake occurred in December 1930: the local quake was not captured by the instrumental array put into place precisely to gather data on such events. Presumably even highly disciplined Jesuits might have employed some salty language at this exasperating failure.

to the side any research into noninstrumentally recorded earthquakes in what seemed to be a stable continental interior.[98]

Seismology at Midcentury

Meanwhile, in the middle twentieth century seismology became part of postwar "big science," defined by massive scale, governmental funding, and priorities of national defense. A range of institutions buttressed theoretical and technological changes within the earth sciences. In 1922 the Scripps Institute of Oceanography was founded, to be joined in 1930 by the Woods Hole Oceanographic Institute. Less formally, but with equal significance, the development of American and international oil industries underscored the importance of seismology to economic and hence national defense priorities. In 1954, the Jesuit seismological network detected the Bikini Atoll hydrogen bomb test: a sixteenth-century religious order revealed the post-atomic arms race. Global seismological networks proved key to the detection and verification of US/Soviet test ban treaty negotiations begun in 1958. The need for global nuclear safety thus contributed to the expansion and linkage of seismological techniques and observation stations: brought of age by these postwar tensions, seismology came to define itself as concerned with the present state of the world, unlike other deeply historical subfields within the earth sciences. Local observations of a particular American backwater several hundred years before were less connected to main modes of seismological research.[99]

Jesuit networks and stations helped establish this modern seismological observation. Until the early 1960s, the SLU Central Station continued to amass and publish data and locate earthquake epicenters, complementing reports by the US Coast and Geodetic Survey and international seismological organizations. Eleven Jesuit stations formed part of the 125-station World-Wide Standardized Seismograph Network established and maintained by the US government beginning in 1962 (this network immediately became known, like all big science in the post-WWII world, by its acronym, WWSSN). Later, in 1974, SLU installed the first stations of the Upper Mississippi Valley Seismic Network, under the sponsorship of the US Geological Survey.[100]

Yet larger changes to seismology slowly made Jesuit networks obsolete. Modern techniques of data capture have rendered multiple independent observational points unnecessary. Since 1970, most Jesuit seismological stations have closed. Moreover, in the late 1960s seismology convulsed with the second of its major upheavals—this one just as profound as the move toward instrumental data.[101]

Plate Tectonics and Revolution in Earthquake Models

In the 1960s, plate tectonics abruptly turned the world of earth science up-side down and put it together in a different way. By the late 1960s, the discipline of seismology had developed key elements that would enact the change to a model of plate tectonics: the World-Wide Standardized Seismograph Network; the increasing availability of high-speed computing; and the ability, through WWSSN seismograms, to study P (longitudinal or, colloquially, primary) waves. Through a series of dramatic publications, debates, and encounters, conventional wisdom decisively shifted between about 1962 and 1973.[102]

The essential insights of what would become the theory of plate tectonics were originally forwarded by Alfred Wegener as a theory of continental drift. Recognized for his work in meteorology, paleoclimatology, and polar exploration, Wegener took inspiration from the jigsaw-puzzle look of a globe in a colleague's office to argue beginning in 1910 that the continents had once formed part of a whole landmass. By midcentury Wegener's theory had gained some adherents, especially among biogeographers seeking to explain close connection between species on otherwise far-separated areas such as South America and southern Africa, but for decades it remained a debated and sometimes ridiculed theory. Studies of land-based paleomagnetism gained the theory of continental drift some adherents in the early 1950s, but it was primarily studies of the seafloor that made earth scientists welcome and indeed embrace Wegener's fundamental idea.[103]

Plate tectonics coalesced over a very short period through the synthesis of instrumental data from several formerly undeveloped and unconnected subdisciplines within earth science. World War II and the funding and equipment of national defense in the United States proved crucial. Wartime developments in ocean exploration and remote sensing created a boom of data after the war. In the first years of the 1960s, several researchers advanced theories of how new seafloor was being created in a slow process of spreading out from midocean trenches. These approaches linked midocean ridges and trenches in a dynamic system and explained these features as part of a cycle of convection currents which carried continents with, not through, the ocean crust. By the mid-1960s, researchers had a much more complex image of the dramatic ridges and trenches on areas of the oceans' floors.[104]

Moreover, data on the seafloor collected during the recent global war and declassified in the 1950s yielded increasingly detailed maps of magnetic anomalies, especially the curious pattern of an area off the Juan de Fuca

Strait that seemed to be striped with one magnetic pattern and then another in parallel and repeated rows that were roughly symmetrical across an apparent seam at a midocean ridge. In 1963, Fred Vine and Drummond Matthews at the Cambridge Department of Geodesy and Geophysics tied together seafloor spreading and paleomagnetism in a speculative explanation of these parallel stripe patterns: the stripes were, they argued, evidence of new crust having been created from seafloor ridges and flowing symmetrically toward each side. In 1965, Vine and Wilson applied these new theories to the patterns and structures of the Juan de Fuca Ridge, arguing that the magnetic zebra stripes were an artifact of successive episodes of seafloor creation and spreading that took on the successive magnetic directionality of the earth's poles in the eras of their creation and then "froze" in that direction.[105]

The meeting of the American Geophysical Union in April 1966 was electrifying for advocates of the new understanding of crustal development and movement. Many researchers became convinced of this newly elucidated theory of continental drift as shown by evidence from the ocean floor. In 1967 Dan McKenzie articulated this new understanding as the "paving stone theory" of "plate tectonics," in which massive, jagged tectonic plates fit together more or less like paving stones, and evidence from a variety of subdisciplines soon began to provide confirmatory evidence.[106]†††††††

By the early 1970s, plate tectonics was becoming the new orthodoxy. This new near consensus was most powerful and most important within the newly named and interdisciplinary "earth science" community, but it is perhaps best glimpsed in its broader impacts. In 1972, Golden Press— publisher of the brightly colored, deeply beloved series of "Little Golden Books" that have introduced generations of American children to simple, illustrated versions of *Heidi*, *Black Beauty*, and the *Pokey Little Puppy*—published a "Golden Science Guide" to geology. This version of earth science included paleomagnetism and seafloor spreading. A revolution was made complete.[107]

†††††††Plate tectonics is a fascinating example of just how instrument-driven seismology had become: instrumental readings of ocean floor magnetism became "seeable" as evidence of seafloor spreading, allowing a number of smart experts to accept an account of the world that differed radically from their own sense perception—who after all can see or feel the earth's tectonic plates moving, except in dramatic moments of fault rupture?

Suddenly and abruptly, scientists understood what caused most earth-quakes in most parts of the world: vast tectonic plates, grinding, colliding, and bumping up over each other. Eureka! Yet the exciting, rapid development of the plate tectonic model left little interest in seismicity far from a plate boundary. The Mississippi Valley is in the middle of a tectonic plate. The theory of plate tectonics does nothing to explain why earthquakes happen there.[108]

From a contested, even scorned, theory, the notion of moving continents became—relatively suddenly—hegemonic knowledge. By the late 1970s, the theory of plate tectonics had become the dominant model to such an extent that researchers had difficulty seeing evidence that failed to fit the plate tectonic model—such as major earthquakes smack in the middle of a tectonic plate. Attention, funding, and research turned to all that could be explained and revealed with the revolutionary new model of tectonic plates. The swamps of the middle Mississippi remained a scientific wasteland.

Iben Browning and the Collective Sheepishness of False Prediction

In 1990, a specific episode of false prediction raised New Madrid earth-quake risk into the news only to drop it far and fast, making New Madrid science even more of a fringe endeavor. The elements of credibility of this episode stem from long-standing popular and scientific interest in earthquake prediction. In 1925, erroneous reporting led to the widespread notion that Bailey Willis had successfully predicted the Santa Barbara quake. (He had certainly given general warnings about the high likelihood of California earthquakes, and he happened to be in Santa Barbara the day the quake struck—all of which led several newspapers to report that he had specifically predicted the tremor.) By the 1970s, international seismology was cheered by optimism that researchers would soon start to predict and understand, even if not control, seismic upheaval. Such work was given considerable impetus by the apparently successful prediction in 1975 of the Haicheng earthquake in China. From the point of view of serious seismological research in much of the West, this optimism with respect to earthquake prediction was a brief period that soon passed.[109]

But within larger American culture, 1990 was a moment at which anxiety to predict earthquakes was at a high point. In October 1989, as Americans across the country were settling onto sofas and into sports bars to watch game 3 of the World Series of Major League Baseball between the Oakland A's and the San Francisco Giants, a 7.2 earthquake tore up the coast toward San Francisco, rippling the bleachers of the stadium where the ballplayers were warming up. The Loma Prieta quake, as it was subsequently named, shut down the Bay Bridge in San Francisco, brought down the elevated Embarcadero freeway along the San Francisco waterfront, killed 63 people, injured close to 4,000, and caused an estimated $6 billion in damages. ╫╫╫╫╫╫ It was the first earthquake whose main shock was broadcast live on national TV—cameras in the stadium just south of San Francisco filmed as the World Series crowds screamed and ducked in the bleachers and then, after surviving the first shock, sent up a defiant cheer. The widespread media coverage and significant damage to a photogenic city made Americans notice when, shortly thereafter, someone with apparent authority began to predict increased risk for a seemingly quiescent earthquake zone.[110]

In 1990, Iben Browning, a one-time meteorologist with a sideline in earthquake prediction, warned of a repeat of the New Madrid quakes for 3 December 1990, a day when a full moon would result in particularly high tides and, he argued, high gravitational pull on the earth's tectonic plates. This prediction became a regional and even national event in part because of surprising support from a seemingly credentialed local expert. David Stewart, a geologist and director of the Center for Earthquake Studies at

---

╫╫╫╫╫╫ One thing about working on earthquake history is that everyone who has been in one has to make sure and say that. Well, I was in this one. Heard it coming up the coast, even. I had just graduated college in the midst of recession and was writing copy for an insurance company on the seventh floor of a building on Market Street in downtown San Francisco. One afternoon, a salesman in a nearby office suddenly hollered into the phone, "What do you mean, it's shaking? What's shaking? WHAT'S SHAKING?!?" As those of us in the office turned to him, suddenly we were shaking, too: he had been on the phone with a client down the coast, who had felt the shocks before we had. I walked home through a sea of broken glass, through one busy intersection snarled when the power blackout knocked out the traffic lights; a homeless man had parked his shopping cart and was directing traffic safely through ruins. Since all the rooftop chimneys had collapsed, I had an excellent view from my apartment on Chestnut Hill of smoke from the burning buildings in the Marina, a district built on landfill where liquefaction had destabilized foundations. The only phone number I could reach through jammed connections was the dorm room of my brother, off at his first semester of college. I wanted to ask him to pass along to our family that I was all right, but worried about my four housemates (thankfully, it turns out they were all fine, but scattered throughout the Bay Area—it was days before we all saw each other again). My brother was in class, but one of his Harvard roommates answered. In an experience of communication utterly the inverse of the nineteenth-century relationship between local and far-off sources of information, I stood looking out my window, speaking across the continent to a person I had never met, asking "OK, I can see smoke to the north and west, what's that?" at which my brother's roommate would flip TV channels around to the various news shows running live coverage to try to explain the columns of smoke I could see rising from the landscape around me, as parts of the Marina district burned under a clear blue sky.

Southeast Missouri State University, announced Browning's prediction at a teachers' earthquake workshop in Arkansas in June and at presentations in Indiana and Missouri later in the year. Though already regarded with skepticism by many within the earth sciences for his unorthodox approaches to seismic study (he had once hired a psychic to fly over the North Carolina coast with him), Stewart had been referenced at a congressional hearing the previous November as an "eminent expert on earthquakes," and his advocacy for more earthquake preparation was introduced into the formal record. He was not a character of the fringe, though he would become a fringe character after the Browning episode.[111]

David Stewart's advocacy played a main role in legitimizing Browning's prediction. Media outlets initially reported that Browning had accurately predicted prior earthquakes. Though these inaccurate reports were later corrected, they shaped public perception. As Stewart told the *Dallas Morning News*, "Here's a man who verifiably has hit several home runs, and he's up to bat. . . . You can't ignore the batting record." Throughout the fall of 1990, Stewart's refrain, as he told the *New York Times*, was the seemingly conservative "Will he hit another on December 3? We don't know, but there's no excuse for not being prepared." Stewart was the only scientific figure to speak in even partial endorsement of Browning, but his credibility lent him authority: he was quoted more often than Browning himself. In media coverage, a well-spoken scientist speaking for the *pro* side seemed a reasonable balance for geologists on the *con* side: the public saw a disagreement between experts, not an outlier opinion ranged against a bulwark of scientific consensus. Later surveys found that residents in the New Madrid region did not understand that scientists overwhelmingly rejected Browning's warnings: nearly half thought that scientists were divided in their reception of this prediction.[112]

Meanwhile, legitimate experts struggled to find the vocabulary in which to speak in general terms about earthquake preparation without giving undue weight to this specific alarmism. Most earth scientists in the region chose simply not to respond to Browning's advocacy, or tried to emphasize the need for earthquake preparation in a more general sense. A working group of the National Earthquake Prediction Evaluation Council convened to consider Browning's prediction and issued its report on 18 October, clearly and unequivocally repudiating his predictions.[113]

Such measured words were swept aside as a flurry of fear and preparation swept central and southeast Missouri.§§§§§§§ The *St. Louis Post-Dispatch*

§§§§§§§ A usually sober and reliable academic colleague of mine claimed to have actually seen in the back of a departmental coat closet the crowbar that the administration of

ran ads for "earthquake kits" in preparation, and government agencies found themselves overwhelmed with calls from worried citizens. Media vans crowded the several blocks of the New Madrid downtown, jostling for something to photograph. This was, after all, sweeps month, when stations were desperate for high viewership ratings. New Madrid itself went from small town to circus. The New Madrid Museum created commemorative t-shirts, while a local diner offered "quakeburgers." National reporters scoured streets looking for people with sufficient local color or sufficiently convincing rural accents. St. Louis supermarkets reported runs on dry goods, while hardware stores ran out of generators. Schools in the area closed or made attendance optional.********[114]

Then, on the appointed day, in as dramatic a way as nothing can possibly happen . . . nothing happened. Media vans decamped unceremoniously; officials looked sheepish. The aftermath: an angry public demanded answers from scientists who tried awkwardly to emphasize that they were not the ones who had said Browning was right in the first place. Earthquake risk in the flat floodplain of midcontinent became something ridiculous, a risible punch line.[115]

Within seismology, there was immense concern and frustration among regional experts that the job of figuring out actual risk and urging reasonable preparation had just gotten immensely harder. Earth scientists in the region registered profound frazzlement that someone who had not actually been an expert had been taken for one by the public, with resultant chaos. As John David McFarland, who was then the Arkansas Geological Survey's contact person for New Madrid seismicity, explained twenty years later, "After Browning, it was hard to get anyone to listen about New Madrid earthquakes."[116]††††††††

Much of the story of the forgetting of the New Madrid quakes is a story of attention turning elsewhere. The quakes became buried under other forms of knowledge and conflict: the middle Mississippi was a site of river

---

Washington University in St. Louis issued to every department to enable office staff to free themselves from the earthquake debris. Like many good stories, this one does not, I am afraid, seem to be true, but the retelling of such stories, with evident relish and elaboration, speaks to the hysteria of the moment, as well as to the collective sheepishness of the region when it became clear that the prediction would not in fact bear out.

******** Frequently, people who grew up in the midcontinent tell me they knew about the New Madrid earthquakes only because they remember getting out of school that day.

†††††††† That is, in terms of disaster preparation. The Illinois alt-country band Uncle Tupelo got quite a hearing for "New Madrid," a love lament based around the Browning failed prediction, on their 1993 album *Anodyne*.

battle and guerrilla warfare, not cross-cultural trade; a zone of floods, not earthquakes; a region of black sharecroppers, not Indian traders or French farmers. In the reaction to Iben Browning's prediction, however, is another element: a suppression of knowledge about earthquakes because that knowledge had become tainted by quack science. The quakes were actively forgotten as well as submerged by changes to the earthquake zones.

Conclusions: Submerged Knowledge

Transformations in the social geography and environment of the former New Madrid hinterland erased evidence of the New Madrid earthquakes and of the polyglot world they helped displace. A regularized geometry of irrigation ditches and industrial farming overwrote earlier environments and, to a large extent, environmental knowledge. The barely visible traces of past environmental history were more likely to be the rusty boilers of abandoned timber mills than fault scarps or sand blows.[117]

In the long period from the Civil War to the late nineteenth century, changes in seismology, in the environment around the epicenters, and within the social history of the middle Mississippi Valley allowed the 1811–12 tremors to subside beneath the surface of attention and memory. Beginning in the middle nineteenth century, a few large landowners came to control most of what had been the New Madrid hinterland. From a multiethnic trading zone of the late eighteenth and early nineteenth centuries, the area became a zone of monocrop hegemony, agriculturally rich and culturally riven. Rigid hierarchies of economy and race controlled social life. Southeastern Missouri witnessed such extreme inequity that it was the site of nationally publicized protests by the Southern Tenant Farmers' Union in the desperate years of the Great Depression. Prior histories of people or place became hard to see.

Only a few observers glimpsed evidence of the quakes beneath the surface. Yet these researchers worked against the vastly dominant regional trend within American seismology and American popular perception in which California became American earthquake country. Within twentieth-century earth science, the development and deployment of networks of seismometers turned scientists' attention elsewhere. Instrumental evidence, not the wide-eyed narratives of frightened settlers, seemed to promise insight into the nature of seismic upheaval. In the environment itself, erosion by wind and water, shifts in land use, and scientific inattention rendered invisible many environmental effects of the quakes. By the middle twentieth century, drainage canals, irrigation districts, railroads, highways, and industrial agriculture had flattened, regularized, and imposed a new geometry on the New Madrid seismic zone, erasing evidence of the quakes. In part because the environmental record seemed so unremarkable, the quakes receded from scientific as well as popular understanding. The exciting, rapid development of the plate tectonic model in the late 1960s and early 1970s left little interest in seismicity far from a plate boundary.

Ironically, one remarkably effective call for attention to New Madrid seismicity proved a grave hindrance to serious research as earthquake preparation. In the fall of 1990, when few people in earth sciences circles or in the larger public had heard much about the New Madrid quakes, national and even international attention became focused on the possibility of a recurrence predicted by Iben Browning and supported by a few local scientific figures. A frenzy of earthquake fear and media hubbub swept Missouri in the weeks leading up to his predicted December earthquake date. When nothing at all seismically interesting took place on that date, many in the area felt fooled and resentful, resistant thereafter to further alarm. Browning's pseudo-science made serious study of the quakes that much harder. The New Madrid events seemed best suited for dusty exhibits of local history or the promotion of "quake specials" at area diners.

By the early 1990s, the New Madrid quakes occupied a curious cultural place of widespread recognition but equally widespread disregard. Many people had heard of these long-ago earthquakes, but few people—including few experts—knew much about them. What had been common knowledge and the subject of common curiosity in the 1810s and '20s became by the late twentieth century simply a curiosity—a trivia question but not a subject for serious research or policy planning.

The earth itself hid its evidence. Earthquakes in a floodplain leave no dramatic escarpments or ragged lines of rock: instead, they produce strange circles of sand, snakelike fingers of white on the occasional riverbank, subtle changes like long and low rises that are dramatic only to people who can see past the topographic subtlety to the forces that must have produced them. By the time of the centennial of the New Madrid earthquakes, much of the evidence that they had ever occurred had been plowed under or washed away by a century's worth of spring floods. In the 1970s, Associated Electric Cooperative built a coal-fired electricity plant in New Madrid, more or less on top of a fault that had destroyed the town in 1811–12.[118]

Yet when people have forgotten, the landscape can still remind them. "During spring floods," noted a 1979 Arkansas Geological Survey study of earthquakes in Arkansas, "much of the old sunken lands return to their former state in spite of levees and drainage ditches."[119] Human effort and inattention can elide only so much: places resist and assert, complicating narratives by the stubborn insistence of the material, the fluid, the corporeal.[120]

One of the few experts still to see evidence of New Madrid, US Geological Survey geologist Myron Fuller, insisted on the centennial of the New Madrid quakes that the "watchful eye" could still detect earthquake evidence, in the form of sand blows and topographic changes throughout the region. The continuing story of the New Madrid quakes over the very late twentieth century and into the twenty-first is one of the development and articulation of regimes of watchfulness by an increasing range and background of observers, and the re-emergence of New Madrid seismicity into the realms of scientific and public dialogue—as well as intense scientific and public debate.[121]

7 ❋ The Science of Deep History: Old Accounts and Modern Science of New Madrid

In 1970, it might easily have seemed that the New Madrid earthquakes of 1811–12, so long submerged beneath the surface of popular and scientific memory, would forever remain half-buried and forgotten. From the Civil War through the American war in Vietnam, changes in science, the environment, and the social order of the earthquake zones allowed the New Madrid tremors to subside beneath the surface of attention and memory. By the middle twentieth century, drainage canals, irrigation districts, and large-scale agriculture had flattened, regularized, and imposed a new geometry on the New Madrid seismic zone, erasing evidence of the quakes. In part because the environmental record seemed so unremarkable, the quakes receded from scientific as well as popular understanding. Meanwhile, the enduring and ugly social and racial divisions of the late nineteenth- and early twentieth-century South made many kinds of local scientific investigation all but impossible.

Instrumental data from plate boundary regions like Japan and the American West Coast gave researchers powerful and abundant material with which to work, even as plate tectonics suggested research questions far removed from narratives of the American frontier. James Macelwane's efforts to make New Madrid the midcontinent cognate of the West Coast failed to convince scientific or public supporters. Written accounts of the New Madrid earthquakes faded into obscurity as frontier tall tales, and the earthquakes themselves became regarded as a long-ago curiosity—indeed, barely regarded at all.

A hundred years after the earthquakes, a USGS geologist named Myron Fuller demonstrated the textual and environmental evidence for such mid-Mississippi Valley tremors. Yet even as he did so, engineer Otto Kochtitzky was draining and dredging the New Madrid epicenters and along the way redefining them as not epicenters at all. Only much later, starting gradually in the 1970s and gathering momentum into the '90s, did other researchers begin to heed Fuller's assessment. In 1973, Saint Louis University seismologist Otto Nuttli used both textual evidence and mathematical modeling to push mainstream seismology to re-examine the evidence for powerful quakes in 1811–12.

Slowly, other researchers began to insist that the events at New Madrid were significant and worthy of study—and that the aged written record might pose interesting questions for the new plate tectonic model. Earthquakes powerful enough to be felt across the country? Earthquakes that had spawned enormous sand blows? Earthquakes of such apparent power, hundreds and hundreds of kilometers from a plate boundary, in what was thought to be a seismically stable continental interior? The written record of these otherwise obscure nineteenth-century accounts insisted upon this strange environmental truth.

The New Madrid earthquakes re-emerged over the late twentieth and early twenty-first century into both public and scientific visibility because of the renewed salience of such "old knowledge"—though sometimes in surprisingly altered form. Researchers began to re-evaluate texts such as William Leigh Pierce's, finding them factually indicative rather than merely fanciful. Geologists began to investigate streambeds and sand blows, rediscovering environmental evidence identified in the early nineteenth century but ignored since. At the same time, researchers argued not only over the magnitude of New Madrid's past earthquakes but also over the potential of future tremors. As in the early nineteenth century, debate took place in public as well as scientific venues. By the second decade of the twenty-first century, professional consensus warned of the possibility of very large future earthquakes in the New Madrid seismic zone, and a new push for earthquake preparation unfolded in the context of the New Madrid Bicentennial. This new public awareness placed the middle Mississippi Valley once more in a global context, as public commentators as well as earthquake engineers drew lessons from Haiti and Chile, New Zealand and Japan. Americans of the midcontinent began to regard themselves as part of a seismically active and globally connected world—much as Americans of 1812 did. Finally, in sometimes subtle ways, those researching seismic evidence paid renewed attention to both local knowledge and bodily experience. Modern science and modern earthquake awareness bear surprising parallels to the seemingly antiquated forms of knowledge of the American frontier.

Following Bits of Paper through Time

In the era of instrumental seismology, one lone serious investigation of the historical quakes of the middle Mississippi Valley emerged. On the centennial of the New Madrid tremors, USGS geologist Myron L. Fuller tramped through the woods and sloughs of north Arkansas and southeast Missouri to document evidence of major earthquakes. In a 1912 report, he used the firsthand reports from the months and years immediately after the tremors to argue for the historical reality and current scientific interest of these massive middle Mississippi Valley earthquakes. His work is now cited throughout contemporary science as the beginning of serious seismology of the quakes, but at the time he was quite literally a voice in the wilderness. For many decades Fuller's report languished in obscurity.[1]

The late nineteenth-century submergence of knowledge of the New Madrid quakes had many elements: changes in the racial and social composition of the middle Mississippi Valley, the searing conflict of the Civil War, and revolutions in the practice and theory of seismology all contributed to the elision of the New Madrid quakes as important historical or scientific events. The submergence of knowledge about the earthquakes was also textual, created through the dismissal, selective use, and successive reinterpretation of writings about the quakes. To trace the slow resurgence of knowledge about New Madrid earthquakes is to follow bits of paper across time.

Myron Leslie Fuller was a government man willing to think outside the usual parameters. He was one of the first researchers in the newly formed US Geological Survey to study groundwater, and he was responsible for a brief, significant period of attention to groundwater pollution within USGS—and within American science broadly—in the first decade of the twentieth century. Trained at MIT (he graduated in 1896), he published eighty-two articles and studies on topics including glacial geology, fossils and various formations, and earthquakes in Jamaica and continental North America, but was best known for his studies of underground oil and water. In his later career, he became a frustrated private "oilogist," always searching for funding to explore further sites in North America and northern China.*[2]

In his work on earthquakes, as on underground fluids, Fuller was a pioneer. His work on the New Madrid quakes came about, in fact, because a colleague, Edward M. Shepard, became intrigued by the role of earthquakes in changing wells and underground water sources. Shepard urged Fuller to investigate the New Madrid events, and the two collaborated on investigations through the sunk lands of Arkansas and Missouri.[3]

Myron Fuller's research reflects the way in which geology was done from the eighteenth century until the recent advent of computer-enabled

---

* His will also founded a museum in his home community of Brockton, Massachusetts. The Fuller Craft Museum is now a renowned center for handicraft art ranging from delicate sculptural metalwork to fabric representations of weather systems. Dedicated and careful, Myron Fuller clearly valued creativity.

analysis and modeling: he tromped around. In all, he made three autumn trips to the regions where the quakes' effects had been most intense. On his first trip, in the fall of 1904, Fuller traveled with Shepard by dugout canoe and horseback "up the old de Soto trail." The fact that later researchers on Hernando de Soto debate the Soto expedition's precise route does not change the historical importance of Fuller's description: he and his colleague were traveling areas that had first been traversed by Europeans in the mid-fifteenth century and about which formal European and American knowledge had in many ways not significantly deepened.[4]

The second year, Fuller went out on horseback with a Mr. C. B. Baily, the city engineer of the small town of Wynne, Arkansas, and a veteran timber surveyor. Baily had earlier helped Shepard on his trips, and Fuller was gratified by Baily's observations on many "earthquake features in the still almost untouched forests." As always, a few locals knew about features of a landscape that was still almost as unknown to outsiders as it had been to a would-be conquistador.[5]

A Centennial History of New Madrid

In the plainspoken prose of a government agent, Myron Fuller wrote a surprising and new centennial history of earthquakes based on entirely old-fashioned sources: aged writings and half-eroded environmental traces. The only scientific publications on the New Madrid quakes at all contemporary to his, a pair of articles from 1902, were simply compilations of extended quotations from historical narratives. Such work was fully in the tradition of late nineteenth-century seismology, but Fuller attempted an in-depth integration of historical accounts with reconstruction and analysis of environmental damage. Fuller addressed both scientists and the broader public. He published several articles in popular journals, and then compiled his findings into a USGS report dated 1912.[6]

Fuller's interest in earthquakes was part of a larger explosion of seismic investigation following the devastating 1906 San Francisco earthquake and subsequent fire. His illustrations explicitly compare hundred-year-old sand blows in the New Madrid region with much more recent smaller sand blows produced by the San Francisco temblor, and his report came out just after the first issues of the *Bulletin of the Seismological Society of America*, the journal of the Seismological Society inspired in part by the earthquake.[7]

Yet Fuller's report reflects very little of the instrumental focus of seismology taking hold in that first decade of the century: instead, he documented what must have happened by comparing what earthquake narratives *said* had happened with the evidence that he could decipher from the heavily wooded swamplands of southeast Missouri and northeast Arkansas. Pulling those together, he documented that the 1811–12 quakes had been felt at very large distances, that their local effects were extremely powerful, and that they were far from unique in the environmental history of the region.

Fuller framed his book as a synthesis of observation and textual investigation, made necessary by the relative absence of knowledge of the quakes and by their potential significance as evidence of "the geologic effects of great disturbances upon unconsolidated deposits," that is, the loose soils of the Mississippi Valley. Although scientific literature usually included the New Madrid events in lists of significant tremors, he observed in his introduction, "the memory of it has lapsed from the public mind." As a result, he explained, "the story of the earthquake is told in two ways—in the quaint, picturesque, and graphic accounts of contemporaries, and in the equally striking geographic and geologic records." Throughout his slim volume, he integrated these two kinds of sources.[8]

Like early nineteenth-century researchers, Fuller collected and collated what historians would term primary sources of the quakes, and he used them to great effect. He opened with a summary of the events of the quakes drawn from historical narratives, and then used historical accounts at every section to buttress, detail, and ground his subsequent environmental investigations. In an extensive investigation, Fuller documented that much of the swampy area of northeast Arkansas and southeast Missouri was not, as those in the region by then assumed, the primeval condition of unchanged land, but was instead "sunk land" created by the earthquakes. Fuller asserted new arguments for familiar environmental evidence. He noted that although trees did not seem to have died wholesale in 1812, cypress trees whose roots are disturbed can live for many years before finally dying (observations confirmed by later dendrochronologists). Fuller's evidence was both observational—deadened trees still erect in standing water—and textual. In one passage that combines his own investigations with references to several narratives of the earthquakes' effects, Fuller quoted a hunter from the late nineteenth century on the recent earthquake origin of the "sunken land": "In some of the lakes," wrote the old trapper, "I have seen cypresses so far beneath the surface that with a canoe I have paddled among the branches." Fuller himself had canoed many of the same bayous: he and this grizzled trapper had the same experience, both brought forward as legitimate evidence for seismological history.[9]

Many of Fuller's forms of research not only called upon nineteenth-century sources but hearkened back to nineteenth-century forms of exploration and potential explanation. In an attempt to investigate causes and correlations of the tremors, Fuller created a chart and diagram listing the relation of the major shocks to phases of the moon and time of day, as well as to weather conditions. He attempted to use reports of felt sensation to figure out the directionality of the earth's vibrations. Fuller's report thus collected and recapitulated certain older forms of knowledge and investigation, even while transmitting knowledge and analysis of the quakes into a new century.[10]

Fuller reintroduced the notion of the New Madrid quakes as very large—larger than recent Charleston and San Francisco quakes, he insisted. In a portion of his account that was completely ignored until very recent investigations, Fuller insisted that the 1811–12 events were only the latest in series of very large earthquakes to hit the region. As evidence, he pointed to trees of several centuries' growth emerging from fault scarps along the St. Francis River in northeast Arkansas, to subtle topographic evidence of former uplift prior to 1811, and, significantly, to accounts of quakes from the 1770s and '90s recorded by Cincinnati physician and natural history observer Daniel Drake after the 1811–12 sequence, as well as to an account of Indian knowledge of a prior great earthquake referenced (and discounted) by Charles Lyell.[11]

In little more than a hundred pages, Myron Fuller demonstrated the serious extent and environmental consequences of the historic earthquakes, documented prior seismicity in the region, and warned of the effects of future quakes. A few serious researchers such as James Macelwane read Fuller's report, and later, during the gradual resurgence of interest in historical New Madrid seismicity that began in the late 1970s, Fuller's book was the main source of information. Yet for more than half a century, Fuller's report lay apparently unknown.[12]

Even those familiar with the book did not apparently absorb or accept its more challenging conclusions. Arkansas Geological Survey earthquake expert Scott Ausbrooks points out one mid-1920s geography of the area around the St. Francis River which blandly noted that "sandy spots or streaks occur commonly in the prevailing heavier soils," without identifying these as likely sand blows and hence a likely earthquake feature—despite listing Fuller's report on the New Madrid quakes in the bibliography. The 1920s author notes that there is dispute over whether the "sunk lands" had much at all to do with earthquakes: the subject was clearly one to be approached with care.[13]

Over the first two-thirds of the twentieth century, Myron Fuller's report faded into the obscurity of a little-read, earnest government report. Perhaps a few old-timers claimed the Mississippi had flowed backward, just as legendary keelboatman Mike Fink had claimed to wrestle alligators barehanded: the river and those who worked it were always good for colorful exaggeration, but that was a far cry from sober evidence. As researchers Arch Johnston and Eugene (Buddy) Schweig observed in the mid-1990s, accounts of the quake elicited "widespread modern disbelief." The New Madrid quakes became something of a crackpot science.[14]

The Complicated History of Myron Fuller's New Madrid Report

The reception of Fuller's work is bound up in the tangled mid-twentieth century history of earthquake risk and earthquake alarmism, and especially the false prediction of a repeat New Madrid quake by Iben Browning in 1990. Fuller's history of the New Madrid earthquakes—like many of his publications—was published by a government agency. It came out in 1912, as a staid publication of the Department of the Interior. Fuller's report was reprinted in the 1950s by a small local firm which promoted southeast Missouri history. It was then reprinted several more times in small editions by David Stewart's Center for Earthquake Studies beginning in 1989, and then came out from Stewart's private press in 1995, in a paperback edition with a lurid cover.[15]

The 1995 edition of Myron Fuller's *The New Madrid Earthquake* is a complicated, multilayered document. Though the bulk of the text is simply a facsimile reprint of Fuller's 1912 report, the book was published by and has a foreword by David Stewart, who by that point had been disavowed by most in the scientific community for his promotion of Iben Browning's unfounded 1990 prediction. Stewart's press (Gutenberg-Richter Publications, named for Beno Gutenberg and Charles Richter, leading seismologists who among other achievements developed the earthquake magnitude scale that bears Richter's name) was based in a small town in Missouri and existed as form of protest against scientific researchers who would have nothing to do with Stewart or his view of New Madrid. Yet Fuller's original text is cited and referred to as a touchstone by all who speak of New Madrid—and Stewart's edition of the book, lurid cover drawing and all, represents a main way most interested readers could find a handy copy.[16]

Ironically, Stewart's view of the public risk from future earthquakes in the New Madrid region (if not his methods of understanding or predicting such risk) was much closer to the officially promulgated view by the 2010s than it was in the late '80s. In a further irony, the *Fault Finders* guides published in the 1990s by Stewart and his few remaining colleagues are an accessible and engaging way for most laypeople to understand the impact of the earthquakes on the terrain surrounding the epicenters. Containing photographs of sand blows and other still-visible earthquake features, the guides are far more immediate and comprehensive than anything available today from official earthquake-related agencies. Yet their provenance makes them unpalatable, so these earthquake guides are not available in any credible or government-related agency studying New Madrid. (For many years they were only available at idiosyncratic local institutions like the tiny New Madrid Historical Museum just off the levy on the small main street of present-day New Madrid, or, more recently, online.)[17]

In his 1912 report, Myron Fuller used original New Madrid accounts to support his own field investigations. His was a seismology drawn in large part from textual accounts, buttressed by on-site observations. The history of Fuller's report is similarly textual, bound up in how the book was published and made available. What science gets read may depend in part upon how it is packaged.

Midcontinent Seismic Study as Instrumental, Rather Than Textual

Myron Fuller's report was also decidedly noninstrumental. Written just as instrumental seismology was beginning to gain intellectual sway and institutional heft, Fuller's report would soon seem the product of a former time, informed by archaic vocabulary rather than up-to-date seismograph readings. Fuller's 1912 research on the quakes languished compared to the abundant and compelling seismographic evidence of more active world hotspots. After Myron Fuller, study of Mississippi Valley earthquakes—to the extent that it took place at all—took place through instrumental records.[18]

In the late 1920s, Saint Louis University had just put in place its initial array of seismometers. Knowledge hard-earned by climbing up and around tree limbs and wading through morasses alive with mosquitoes and water moccasins was no longer the only way to comprehend local territory. Seismographs gave researchers access to tremors far from them, even on the other side of the globe, even as seismograms made precise information portable, a medium of exchange.[19]

Gradually, the few instruments operated by Saint Louis University were joined by a small number at other universities and governmental institutions in adjacent states. Earthquake listings for Mississippi Valley tremors that had been purely textual became more quantitative over the course of the twentieth century, as the number of observations stations grew and more sensitive instruments not only detected smaller and smaller tremors, but (by the 1930s) timed them to the hundredth of a second.[20]

These changes were not simply in the number or range of instruments, but in the central intellectual importance of the quantified data they produced. Instrumental evidence of very small, current tremors became the way to understand the potential of the region—not wild reports of huge long-ago upheaval that seemed just a part of regional tall tales. Historic narratives of a frontier-era earthquake ceased to be central to understanding the middle Mississippi environment. Instead, SLU's seismometers, along with those in a few other regional stations, became the primary testimony to seismicity in mid-America.

Bits of Paper Return: Otto Nuttli and the Seismology of the Midcontinent

The career of Saint Louis University seismologist Otto Nuttli is a landmark in the rediscovery of historical narratives and thus of New Madrid seismicity. Nuttli initiated a refocusing of attention on New Madrid–area seismic history in a 1973 paper published in the *Bulletin of the Seismological Society of America*. In this piece, Nuttli resurrected and resumed Fuller's practice of combining textual research on the 1811–12 events with other forms of seismological evidence. Further, he attempted to translate narrative evidence into quantified forms such as magnitude estimates accessible with the techniques and conventions of instrumentally based seismology.[21]

Otto Nuttli was thoroughly of the mid-Mississippi. Reared in St. Louis and trained and then employed by Saint Louis University, Nuttli was steeped in SLU's focus on seismology and rose to become a main figure in the field. Nuttli contributed to studies of the core-mantle boundary, shear waves near the earth's surface, and comparisons between plate-boundary quakes and what are now termed stable continental region or intraplate quakes. He developed what came to be called a Nuttli magnitude scale, which measures small events at short to moderate distances in eastern North America. Local environments, though, shaped his intellectual legacy: Nuttli was positioned to investigate and insist upon the historical reality and scientific validity of seismicity in the midcontinent. Throughout his research, he observed and insisted upon different models for central and eastern North America than for the West Coast.[22]

In his 1973 article, he argued for a reassessment of the New Madrid earthquakes. Because there was no instrumental record, he argued, researchers have to consider "indirect sources of information, such as published accounts of the effect of the earthquakes upon the landform and upon people."† While remaining interested in deciphering the earthquakes through

---

†This hardly seems a ringing endorsement of historical method: to term firsthand accounts of people's own experiences "indirect sources of information" goes against most historians' sensibilities. Nuttli's essay was, nonetheless, a substantial redirection of seismological research toward historical sources.

evidence of physical disruption, Nuttli argued that written accounts of the quake could indeed help reveal those environmental truths.[23]

Setting himself the task of recovering physical events from written sources, Nuttli practiced an essentially nineteenth-century form of seismology. "In an attempt to obtain as complete a set of information as practical," he explained, "and to avoid possible errors resulting from using secondary or tertiary sources, the present author has gone back to the original newspaper accounts wherever possible." Newspaper accounts were the heart of Nuttli's reassessment of New Madrid—and his paper literally carried them. The issue of the *Bulletin* that published Nuttli's essay included a microfiche card tucked into a pocket in the back cover, containing a "a retyped copy of the newspaper accounts of the earthquakes."[24]‡

Nuttli read newspaper accounts carefully. He dismissed and did not reprint in his typescript a series of letters about an event later proven to be a hoax (though he does not give these the dignity of explanation, they were likely the series of reports about the collapse of Virginia's natural bridge). He acknowledged that not all that occurred appeared in newspapers: "Although aftershocks were felt at least through 1813," he observed, "they apparently were not considered newsworthy, especially when compared with such events as the War of 1812 and Napoleon's adventures in Europe." Nuttli's analysis reflects a careful reading of newspaper accounts against one another, checking for internal consistencies. Nuttli's essay reintroduced to North American earthquake study the examination of historical accounts as a legitimate technique of research into this earthquake sequence.[25]

His conclusions were surprising to his many readers who regarded the middle of North America as seismically quiet. First, he used newspaper accounts of the earthquakes to construct an isoseismal map of the first main shock, a map of where similar effects had been felt, which was "characterized by an unusually large felt area," with significant felt intensity "as far away as the southeast Atlantic coastal area." Nuttli thus reinforced claims asserted by Myron Fuller about the extent of the earthquakes and their status as an actual, historical event worth investigation. Nuttli then offered estimates of the magnitude of the shocks, estimating the 16 December shock as a body-wave magnitude of 7.2, 23 January as 7.1, and 7 February as 7.4. These estimates would ultimately draw other researchers into the question

---

‡ Present-day online archives of the *BSSA* include alongside Nuttli's paper a reproduction of this microfiche typescript, the strokes of a manual typewriter now digitized but still carrying his earnest commitment to uncovering and sharing the most direct evidence of the quakes.

of New Madrid seismicity and spark wide-ranging debate about the size of the quakes.[26]

The Textual Approach Develops at CERI

Much of the initial, halting follow-up to Otto Nuttli's publication and eventual renewed attention to textual study of the New Madrid earthquakes came through the founding of a new institute for earthquake study in mid-America. Just as Saint Louis University was central to the instrumental understanding of midcontinent seismicity in the first two-thirds of the twentieth century, the Center for Earthquake Research and Information (CERI) became central to debate over the New Madrid quakes over the last third. Founded in 1977, with over fifty-five staffers by the early 2010s, CERI is an unusual hybrid entity affiliated with the University of Memphis but also housing the local field office of the US Geological Survey. From the outset of the program, historical narratives have been central to many CERI researchers' work.[27]

As with most developments having to do with seismology, the key moment for the creation of CERI was an earthquake: a magnitude 5 event in September 1976, centered at Marked Tree, Arkansas, that in the words of founding director Arch Johnston "scared the hell out of people." The quake brought down telephone lines, damaged roofs, broke small objects, blew out a few windows, and frightened state legislators into action. The State of Tennessee created the Tennessee Earthquake Information Center, soon renamed the Center for Earthquake Research and Information, centered at the University of Memphis, the school closest to the tremors.[28]

Arch Johnston reflects conventional thinking in his own approach to earthquakes in the 1970s. A young geophysicist just finishing a PhD in Colorado after a five-year term in the air force during Vietnam, Johnston got the job as director of TEIC without thinking much about historical quakes or lower-level seismicity. As he reflected in 2009, as a geophysicist, he felt that on a global scale, anything under about magnitude 5 "did not register"—and the seemingly casual term is quite telling: certainly seismicity that did not instrumentally register was not useful data for Johnston or most of his contemporaries. Like many recent grads, he found his interests radically reoriented by a new job: "There I was, didn't really care about small-magnitude earthquakes, much less historical earthquakes"—and he then got funding for a seismograph network to study small-magnitude earthquakes in a zone that had once been the site of large but preinstrumental quakes.[29]

Catching up quickly on this little-known field, Johnston read Otto Nuttli's 1973 piece and contacted him for advice and critiques. After turning only "grudgingly" to narrative accounts of quakes, Johnston quickly realized that "we had to use historical accounts—we had to see just how quantitatively we could use qualitative accounts." Slowly, and with a certain degree of reluctance, Johnston found himself turning to what he began to see as "a very worthy scientific problem, and one that was not appreciated in the scientific community": the nature, extent, and recurrence of midcontinent seismicity.[30]

In 1982, Johnston published his first major piece on New Madrid–area seismicity, an article in *Scientific American*. At around the same time, other fields were beginning to pick up on long-buried evidence of the 1811–12 quakes. In 1976, historian James Lal Penick of Loyola University in Chicago published a well-regarded account of the social and historical impact of the quakes.[31]

Yet despite the growth of interest among a few professionals (mostly professionals with institutional links in the midcontinent), larger currents ran against any consideration of historical narratives as anything but sideline curiosity. In 1988, Arch Johnston enquired with the Saint Louis University earth sciences program about where Otto Nuttli's work had been archived. It had not been. After Nuttli's death, the department chair at SLU had cleaned out his office and thrown everything away. As Johnston concluded, in a wry mid-South drawl, "That gives you an idea of the value placed by a pretty high-level seismologist on historical work."[32]

At CERI, however, one of Nuttli's essential projects—the compilation of primary source accounts that had appeared as supplementary material for his 1973 piece—has been preserved and vastly expanded. The current website for CERI features the New Madrid Compendium, a collection of first-person nineteenth-century accounts of the quakes. Some are retyped, others (more recent) are scanned in; all are indexable by specific effects as well as source and date. The compendium is essentially a higher-tech web-based version of Otto Nuttli's 1973 database. It began with Arch Johnston's research into the bibliographies listed by Penick and Fuller, but was carried forward when in 1999 he hired recent University of Memphis history PhD Nathan Kent Moran.[33]

For years, in a back room of the CERI offices in Memphis, Kent Moran has marshaled undergraduate research assistants and balky microfilm machines in his mission for archival completeness, reading the entire run of every American newspaper he can locate for the few months surrounding the main 1811–12 quakes. All these sources are available through the New Madrid Compendium on the CERI website, alongside the scientific context such as an earthquake catalog of known midcontinent quakes. Present-day disaster planners, seismologists, and historians turn to the CERI website to find instrumental data on the past hundred years of seismicity in the region, as well as to research particular aspects of the experience of the New Madrid quakes.[34]

The New Madrid Compendium is a project of straightforward historical data-gathering, yet it is at once consistent with and distinct from older historical approaches. The goals of the project are not far from Samuel Mitchill's or those of other early nineteenth-century compilers: comparison of myriad accounts can help reveal useful connections and information. The project's overall parameters, however, are far distant from Mitchill's world. The original New Madrid accounts were deeply embedded in a context of newspaper reporting and of narrative accounting for changing processes—whether processes of body or of land. Little of that context appears today in the CERI Compendium: various entries appear either as disembodied, retyped text, or more recently as scanned-in snippets separate from their commercial or political context.

This fracturing of narratives into snippets of text—a fracturing perhaps necessary in a searchable online database—does seem related to the lack of connection between the compendium and broader public conversation about the quakes. There is little sense of whether the compendium saw an uptick of interest with the bicentennial of the New Madrid quakes, or how the compendium might be promoted as an educational resource, perhaps for school syllabuses that might use the rich and detailed source material.[35]

And yet—it is nothing to dismiss, this encyclopedic array of sources on an important American quake, made available, freely and publicly to all. Certainly the New Madrid Compendium reflects the most broadly accessible set of the quake accounts created over the past two hundred years. At the centennial of the quakes, the director of CERI, Charles Langston, remained enthusiastically committed to the project, citing its use by disaster planners as well as earth scientists, and hoped to expand the educational reach and mission of the compendium alongside other CERI efforts. This twenty-first-century compilation of New Madrid accounts is thus no more static than were the flurry of hastily reprinted newspaper accounts of the nineteenth century: the bits of paper that carry information about New Madrid continue their travels.[36]

Seismic Networks, Policy Questions: Reaction to Nuttli

In the decade and a half following Otto Nuttli's original 1973 publication, interest in New Madrid seismicity and information about midcontinent seismic activity began slowly to grow. Concern for safe placement of nuclear power stations impelled a brief burst of monitoring. Starting in 1974, the US Geological Survey and the US Nuclear Regulatory Commission supported seismographic stations in the midcontinent, including a regional network anchored at Saint Louis University—James Macelwane's vision had at last come to fruition, nearly half a century after he first began to advocate for it. Monitoring networks produced a catalog of over four thousand midcontinent earthquakes from 1974 through 1992 ... when the monitoring network ceased.[37]

In 1992, facing a climate of public opposition to new nuclear power plants, the Nuclear Regulatory Commission withdrew its support for seismic monitoring of New Madrid and the eastern United States. Instead, the funds would be used to implement the USGS-run United States National Seismographic Network, or USNSN. Advocates for New Madrid argued that the USNSN would be sparse and lack the coordination provided by previous regional centers. Thus despite a period of renewed seismic networking, at congressional hearings in 1989, Arch Johnston, director of the Center for Earthquake Research and Information in Memphis, gave essentially the same plea as James Macelwane decades before: reporting on a recent US Geological Survey workshop to plan research and mitigation efforts, he underscored that "one of the highest priority items was for a modernized earthquake monitoring network to give more and better quality information about the New Madrid seismic zone."[38]

A renewed period of networked monitoring began in the late 1990s, pushed both by technological developments—especially digital instruments available in the late '90s—and better institutional coordination. In 1997 Congress authorized the development of what became the Advanced National Seismic System, or ANSS. By the first decade of the twentieth century, local institutions including the University of Arkansas at Little Rock and the Arkansas Geological Survey began to set up additional seismic arrays, linked to the regional networks of Saint Louis University and CERI and therefore to the ANSS. Yet for an area known two hundred years previously as an earthquake zone, such arrays were long in coming.[39]

At a national level, concern for seismic threat and hope for the possibility of eventual earthquake prediction came together in the late 1970s in the passage of the 1977 Earthquake Hazard Reduction Act. This act carved out an official federal role for earthquake hazard mitigation, creating the National Earthquake Hazards Reduction Program, a coordinative effort to mitigate earthquake hazards to American people and property.§ A main concern of

---

§ Abbreviated, of course, as NEHRP, and pronounced, with a straight face, as "nee-herp." NEHRP is primarily administered under four agencies: the Federal Emergency

this national as well as regional interest was to estimate the consequences of the newly defined earthquake risk of the middle Mississippi Valley. In 1974, a team of engineers carried out an evaluation for the US Department of Housing and Urban Development of seismic hazards to Memphis. Their analysis concluded that an 1812-type event that occurred in the year 2000, during the daytime, would kill more than three thousand people and cause more than $1.3 billion worth of damage (in 1973 dollars).[40]

People and governments in the New Madrid region began—slowly, very slowly—to pay attention.** In 1983 emergency management agencies from states of the New Madrid seismic zone formed CUSEC, the Central United States Earthquake Consortium, with funding from the Federal Emergency Management Agency, or FEMA. Regional state geologists began meeting in the early 1990s and undertook planning workshops under NEHRP funding. CUSEC coordination helped create partnerships with organizations of geologists, disaster planners, and engineering and transportation. The Mid-America Earthquake Center at the University of Illinois, Urbana-

---

Management Agency (FEMA), the National Science Foundation (NSF), the US Geological Survey, and the less well known but important National Institute of Standards and Technology (NIST). In addition to behind-the-scenes research and analysis, the public information and interpretation found on USGS websites is one direct result of NEHRP funding. Such resources include e-mail alerts for worldwide tremors, apps that report and estimate magnitude almost instantaneously, and in-depth reports on historic and recent earthquakes around the earth. Anyone with access to the web can today have faster and more complete information about global seismicity than anyone in the world could possess until well into the twentieth century.

** In the late 1970s, my step-father bought several items to store in the basement of our home in downtown Little Rock, Arkansas, against possible, if unlikely contingencies. He was equally sheepish about both purchases. The first was a six-pack of "Billy Beer," a bicentennial brew produced in 1976 by president Jimmy Carter's outspoken but not-very-well-informed brother Billy (his politics and his beer drinking were both embarrassing for his older brother). My step-father put some aside in case it might increase in value as a collector's item some day. Over several decades, it didn't, and our household supply did not survive later basement cleanings. It may in fact have provided fuel for hardworking basement-cleaner-outers. The second item was a couple of gallons of water as backup supply, since my science-minded dad had recently been reading about the possibility of earthquakes around Memphis. His embarrassment at possibly ridiculous-seeming preparation for such a far-fetched disaster reflected how little was known about area earthquakes (as well as how little was known about disaster preparation: two gallons is a drastically insufficient emergency supply for a family of five people and a German shepherd). At the time, the very idea of area earthquakes seemed vaguely preposterous, yet the water remained in the basement for many years, a cautious hedge against a possibly dubious seismic calamity.

Champaign, was established to coordinate efforts between interested universities in the region. By the early 1990s most of the middle Mississippi Valley states had some form of seismic building codes or other planning. Still, in 1999, the Center for Earthquake Studies at Southeast Missouri State University—David Stewart's former institution—consisted primarily of two part-time work-study students who answered phones after tremors. By the turn of the century, area experts rued that most people living in states surrounding New Madrid remained "unaware or unconcerned" about seismic hazards.[41]

Other natural disasters began to sway some in the midcontinent. The calamity of the massive Mississippi River flood of 1993, which took out several main bridges, reinforced concerns for highway overpasses and Mississippi River levees and bridges. In 1999, Missouri disaster management official Ed Gray warned, "If you have an earthquake in New Madrid, I can't guarantee you that any of those bridges from Vicksburg, Miss., to St. Louis, Mo., are still going to be functioning."[42] Such concerns were heightened by levee failure following Hurricane Katrina in 2005 and massive Mississippi Valley flooding in 2011. Civic planners began slowly to take earthquakes seriously, especially their threat to natural gas pipelines and fiber optic cables.[43]

Scientific research undergirded these policy developments, as earth scientists began to bring a variety of analytic tools to bear on the problem of New Madrid seismicity. This interest formed part of a late twentieth-century resurgence of interest in historical seismicity more broadly. By the early 1980s, researcher Ronald Street created a massive catalog of reports of historical seismicity in the midcontinent under contract with the US Geological Survey. Researchers in the early twenty-first century began to model wave velocities in soils of the New Madrid seismic zone, including sites in western Tennessee like those into whose seismic fissures Davy Crockett had stumbled.[44]

Initial stages of re-evaluation surrounding New Madrid seismicity focused on the geological structures that might have caused such alarming activity. Researchers immediately began to look for the structural implications—and fossil fuel potential. Beginning in the 1990s, some researchers argued that the New Madrid quakes may have been a response to the new freedom of land liberated from the crushing weight of the latest round of glaciers, a theory known as "glacial rebound," or, similarly, release of the earth's crust from the topsoil eroded away in the past history of the Mississippi Valley. Other research suggested that the structure of the Mississippi embayment may have resulted from continental rifting millions of years ago. Some think that the New Madrid seismic zone may have passed over a "hotspot" that weakened the area and left it prone to faulting.[45]

New research suggested new histories for the landforms of the New Madrid seismic zone. Crowley's Ridge had long posed a puzzle for many fields. The long, narrow ridge stretches more than 150 miles from southeastern Missouri down along the Mississippi into east-central Arkansas. Though generally only about two hundred feet higher than the surrounding floodplain, it is striking in its difference from the topography around it, and with distinct creatures and loess soils (in Arkansas farmers' markets, "Crowley's Ridge peaches" fetch a higher price than lowland fruit because of their flavor). Crowley's Ridge has long been regarded as a leftover of long-ago erosion from an ancient grandparent of the Mississippi, but a team of researchers published results in 1995 suggesting the ridge is a remnant of ancient earthquake-related rupture. Active faults may have shaped the topography even of this midcontinent floodplain.[46]

By the 1990s a research team led by Roy Van Arsdale and David Stahle investigated the dendrochronology of the region—the history told in tree rings of long-lived species.[††] Van Arsdale and colleagues used cores from baldcypress trees to document the substantial effects of the sudden creation of Reelfoot Lake in 1812, as well as change to the St. Francis sunk lands. This dendrochronological work harkened back to observations made by Charles Lyell on his visit to New Madrid and by Myron Fuller on his: throughout much of the late twentieth-century research of New Madrid runs the sense of picking up former, dropped strands of research.[47]

Seismic History through Muck, Soils, and Pot-Shards: Paleoseismology

In similar ways, researchers followed Myron Fuller, not only through bits of paper, but through environmental evidence. Deep in the woods and rivers of southeast Missouri and northeastern Arkansas—and more recently in adjoining areas such as west Tennessee and southern Illinois—researchers

[††] Partly because it involves so many different kinds of intellectual, physical, and interpersonal skills—navigating drainage ditches by canoe, recognizing earthquake features beneath plow land, coaxing delicate machinery, programming complex computer languages, negotiating with suppliers of dynamite, purchasing plane tickets in a hurry—geological research tends to have a much more informal, team structure than, say, the large and hierarchical labs of contemporary biology. It is not uncommon in geology, including in studies of the New Madrid seismic zone, for teams of researchers to include graduate students as well as senior scholars, and often people from seemingly disparate disciplines. Paleoseismology in particular is characterized by this sort of broad team approach.

began by the late twentieth century to coax a narrative of deep history out of seemingly unremarkable features on the terrain.

Through the many forms of investigation that make up paleoseismology, researchers seek to understand how seismic events that took place far before recorded human history leave marks upon the geologic record. Paleoseismology has deepened the time and widened the geography of New Madrid seismicity. Seismologists now understand the 1811–12 New Madrid earthquakes as only the most recent: other, similarly large sequences of earthquakes shook the middle Mississippi Valley sometime around 1450, 900, and 300 CE, and 1100 BCE, as well as in even longer-ago quakes less precisely dated.[48]

Much of the paleoseismic investigation of the New Madrid region focuses on liquefaction features—sand blows and related sand dikes (the sand-filled fissures and conduits that fed sand blows)—and similar effects of strong ground shaking on the soil and water of the alluvial landscape. Myron Fuller researched these features a hundred years after the tremors, but his work was not picked up by others until after Otto Nuttli's publication. Even then, as in many aspects of New Madrid research, the results took decades to impact broader understanding.[49]

Beginning in the last two decades of the twentieth century, and pushing against substantial professional as well as public skepticism, researchers assembled convincing evidence that the earthquake history of the middle Mississippi Valley dated to long before 1811. Initial efforts began through archeological sites and through the study of liquefaction features, especially along the region's many waterways.

In 1989 area archeologist Jim Price was called to what he expected to be a routine investigation: to excavate a hump on the side of a sixteen-foot mound at a site named Towosaghy, near New Madrid. He found not the expected staircase, but a fire pit stuffed full of debris: around 1400 CE, the people using the ceremonial mound had burned their temple and dumped the trash next to the mound it was on. Subsequent research established that in at least four sites, residents set fires, capped sand fissures with clay, and may have built mounds over them, apparently sealing and abandoning them.[50]

Still, conventional wisdom held that most of the sand blows abundantly in evidence around the region were from the 1811–12 earthquakes. Over the next two decades, innovative cross-disciplinary work conclusively demonstrated that many area sand blows were from very large events long before 1811. Geologist and soil scientist Martitia (Tish) Tuttle had in her own words "read Fuller cover-to-cover," and she used modern research methods to validate many of his conclusions. Tuttle, along with geologists Buddy Schweig

and John Sims, geophysicist Lorraine Wolf, and archeologists Robert Lafferty and Marion Haynes, expanded the history of seismicity in the region
by investigating sand blows and related sand dikes.[51]

The drainage of the sunk lands in the early twentieth century had hidden and effaced much of the earthquake evidence on local terrain. Tish
Tuttle and her colleagues used that same swamp drainage system as a way
to rediscover the history of the land. Starting with aerial photographs and
proceeding by canoe and motorboat along more than seven hundred kilometers of rivers and large drainage ditches, they searched laboriously for
liquefaction features, finding more than five hundred of them in the banks
of local waterways. In particular, they found numerous sand blows in which
thick soils had formed that clearly predated the 1811–12 paroxysm. At some
of the liquefaction sites, they found evidence of Native American occupation in the soils developed in or buried by the sand blows: time capsules
left by the massive earthquakes. Radiocarbon dating and archeological ty-
pology techniques allowed the researchers to date organic material such
as charcoal and corn kernels and artifacts such as bits of pottery or broken
arrowheads. When successive eras of these seeming bits of organic detritus
and remnants of long-past cultures occurred above and below a set of sand
blows, the researchers could derive bounding dates for the sand blows and
therefore the earthquakes that had caused them. Geological techniques and
archeological analysis made for successful seismic sleuthing.[52]

In 1999, Tuttle compared the multiple sand blows successively formed
during the 1811–12 sequence of tremors to the same kind of compound sand
blows formed during long-ago earthquakes—so-called paleoliquefaction
features—documenting that earthquakes in the New Madrid seismic zone
have been clustered in time for at least the past four thousand years. She
and colleagues eventually identified hundreds of paleoliquefaction sites, including several dozen sand blows associated with cultural relics from past
settlements. These investigations have demonstrated that series of large
quakes were characteristic of seismic activity in this region, in a way very
different from the pattern of single main shocks usual in many other earthquake zones. Major earthquakes—indeed, major *sequences* of quakes—characterized the New Madrid seismic zone over millennia.[53]

Such muddy but intellectually exciting investigation led to what Susan E.
Hough terms a "modest geologic goldrush." Just as these researchers had
expanded the time horizon of New Madrid seismicity, successive research
expanded the geographic study of midcontinent seismicity. Inspired by paleoliquefaction findings in the New Madrid seismic zone, teams began to
identify earthquake features near the small Arkansas town of Marianna, in
southern Illinois, and in the St. Louis region. In cotton fields throughout

the middle Mississippi Valley, investigators engaged in a series of "trench-ings" to delve into the seismic history beneath the earth.‡‡ Near Marianna, Tuttle worked with Haydar Al-Shukri, Okba Al-Qadhi, and Hanan Mahdi to investigate sand blows two meters thick, a hundred meters long, and sixty wide: one area of study was known to local geology students as "Daytona Beach" for its sandy deposits. They found that these features and their re-lated sand dikes were created sometime between five and seven thousand years ago—outside the New Madrid seismic zone, in east central Arkansas. Such impressive and sobering earthquake evidence further widens the zone of past and possibly future seismic territory.[54]

To the north, paleoseismologists have demonstrated significant moder-ate to large earthquake activity over the past millennia in the Wabash seis-mic zone—what most Americans would probably think of as the seismically quiet region of southeastern Illinois and southwestern Indiana. One group of researchers has argued that the 23 January 1812 New Madrid earthquake was in fact a shock along the Wabash seismic zone, with an epicenter in what is now an Illinois soybean field. A June 1987 magnitude 5 quake near Law-renceville, Illinois, may have served to do for public awareness what this re-search has done in seismological terms: suggest yet another new geography of American earthquake hazard.[55]

Over several decades, research demonstrated that far from being a freak, one-time event, the New Madrid earthquakes of 1811–12 were one of many seismic episodes to shape the landscape of midcontinent. Such careful and extensive work set the standard for studies in paleoliquefaction. Ironically, given all the skepticism applied to accounts of the New Madrid earthquakes, the New Madrid tremors have—recently and swiftly—become a main and comparatively well studied example in the field of paleoseismology. Simi-larly, the New Madrid earthquakes are now a main example in the study of intraplate or stable continental region earthquakes—that is, earthquakes that do not occur on a boundary zone between tectonic plates.[56]

In the late twentieth century, researchers began to reinvestigate histori-cal seismicity in the middle Mississippi Valley. They did so by following leads provided by Myron Fuller almost a hundred years before: they dug

---

‡‡ Trenching is exactly what it sounds like: researchers dig a large, rectangular trench and climb inside to look at a cross-section of soil. Paleoseismology is sometimes termed "science by backhoe."

into archives and dug into the earth. Through a renewed attention to early nineteenth-century accounts of the New Madrid earthquakes and through new interdisciplinary techniques, researchers developed insights into New Madrid's seismic history. Modern science thus harkens back in surprising and striking ways to the vernacular science of the nineteenth century. Researchers in diverse arenas, with diverse skills, worked to create an understanding of the New Madrid earthquakes consistent with the instrumental and quantified rigor of modern science.

Or at least they tried to. Now that the seismological community is taking seriously the New Madrid quakes, and intraplate quakes more generally, the facts of the matter have become worthy of scientific debate. And a lively debate it is: serious and well-credentialed researchers present dueling theories of the size, measurement, movement, recurrence, and future hazard of the New Madrid earthquakes and associated midcontinent faults.

This scientific debate has meant even greater conflict in the public sphere. Despite the opposition of a vocal minority, modern seismology has wrestled itself into consensus around the potential threat of future strong series of earthquakes in or near the New Madrid seismic zone. Yet bringing the New Madrid earthquakes up from beneath the surface of public attention and historical memory has forced Americans of the midcontinent to confront seismic threats both powerful and diffuse. How exactly to respond has made for intense arguments at city council meetings and state legislatures. Highway engineers and disaster planners, school officials and insurance executives, homeowners and voters—all those concerned with making homes and lives in the middle Mississippi Valley must now consider the possible implications of even moderate earthquakes beneath their terrain.

The New Madrid quakes have become the flashpoint for conflict over conservatism in earthquake engineering: How prepared for a devastating earthquake do Memphis or St. Louis or Little Rock really need to be? Should critical facilities such as fire stations and hospitals built of unreinforced masonry (a killer during earthquakes) be retrofitted to a higher standard? Should states facing high rates of teen pregnancy, low rates of childhood vaccination, and substantial illiteracy spend daunting sums for seismic retrofitting? Or make even the small increases necessary for seismic design? In building code hearings in early 2008, Kay Brockwell, the economic development officer for a small town in east Arkansas, contrasted "the possibility of earthquake damage of buildings" with "the near certainty of people never having decent jobs if we don't get economic development." He spoke for many in what some would regard as an earthquake zone: towns clearly need jobs, but do they really need seismic stabilization rods? And as others asked—might one, ultimately, bring the other?[57]

At stake in emerging early twenty-first-century understandings of the 1811–12 quakes are at once history and future. Archeologists' interpretations of long-ago settlements and trash piles impact engineering requirements for twenty-first-century industries. Researchers use New Madrid evidence to debate frameworks for large quakes not accessible to instrumental records, not based at plate boundary zones, and not frequent enough to create social consensus around the need for damage prevention. Even as they do so, disaster planners and city officials struggle to balance new and often disturbing scientific evidence with a variety of other civic priorities. How scientists read the environmental past and how that past shapes future priorities are not abstractions but pointed questions of local governance in the middle Mississippi Valley. These stakes are intensified in the struggle to commemorate past earthquakes in the New Madrid Bicentennial and to plan for the region's seismic safety: all these efforts involve conflict over new views of the history of the New Madrid seismic zone, and over its likely future.[58]

Narrative Accounts and Debate over New Madrid Magnitudes

Since 1973, historical sources have taken an ever more important role in the active seismological debate over the New Madrid quakes, especially the heated disputes over the magnitude of the 1811–12 events. Certainly, for disaster planning and forecasting, the size of large tremors in the past is extremely significant. On this crucial question of analysis there is a high degree of disagreement, conducted at high volume.[59]

In 1996, Johnston and USGS colleague Buddy Schweig, then also based at CERI, opened the debate. In an article that summarized work done since Nuttli's 1973 article, they challenged seismologists to unravel the "enigma" of the quakes, and in particular they emphasized the sheer size of the tremors. Noting that geophysical and geological work had largely verified historical accounts of tumultuous upheaval, they asserted that "the sequence included at least six (possibly nine) events of estimated moment magnitude M.7 and two of M ≈ 8."[60] Johnston and Schweig argued that the New Madrid quakes were probably the greatest sequence to strike the continental United States.§§[61]

_____

§§ Many similar but garbled claims echo throughout our popular media. Some of the highly varying ways in which the New Madrid quakes are characterized probably stem from lower-forty-eight vs. continental US confusion. The largest earthquake experienced in modern North America is, as of this writing, the 1964 Good Friday earthquake

In seismological terms, them's fightin' words. Since then, geophysicist Seth Stein at Northwestern University has assailed the magnitude assessment, causal mechanisms, and unnerving future implications of the New Madrid advocates. Significant parts of this debate have played out in the public sphere as debates about public regulatory priorities—as Memphis contemplated a more rigorous seismic building code, Stein decried additional regulation and urged different priorities in seismic preparation. In the seismological literature, debate has ranged over almost all aspects of the tremors, including whether GPS systems show movement along a fault, whether models applied to other quake systems necessarily apply to intraplate quakes, what counts as a main shock versus an aftershock, and what recurrence interval paleoseismology reveals. But much of the debate over the New Madrid quakes, and particularly about their magnitude, rages about how to read old texts.[62]

In 2000, Hough and three colleagues—one a political scientist—offered a landmark revision of original magnitude assessments, hinging on careful reading of historical accounts as well as a reassessment of settlement patterns. Hough is a USGS geophysicist based in California who has written widely and engagingly for a popular audience. As she recounts in her 2002 introduction to seismology, *Earthshaking Science*, it was only because of her research for that broadly accessible science survey that she realized that certain interpretations of historical accounts of New Madrid did not take adequate account of "site response," the way in which soft sediments amplify seismic waves.[63]

In subsequent research following up on that insight, Hough and her coauthors built on an acknowledged feature of the "felt reports": the quakes were experienced powerfully in certain places far away, such as lowlands in North Carolina, and had damaged buildings in Cincinnati, but had not wakened sleepers much closer to the epicenters in Kentucky. Researchers had long recognized that the loose soils of coastal plains and bottomlands along rivers shake much more than, say, the central hills of Kentucky, where a thin layer of soil covers well-anchored bedrock. Hough and her coauthors

---

that struck Port William Sound, Alaska, with 9.2 moment magnitude. The quake devastated parts of Anchorage and the surrounding area, and the resulting tsunami peaked at ninety-seven meters high and registered on tide meters in Cuba and Puerto Rico. Researchers now agree that the New Madrid quakes are *among* the most powerful to hit anywhere in the United States, along with a number of lesser-known Alaskan quakes, the 1906 San Francisco earthquake, and several nineteenth-century Southern California quakes.

were the first to argue that patterns of settlement on the frontier dictated that very few people would be present to report lack of damage in places such as those central hills, while many would be concentrated in river bottom areas, contributing to an atypically high site response there.[64]

The content of historical narratives was important to the downgrading of New Madrid intensity by Hough and colleagues. In their 2000 piece, the coauthors reexamined felt reports and translated their assessments of damage differently than had Otto Nuttli. Further, they argued that Nuttli simply misread his own primary source in assigning one important moment magnitude reading—and they pointed out a likely transcription error in the date for another shock. Poor historical analysis, in other words, led to poor seismology.[65]

In a 2004 review article on the use of historical texts, Hough argued that "overall, many of the accounts do not appear to support values as high as those originally assigned." To make this argument, she carefully dissects previous analysis:

> In St. Louis, Missouri, Fuller (1912) describes reports (from the *Louisiana Gazette*, 21 December 1811) of people having been wakened by NM1 [that is, the first major tremor, 16 December 1811] and furniture and windows having been rattled. He notes that "several chimneys were thrown down," and a few houses "split." To understand such accounts one must be familiar with the vernacular of the time; in this case, the word "split" seems to have been used in a number of accounts to mean "cracked" rather than destroyed. Consistently, as in the above example, the phrase "thrown down" is used to describe catastrophic damage to chimneys, walls, or houses. The *Louisiana Gazette* account goes on to note that "no lives have been lost, nor has the houses sustained much injury." This observation also suggests that the word "split" does not imply substantial damage. On the basis of these reports, a MMI [Modified Mercalli Intensity] of VI–VII appears to be more appropriate than the value of VII–VIII that Nuttli (1973) assigns for NM1.[66]

Here, to understand the past environment of the Mississippi Valley and therefore the future seismic risk of a large American city, geologists argue about early nineteenth-century vernacular.

In such debates researchers return to earlier, preinstrumental methods of evaluating ground shaking. Early seismological researcher Robert Mallet, for instance, advocated the careful, comparative use of wall cracks and earth fissures to study earth motion during the 1850s. Such distinctions are still part of the diagnostic criteria for evaluating the intensity of ground motion, though generally in conjunction with instrumental evidence. In this passage, as in other contemporary discussions of New Madrid, modern researchers dissect two-hundred-year-old texts to make what is in essence a preinstrumental assessment of ground motion.[67]

In a 2008 summary article, Susan Hough argued that the New Madrid earthquakes, once overlooked and seen as unimportant, then rediscovered as the largest earthquakes to rock the continental United States, should instead be understood as roughly the same as the 1886 Charleston quakes. Unlike past dismissals of the New Madrid quakes, though, Hough's assessments were based in part on careful analysis of archival records. In the work of Susan Hough and her collaborators, the very texture, siting, and character of nineteenth-century settlement and writing are at issue in reconsidering the New Madrid quakes.[68]

In much of this resurgence of scientific work on New Madrid magnitudes, researchers bring keen tools to the meanings of words, phrases, locations, or building damage. They have used nineteenth-century evidence to create twenty-first century science. Yet even so, such approaches have ignored in these nineteenth-century accounts the essential qualities of narrative—the ways in which letters or reports on illness told a story about processes that were interconnected with a vast web of often ill-understood correspondences and states. In this move from narrative to text—from readers contemplating interconnection to readers astutely extracting snippets of usable, comparable, and at times even quantifiable information—the seemingly same documents appear in fundamentally different, altered form. In similarly subtle forms, emerging scientific study of the New Madrid earthquakes has knit back in older strands of research and methodology.

Recent estimates of New Madrid magnitude draw not only on historical accounts but on comparison with other, similar world earthquakes and on extensive analysis of liquefaction features. Some researchers, including Hough, argue for upper magnitude bounds of 7.1–7.3 for the strongest New Madrid shock in late February 2012. Recent surveys by W. H. Bakun, Arch Johnston, and M. G. Hopper place the upper magnitude of the February shock in the range of 7.4–7.8.*** While lower than the original estimates by

---

*** Careful seismological researchers try to discourage using earthquake magnitude like a trading card: an earthquake in the middle Mississippi Valley, for instance, would likely have a far greater effect than an earthquake of the exact same moment magnitude in California, because of how soils shake, how buildings are made, and how far seismic waves travel. Still, most of us laypeople need a certain amount of trading card simplicity. A range of 7.1 to 7.8 might best be imagined as somewhere between the 1989 Loma Prieta quake and the 1906 San Francisco earthquake. I find this a useful comparison because I lived through one and have seen pictures of the other, and I have no interest in getting close to anything resembling either: in other words, while these magnitude differences

Johnston and Schweig, this range is in keeping with warnings by disaster preparation advocates that the middle Mississippi Valley needs to be ready to shake—and recover.[69]

Hazard Maps, Building Codes, and Lots of Shouting

In 1994, the magnitude 6.7 Northridge quake in Southern California struck directly under a northern suburb of Los Angeles and caused more damage than any tremor since 1906: sixty people dead, more than seven thousand injured, over twenty thousand left without a place to live, and between $13 and $20 billion in financial losses. Such totals demonstrated that even a moderately large quake could cause extensive damage—even to earthquake-resistant construction. In the late twentieth century, lawmakers in regions surrounding the 1811–12 epicenters began to debate building codes and spending priorities, especially how much seismic safety should be mandated in area structures. Some important institutions took earthquake risk seriously. In 1997, the Missouri Botanical Garden opened its new research center and herbarium, the Monsanto Center, specially designed to withstand potential future earthquakes. As one of the world's hubs of plant classification, with delicate samples consulted by experts around the world, the Monsanto Center could perhaps warrant an abundance of caution. Could ordinary buildings, housing people?[70]

Debate over seismic building codes riled many cities and highlighted areas of scientific as well as societal disagreement. Seth Stein in particular waged a war of New Madrid skepticism. In 2004, he and Joseph Tomasello inveighed in the *New York Times* against what they regarded as the high and unreasonable cost of seismic safety in New Madrid, editorializing on "When Safety Costs Too Much." Myron Fuller might well have been amazed to read of this forgotten earthquake in the pages of the *New York Times*.†††[71]

"You can build a school with steel and make it withstand any earthquake within 2,500 years," Seth Stein argued at a 2008 Arkansas hearing, "or you

---

are extremely important in seismological terms, as a practical matter they all suggest the need for safety plans, water heater bolt-downs, and resilient construction.

††† Myron Fuller would likely have been just as dismayed at some of the ways awareness of midcontinent earthquakes has entered American life. A 2006 made-for-television mini-series, *Apocalypse 10.5*, follows a catastrophic 10.5 earthquake that destroys Los Angeles and triggers ancient rifts, splitting North America in two along the Mississippi embayment. In wildly inaccurate ways, the recent three decades of research on the subsurface structures of the New Madrid seismic zone bubble up to the surface of popular awareness in the improbable and overstated graphics of this unabashedly trashy bit of pop culture.

can hire teachers with that money." Such objections struck a chord with many. In 2008, mayor Barrett Harrison of Blytheville, in northeast Arkansas, spoke out against new seismic building regulations adopted in his state. He argued that if a company was considering moving to his region rather than one without a seismic code, "and it's going to cost a million more dollars to build a facility here that's earthquake proof, we may very well lose the industry." ‡‡‡ [72]

Building codes in Memphis, Tennessee, were a particular flashpoint, as citizens and policy makers debated code upgrades in the early 1990s and through the first decade of the 2000s. Against criticism from some in the public, the scientific and engineering community increasingly reflected new understandings of New Madrid in strengthened code recommendations. In 2000, the International Code Council used the revamped USGS estimation of seismic hazard in setting guidelines that buildings in the New Madrid seismic zone should be built or retrofitted to essentially the same standard as those in Southern California. [73]

Such recommendations struck at a key element of New Madrid skepticism: a deep sense from many in the midcontinent that California, not their own terrain, was American earthquake country. Whether California was an appropriate model for mid-American earthquakes resonated as a crucial question, from Myron Fuller's 1912 report to James Macelwane's early twentieth-century advocacy to Otto Nuttli's 1973 work. Through the subsequent resurgence of interest in seismicity in the American interior, California has been a constant reference point. [74]

Increasingly, public information reflected growing scientific consensus that earthquakes in the eastern United States *were* different from those "out West": felt over a broader area and able to create more damaging shaking even at lower magnitudes. In 2006, Gary Patterson, information officer for CERI, warned a public meeting in northeast Arkansas about underestimating damage from future New Madrid tremors. "This is not a California earthquake," Patterson stressed, emphasizing repeated shocks and the impact of seismic shaking on loose soils. "There are some basic differences here that drive the hazard level up." Yet without the frequent midrange earthquakes that can break pottery and jangle nerves and in the process create social consensus about the need for seismic preparation for bigger quakes, middle Mississippi Valley regulatory advocates have to struggle against the challenges of what CUSEC director Jim Wilkinson termed a "low probability, high consequence hazard." [75]

---

‡‡‡ It would not, in fact, cost a million more dollars—but even a few percentage points of added cost can be significant.

Despite increasing scientific agreement, skeptics seized upon California comparisons. In a 2003 editorial in the journal *Eos*, New Madrid critics Stein, Tomasello, and Andrew Newman asked, "Should Memphis Build for California's Earthquakes?" Their answer: a resounding no. In the past century, they observe, "earthquakes typically have been more of a nuisance than a catastrophe." Stretched resources should be spent wisely, not stretched further: proposed building codes, they acknowledged, "might, over time, save a few lives per year, whereas the same sums invested in public health or safety measures (flu shots, defibrillators, highway upgrades, etc.) could save many more."[76]

Debate and Growing Consensus

The 2003 Stein, Tomasello, and Newman *Eos* editorial was a shot across the bow, and others in the field responded. USGS geologist A. D. Frankel, a lead author of the USGS seismic hazard maps, published a response in *Eos* criticizing the methods of hazard assessment used by Stein, Tomasello, and Newman and assailing the "unrealistic calculation of seismic loss in Memphis over the next 650 years." Susan Hough wrote to emphasize the occurrence of large quakes many times in past millennia, as well as the potency of triggered shocks at large distances from main shocks.[77]

Agreement about the need for public planning continued to mount despite further debate over new forms of evidence—debate in popular media as well scientific literature. Beginning in 1991, geodetic data from GPS measurements showed only low rates of strain accumulation in the New Madrid seismic zone, which in turn suggested to some researchers that sizable, threatening earthquakes on the fault would take eons to build up. A team whose authors included Andrew Newman and Seth Stein argued that "the hazard posed by great earthquakes in the NMSZ appears to be overestimated."[78] Late in 2009, Stein and Missouri-based geologist Mian Liu argued in *Nature* that the New Madrid fault system is shutting down. Recent rumblings, they argue, are simply aftershocks.[79]

"It just doesn't work that way," responded CUSEC director Charles Langston to Stein and Liu's *Nature* paper. "It takes hundreds of thousands of years for the Earth to do something—either start up or shut off." By 2005, geodetic data conflicted with the earlier studies of Stein and his team. Robert Smalley and colleagues reported contradictory data of strain rates comparable to those across plate boundaries. The article announcing this finding, by Smalley and colleagues, came out in *Nature* in 2005, alongside a supporting discussion by leading New Madrid researcher Tish Tuttle.[80]

These debates made big news. In Arkansas, researchers' arguments over GPS data and fault loading made the first page of the state news section of the Sunday paper.[81] In sharp contrast to the muddled messages that accompanied the Iben Browning prediction, advocates for seismic planning have increasingly polished sound bites. "It's not like we think it's going to happen tomorrow, but we need to plan," cautioned CUSEC's Gary Patterson in 2010. "Past New Madrid earthquakes were really big, they really happened, and they could happen again."[82]

Far from esoteric disagreements, New Madrid debates carried clear social and policy consequences. Struggles over seismic preparation took place not only in the context of deep global recession, but of the draining away of population and resources throughout the agricultural regions of the middle Mississippi Valley. "The farmers of east Arkansas," wrote regional planning official Rex Nelson in a newspaper editorial in 2009, "are some of the best in the world at producing food and fiber." Even so, loss of jobs to industrial cities in the era of World War II and the increasing mechanization of agriculture meant that, as Nelson noted, license plates in eastern Arkansas driveways around Christmastime were apt to be from Illinois, Michigan, Ohio, even California. Arkansas might be where grandparents lived, but it was not where children or grandchildren worked.[83]

In response, preparation advocates increasingly met the New Madrid skeptics on their own ground, emphasizing the stark societal consequences of earthquake-mitigation choices. In 2009, USGS issued a fact sheet to rebut the arguments of the New Madrid skeptics. The authors emphasized USGS estimates that over the next fifty years, the New Madrid seismic zone stands about a 7–10 percent chance of having an earthquake roughly the same size as the 1811–12 quakes. The chance of a magnitude 6 or larger earthquake in the next half century is much greater—25 to 40 percent. Consequences of even such a seemingly moderate earthquake sequence included impassable roads, extensive flooding, contamination by stored agricultural chemicals, levee failure, and disruption to telecommunication and fossil fuel pipelines. In a resounding rebuttal to New Madrid skeptics, this group of USGS experts emphasized that midcontinent earthquakes "can be expected in the future as frequently and as severely as in the past 4,500 years. Such high hazard requires prudent measures such as adequate building codes to protect public safety and ensure the social and economic resilience of the region to future earthquakes."[84]

Perhaps most ominously, the USGS report warned that "unreinforced schools and fire and police stations... would be particularly vulnerable when subjected to severe ground shaking." Warnings such as these reveal the dark shadow of the 2008 earthquake in the Chinese province of Sichuan. Over all discussion of earthquake hazard in the early twenty-first century falls the knowledge of the grim toll of this catastrophe, in which schools throughout the region pancaked and collapsed, killing thousands of schoolchildren.[85]

A 2010 FEMA-funded report underscored these seismic concerns. Released by the University of Illinois Mid-America Earthquake Center, the report estimated damage from a magnitude 7.7 New Madrid earthquake. Direct losses could reach $300 billion, the authors concluded, with indirect losses at least twice that. Tennessee, Arkansas, and Missouri would suffer the most, with damage reaching into Illinois and Kentucky. About 715,000 buildings could be damaged due to an earthquake, with major destruction to infrastructure like roads and bridges, hampering services from emergency personnel. Such an earthquake could kill over 3,000 people and injure 86,000; 2.6 million households might be left without power, with 215,000 people seeking shelter. The stakes for debates over earthquake history are high indeed.[86]

Increasingly, in planning recommendations as with magnitude estimates, experts have come to consensus around both history and caution: past New Madrid earthquakes indicate future threat. Working groups in 2011 and 2012 for the US Nuclear Regulatory Commission and the National Earthquake Prediction Evaluation Council (NEPEC) all firmly endorsed the notion that past is warning: very large earthquakes (even if there is still debate over exactly *how* large) create future threat that present populations must prudently heed. "The New Madrid seismic zone," concluded the 2011 NEPEC expert panel, "is at significant risk for damaging earthquakes that must be accounted for in urban planning and development."[87]

New Madrid Bicentennial

Achieving scientific consensus—even near consensus—is challenging enough. Planning for earthquakes is far more so. Recent science emphasizes past *sequences* of earthquakes—sets of events that would stymie even the most capable planning. At a 2006 earthquake preparation event in Little Rock, Arkansas, organized in part by the national disaster-preparedness advocacy group Protecting America, emergency planners for the state attempted to raise public awareness of earthquake threat and also inform Arkansans about mitigation and response efforts already in place. Earnest transportation engineers laid out plans, for instance, to repair highway overpasses after the next New Madrid earthquake. A visiting historian raised a hand and asked about plans for repeated shakes—what would happen if, just as new repairs are put into place three months after a tremor, another massive one hits? The response was both honest and pained: new knowledge of earthquake sequences are simply not yet incorporated into disaster plans. It is hard enough to plan for one earthquake, but no one has any real sense of how to plan for three or four in quick succession. At a similar 2012 presentation by a Tennessee highway engineer, the question and the answer were virtually identical.[88]

Still, earthquake planning in the midcontinent has moved well beyond the zone of punch line and into actual emergency preparation. USGS handbooks encourage residents of the middle Mississippi Valley to keep a bag with shoes and a flashlight by the head of their beds, and they supply checklists for relatively easy household remediation such as securing water heaters and installing cabinet latches. "Earthquake weeks" now take place in at least five states of the New Madrid seismic zone. School workshops and "geocaching" projects involve children and families. In exercises that can leave grandparents of the Cold War era both reassured and bemused, schoolchildren in many parts of the New Madrid seismic zone learn that when the earth shakes, they need to "Duck, Cover, and Hold On!"[89]

Concern and planning, research and debate, came together in plans for the New Madrid Bicentennial, a year-long set of programs and events for experts and the public. The 2011 National Level Exercise in disaster response was based on "a catastrophic NMSZ event," the first such disaster-planning exercise based around a fundamentally natural disaster. Following the model of California communities, CUSEC and other planners initiated the Great Central U.S. Shake-Out, a form of regional earthquake drill for businesses and schools as well as public institutions. The Shake-Out was supported by a set of disaster-planning organizations including CUSEC, FEMA, the American Red Cross, NEHRP, Ready.gov, and the USGS. The 2012 Shake-Out involved more than 2.4 million people across the midcontinent.[90]

New Madrid bicentennial commemorations included National and Eastern Section meetings of the Seismological Society of America held in Memphis and Little Rock, respectively, at which seismologists, paleoseismologists, and earthquake engineers presented recent findings and took field trips to view prehistoric sand blows in the New Madrid seismic zone and the Marianna area. The bicentennial also included public talks and workshops, an auction of a "Dream Home" built to withstand earthquake shocks through the St. Jude Dream Home program, and tours of seismically retrofitted buildings throughout Memphis. A National Earthquake Conference in Memphis in April 2012 brought together planners and experts across many disciplines and governmental agencies to share lessons, strategies, and earthquake-preparation products ranging from enormous stabilizing rollers for banks of hard drives to a popular courtyard exhibit colloquially called the "shaky house," which seismologists, engineers, curious tourists, hotel staffers, and one somewhat apprehensive historian could climb inside to experience the sensations of a 7.2 earthquake. Such efforts marshaled history as a resource in conversations among experts and in awareness-building among the larger public.[91]

Public media and popular sources increasingly recognize the historical existence of the New Madrid earthquakes of 1811–12, and even to some extent the prehistoric earthquakes in and near the New Madrid seismic zone. Policy and regional planning increasingly incorporates earthquake potential—or at least is beginning to. That the New Madrid Bicentennial could even exist is one demonstration of this new recognition of the 1811–12 shocks as real historical events and of mid-Mississippi Valley seismicity as part of history and potentially of the future.

A New Geography of Disaster

At the same time, much of what made the New Madrid earthquakes real to many in the midcontinent had little to do with governmental planning or educational efforts, but had to do instead with headlines and world events. In 1811 and 1812, the New Madrid earthquakes were for Americans and Europeans an event of the far West. Two centuries later, future earthquakes were a concern for the poor South. Early twenty-first-century disasters reoriented concerns toward the vulnerability of poor and rural people far from national centers and resources.

The terrorist attacks of 11 September 2001 on the Pentagon, New York City, and (presumably) the White House changed a great deal in American culture. In the United States and abroad, resources and preparation focused on buttressing against large-scale inflicted damage and preparing for disaster and emergency response. Four years later, though, little of this vaunted capacity seemed anywhere in evidence, as a stunned American public stood helpless to alter the unfolding catastrophe in New Orleans and the Gulf Coast on Labor Day weekend 2005. Decades of flood control policies and unheeded warnings crumbled like the levees themselves in the force of Hurricane Katrina and its subsequent storm surge. As Katrina devastated much of the Louisiana and Mississippi Gulf Coast, policy makers and scientific advocates of the middle Mississippi Valley found themselves transfixed by images of poor Southerners trapped on rooftops, pleading for aid, and cut off from assistance for days. Watching, they were helpless and aghast—and they pictured their own citizens.[92]

"After Katrina" became shorthand for people in the Louisiana Gulf and for disaster planners more generally, standing for their awareness that federal help would not arrive swiftly in the aftermath of catastrophe. After Katrina, all knew with grim certainty, local people and local governments have to be prepared to respond and help themselves.

Many in the New Madrid seismic zone also drew sobering lessons from another regional disaster—precisely because of its relative invisibility on the national stage. In early January 2009, a vast winter storm coated many areas of the New Madrid seismic zone and the Ohio Valley with unusual and crippling ice. It led to power outages for over two million households, killed sixty-five people, and required the largest callout of the national guard in Kentucky's history. "This thing was 387 miles wide and 100 miles deep," commented Brig. Gen. John W. Heltzel of the Kentucky Division of Emergency Management, and director of Kentucky's response. "Katrina wasn't that big. This thing was huge." At the same time, the catastrophe barely registered in the national news media. Far-off friends and relatives called to chat, blithely unaware that people in the affected areas were coping with anything like such a disaster. The ice storms of January 2009 were a reminder to many of the invisibility the same region might face in a large earthquake—and their need to mount self-sufficient response after a future disaster.[93]

At the same time, the public view of the New Madrid threat was shaped by an increasingly international understanding of comparisons for New Madrid seismicity—an internationalization that harked back to the reaction to the 1811–12 quakes. A main comparison made in the early stages of the revival of interest in New Madrid seismicity was to a 7.8 earthquake in 1819 in the Kutch region of India that created a sharp uplift known as the Allah Bund, or "wall of God," and which submerged and buried a British fort in the same way that the New Madrid quakes buried baldcypress along the St. Francis River. Later researchers would trace similarities between the New Madrid earthquakes and the devastating quake that struck Bhuj, India, near Kutch, in January 2001.[94]

Early twenty-first-century efforts to understand seismicity in central North America increasingly turn to northern China, where the 1556 Shaanxi and the 1976 Tangshan earthquakes provide horrible reminder of the human cost of intraplate seismicity. Researchers draw similar parallels between the United States and Australia, where in 1989 a 5.6 intraplate earthquake—relatively mild by seismic standards—killed thirteen people and caused a staggering $4 billion in damage. Such high cost and relatively high casualties served as a grim reminder of the potential impact even of moderate earthquakes on human communities unprepared for them.[95]

Similarly, re-evaluation of midcontinent seismicity took place as part of a larger re-evaluation of many areas of the United States once never thought of as earthquake zones and increasingly regarded as threatened areas. Recent research on historical New England earthquakes has begun to establish that region also as a zone of threat, for scientific researchers and increasingly in public media. Disaster planners as well as earth scientists are beginning to take stock of the Charleston, South Carolina seismic zone and the Charlevoix seismic zone in Quebec. Earthquakes may be a concern not simply for those living in the middle Mississippi Valley but for residents of what had seemed to be previously quiet zones of the mid- and eastern continent.[96]

Global Earthquakes, Global Awareness

Recent seismicity in the United States and throughout the globe has heightened scientific and public interest in earthquakes generally and in the particular hazards posed by midcontinent earthquakes. Throughout 2010, tremors rocked parts of the world from the Chilean coast to the deltas of New Zealand. Each provided lessons for disaster planners, seismologists, and ordinary American citizens who witnessed reaction and recovery efforts through the Internet and news—and increasingly through social media and personal interaction.

Two major earthquakes struck early in 2010. One, a large earthquake in Haiti, caused catastrophic and likely nation-changing damage. The other, in Chile, was more than an order of magnitude larger, yet its damage—though severe—was widely viewed as both contained and recoverable. The sobering comparison of these earthquakes signaled to many in the United States the high stakes of earthquake preparation.[97]

The earthquake that rocked the Republic of Haiti in January 2010 had a moment magnitude of 7.0, only slightly larger than the Loma Prieta earthquake that hit San Francisco in 1989. It is among the most destructive earthquakes in the history of humankind. The earthquake hit a country already struggling under the weight of high poverty, limited infrastructure, both recent and historic political instability, and recovery efforts from recent hurricanes and tropical storms. Many homes were constructed by hand, often on steep hillsides out of make-do materials. Few buildings were designed by engineers, much less designed to seismic safety standards: even engineering professionals in Haiti are likely to have been trained under French building codes that do not include provisions for earthquake resistance. The earthquake's damage to homes, businesses, and institutions was widespread and in many places catastrophic. Port facilities and portions of the capital, Port-au-Prince, suffered severe damage related to earthquake-induced liquefaction. Approximately 15 percent of the nation's population was directly affected: out of an overall population of nine million, well over 300,000 Haitians perished and over a million were forced into shelters. Recovery will continue for years.[98]

Six weeks after Haiti, while its port was still largely obstructed and recovery efforts were still meeting emergency needs, a far more massive earthquake struck Chile. At an 8.8 moment magnitude, the Chilean earthquake ranks as the largest ever directly reported by strong motion instruments. Like Haiti, Chile has a long history of powerful earthquakes, most notably the immense 9.5 earthquake whose tsunamis devastated the Chilean coast in 1960. Unlike Haiti, Chilean seismic building standards, which are often modeled on those of the United States, are robust and well enforced. American earthquake engineers evaluated this earthquake as a "natural laboratory." The earthquake was massive, yet its damage was, relatively speaking, contained. Though the earthquake affected an area of over 8 million people, it caused only—if deaths can ever be counted in *only*—approximately 550 deaths. The tremor caused $30 billion of damage, fully or partially closing one in five hospitals in the affected areas, yet most Chileans who felt the powerful tremors walked alive out of partially damaged but still-standing buildings. Compared to Haiti—a far less powerful earthquake—the damage was limited indeed. For Americans, the "take-home message" of Chile's earthquake was that seismic building standards and earthquake preparation made a difference in people's lives and community recovery.[99]

From September 2010 to February 2011, a series of large earthquakes shook southeastern New Zealand, in the coastal plain east of the Southern Alps. Over 180 people were killed, and the central business district of Christchurch, New Zealand's third largest city, had to be entirely sealed off. Ultimately half of its buildings may need to be torn down. Economic losses were estimated by mid-2012 to be at least $20 million NZ, with recovery still drastically incomplete. The region most affected by the earthquakes was in some ways similar to the New Madrid seismic zone: a deep embayment created by a main river. In New Zealand, liquefaction was less severe than in the 1811–12 New Madrid earthquakes, but it was nonetheless stunningly visible. Images of vans stuck upright in sand blows and large streets turned into mud pits fascinated viewers of YouTube worldwide, but gave highway engineers in the mid-American continent reason to shake their heads and look grave. Looking at the evidence from New Zealand, a wide variety of professionals concluded that in areas prone to liquefaction, moderately large earthquakes can produce severe damage and economic consequences even in the absence of large-scale casualties.[100]

The New Zealand quakes would have continued to be major seismic news through much of 2011, except for what happened on 11 March 2011: shortly before 3:00 p.m. local time, the magnitude 9 Tohoku earthquake and resulting tsunami struck off the northeast coast of Japan. The earthquake occurred along the subduction zone of the Pacific tectonic plate that dips beneath the island. It was the largest and most destructive temblor in Japan's recorded history, one that ruptured a stunningly large stretch and created a tsunami both unanticipated and lethal.§§§ The Fukushima Daiichi nuclear plant was flooded, damaged, and shut down—but not before partial melt-downs and breaches of containment vessels leaked radioactive waste into the surrounding environment. Such seismic catastrophe in arguably the world's most earthquake-prepared nation calls into question US resumption of licensing of new nuclear power plants and earthquake planning generally.[101]

The seemingly esoteric observation that earthquakes have more impact and are more felt in the eastern than the western United States was brought home dramatically by the Mineral, Virginia, earthquake felt throughout the East Coast and as far away as Chicago and Montreal. In August 2011, this relatively moderate tremor—moment magnitude 5.8—sent people out into the streets in panic across the central Eastern Seaboard. People in federal agencies in the nation's capital feared it was a terrorist attack. With key systems down, businesses were unable to communicate with their employees milling in the street. The earthquake exceeded the design criteria and triggered an automatic shutdown of the North Anna Nuclear Power Plant located only eleven miles from the earthquake epicenter. Buildings within a hundred-mile radius sustained damage. Because of the earthquake's harm, the Washington Monument remains closed to the public. Now there is heightened concern about the earthquake potential of the Central Virginia seismic zone. Could this or other areas in the developed and populated Eastern Seaboard produce earthquakes larger and more damaging than the 2011 earthquake?[102]

Further fossil fuel development also created seismic anxieties for many in the midcontinent. Recent changes in technology have made the mining practice of hydraulic fracturing, or fracking, both possible and potentially economically rewarding on a large scale. Residents of north-central Arkansas, near the small town of Guy, began to feel clusters of earthquakes

§§§ Though paleoseismologists had warned over the previous decade that previous large tsunamis had hit the coast in the past three thousand years, change in disaster planning and seismic engineering came too slowly to save the lives of the over 15,700 people killed and almost 5,000 missing.

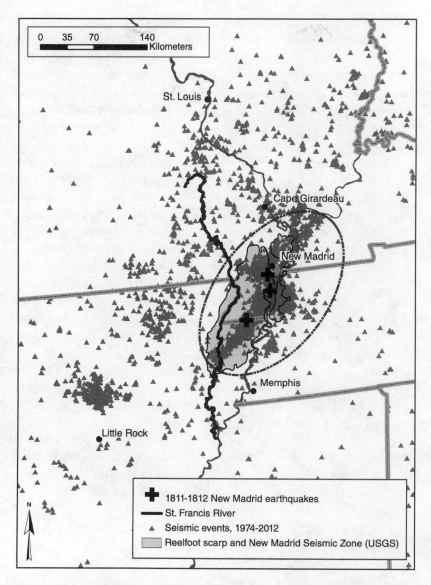

FIG. 7.1. New Madrid-area epicenters, 1974–2012. Seismic monitoring reveals the on-going small-scale seismicity in the New Madrid seismic zone (the dashed ellipse). The clump of events above Little Rock represents an earlier set of small earthquakes, as well as the 2010–2011 Guy earthquake cluster, attributed to the waste-disposal drilling associated with hydraulic fracturing, or fracking. Continuing research into the New Madrid region raises questions about societal preparation as well as our own created seismicity. (Geographer: Bill Keegan)

in late 2010 and 2011, shortly after energy companies began to inject the waste slurry produced by fracking into disposal wells near fracking sites. The largest earthquake of what became known as the "Guy swarm," in February 2011, measured only 4.7 moment magnitude—but it was felt across ten states.[103] This nerve-wracking tremor prompted calls in small local communities for fracking to stop—calls amplified by national media that reported concerns over fracking in many communities, including in Ohio, Pennsylvania, and New York.[104]

People in central Arkansas were unaccustomed to paintings leaping off shelves and houses suddenly shaking. Many were spooked not only by the tremors themselves, but by the fear that they might be connected to even larger earthquakes. Arkansas Geological Survey staffer Scott Ausbrooks is the "go-to" person for seismic concern in the state and was for much of 2010 and 2011 a very busy man. His phone rang constantly, he reported, with people asking, "If we get an earthquake in Greenbrier [a small town near Guy], is that going to trigger one in New Madrid?"[105] Though the scientific answer to that questions appears to be "no," at least not on the scale people are worried about, the political question remains: how much shaking are residents of an earthquake zone willing to feel in order to maintain inexpensive gas prices?

Amid national concerns about energy policy and environmental harm, relatively low-level seismicity in a rural mid-South state—along with similar disturbances elsewhere—raised questions both economic and environmental. The Arkansas legislature imposed a moratorium on new fracking in the region, which apparently "turned off" the tremors, but the question of induced seismicity remains intensely studied and politically fraught.[106]

At the time of these highly contested small earthquakes, earth science researchers noted a surge in seismicity generally in the region between the Rockies and the Appalachians. "Beginning in 2001," warned a 2012 USGS report, "the average number of earthquakes occurring per year of magnitude 3 or greater increased significantly, culminating in a sixfold increase in 2011 over 20th century levels." Together, these unexpected earthquakes, global and North American, some catastrophic and some simply unsettling, created a climate of popular and scientific concern about potentially damaging earthquakes.[107]

Making Old Knowledge Anew

Just as the American public is revisiting collective common knowledge about earthquakes, more expert researchers have likewise called upon older approaches within their fields. Coming to understand the New Madrid earthquakes involves a resuscitation of seemingly moribund forms of knowledge. Such continuities link up-to-date science with approaches from a very distant past.

Scientists have turned to the historical record, taking folk stories seriously and building new arguments based on the accounts of earthquake experiences that were important to nineteenth-century vernacular science. Much of the late twentieth-century creation of knowledge about New Madrid consists of the sometimes sudden transformation of tall tales—a river flowing backward, massive earthquakes, huge volcanoes of hot slurry—into substantiated fact. One team of researchers in 1998 titled a section of their paper on New Madrid earthquake faulting "Extracting Facts from Folklore." They attempted to take seriously stories that had been merely a scientific "curiosity," such as tales of the Mississippi running backward. Scientists have simultaneously brought some of the methods of nineteenth-century vernacular science to bear on twenty-first-century data gathering and research. They have begun to turn to those closest to the evidence, farmers in area fields, as sources of expertise. They have created broad-based modes of experiential and quantitative data collection, inviting input from those who experience quakes.[108]

But scientists' use of the historical record and the methods of nineteenth-century vernacular science has been tentative and selective. Researchers employing these sources and methods, always alert to the longtime marginality of their work to mainstream seismological research, use only the bits of historical data and modes of research easiest to incorporate into the kind of modeling they are comfortable doing.

Extrapolating these research trajectories, with awareness of the full history of the New Madrid earthquakes and the centuries of vernacular and professionalized science surrounding them, suggests possibilities for future research. There is potentially much more to be tapped in the historical record. Nineteenth-century vernacular modes of data gathering, communication, and theorizing might productively inform modern earthquake research more broadly.

Native Knowledge Reappraised

Indigenous peoples' knowledge and understanding of quakes in the midcontinent were the subject of investigation in the early nineteenth century, and continue to lead in interesting directions today. "Since the settlement of our country," wrote the editor of an 1812 collection of accounts, "we have no record of such dreadful convulsions of the Earth as is recounted in the following pages." On the contrary, others insisted that earthquakes were part of both the recent and the longer history of the Mississippi Valley. "The fact is," concluded territorial official Amos Stoddard in 1812, citing earlier reports and geological phenomena such as pumice stone, as well as his own experience of tremors in 1804, "that earthquakes are common in that country, and may be traced to the first settlement of it."[109]

Such fundamental disagreement often came to a head in American and European writers' assertions about Native American knowledge of earthquakes. What was the Indian history of the earth? Indigenous communities' stories of prior quakes in the New Madrid region were frequently cited through late nineteenth- and early twentieth-century literature on New Madrid—though in typical nineteenth-century fashion, rarely with much in the way of specificity or precision about the sources of that knowledge. Charles Lyell, for instance, reported in the 1840s being "informed at New Madrid, that the Indians, before the year 1811, had a tradition of a great earthquake which had previously devastated this same region." Deciding that the environmental evidence said no such thing, he left it at that.[110]

The few hints of Indian knowledge of earthquakes are tantalizing. In an 1819 guide to Louisville, Kentucky, a local booster recounted a conversation seven years before, in 1812: Several leaders, including Black Feather of the Shawnees, were encamped on their way to Fort Meigs along with American troops when they felt the earth shake. Black Feather "observed that the *earth tottered.*" Asked by an American officer if he had ever felt anything similar, Black Feather counted back to reply that he had felt similar tremors three times during the past forty years. "This anecdote," the writer made sure to include, "I have from a gentleman who was present."[111]

Recent scientific views have in many ways circled back to this earlier nineteenth-century view of an active seismic history and of the value of oral traditions from people with a long cultural history in a place. Such reappraisal is a local version of a global story—as, for instance, footsteps of the archangel Michael in a southern Italian village, reinterpreted in 2005 from a seismological perspective as fault features. Recent geological work on Australia's east and west coasts has revealed evidence of massive tsunamis, likely from multiple meteor impacts in the nearby oceans. Such geological reassessment bears out aboriginal legends of a white wave falling from the sky. Around the beginning of the twenty-first century, researchers correlated long-told oral histories among Native peoples of the Pacific Northwest with earthquake records in Japan and with evidence of damage to trees inland of coastal areas, confirming that in 1700 a massive tsunami swept the region.[112]

Perhaps this reassessment of former knowledge or past peoples' experience may go even deeper. Renewed or sustained Indian knowledge may suggest some changes to current scientific models, in particular, the insistence that the New Madrid quakes were only felt east of Mississippi and attenuated strongly thereafter. Edwin James's accounts of Missouri Indians suggest that the quakes were felt strongly along the upper Missouri—an area almost never included in maps of the quakes' effects.

Paleoseismological evidence of earlier quakes may also alter narratives of much earlier communities. Geological evidence of the Mississippian culture of the middle Mississippi Valley has begun to suggest that many of the Mississippian mounds may have been built over sand blows, as if to stop them up or recognize them. (Alternatively, Mississippians may have constructed ceremonial mounds on sand blows simply out of convenience: hummocks of well-drained sand certainly prove useful in areas prone to seasonal flooding.) Yet the chronological coincidence—as far as the imprecision of carbon-dating can show coincidence—is intriguing. Mississippian culture is traditionally dated to around 900 CE, right around the time paleoseismologists like Tish Tuttle have established that powerful quakes shook the region. How might prior large earthquakes have changed the communities of the middle Mississippi Valley? Such questions have implications for the early history of European and indigenous contact. If an earthquake struck the region somewhere around the mid-1400s, was the European explorer Hernando de Soto moving in the 1540s among communities in the generations following a massive earthquake?[113]

A New Vernacular Science?

Throughout research on New Madrid runs the thread of reappraisal of once-dismissed knowledge. In his landmark 1973 study, Otto Nuttli used the tables generated by Jared Brookes and Daniel Drake to estimate intensities—just as Myron Fuller had relied upon their accounts in his own 1912 analysis. Both of these nineteenth-century observers appear in current scientific literature as key points in heated debates over the magnitude of the tremors, such as Susan Hough's reassessments of damage reports from bottomland regions along the Ohio.[114]

Writing in the years leading up to 1930, Bootheel reclamation director Otto Kochtitzky recounted his experience as a younger man listening to a steamboat captain describe a mud-filled channel in the riverbank downstream of New Madrid as an old earthquake feature, a crevice formed by the 1811–12 earthquakes. Noting that "cypress trees more than two hundred year old were standing in plain view on its surface," Kochtitzky scoffed at this version of regional environmental history. Clearly, the old crevice was simply "a channel or an erosion of an ancient overflow bayou," and the captain was victim of regional tale-telling: "We all enjoy romance," he concluded dismissively. Yet recent paleoseismology suggests at least the possibility that the river captain could have been accurately describing a feature created by an earthquake long before 1811.[115]

Over the very late twentieth century and into the early twenty-first, New Madrid researchers have returned to paying attention to first-person accounts and to highly localized, site-specific observation. These techniques—techniques that would have been very familiar to the naturalists and botanizers and busy theorizers of the nineteenth century's vernacular science—are at a center of a revolution in the study of intraplate seismicity. Common queries about earthquakes today echo concerns of the nineteenth century—links across vast distances, animal connections, meteorological storms. At the same time, present-day scientists, who are often intensely interested in nineteenth-century reports for what they reveal about the movement of faults and the travel of seismic waves, generally dismiss out of hand the widespread reports of flashing, burning, glowing, hovering lights. These are for modern researchers of an entirely different order than reports of chimneys cracking or crevasses opening, both of which are read eagerly and carefully. Similarly, interest in earthquake noise is largely absent from late twentieth-century and early twenty-first century seismology, though the noises of earthquakes were a main aspect of earlier earthquake description. As New Madrid research continues, will investigators continue to reappraise older forms of knowledge, or find in previously dismissed insights some unrealized bit of evidence?[116]

Bodily Experience and Changing Seismology

In the New Madrid earthquakes, the lived experience of human and animal bodies became part of the history of the natural environment. In early nineteenth-century reports of the quakes, natural events act in and through living creatures. The disoriented sensations of human beings and the alarm of cattle, pigs, ducks, and geese were as much a part of the earthquakes as huge, spurting sand blows.

For early nineteenth-century observers, to reference animal bodies as a measure and indication of an earthquake was straightforward. Tiger, the Newfoundland dog aboard the steamboat *New Orleans*, was alone among those on board in perceiving the more moderate shocks while underway. He "prowled about, moaning and growling" at the smaller tremors, but when he put his head on his mistress Lydia's lap, "it was a sure sign of a commotion of more than usual violence." Similarly, after the 1885 Charleston quakes, one physician reported that a "very intelligent dog" got up from a hearth one morning, "went into a corner of the room, and remained there trembling and shivering. His master remarking that the animal was mad, his little son, who had a similar experience previously, replied, 'No; it is an earthquake.'" Soon after, the family felt "a well-developed trembling" bearing out the dog's presentiment.[117]

As earthquake science generally became formalized, instrumentalized, and quantified, trembling dogs retreated from the picture. In the early twenty-first century, many seismologists are extremely dismissive of animal behavior, even though many nonscientists take for granted that their household animals behave differently before storms or even earthquakes.**** Scientists' skepticism may in part reflect how profoundly most people in developed countries are separated from daily knowledge of large numbers of birds and domestic creatures. Few modern North Americans live in environments surrounded by horses and cattle, pigs and goats, domestic fowl, dogs, and creatures in streams, woods, and fields. Our expertise and familiarity with many animals and birds and our use of bodily referents have become estranged from the processes of demonstration and proof of instrumentalized and objectified seismology.[118]

Only in Chinese science have reports of animal behavior—as well as other forms of folk knowledge about earthquakes—been a part of modern mainstream earthquake science. Under Mao's dictum "Science walks on two legs," expert knowledge after the Cultural Revolution was to work alongside folk knowledge. Folk knowledge of snakes coming out of the ground, wells being disturbed, and household fowl behaving strangely before earthquakes became the subject of serious study, as researchers tried to establish mechanisms and also understand how widespread such phenomena might be. Seismologists established a program of folk observation that involved youth groups, schoolteachers, and community associations. These groups would go out into the countryside to check devices measuring electrical differential across expanses of soil, take down reports of odd animal behavior, or record various measurements about local wells.††††

**** Some of this dismissal stems from the countless good stories that come after a quake; some too comes from the undeniable power of post hoc explanations: Susan Hough explains that house cats, for instance, are often later reported as behaving peculiarly before a quake—but on any given day, quite a few housecats are likely to be behave peculiarly, and might cat owners simply remember more vividly the ones that do so before a major quake?

†††† During casual conversation one morning, I discovered quite to my surprise that my daughter's Chinese teacher was part of this program as a young teacher-in-training. I should not have been surprised: this was a massive program involving people of many regions, ages, and professions.

Though much of the political excess of Chairman Mao and the Gang of Four was subsequently repudiated, Chinese seismology has continued to possess a strong strain of respect for vernacular knowledge of earthquakes. Such approaches were involved in what some regard as the only successful—if still not fully explained—lifesaving evacuation before a quake, in 1975. Folk traditions of interest in animal, bird, and reptile behavior continue to inform Chinese earthquake research.[119]

In European and American science (and indeed in most non-Chinese seismology), investigation into the possible extent of animals' precursor behavior has been slow and halting. As of the early 1980s, research suggested that birds, animals, and fish might well be sensitive to sound frequencies and electrical displacements that would render them much more responsive to the environmental signs of coming quake. More recent researchers have cautioned that abundant anecdotal evidence of animal precursors does not seem to be borne out by careful study. In one study, California ants were unbothered by a coming quake, and in another, the "seismoometers" mounted on the backs of cows in Sweden registered tremors that left the grazing cows unbothered. #### The use of sensing bodies to identify possible earthquake precursors is an area of contested work, fraught with the anxieties of scientists on the fringes of recognized science.[120]

Attention to animal bodies thus indicates how bodily knowledge functions differently in different scientific regimes. In early nineteenth-century North America and mid to late twentieth-century People's Republic of China, the sensitivity to unusual behaviors characteristic of cultures that live with and among large populations of nonhuman creatures led to detailed reporting and widespread interest in bodily reaction and bodily prediction of seismic events. In the twentieth- and early twenty-first-century West, science has been largely divorced from such background bodily knowledge—how many contemporary Americans rely for the winter on the meat of deer or ducks we have hunted or the milk of cows we tend? As Americans' daily experience has grown separate from that of large groups of animals, birds, fish, and other creatures, Western science has asserted skepticism about the ability of living bodies to supply useful and actionable information about coming quakes.

One seemingly fanciful account of the mass migration of squirrels suggests that the older accounts might actually have something to offer to modern science. Charles Latrobe—a relative of steamboat pioneer Lydia Latrobe and one source on her family's journey—also wrote an account of

Seismologists—and earth scientists generally—seem to be unusually fond of terrible puns.

the New Madrid earthquakes in an entertaining 1835 memoir, *The Rambler in North America*. His memoir was widely read, and it has been regarded as somewhat fancifully embroidered. In one passage, for instance, Latrobe noted many portentous omens of the fateful year of 1811, including that a "countless multitude of squirrels" spontaneously migrated south, "by tens of thousands, in a deep and sober phalanx," many of them drowning in the Ohio River in the process. Early compilations quoted his account of this apparent mass migration, but modern historical and scientific accounts have dismissed it as one of the local myths which emerged from the actual event of the earthquake.[121]

Yet from the point of view of modern biology, Latrobe may well be a truth-teller as well as storyteller. In heavily forested earlier eras of North American history, gray squirrels would indeed engage in local mass migrations brought on by a boom-and-bust cycle of food resources. When oak trees had a particularly healthy season and produced massive numbers of acorns, gray squirrel populations exploded the following year. In the autumn, these overpopulated and resource-stressed squirrels engaged in mass movements toward more resource-rich territory. Though these migrations have been episodically reported into the mid-twentieth century, they were far more common and much, much larger in preindustrial North America. The numbers could be simply astounding. Armies of squirrels on the move ate over farmers' fields like locusts, and dead squirrels collapsed by the thousands on riverbanks. In 1747, during one such squirrel migration, the State of Pennsylvania paid bounties for the killing of 640,000 squirrels; during a Michigan gray squirrel migration in 1866, one settler counted 1,400 squirrels in the course of a two-mile drive. Latrobe's seeming exaggeration was likely a description of an actual local phenomenon in 1811—not an earthquake precursor, but a periodic ecological event. Might other seemingly far-fetched stories of animal behavior from 1811–12 be likewise borne out in modern frameworks?[122]

Bodily Knowledge in Modern Seismology

Sensations of the body were ever-present in early nineteenth-century earthquake accounts. One characteristic of modern seismology has been the effacement and replacement of bodily signs with indicators on instruments. While once many people regarded the ordinary sensations of ordinary people and ordinary animals as important evidence, in the current day, a few highly trained people make highly disciplined and highly constrained observations of a very specific set of human or animal bodies. The physical form still gives evidence, crucial evidence, but the subjective physical experience of ordinary people is in most modern science no longer considered a crucial way of registering and managing information about the natural world. Instead, researchers disdain and even excise the particularity of bodily experience from scientific evidence. §§§§[123]

Nineteenth-century interest in how an earthquake sounded contrasts strongly with contemporary science. Though earthquakes today are frequently experienced as extremely noisy, current seismologists have no interest in the characteristics or apparent directionality of that sound. Such aural experience is regarded as too dependent on local situation (noise may be reflected back from nearby ridges or buildings, for instance).[124] That an earthquake sounds as if it is coming from the southeast or the west is no longer a scientifically meaningful observation.

This difference speaks to the gulf across two centuries in the faith in one's own body as a meaningful and significant guide. Careful personal observation and recording of one's bodily experience were in 1811 and 1812 among the few ways to investigate earthquakes. Now such evidence is discarded in favor of the multiplicity and exactitude of GPS measurement readings or satellite photos.

Since 2005, someone named Tiempe has organized an online Yahoo group for "Earth Sensitives who daily feel sensations in muscles or hear tones in their ears, etc. for upcoming quakes." Tiempe's website maps specific bodily symptoms onto the world map of seismicity in order to chart earthquake prediction. One day in February 2007, for example, Tiempe reported for the Hawaiian islands "2 symptoms including a charley horse— Very Strong." Five years later, Tiempe charted strong probability of earthquakes in several zones of the globe, including the New Madrid region, based on symptoms correlated with that longitude and latitude.[125]

Mapping the world onto one's own body has no place in modern understanding of seismology, and in most respects Tiempe's concern to keep the globe safe through intense self-monitoring bears no relationship to the somatic frameworks through which nineteenth-century Americans experienced the New Madrid tremors. Indeed, perhaps Tiempe's extreme attention to the role of the individual body in knowing earthquakes helps point toward the deep gulf in understandings of the world separating 1811–12 and the twenty-first century. Where early Americans viewed bodily experience as a straightforward source of insight into unusual environmental tumult, expressing their somatic histories as a taken-for-granted part of accounting

---

§§§§ A common assumption made by seismologists is that the duration of any shaking was far less than scared people reported—in much the same way as people who brandish guns are usually much shorter than reported by people frightened of those guns.

for an earthquake, Tiempe's concern would likely strike most Americans today as both misplaced and self-referential: a kind of fetishistic attention to bodily experience not well related to a scientific understanding of the world.[126]

The USGS "Did You Feel It?" website stands as an intriguing exception to the disembodiment of modern earthquake study. This site invites ordinary people to record what they feel, creating a register of seismic events through seismic sensation. Such an effort is not simply crowd-pleasing outreach or positive public relations for an impersonal federal bureaucracy: especially in the context of reduced funding for seismic monitoring, researchers increasingly turn to the site for accurate reports of earthquake intensity. Intensities from the August 2011 Virginia earthquake come directly from "Did You Feel It?" In regions such as the East Coast, without closely spaced seismic monitoring arrays, the site can be the fastest and most accurate means to site epicenters and gather information about intensities.[127]

Local Knowledge, Remade

A Mississippi River traveler named William Leigh Pierce helped shape the understanding of the New Madrid earthquakes in the early nineteenth century. Not well known before the quakes, Pierce became well known because of the accounts he wrote about them. Similarly, a cotton farmer from eastern Arkansas helped shape the understanding of the New Madrid earthquakes in the late twentieth century, being published in the *Bulletin of the Seismological Society of America* because of his expertise in interpreting his farmland environment.[128]

Marion Haynes of Haynes Farm in Blytheville started his earthquake geology career with curiosity and a morning's amble, and has proceeded to work as part of innovative research teams investigating the paleoseismology of his home region. Once a cotton farmer, he is now employed at the Arkansas Archeological Survey's Blytheville station and as a consultant for the paleoseismological research firm M. Tuttle & Associates. His story, in bare outline, sounds like those of the polymathic amateur enthusiasts who created early nineteenth-century American science, but his career reveals how New Madrid investigations have emerged as a twenty-first-century science—in large part through a surprising return to nineteenth-century sources of information and forms of expertise.

Haynes grew up on Pemiscot Bayou after World War II, a time when many families had outhouses and when, as he says, "they were still cutting a few trees and killing hogs." Like many in the middle Mississippi Valley, Haynes learned about his home region's history through the arrowheads and pottery shards in his father's overalls pockets after a day's work in field and ditches. Haynes remembered his childhood curiosity about the sand blows that made the cotton grow poorly—because of the name, he assumed that somehow that they had been blown there by the wind, never picturing them erupting from beneath the earth.[129]

One day in the late 1980s, Haynes saw archeologist Robert Lafferty preparing some archeological investigations in a nearby field and went to see what he was doing. As they talked, Lafferty recognized Haynes's knowledge of the local soils, and as Haynes recounted, got him "down in a trench looking." During their work, Haynes mentioned to Lafferty that there were some sand blows that might be near archeological remains near his own house. One morning soon after, as he said, Bob Lafferty and his coinvestigator, Tish Tuttle, "came knocking on my back door." They had been denied access to the land they had planned to trench and asked Haynes to show him the sand blows he had mentioned on his own land. That site ended up being one of several in the area central to the team's significant findings about the paleoseismology of the New Madrid seismic zone.[130]

Marion Haynes would not be a country farmer if he did not downplay his own skill and his own work. "I am pretty good with a backhoe," he says, in explanation of his own involvement with this pioneering paleoseismological team. More than that, Haynes had a dedicated farmer's knowledge of the soils and signs of his beloved countryside. He had long observed the slight color differences in area sand blows that he mentioned to the team: as paleoseismologist Tuttle would confirm with radiocarbon dating, the darker-colored sand blows are older deposits, into which more organic matter has worked. Haynes's deep tenure in the region, his ability to read the local terrain, and, yes, his skill with a backhoe were all central to investigations that resulted in a delta farmer being published in a leading scientific journal.[131]

Multidisciplinary and multicausal science hearkens to a much earlier, less specialized era.[132] Local knowledge and local involvement increasingly characterizes New Madrid research and expertise.***** In the summer of 2009, researchers sought out area residents' participation in a latest

***** When I once asked Tish Tuttle how her team located sand blows to study, she talked first about techniques that were satisfyingly impressive to a nonscientist: taking aerial photographs and combing satellite images to look for visual clues, combined with "ground truthing," trenching, and in other ways checking that a given deposit is in fact a sand blow. She then surprised me by talking about what sounded like a very unscientific method: talking with people in the area about what they'd seen. "Never underestimate the knowledge of the local people," she told me, "especially those who have farmed the land." I learned a lot about modern science in that conversation.

round of research into New Madrid seismicity, seeking southeast Missouri landowners willing to volunteer space for new seismic monitoring stations, part of the EarthScope monitoring project.[133]

In this resumption of interaction with a broad public, work in New Madrid is consistent with national and international efforts to promote "citizen science." A key example is the Quake-Catcher Network. This innovative program tracks seismic movement by networking individual laptops, desktop computers, and even Nintendo Wiis through accelerometers present in many devices already (the motion detectors that allow a smartphone to switch from horizontal to vertical or an older cell phone to turn on when opened). In many ways and places, ordinary people could increasingly, and once again, participate in scientific investigations. Such work may ultimately begin to reshape expert knowledge as well.[134]

The first response of early nineteenth-century North Americans who felt a quake was to immediately write about it to see what other people had felt. The first response of early twenty-first-century North Americans is, in technologically adjusted terms, the same: after a 3.8 tremor in Chicago early in 2010, one resident reported her reaction upon being awakened suddenly: Sarah Evans, of the northwest Chicago suburb of Elgin, "popped out of bed" at 4:00 a.m. when she felt the "extreme shaking." She then reached for her phone to post a message on the social networking site Twitter. "Seriously weird," she wrote. "Something that felt like a minor quake just woke us up. Can anyone else in the CHI area/burbs confirm?" Evans reported that very soon Tweeters throughout her region were trading information about the earthquake. An online blog of the *New York Times* immediately solicited and began to republish accounts of the two major earthquakes in early 2010, in Haiti and Chile. Such conversations and compilations express a form of immediate and collective information gathering consistent with the early nineteenth-century compilers who began to construct knowledge of earthquakes in their time.[135]

Clearly, earthquake science of the early twenty-first century is not the same as that of the early nineteenth. Seismic study has transformed completely over those two centuries, through precision instruments, rigorous training, computer-aided modeling, global exploration, and new theoretical models. What is perhaps surprising is to find in technologically advanced, theoretically sophisticated, professionally trained modern seismology *any* trace of its nontechnical, theoretically diverse, and decidedly nonprofessional early nineteenth-century roots. In making sense of a set of earthquakes from the early nineteenth century, modern experts have realized that they need to delve into early nineteenth-century forms of knowledge.

Tensions abound in the uneven project of the resuscitation of formerly disdained accounts and expertise. Certain aspects of those early nineteenth-century New Madrid accounts are of interest to modern seismologists: the length of an earthquake crack but not the bellows of nearby cattle, a cracked chimney but not the hopes for eternal salvation of frightened people huddling nearby watching the crack spread. What more might present-day seismologists see, however, if they included somewhat more of the behavior of cows or the sensations of scared people? To even pose such questions risks not just raised eyebrows but the slammed door of (understandable) professional disdain: many experts in the earth sciences have had their fill of kooky science and have no wish to confuse rigorous models and testable hypotheses with fuzzy what-ifs. Perhaps one role, though, of histories of long-ago earthquakes is to suggest not only the richness but even perhaps the potential utility of past knowledge in writing the future of earthquake knowledge.

Mapping Uncertainties

The participation of a nineteenth-century-style public is apparent not only in the collection and production of scientific knowledge about current and future seismic activity in the New Madrid region, but even more so in political debates about the practical implications of earthquake science. In June 2008, researchers from CERI and the University of Memphis and the University of Texas at Austin set out to map faults in the New Madrid seismic zone by towing a seismic source machine along the Mississippi River with the help of a US Army Corps of Engineers tugboat. Their enterprise made front-page news in the main Arkansas paper. Their research was, admittedly, more photogenic than most basic science. (The newspaper's coverage featured a photo of attractively athletic sunbathing seismologists; moving at two miles an hour to ensure a detailed set of data, there is a lot of time to sunbathe.) Still, the fact that scientific efforts would garner a front-page leader and a full page of coverage speaks to the salience of such geophysical investigation to the lives of ordinary people in the early twenty-first-century middle Mississippi Valley.[136]

In the first years of the twenty-first century, traffic often slowed to a crawl between Arkansas and Tennessee on the grand steel arches of the Hernando de Soto Bridge, the span carrying the American transportation artery Highway I-40 across the Mississippi between Arkansas cropland and the city of Memphis. Work began in 2000 on a multistage seismic retrofit to allow the bridge to withstand up to a 7.7 earthquake. Emergency preparedness conferences in states along the middle Mississippi River emphasized that proper preparation would allow a rapid return to a "new normal" as soon as possible after a disaster. Washington University in St. Louis held workshops on "Disaster Reduction through Preparedness." A national insurer included the New Madrid earthquakes in an e-newsletter to policyholders in 2002, using damage from 1811–12 as a preface to checklists of how to secure water heaters and shelve heavy objects carefully. Earthquake preparation and earthquake knowledge reach local residents through many channels in the early twenty-first-century middle Mississippi Valley.[137]

At the same time, many Americans, in the area and further afield, resisted this resource-heavy redefinition of the midcontinent as an earthquake zone. Arizona senator John McCain listed funding for seismic study of Memphis as one of his October 2009 "Top 10 Earmarks for the Day" representing wasteful federal spending. Headlines in the fall of 2010 positively reviewed Seth Stein's critique of New Madrid alarmism and ridiculed "Doomsday Midwest Quake Predictions." Despite two decades of increasing, if debated, seismic regulations in building codes in Shelby County, critics pointed out that early twenty-first-century renovations of schools in Memphis did not take earthquake resistance into account—even when engineers in the community volunteered to do a pro bono evaluation of seismic stability of the $20 million 2006 renovation of a local high school.[138]

Knowledge of New Madrid seismicity was not complete in the early twenty-first century, but there was a great deal more of it, and more kinds of it, compared with investigations even two decades before. Seismologist Susan Hough could in 2008 make easy reference to New Madrid as the "best-understood intraplate source zone in the world." Yet the same year, Cliff Chitwood, an economic advocate from eastern Arkansas, could editorialize—equally accurately—that "at present, the very science of the New Madrid Fault Zone is in flux." For Chitwood, the specific conclusion was clear: building standards in his region were being unfairly and inefficiently decided by "FEMA bureaucrats" rather than scientific clarity. His particular conclusion could be—and was—contested by many engaged in debates over New Madrid history, science, and policy, but Chitwood's assessment of the state of scientific knowledge was undeniable. What Chitwood may not have realized is that his statement held also for the previous two hundred years of knowledge about seismic activity in the American heartland. Powerful historical continuities reverberated through his advocacy. In the early twenty-first century as in the early nineteenth, understanding of the New Madrid quakes was debated in the pages of newspapers and at civic gatherings, the content of scientific knowledge up for discussion by an interested and engaged public.[139]

Once, the New Madrid earthquakes were a punch line. Now, they are a line item. Those with expertise increasingly agree that large earthquakes of the past give reason for thoughtful planning and preparation for the future. Debate roils scientific journals—and occasionally news websites—but by the early 2010s the overall perception of hazard from the New Madrid seismic zone was utterly transformed from the 1950s, 1970s, or even early 1990s. *How* exactly to prepare for "the next big one" might be up for argument, but that communities need to prepare is widely, if ruefully, agreed.

The creation of knowledge about New Madrid continues in the early twenty-first century. As investigators continue to take seriously the knowledge produced by vernacular science from the nineteenth through the twenty-first centuries, the process of research and discussion will enrich and inform public debate and has the potential to ground political decisions in increasingly more robust (and perhaps more conclusive) science. The creation of knowledge about the New Madrid earthquakes has been interwoven with the debates over policy, symbolism, and larger cultural meaning of science throughout the two centuries in which people have puzzled over the power so suddenly manifest in the earth beneath them.

Conclusion: Memory and Earth in the Mississippi Valley

When the subterranean geography of this country shall become better known, it will probably be found to be one of the most interesting in the world.

JOHN BRADBURY, describing his travels in the Mississippi Valley just before the New Madrid earthquakes[1]

*

In 2004, two friends and fellow academics in St. Louis took me on a single-engine Piper aircraft flight across the Missouri Bootheel. I had just started to research this book, and I wanted to get a better understanding of the area's topography—a sense for the big picture. (Buying that tank of gas remains the best single use of research funds of my entire academic career.) The three of us in the plane—a Japanese medievalist, a Chekhov scholar, and a US historian—strained at the windows as our plane buzzed loudly over the region of the middle Mississippi Valley that had been shaken by tremors in 1811–12. I knew that there should be sand blows visible below, even on such a relatively hazy day. Yet as we crossed huge curves and whorls of the Mississippi River, as our engine propelled us over soybean and cotton fields, we squinted and frowned.

After a long and discouraging while, we came to a more or less simultaneous realization: the splotches of white everywhere below us, the blotchy patches easily visible across many fields, were in fact exactly what we were straining so hard to find. Each of us had taken that blotchiness as some sort of drainage difficulty in local fields, or a remnant of planting or irrigation. There was so much of this circular discoloration that we did not think it could be the sand blows we were looking for—traces of two-hundred-year-old earthquakes could not possibly be so obvious in the farmland of rural Missouri. We had not seen this evidence of seismic upheaval because it was too obviously in sight.[2]

Arkansas farmer and archeological authority Marion Haynes told me that back in the 1960s, a farmer some ways south of him bought two crawler tractors and a breaking plow, with the plan of once and for all churning up the sand blows that frustrated the even growth of his cotton. Haynes chuckled quietly as he told the story: the sand blows were, of course, too much—too big, too deep—for the man's equipment. From Haynes's perspective, the endurance of the sand blows on those Arkansas fields was good because they protect archeological remains increasingly threatened by industrial agriculture. They are, moreover, simply an integral part of the terrain that he loves and in which he is deeply rooted.[3]

Yet such signs on the earth—so obvious, once we in the plane understood what we were looking at, and so intractable to frustrated Delta cotton farmers—proved invisible to generations of people interested in the middle Mississippi Valley. Sand blows were for decades after 1812 noted, discussed, measured, and marveled at. But that did not stop them from disappearing from view in the late nineteenth century. Swamp reclamation advocate Otto Kochtitzky was only the most vocal and active of local improvers in the late nineteenth and twentieth centuries to reconsider and dismiss environmental evidence of earthquakes along with long-ago tales of earthquake upheaval. Kochtitzky set out to change his local terrain, and he did, altering its history just as he and his many workers altered its drainage patterns.

What seems to be so may not be. What is obvious to one generation of practitioners, or to people with one set of knowledge, may be invisible to those directing a dredging machine or flying high above the earth's surface. "Seeing" is no simple matter.

Coming to Terms with Earthquake Territory

Knowing earthquake territory has been part of my task in writing this book. At the outset I expected to read books and articles on the history of the middle Mississippi Valley, but I have also spent much more time that I ever thought I would talking with cotton farmers and sharecroppers' grandchildren, poring over my highway map for small routes traced in thin lines and light colors, listening to the evangelical Christianity of American Family Radio on small roads of northern Arkansas and southern Missouri. I never expected to hear someone speaking in tongues as part of my historical research, but late one night on a highway in southeast Missouri I was surprised to hear a local religious radio program turn into on-air prayer—and I was just as surprised to realize with a start that I was familiar with what speaking in tongues sounded like. I was not aware of having ever been around that form of devotion before, but somewhere in a childhood in the

Bible belt, I had. This project has surprised me not only in its methodological complexity, but also in the ways in which it has forced me to be what I used to be—a kid from Arkansas—as well as what I am now, a PhD more at home in library stacks than in any cow field.

Over the course of this project, I have crouched in an east Arkansas cotton field alongside a landowner and his two improbable miniature poodles, crunching tart, crisp turnips fresh from his side garden while we watched as a team of researchers tried to figure out the story of sand blows and ancient earthquakes told in cross-section on the walls of an eight-foot trench dug by a backhoe rented through a grant from the National Science Foundation. In ways that have stretched me and taught me a new understanding of terrain and history I thought I knew, I have tried to understand the conflicts and the geographies and the social worlds in which people across two centuries have bounded, explained, or kept at bay a tumultuous rebellion of their earth.

In these many ways, this book has sought to remember. Not simply to remember these particular earthquakes, though they are interesting and may hold cautionary lessons for the future. Rather, this book has sought to remember how knowledge changes, to trace a process of submerged and re-emerging certainty. Sometimes even cracks running across the earth in front of us can go unseen because they fail to fit a current model. Across many sciences, events without apparent explanation can at times go ignored. We commit such unacknowledged forgetting at our peril.

✳

The history of this flat and seemingly uninteresting floodplain, the continental interior that young people of ambition tend to flee and that Americans from the coasts regard with a mix of bemusement, condescension, and horror, turns out to be far from obvious. Much depends on perspective, on a few governing assumptions, on the choice to see or not see.

After the 1994 Northridge earthquake, power was knocked out over a wide swath of the Los Angeles area—normally a region of extreme light pollution. Stumbling outside at 4:00 a.m., hundreds of thousands of Los Angelenos saw a dark night sky for the first time. That night and for several weeks after, observatories, emergency services, and radio stations in the LA area received hundreds of calls from people asking whether the sudden brightening of the stars and the "silver cloud" spanning a portion of the sky had something to do with causing the quake. The director of the Griffith Observatory in LA reported that when staffers explained that the nebulous silvery formation was the Milky Way and that the stars always shine that brightly, many people refused to believe them: what LA residents were seeing did not in any way resemble a normal night sky, as they had always experienced it from the midst of their brightly lit, glittery megapolis. The sky looked so different that what these experts said was the Milky Way simply had to be something new and possibly ominous visible there in the night sky.[4]

Over the past two centuries in the Mississippi Valley, sand blows that were once obvious and everywhere remarked became invisible and unexplained. They came into visibility again, but now may be increasingly effaced—in reality as well as in arguments and scientific frameworks—by the economic demands of a land rich in agricultural fertility and little else.

Throughout the regions of some of the world's greatest liquefaction features, the extremely tight profit margins of modern agriculture cut with a new and threatening edge. Today, seismologists use GPS systems to study rates of strain accumulation on faults in the New Madrid seismic zone. Rice farmers in what was once the New Madrid hinterland also use GPS systems, to precision-grade large tracts of land so that they are perfectly flat and tilted ever so slightly to one side or another in order to drain smoothly. Landowners will completely scrape off the top layer of soil, grade the remaining layers to the desired angle, then respread the surface. Such land grading may allow for efficient cultivation of rice, but it also completely destroys the cultural, seismic, and geographic record. The large-scale machinery of the early twenty-first century can accomplish the effacement of surface terrain that Marion Haynes's neighbor wanted fifty years before. Just as in the early twentieth century, early twenty-first-century land use threatens to obliterate the seismic history of New Madrid terrain.[5]

Meanwhile, commemorating seismic history on the land proves just as slippery, even if all agree on what that history is. In China, sites of the devastating 2008 Sichuan earthquake are being developed for earthquake tourists. Office workers can now play laser tag in a park next to destroyed homes, and in 2010 the famous "earthquake pig" who survived for thirty-six days in a home's rubble (nicknamed Three-Six, and later, Strong-Willed Pig) went on honored display. In 2011, the much-lauded but castrated Strong-Willed Pig was cloned: six clonelets will presumably carry his sturdiness on to future generations. Redevelopment plans include tea shops and recreational boating near the horrifically created earthquake lake, along with an earthquake simulator that will enable visitors to feel lurching tremors. Many in China and elsewhere question how laser tag and a photogenic boar can coexist with unbearable and recent tragedy.[6]

How might an emerging sense of earthquake history similarly shape resource use in the areas around the Missouri Bootheel? In Arkansas, the

Crowley's Ridge Nature Center features an earthquake movie, and Reelfoot Lake is regionally known as a good fishing spot. How might the re-envisioning of nearby rural areas as earthquake country change local billboards, as well as local disaster planning? Will the New Madrid earthquakes come to be an economic resource for some of the country's most depressed rural counties, or just an economic drain on already resource-stressed mid-continent states? The earthquake history of mid-America suggests more questions than answers, but the stark changes in how these earthquakes have been regarded in American history—indeed, in whether they have been seen at all—suggests that the future place of earthquakes in the cultures, economies, and histories of Mississippi Valley states could take many forms.

✳

The past two hundred years have witnessed successive waves of effort to understand what happened in the winter of 1811–12 in the Mississippi Valley. Early nineteenth-century Americans tried to understand the events with the resources they had to hand. For many in North America, the massive quakes were a powerful portent. American boosters feared that they would cripple Western expansion. These fears proved groundless: American territorial encroachment was far too robust for even long-tapering quakes to subdue. Prompted by shaking ground, many Native Americans rallied to Tecumseh and Tenskwatawa's cause of Indian unity and resistance to American encroachment in the struggles surrounding the War of 1812, while Indian communities such as the eastern Cherokees turned instead to rituals of spiritual and cultural rebirth. Americans likewise experienced crises of spiritual faith intensified by the earth's upheaval. In American as well as Indian communities, those who had felt the quakes interpreted them through the physical sensation, linking the experience of earthquakes with physical, spiritual, and cultural well-being.

The quakes were also an intellectual puzzle, one attacked vigorously by backwoods doctors and Philadelphia savants alike. The New Madrid earthquakes were part of a lively and widely shared vernacular science long hidden in the seemingly prosaic pages of commercial journals and folksy letters. These quakes show us moderns what was everywhere apparent in the early nineteenth century: that lively discussions of basic causes and fundamental mechanisms took place in the midst of everyday life, scientific discussion wedged in alongside columns of crop prices, parochial politics, and homespun horse remedies.

However wildly varied, many of these early scientific explanations knit the New Madrid earthquakes into global and holistic frameworks. Whether through the new and exciting science of "electrical fluid" or through as-yet-undetermined subterranean channels, the New Madrid quakes were connected by early American writers and thinkers with events and causal chains reaching far across the globe. Observers throughout the early United States—many quite ordinary people, as well as the few more intellectual elites—made clear their sense that disturbance in their western borderlands would someday be fully explained as a significant aspect of global science.

For Americans of the late nineteenth and early twentieth centuries, such assumptions seemed as quaint and ill-founded as the religious and bodily enthusiasm of backwoods revivals. The New Madrid earthquakes became submerged beneath changes to environments and in national priorities, no longer visible beneath accreted layers of racial conflict and drainage sludge. The middle Mississippi Valley was remade by war, by cotton farming, and by John Deere. For most of the twentieth century, the New Madrid quakes existed as little more than a historical footnote.

Only tentatively, in sole efforts that took decades to gain broader hearing, did scientific researchers begin to once again cast the New Madrid quakes as a significant set of events, with significant consequences for broader scientific understanding. This remembering took place as a process of textual reconstruction, as researchers once again gathered and pieced back together firsthand accounts of the quakes. The reassessment of New Madrid also took place as a blossoming of post-plate-tectonic reevaluation within seismology, as the very question of seismicity far from a plate boundary zone became both more globally intriguing and more locally intense, of interest to researchers in north-central China and to contractors in Memphis, Tennessee.

Technology—seismographs, GPS monitors, even the techniques of rapid communication and swift travel to far-flung parts of the world where intraplate seismicity occurs—is a crucial theme in these changes. Upheavals in technology mark the crucial eras in seismology, as the introduction of instrumentation around the beginning of the twentieth century created a revolution in the study of earthquakes, and then, over sixty years later, the mapping of magnetic anomalies on the ocean floor helped impel a sudden, dramatic shift in worldview, from a more or less static globe to a world of plate tectonics in which vast chunks of the earth's crust shift and collide in ongoing movement.

Yet neither textual study nor global positioning systems solved the questions of New Madrid seismicity. Despite the insights of instrumental knowledge, and despite intense seismological research, intraplate quakes remain poorly understood, their basic mechanisms debated and their causes still unknown.

Moreover, we do not yet fully understand some of the most basic aspects of these earthquakes' history. Leading New Madrid scientist Seth Stein asserted in a 2004 *New York Times* editorial and in a 2010 book that "as far as we know, no one has ever died in an earthquake in the New Madrid zone." It is indeed clear that people died in the winter of 1811–12, along and in the Mississippi River, though their specific numbers and stories will likely remain obscure. This insistence upon the lack of historical certainty about the New Madrid earthquakes indicates the ways in which their nature is still held at arm's length in scientific as well as historical sources, their unknowability insisted upon across two hundred years.[7]

For most people in the early nineteenth century, accounts of what individuals had felt or heard were the only way to understand and apprehend what had happened. Earthquake reports were not evidence for science: they were the science itself. Exchanging accounts—accounts laden with implicit suppositions about correlations and causes—was the mode of conducting scientific conversation in the far-flung, decentralized early United States.

These earthquakes show us that people in early American created knowledge about the natural world in a number of different ways that would now seem to us completely separate: through faith, through political action, through settlement, and through spirited discussion of scientific causation in ordinary newspapers and other everyday venues. These earthquakes further show us how knowledge in the United States continues to be made—in and through social and public policy conversation, as well as through scientific investigations of New Madrid seismicity that are as contentious in the early twenty-first century as they were two centuries before.

In the late nineteenth and much of the twentieth centuries, personal accounts came to be regarded as unscientific and to a very large degree seismologically uninteresting. Though modern workers have rescued from obscurity the efforts of one researcher, Myron L. Fuller, who did find historical narratives useful, his work languished compared to the abundant and compelling seismographic evidence of more active world hotspots.

Now, accounts of the New Madrid earthquakes are some of the key evidence for heated debates within studies of mid-American quakes, and indeed of intraplate seismicity generally. Yet ironies abound. One of the leading New Madrid researchers, Susan Hough, has used fine-grained analysis of the sources and locations of historical narratives in order to make the case that the 1811–12 temblors were likely less fierce and even less destructive than recent scholarship had recently begun to agree.

In a beautiful and massive recent volume, historian of geology Martin Rudwick carries readers along on the fascinating story of how and why European earth scientists came in the very late eighteenth and early nineteenth centuries to think *historically*.[8] This narrative of New Madrid earthquake history points to the continuing centrality of historical texts, historical records, and historical thinking in earth sciences research on this particular set of dramatic events—a centrality that has endured despite the slippery and changing readings and uses of the texts themselves over two hundred years.

As I worked to investigate this history, I had the chance to present at CERI in Memphis. I was nervous: would earth scientists resent how I would tell them the history of their own field? In the course of speaking about nineteenth-century scientific approaches to the New Madrid earthquakes, I read a nineteenth-century letter illustrating how familiar objects could help show the nature and extent of disturbances. I quoted Alexander Montgomery's description to Benjamin Smith Barton that in Frankfort, Kentucky, "parts of some Houses of Brick have been Shaken down; But in Louisville at the distance of about eighty miles below us and on the banks of the Ohio, the Damage done Houses has been considerable. There is scarcely a brick House in the town that has not lost Chimnies—the walls of others have been very much *Cracked*."[9] I saw heads nodding and was pleased that no one appeared poised to lob research binders at my head as I answered questions. Afterward, one seismologist came rushing up. He said he appreciated my arguments, but was very anxious to hear more about the chimney damage. Where exactly were those chimneys? How long were the cracks?

That exchange demonstrated to me both the continued centrality of earthquake narratives to understanding this preinstrumental set of earthquakes, as well as the many facets to what reading or using an earthquake account can mean.[10] Texts in this story serve as historical trace markers, showing us what changes and what stays the same.

As with Myron Fuller's 1912 report on the quakes, there is no standing outside the complicated textual history of these quakes. Even as I have traced the debates over scientific standing that emerged after seismologist David Stewart's support for Iben Browning's highly untraditional quake prediction, I find myself both citing and placing in historical context Stewart's reprinting of Fuller's foundational text. As I try to comprehend how these quakes have been understood, I find myself referring for technical information to the same articles and books whose origins I am attempting to narrate. Tagging along on a field expedition to an east Arkansas cotton field, I asked soil scientist and geologist Tish Tuttle about her very recent research into prehistoric New Madrid quakes. "Go back and look at Fuller again," she urged me.[11] As a historian of the nineteenth century, I have grown accustomed to having the past lie flat in dusty boxes. I can take each file out as I need it, then put it back again and move on. In studying the

New Madrid quakes, the narratives of past and present swirl like leaves on a street corner whirled by autumn wind. Reading and rereading texts, collecting them in various eras but using them for very different ends, those struggling to understand and explain the New Madrid earthquakes keep past sources always in present mind, in ways that sometimes conflate and sometimes highlight different understandings of the heavings of a tumultuous earth.

Back to my original question: how do we know what we know about the New Madrid earthquakes of 1811–12? Just like the three of us confused in that airplane, staring at sand blows we could not see, we are, in the present day, understanding the New Madrid earthquakes still. We remain very much a part of that story. Our sciences have not completely left behind their nineteenth-century ancestors. Our notions of the causality of earthquakes have changed considerably, but we still have not completely solved their riddles—nor have we escaped their challenges. Coming to terms with their many forms of inquiry and understanding is therefore deeply intermixed with our continuing efforts to comprehend seismic disturbances of long ago that may indeed have impact well into the future.

NOTES

Note: Numbers in bold refer to sections in the bibliographic essays.

Introduction

1. Crockett: *A Narrative of the Life of David Crockett*, 185–92 (**1.1**); Jonah 1:17.
2. Please see **1.1**.
3. Runny egg casserole: conversations with Tish Tuttle, fall 2003.
4. Tabletop liquefaction: Gerstel, "Earth Connections" (**1.22**).
5. My thanks to Martitia (Tish) Tuttle, M. Tuttle and Associates, and Haydar Al-Shukri, University of Arkansas at Little Rock Earthquake Center, for allowing me to follow them about. On the nature and impressive size of New Madrid liquefaction features, see Tuttle, "The Use of Liquefaction Features in Paleoseismology," esp. 366 (**7.5.5**); I also appreciate conversations with Eugene Schweig, USGS.
6. Murder of slave: Feldman, *When the Mississippi Ran Backwards*, 11, 190–93 (**1.1**). See also **1.18**. New Hampshire: "Earthquakes," *National Intelligencer* (Washington, DC), 20 February 1812. Other effects: see Smith, *An Account of the Earthquakes* (**1.2**); and Smith, "On the Changes Which Have Taken Place in the Wells of Water Situated in Columbia, South-Carolina, since the Earthquakes of 1811–12" (**5.6.1**).
7. Even as researchers' estimates have moved downward, sources accessible to the public swing widely. A 2011 *Huntington Post* editorial, for example, called attention to the New Madrid quakes as the "Mother of All American Earthquakes," the "largest ever in North America." Yet as the US Geological Survey documents, quakes in Alaska, California, the Cascadian coast, and Hawaii have been larger. Landon Jones, "The New Madrid Earthquake: 200 Years Later," 11 August 2011, *Huffington Post*, Huff Post Green, http://www.huffingtonpost.com/landon-jones/earthquake-new-madrid_b_934803.html; USGS, "Largest Earthquakes in the U.S.," http://earthquake.usgs.gov/earthquakes/states/10_largest_us.php, updated May 2010. Debate on New Madrid magnitude: see Hough and Bilham, *After the Earth Quakes*, 79 (**1.6**). Summary of magnitude estimates as of 2012: NMSZ Expert Panel Report to NEPEC, p. 6 and table 1 (**7.5.1**). The USGS website explains magnitude under "Earthquake Topics" and "Earthquake Glossary." See **7.5.2**.
8. For an excellent, accessible introduction, see Hough, *Earthshaking Science* (**1.6**).
9. Seth Stein, "Approaches to Continental Intraplate Earthquake Issues," in Stein and Mazzotti, eds., *Continental Intraplate Earthquakes* (**7.8**).

10. Recurrence interval: Tuttle et al., "The Earthquake Potential" (1.1). Butler, Stewart, and Kanamori, "The July 27, 1976, Tangshan, China Earthquake"; and Liu et al., "Active Tectonics and Intracontinental Earthquakes in China," esp. 300 (both in 7.8.2).

11. See citations at I.1. See Coen, *Earthquake Observers*, 3 (1.6).

12. "Barren spots upon which little will grow": Fuller, "Our Greatest Earthquakes," 79 (1.1).

13. Bryan letter, 344 (1.2).

14. Disasters can provide a particularly good way to investigate such events (1.34).

15. See 1.5.

Chapter 1

1. "New-York, February 11, Earthquake," February, listed in citations to 1.3, and 1.3.1.

2. "Often exaggerated": Saucier, "Origin of the St. Francis Sunk Lands," 2847 (2.6).

3. See 1.3.1. Pierce leaves his role in the convoy—owner of the goods? hired help?—unclear. Either way, a young Yankee man trying to make good through Mississippi trade would not have been unusual in his era.

4. See citations in 1.2 (esp. Smith, *An Account of the Earthquakes*, and *An Account of the Great Earthquakes*); 1.3.1; and 1.9. Poems: see 1.4; Schoolcraft, *Transallegania*, quotation on p.141.

5. See 1.9.

6. See citations on 1.10. Pseudonyms common in publishing: O'Brien, *Conjectures of Order*, 553 (1.8).

7. On one particularly telling journal, see 3.17. Also 3.2.

8. Travel time, pre-steamboat: Cincinnati to New Orleans, 25 days; return, 65 (Cutler, *A Topographical Description of the State of Ohio*, 45 (1.17); in the late eighteenth century, New Orleans to St. Louis took a month by land and at least two by upstream barge (Bolton, "Jeffersonian Indian Removal," 78) (1.11). Crucial role of Mississippi and other rivers at the time of the quakes: Johnston and Schweig, "The Enigma of the New Madrid Earthquakes," 342 (1.1); Aron, *American Confluence* (1.12).

9. Aron, *American Confluence* (1.12).

10. Mississippi embayment: Johnston, "A Major Earthquake Zone on the Mississippi," 60–63 (7.3); Snow, *From Missouri*, 3–4 (6.11.1); "Formation of the Mississippi Alluvial Plain" in Foti, "Geography and Geology" (7.6); Soil thickness: Street et al., "NEHRP Soil Classifications," 123 (in 7.5.12). I had never thought of myself as a child of the floodplain until I drove through the northern New England countryside as a young adult and was astonished at row after row of stone walls: I had never realized farmed land could have that many rocks in it.

11. DuVal, *The Native Ground*, chapters 4, 6 (1.12); Morrow, "New Madrid and Its Hinterland" (2.1); Foley, *The Genesis of Missouri*, chapter 2, 3, 7; Foley, *A History of Missouri*, 1:71 (both in 1.11).

12. Anxieties over Louisiana Purchase: Kastor, *The Nation's Crucible* (1.12); Flores, "Jefferson's Grand Expedition," 25–26 (2.14); West, "Lewis and Clark: Kidnappers," esp. 6–13 (1.15); Howe, *What Hath God Wrought*, 108; Wood, *Empire of Liberty*, 374 (both in 1.8). Wilkinson: see 2.17.

13. Concerns for free traffic on Mississippi: Foley, *History of Missouri*, 64 (1.11); lifting of restrictions: Morrow, "New Madrid and Its Hinterland," 243 (2.1).

14. Chambers, "River of Gray Gold" (1.11); Latrobe, *The First Steamboat Voyage on the Western Waters*, 5 (1.31); Meinig, *Continental America*, 214 (1.8).

15. Tensions 1811–12: Sugden, *Tecumseh*, 218–19 (3.10); Warren, *The Shawnees and Their Neighbors*, 38 (2.1); Foley, *History of Missouri*, 126–27 and chap. 8 (1.11). Rumors: Ross, *Life and Times of Elder Reuben Ross*, 200–201 (1.17); "formidable combinations of the savages": Ninian Edwards, "Copy of a Letter from Gov. Edwards to Governor Scott," *The Supporter* (Chillicothe, OH), 14 March 1812, Archive of Americana: America's Historical Newspapers. Osage attack: "On Wednesday Last, an Express Came to Governor Howard . . . ," *National Intelligencer* 1812, InfoTrac: Nineteenth-Century Newspapers; American abandoned farms: William Russell, St. Louis, to William Rector, Surveyor, in Carter, ed., *Territorial Papers*, 14:752 (2.16); also Shaw, "Collections of the State Historical Society of Wisconsin, 204 (1.2).

16. On "commotion": "Indigena" [pseud.], "For Liberty Hall," *Liberty Hall*, 1 January 1812, p. 2, cols. 2 and 3, CERI Compendium (1.2). See 1.19. Noisy fowl in Creek nation: Smith, *An Account of the Earthquakes*, 38–39 (1.2). The woman was likely Creek, but the source does not make that clear. Major Burlison: Griffith, *Letters from Alabama*, 192 (1.17). On modern experience of earthquakes, see 7.20: Australians tend to worry most about stampeding possums.

17. Smith, *An Account of the Earthquakes*, small brook: 16; "skaiting": 44; compass, 18 (1.2).

18. See 1.18. Smith, *An Account of the Earthquakes*, "similar to a cradle": 38, like waves: 51; *An Account of the Great Earthquakes*, 14 (both in 1.2). Traveler in Creek nation is likely trader M. Terrall, who also wrote back to his Rhode Island trading firm (Brown family business papers, John Carter Brown Library, Brown University).

19. Sargent, "Account of the Several Shocks," 355. "Conflict of the elements": Smith, *An Account of the Earthquakes*, 58 (both in 1.2).

20. Long history of river raiding, piracy: DuVal, *The Native Ground*, 158, 166 (1.12).

21. "Smart shock": Pusey, "The New Madrid Earthquake," 286 (1.2). "the great shake": Lyell, *A Second Visit*, 179 (5.16); Hudson, "A Ballad of the New Madrid Earthquake," 149 (4.6). "terrible heavings": Flint, *Recollections*, 215 (1.17). "*Shakes*": Ross, *Life and Times of Elder Reuben Ross*, 206 (1.17).

22. See 1.19. "Distinct reports": Pusey, "The New Madrid Earthquake," 286. Rattling of carriages: Mitchill, "A Detailed Narrative," 289, 97; Smith, *An Account of the Earthquakes*, 9, 55; "burning out of a chimney": 12; "loads of small stones": 17; "the ringing of the bells": 58 (all in 1.2) "violent tornado": Audubon, ed. *Audubon and His Journals*, 234 (1.18).

23. Bellows: Smith, *An Account of the Earthquakes*, 8; Mitchill, "A Detailed Narrative," 285; and fire: 291. Steam: Broadhead, "The New Madrid Earthquake," 82 (all in 1.2). See 5.17.

24. Foster, *The Mississippi Valley*, 18 (1.17).

25. See 1.22. "Sand slews": Fuller, "Our Greatest Earthquakes," 362, 366 (1.1). Taloni sinkholes; McLoughlin, "Appendix," 341 (3.17).

26. See 1.23. "very much crack^{d}": Austin, "Diary," 207; "for the ground is cracked": Smith, *An Account of the Earthquakes*, 26 (both at 1.2). Austin as colonization agent for New Spain: Howe, *What Hath God Wrought*, 658–59; Robert A. Calvert, "Austin, Stephen Fuller," in *American National Biography Online*, 2000 (both in 1.8). Cracks and crevices at Fort Massac: Matthias Speed, "From The Bairdstown (KY) Repository," *Western Spy* (Cincinnati, OH), 28 March 1812; "Summary," *Western Spectator* (Marietta, OH), 1 February 1812, both in CERI Compendium (1.2).

27. "A number of ravin[e]s": Notes from (or slightly before) 21 December 1853, "Field Notes of Missouri Surveys, Vol. 643," Land Survey Repository, Missouri Department of Natural Resources, Rolla, Missouri, 139. The surveyor was in township 18N, range 11E, near intersection of sections 22, 23, 26, and 27 (roughly three curves of the Mississippi downriver from New Madrid, close to the Arkansas border). In the present day, that location is in the very small town of Braggadocio, which is east and slightly south of the Mississippi River town of Caruthersville. Map of T.18N, R.11E. in *Plat Book of Pemiscot County, Missouri* (W. W. Hixson, 1930). Fuller, *The New Madrid Earthquake*, 51 (1.1).

28. Castings: Foster, *The Mississippi Valley*, 21 (1.17). Cow: James Fletcher, "From the Nashville Examiner, January 24," *Kentucky Gazette* (Lexington), 18 February 1812, CERI Compendium (1.2).

29. Bradbury, *Travels in the Interior of America*, 199–201 (1.2).

30. Booming floods (from usual spring overflow): Valencius, *The Health of the Country*, 142 (1.13).

31. Smaller islands destroyed: Fuller, *The New Madrid Earthquake*, 93–94 (1.1). "There is a certainty": "Earthquake," *The Supporter* (1.2). "Sunken islands" still visible: Evans, *A Pedestrious Tour*, 195 (1.17). Great interest: Montgomery letter to Barton; Broadhead, "The New Madrid Earthquake," 83 (both in 1.2).

32. Penick, *The New Madrid Earthquakes*, chap. 1 (1.1); Latrobe, *The First Steamboat Voyage on the Western Waters*, 28–30 (1.31).

33. Holmes, *Age of Wonder* (1.26).

34. Earthquake cracks spew, swallow material: Penick, *The New Madrid Earthquakes*, 96–99 (1.1); Bringier, "Notices," 21 (2.4). "Channel coal": interview with Marion Haynes, January 2011. "Hard, jet black substance": Shaw, "Collections of the State Historical Society of Wisconsin," 203 (1.2). Lignite in worm castings: Shepard, "The New Madrid Earthquake," 56, 51 (1.1). Material spewing through nineteenth century: Foster, *The Mississippi Valley*, 25 (1.17).

35. Hymn 34, Evangelical Lutheran Church, *A Collection of Hymns*, 26 (4.15). "Weakness and dependence": "Letters from Honorable L. F. Linn," 5 (1.17).

36. Austin, "Diary," 207 (spellings of *puting*, *dessolation*, and *dispair* are from the original) (1.2). On Christianity as pervasive heuristic for nineteenth-century American life, see for instance Stowe, *Doctoring the South*, 29 (4.5).

37. See the discussion at 1.6, esp. Coen, "Introduction: Witness to Disaster"; Oldroyd, *Thinking About the Earth*, 224–25; Lisbon, Calabria, Sicily: Udías, "Jesuits' Studies of Earthquakes and Seismological Stations," 137; "year of earthquakes": Davison, *The Founders of Seismology*, 1–2; Coen, *Earthquake Observers*, 7–9, 106–7.

38. Mixed-culture men as boatmen, travelers, messengers: Thorne, *Many Hands of My Relations*, 147, 151 (1.11). Death toll: see 1.28.

39. "The violence of the jar": Smith, *An Account of the Earthquakes*, 40, capitalization and punctuation true to the original (1.2).

40. The story may even elide another death of an African American resident: see 1.28. African American history of the middle Mississippi valley: Robinson, "The Louisiana Purchase and the Black Experience" (1.11). Governmental history of the mid- and lower Mississippi: Howe, *What Hath God Wrought*, 147–60 (1.8); Foley, *History of Missouri*, 151–52 (1.11); Bolton, "Jeffersonian Indian Removal," 82–83 (1.11); Kastor, *Nation's Crucible* (1.12). French heritage: Ekberg, *François Vallé*; Ekberg, *French Roots in the Illinois Country* (both in 1.11). Racial and cultural "borderlands" of early Missouri: Thorne, *Many Hands of My Relations*; Winch, *The Clamorgans*, esp. 4–5 (both in 1.11).

41. Shaw, "Collections of the State Historical Society of Wisconsin," 203 (1.2).

42. Lyell, *A Second Visit to the U.S.*, 178–79 (5.16).

43. Stevenson, "Autobiography of the Rev. William Stevenson," 41 (1.2). Bringier, "Notices," 21 (2.4). Topographic change in epicentral regions is not uniformly acknowledged in scientific literature, but is amply represented in accounts of the time: Beck, *A Gazetteer of the States of Illinois and Missouri*, 300; Broadhead, "The New Madrid Earthquake," 82; Featherstonhaugh, *Excursion through the Slave States*, 78 (all in 1.2); Flint, *Recollections of the Last Ten Years*, 218 (1.17). Fuller, *The New Madrid Earthquake*, 57–88; and Van Arsdale, "Seismic Hazards of the Upper Mississippi Embayment," 21 (both in 1.1).

44. See 2.5. *Callé*: de Montulé, *Travels in America*, 113. "Sunk Lands": typical report: Sargent, "Account of the Several Shocks," 355. The effects of land subsidence in creating the "sunk lands" of northeastern Arkansas are discussed in chapters 2 and 6. Tennessee: Smith, *An Account of the Earthquakes*, 23 (all in 1.2).

45. Flint, *Recollections of the Last Ten Years*, 215 (1.17); see 1.24.

46. See 1.25.

47. "Offered, at New Madrid": Smith, *An Account of the Earthquakes*, 83 (1.2). Fleeing to a hill: Ross, *Life and Times of Elder Reuben Ross*, 206 (1.17); and "Personal Narrative of Col. John Shaw," 91 (1.2); Penick, *The New Madrid Earthquakes*, 41–42 (1.1).

48. 1817 New Madrid: Morrow, "New Madrid and Its Hinterland," 246 (2.1). Hendricks, "Sunken Lands" (2.5). Chapter 2 documents depopulation by recent Native emigrants. Area depopulation: Silas Bent to Josiah Meigs, 22 June 1813, in Carter, *Territorial Papers*, 14:684 (2.16).

49. See 1.29.

50. William Rector to Josiah Meigs, in Carter, *Territorial Papers*, 14:700 (2.16). New Madrid repopulation in 1830: Penick, *The New Madrid Earthquakes*, 48 (1.1).

51. "Forty-one shocks": Smith, *An Account of the Earthquakes*, 26 (1.2). Shocks continuing: McClinton, ed., *The Moravian Springplace Mission*, 2:190, 10 December 1817 (3.17).

52. On the sheer mess of nineteenth-century urban animals: Opie, *Nature's Nation*, 272; monumental sights, sounds, and mess of now-extinct migrating passenger pigeons: "Missed Connections," in Price, *Flight Maps* (1.13).

53. See 1.27.

54. Mitchill, "A Detailed Narrative," 281 (1.2). Mitchell: Hindle, *The Pursuit of Science in Revolutionary America*, 314–15 (5.3). Comets and epidemics: "War, Indians, Comets, Earthquakes": Ross, *Life and Times of Elder Reuben Ross*, chap. 20 (1.17). Nat Turner: Howe, *What Hath God Wrought*, 324 (1.8). "When the stars begin to fall": from "My Lord, What a Morning" (4.15.1).

55. See 1.26.

56. "Earthquake," *The Supporter* (1.2).

57. "There was a back current": "Earthquake," *The Supporter* (1.2). The section of one 1998 article dealing with this waterfall is entitled "Extracting Facts from Folklore." Odum et al., "Near-Surface Structural Model for Deformation," 153 (7.5.3). Hough, *Earthshaking Science*, 181; Tachia River: 183 (1.6).

58. Though a separate letter, this was reprinted alongside his initial account: Pierce, "New-York, February 11, Earthquake" (1.3).

59. Trying to figure out local vs. widespread disturbance: "A Subscriber," "Earthquake!," *Poulson's American Daily Advertiser*, 23 December 1811. Estimating extent: Smith, *An Account of the Earthquakes*, 8 (1.2). "In Georgia": "Several Shocks by Earthquakes," *Niles' Register*, 4 January 1812. "From the accounts already received": "Earthquakes," *Lexington American Statesman*, 21 January 1812. See 3.2.

60. Curiosities of the West: Berry, Beasley, and Clements, *The Forgotten Expedition*, 171–72 (1.15). Mammoth bones; Bringier, "Notices," 22–23 (2.4); Hindle, *The Pursuit of Science in Revolutionary America*, 184–85, 323 (5.3); Marshall, *The History of Kentucky*, 1:6 (1.17); Wood, *Empire of Liberty*, 393 (1.8). Reptile and other fossils: Abney, *Life and Adventures of L. D. Lafferty*, 115 (1.17); Mayor, "Suppression of Indigenous Fossil Knowledge" (2.3); Flores, "Jefferson's Grand Expedition," unicorns, giant water serpents: 28; silver deposits, meteors: 28 (2.14); also Flores, ed., *Journal of an Indian Trader*, 94–98. Fawns, paroquets: Cutler, *Topographical Description of the State of Ohio*, 48–50 (both at 1.17).

61. Berry, Beasley, and Clements, *The Forgotten Expedition*, "Editors' Introduction"; Malone, "Everyday Science"; Ronda, *Lewis and Clark among the Indians*; West, "Lewis and Clark: Kidnappers," 6 (all in 1.15); Flores, "Jefferson's Grand Expedition," 22 (2.14); Whayne, "A Shifting Middle Ground" (1.11).

62. See 1.16.

63. See 1.31.

64. Latrobe, *The First Steamboat Voyage on the Western Waters*, 13–14; Dohan, *Mr. Roose-*

velt's Steamboat, 41; Grant, "Roosevelt, Nicholas J."; Description of boat: Grant, "Roosevelt, Nicholas J."; and Latrobe, *The First Steamboat Voyage on the Western Waters*, 11 (all in 1.31).

65. New sort of sawmill: Dohan, *Mr. Roosevelt's Steamboat*, 60–61 (1.31).

66. Grant, "Roosevelt, Nicholas J."; Latrobe, *The First Steamboat Voyage on the Western Waters*, 17–18, 20–21 (1.31).

67. Shock: Latrobe, *The First Steamboat Voyage on the Western Waters*, 23, 26, 29–30 (1.31); Penick, *The New Madrid Earthquakes*, 75 (1.1). Pierce was further downstream when the quakes hit, but he did not have a steam engine: both vessels docked in New Orleans early in January 1812. Smith, *An Account of the Earthquakes*, 27 (1.2).

68. Latrobe, *The First Steamboat Voyage on the Western Waters*, comet dropped into water: 17; Penelore: 27–28. Charles Latrobe says the word was "Pinelore" and was Choctaw in origin: Latrobe, *The Rambler in North America*, 86. The *Comet* made a voyage to Louisville in the summer of 1813 and was later the first steamboat to reach Arkansas, docking in Arkansas Post in 1820. "Steamboats Employed on the Western Waters from 1812 to 1819," in Casseday, *The History of Louisville*, 128 (all in 1.31); Stewart-Abernathy, "Steamboats" (1.33).

69. Latrobe, *The First Steamboat Voyage on the Western Waters*, 27–28 (1.31).

70. Latrobe, *First Steamboat Voyage on the Western Waters*, 32 (1.31).

71. Terrell, letter to Brown & Ives (1.16); Bradbury, *Travels in the Interior of America*, 263 (1.2).

72. See 1.33. Jacksonport: Holder, "Historical Geography of the Lower White River," 136 (1.14).

73. See 1.31 and 1.33.

74. Roosevelt financial trouble: Grant, "Roosevelt, Nicholas J."; and Harris, "Fulton, Robert" (both in 1.31).

75. Mitchill, "A Detailed Narrative," 295 (1.2).

76. Torrent of westward immigration, 1800–20: Howe, *What Hath God Wrought*, 140 (1.8).

77. "Counteract every tendency to disunion": Howe, *What Hath God Wrought*, 87–88 (1.8). Some limited federal "improvements," including snag boats, did soon follow: Hill, *Roads, Rails, and Waterways*, chap. 6 (1.32).

78. New Madrid is in some ways the flip side of story of Chicago told in Cronon, *Nature's Metropolis* (1.13). I appreciate comments by Jenny Smith, Program in Agrarian Studies, Yale University, work-in-progress discussion, 19 January 2007.

Chapter 2

1. Bringier, "Notices" (2.4).

2. See 2.1.

3. On the rhetorical strategies and historical impact of the dismissal of Native American history, see 2.3. Bringier's visit: Williams and Bringier, "Louis Bringier and His Description of Arkansas in 1812," 111 (2.4).

4. Williams and Bringier, "Louis Bringier and His Description of Arkansas in 1812," 108–13 (2.4).
5. Bringier, "Notices," esp. 15, 16 (2.4).
6. Silliman and his journal: Daniels, *American Science in the Age of Jackson*, 18; Greene, *American Science in the Age of Jefferson*, 182, 242–43; Judd, *The Untilled Garden*, chap. 3, esp. 94 (all in 5.3). See 5.4.
7. Bringier, "Notices," 21 (2.4).
8. Bringier, "Notices," 21 (2.4). Modern researchers cite details such as water rising to the belly of Bringier's horse, but do not dwell on the Indian prophecy he reports: for instance, Johnson and Schweig, "The Enigma of the New Madrid Earthquakes," 347 (1.1); Williams and Bringier, "Louis Bringier and His Description of Arkansas in 1812," 134 (2.4).
9. See sources on 2.7. Bringier, "Notices," 39–40 (2.4). The clarification "comet" for "blazing star" was apparently an addition by Silliman.
10. Bringier, "Notices," 40 (2.4).
11. Bringier, "Notices," 41 (2.4).
12. Usual story: for instance, Foley, *The Genesis of Missouri*, 218–19 (1.11); Hough and Bilham, *After the Earth Quakes*, 81–82 (1.6); Ogilvie, "The Development of the Southeast Missouri Lowlands" (6.9).
13. See 2.10.
14. See 2.11; also Mainfort, "The Late Prehistoric and Protohistoric Periods in the Central Mississippi Valley"; Morse, "Protohistoric Hunting Sites" (all in 2.9). Confluence region: Usner, "American Indian Gateway," 44–45 (2.1); Aron, *American Confluence* (1.12).
15. Usner, *Indians, Settlers, and Slaves*, esp. 103 (1.12); DuVal, *The Native Ground* (1.12); Warren, *The Shawnees and Their Neighbors* (3.10).
16. "Principal river": Evans, *A Pedestrious Tour*, 208. St. Francis at time of Civil War: Parker, *Missouri as It Is in 1867*, 35 (both in 1.17). Choctaw name for St. Francis: Bringier, "Notices," 25 (2.4).
17. See 2.7.
18. Arkansas Post: Arnold, *Colonial Arkansas*, 5 (2.16); French immigrants: Ekberg, *French Roots in the Illinois Country* (1.11); Bolton, *Territorial Ambition*, 13–15; Foley, *The Genesis of Missouri*, chap. 2 (both in 1.11); Cleary, "Contested Terrain" (1.14). St. Francis as resource zone for lower Mississippi Valley: Usner, *Indians, Settlers, and Slaves*, 174 (1.12).
19. See sources on 2.18.
20. Reshuffling of lower Mississippi Valley: Usner, *Indians, Settlers, and Slaves*, 102 (1.12).
21. Foley, *A History of Missouri*, 1:20, and chap. 2, esp. 33–34 (1.11). DuVal, *The Native Ground*, 137, 178 (1.12). Daniel Boone: Foley, *The Genesis of Missouri*, 77 (1.11).
22. Foley, *History of Missouri*, 64, 71 (1.11); Thorne, *The Many Hands of My Relations* (1.11). Livingston: Grant, "Roosevelt, Nicolas J." (1.31).
23. See 2.12.
24. See 3.10.1. Shawnees overall: Warren, *The Shawnees and Their Neighbors*, 24 (3.10).

Bowes, *Exiles and Pioneers*, 25 (2.12); Morrow, "Trader William Gillis and Delaware Migration," 150 (2.1); Creek village on the St. Francis: Lankford, "Shawnee Convergence," 395 (2.12); Usner, "American Indian Gateway," 46 (2.1); Bowes, *Exiles and Pioneers*, 25, 30; Faragher, "More Motley Than Mackinaw," 305–6 (both in 2.12); Ogilvie, "Development of the Southeast Missouri Lowlands," 9 (6.9).

25. Bear oil: Holder, "Historical Geography of the Lower White River," 134 (1.14). See also 2.2.

26. Arnold, *Colonial Arkansas*, 70 (2.16); Foley, *The Genesis of Missouri*, 61–63 (1.11); Ogilvie, "Development of the Southeast Missouri Lowlands," 13, 16 (6.9); Morrow, "New Madrid and Its Hinterland," 243 ("a considerable trading point," 245) (2.1). Even so, game began to be exhausted in the New Madrid region by about 1784—but it was still more plentiful than in the Southeast. Stoddard, *Sketches, Historical and Descriptive*, 212 (1.17).

27. Ogilvie, "Development of the Southeast Missouri Lowlands," 15 (6.9); Arnold, *Colonial Arkansas*, 20 (2.12); Schroeder, *Opening of the Ozarks*, 345–47 and chap. 4 (1.14).

28. Key, "The Calumet and the Cross," 153 (2.9); DuVal, *The Native Ground* (1.12); Arnold, *Colonial Arkansas*, 124 (2.12).

29. See 2.12.1.

30. (His name was also sometimes spelled Connitue). Myers, "Cherokee Pioneers in Arkansas," early wave: 133–34; Connetoo: 129 (2.12.1); Connetoo-led wave: Bolton, "Jeffersonian Indian Removal," 85 (1.11); Foster, "The First Years of American Justice," 137–38 (2.14). Spanish/Quapaws/Cherokees: "Statement of the Former Spanish Commandant, Arkansas," 5 June 1816, 179, in Carter, *Territorial Papers*, vol. 15 (2.16); DuVal, *The Native Ground* (1.12). "Choctaws and Chickasaws from over the Mississippi": John B. Treat, Arkansa, to Henry Dearborn, Secretary of War, 31 December 1806, in Carter, *Territorial Papers*, 14:56 (2.16).

31. White River: Abney, *Life and Adventures of L. D. Lafferty*, 17–18 (1.17); Myers, "Cherokee Pioneers in Arkansas," 143 (2.12.1). Frequent movement among settlements: Faragher, "More Motley Than Mackinaw," 307 (2.12).

32. Return J. Meigs to the Secretary of War, 17 February 1816, in Carter, *Territorial Papers*, 15:121–23 (2.16).

33. Myers, "Cherokee Pioneers in Arkansas," 156 (2.12.1). Also McLoughlin, "Thomas Jefferson and the Beginning of Cherokee Nationalism," 109 (2.13). "One hundred gunmen": McLoughlin, *Cherokee Renascence*, 180–81, also 56 (2.12.1). "The country between the mouth of the St. Francis": *Biographical and Historical Memoirs of Northeast Arkansas*, 450 (1.17). James Wilkinson, Governor of Louisiana, to H. Dearborn, Secretary of War, 22 September 1805, in Carter, *Territorial Papers*, 13:228–29 (2.16). Also 2.17.

34. "Western Cherokees": Bolton, "Jeffersonian Indian Removal," 86–87 (1.11); Sugden, "Early Pan-Indianism: Tecumseh's Tour of the Indian Country," 277 (3.11); McLoughlin, "Cherokee Ghost Dance Movement," 115 (2.13); Dowd, *A Spirited Resistance*, 162–63 (3.3.1); McLoughlin, *Cherokee Renascence*, 162 (2.12.1); on political machinations (from all sides): McLoughlin, "Thomas Jefferson and the Beginning

of Cherokee Nationalism" (2.13). McLoughlin later characterizes the number as closer to 800: McLoughlin, *Cherokee Renascence*, 164 (2.12.1). Unclarity of numbers may indicate the unclarity of who was permanently moving and who was visiting, hunting, or simply scouting territory: certainly there was a great deal of fluidity in these settlements; certainly there were substantial numbers involved, and certainly no officials at the time were able to keep very close track.

35. "Rally point": Return J. Meigs to the Secretary of War, 17 February 1816, in Carter, *Territorial Papers*, 15:121–23 (2.16).

36. See citations on 2.1. Polk Bayou: Holder, "Historical Geography of the Lower White River," 135 (1.14). Arkansas Post: Key, "Indians and Ecological Conflict," 129 (2.12); Arnold, *Colonial Arkansas*, 20, 186–87n3 (2.12). St. Francis river trading zone: Markman, "The Arkansas Cherokees," 16 (2.12.1); Myers, "Cherokee Pioneers in Arkansas" (2.12.1). Mixed-culture trading households: Morrow, "Trader William Gillis and Delaware Emigration" (2.1); Thorne, *The Many Hands of My Relations* (1.11). Indeed, language of race is inexact: many French Creoles would have indignantly called themselves "white," while others would likely have casually called themselves "French" as well as "Delaware," "Illinois," or some other tribal group. In this account as in their historical era, they move mostly freely and sometimes uneasily between the necessary but problematic categories of "white," "black," and "Indian." Contrast with lower Mississippi: Usner, *Indians, Settlers, and Slaves*, esp. chap. 3 (1.12).

37. Bolton, "Jeffersonian Indian Removal" (1.11). Opie, *Nature's Nation*, 99–108; Nobles, "Straight Lines and Stability" (both at 1.13).

38. Stiggins Narrative, in Nunez, "Creek Nativism," 134 (3.19); Dowd, *A Spirited Resistance*, 162–63 (3.3.1).

39. Hawkins: Dowd, *A Spirited Resistance*, 156, 162–63 (3.3.1); Stiggins Narrative, in Nunez, "Creek Nativism" (3.19); Bolton, "Jeffersonian Indian Removal" (1.11). "The Cherokees on the River St Francis": John B. Treat, Arkansas, to Henry Dearborn, the Secretary of War, 26 July 1808, in Carter, *Territorial Papers*, vol. 14 (2.16).

40. Warren, *The Shawnees and Their Neighbors*, 37 (3.10).

41. "Numberless Indian stragglers": "Representation of the Inhabitants of New Madrid," St. Louis, 19 September 1804, 61–62; "The insults and depredations" and "military establishment," in Major James Bruff, St. Louis, to Governor James Wilkinson, 29 September 1804, 59–60, in Carter, *Territorial Papers*, 13:61–62 (2.16).

42. "Vagabonds": "An Account of the Indian Tribes in Louisiana," sent to James Madison by Daniel Clark, 29 September 1803, in Carter, *Territorial Papers*, 9:64 (2.16). Travel back and forth: McLoughlin, *Cherokee Renascence*, 94 (2.12.1).

43. See 3.10.1.

44. "The great number of Indians resident here": Militia Officers of New Madrid to Gen. William Clark, no date, 1809, in Carter, *Territorial Papers*, 14:270 (2.16).

45. US official policy was also complicated by the mixed loyalties of territorial governor James Wilkinson: see 2.17. Squatters pressing settler Cherokees: Markman, "The Arkansas Cherokees," 31 (2.12.1). "The depopulation of the feeble scattered

settlements": Governor Wilkinson, St. Louis, to the Secretary of State, 24 August 1805, 189; edict cutting off trade: John B. Treat, Arkansa [Post], to the Secretary of War, 15 November 1805, 278, in Carter, *Territorial Papers*, vol. 13 (2.16). Removal of squatters: Myers, "Cherokee Pioneers in Arkansas," 149 (2.12.1). "Depopulation of our loose settlements": Bolton, "Jeffersonian Indian Removal," 83 (1.11).

46. "Numerous Credits": John B. Treat, Arkansa [Post], to the Secretary of War, 13 July 1806, in Carter, *Territorial Papers*, 13:544 (2.16).

47. "I may find some difficulty": James Wilkinson, Louisiana Governor, to James Madison, Secretary of State, 21 September 1805, in Carter, *Territorial Papers*, 13:220 (2.16). Parallel with lower Mississippi: Usner, *Soldiers, Settlers, and Slaves* (1.12).

48. "The hunting life here is at an end": McLoughlin, *Cherokee Renascence*, 172 (2.12.1). Dowd, *A Spirited Resistance*, 120, 160, 164–65 (3.3.1); Key, "Indians and Ecological Conflict," 127 (2.12); McLoughlin, "Cherokee Anomie," 7, 17 (2.13); Usner, "American Indian Gateway," 45–46 (2.1). Tradition of "long hunts": Aron, "Pigs and Hunters," 182 (2.14).

49. Parallels from the 1820s: Key, "Indians and Ecological Conflict," 141–44 (2.12). Bull, cow, calves: "Inventory of the Property Belonging to and Found in the Possession of Elisha Evans" (no date), Elisha Evans Probate, 1808, microfilm reel c50575, MJR (1.11). William Kincheloe: Probate Records for Pierre Dumay, 1909, microfilm reel c50575, MJR. Conflicts over livestock a broader theme: James McFarland to Governor Lewis, Arkansa, 11 December 1808, in Carter, *Territorial Papers*, 14:266–68 (2.16); Myers, "Cherokee Pioneers in Arkansas," 152 (2.12.1); Key, "Outcasts upon the World," 96–97 (2.14); Aron, "Pigs and Hunters" (2.14).

50. "Forced from his savage cruelty": Samuel G. Hopkins, Point Pleasant, to Major Richard Graham, Agent for Indian Affairs, St. Louis, 6 August 1825, Richard Graham Papers, Missouri Historical Society. Abney, *Life and Adventures of L. D. Lafferty*, 21 (1.17). My thanks to Lynn Morrow for this and so many other sources. Morrow, "New Madrid and Its Hinterland," 250 (2.1). Parallels in survival strategies for people like escaped slaves: Usner, *Indians, Settlers, and Slaves*, chap. 4, esp. 108 (1.12); Dowd, *A Spirited Resistance*, 188 (3.3.1).

51. "Niggerwool Swamp": Kochtitzky, *Otto Kochtitzky*, 139 (6.9).

52. Valencius, *The Health of the Country*, 147–52, photograph of African American family in dugout canoe (1.13); Usner, *Indians, Settlers, and Slaves* (1.12); Stewart, "Rice, Water, and Power"; Stewart, *What Nature Suffers to Groe*, chap. 4 (both in 1.13).

53. American official Amos Stoddard dismissed land between the St. Francis and the Mississippi as "periodically inundated" and "insalubrious"—but he was evaluating land as an American. For Creole rivermen, traders, and eastern Indian settlers, seasonal overflow made the area resource-rich and usefully distant from farming settlers. Bolton, "Jeffersonian Indian Removal," 79–80; later ecological changes: 80–81 (1.11). "They would hunt bear and deer and turkeys": Keefe and Morrow, eds., *A Connecticut Yankee in the Frontier Ozarks*, 129 (1.17). Also: Key, "Indians and Ecological Conflict" (2.12).

54. Trade and bullying: Report of James McFarland to Governor Lewis, Arkansa, 11 December 1808, 266–68; "intruders": Proclamation by Governor Meriwether Lewis, Territory of Louisiana, 6 April 1809, 261; "Cherokee Towns on the Sᵗ Francies": William Clark, St. Louis, to Henry Dearborn, Secretary of War, 29 April 1809, 264, in Carter, *Territorial Papers*, vol. 14 (2.16). Clark's spelling was not improved by his extensive journal-writing during his and Lewis's Corps of Discovery to the Pacific. Also Myers, "Cherokee Pioneers in Arkansas," 152 (2.12.1); broader context: Foley, *History of Missouri*, 133 (1.11). "Big Red Head": DuVal, *The Native Ground*, 202 (1.12).

55. "Usual crossing place": John B. Treat, Arkansa, to Henry Dearborn, Secretary of War, 31 December 1806, in Carter, *Territorial Papers*, 14:56 (2.16). Other crossings: Parker, "Missouri as It Is in 1867," 35, suggesting a crossing twenty-eight miles above Chalk Bluff, a site at the northern end of Crowley's Ridge (32) (1.17). Connetoo's trading post: vagaries of backwoods grammar and spelling make this somewhat speculative, but the US agent James recommended in 1808 that the best site for a trading establishment would be "the pine hills, say coniture on the river Sᵗ Francis." Report of James McFarland to Governor Lewis, Arkansa, 11 December 1808, in Carter, *Territorial Papers*, 14:266–68 (2.16).

56. Myers, "Cherokee Pioneers in Arkansas," 142 (2.12.1). Later village of St. Francis: My thanks to George Anne Draper for sharing documentation from the National Register of Historic Places application for Mt. Hope Cemetery (e-mail, 13 October 2006). Connitoo's trade: Key, "Indians and Ecological Conflict," 130 (2.12). Exchange of $4,800: Myers, "Cherokee Pioneers in Arkansas," 150–51 (2.12.1).

57. "Deposited … in the hands of Wm Weber and Kanetou": Statement from Joseph Michel to Probate Judge Andrew Wilson (spelled Wellson in this document), 3 May 1806, Probate Records for Peter Saffrey and Francis Contelmy, 1806, New Madrid County Probate Court Case 13, microfilm reel c50575, MJR (1.11); "Lately resident at the river St. Francis": Statement by A. Campbell, 2 May 1806; and Certificate of A. Campbell regarding property of Francis Contelmy, deceased, to Andrew Wilson, Judge of Probate, 3 May 1806, Saffrey and Contelmy Probate.

58. A woman named Peggy Chisolm employed a similar strategy—so perhaps did other multicultural people as yet undiscovered in legal records: Foster, "First Years of American Justice," 137–38 (2.14).

59. "In a deranged state of mind": Order by Joseph Charpenter, Judge of Probate for District of New Madrid, 8 July 1807, Probate Records for William King, 1807, New Madrid County Probate Court Case Number 27, microfilm reel c50575, MJR (1.11). Value of estate: Francis Clopper, statement of the estate of William King, [November] 1807, King Probate. Barsaloux estate: Statement of account of estate by Joseph Charpantier, 1807, Probate Records of Jean Baptiste Barsaloux, 1807, microfilm reel c50575, MJR (1.11). "Claimed by Indians": Jean Baptiste Barsaloux, "Inventory of Apraisment for Jean Baptiste Barsaloux Deceased 1807," hand-bound book listing appraisals and sales of the Barsaloux estate in spring 1807, Barsaloux probate. Problems of mobile currency in early republic: Kamensky,

The Exchange Artist (1.8); in early Missouri: Foley, *History of Missouri*, 177 (1.11); of early Arkansas lawyers: Foster, "First Years of American Justice," 130 (2.14).

60. Dumay Probate, MJR (1.11). Orphans' courts apprenticed out children whose parents or fathers died. Foster, "First Years of American Justice," 126–27 (2.14).

61. Statement of Estate, 17 December 1807; "Inventory of the Goods & Chattles of John Henry Merckel," 30 September 1805; "Account of Sale of the personal estate of John H. Merckell," 16 November 180?, John Henry Merckle Probate Case Files, microfilm reel c50575, MJR (1.11). His name may be Merkle, or perhaps Merkell— orthographical consistency is not a hallmark of these early records. His personal effects included a statement of naturalization, and he identified himself as a merchant in depositions connected with the Saffray and Contelmy probate. Probate Records for Benjamin Louis Vanamburgh, 1807, microfilm reel c50575, MJR (1.11). Booze: Rorabaugh, *The Alcoholic Republic* (1.8); Morrow, "Trader William Gillis and Delaware Migration in Southern Missouri," 157 (2.1). Missourians distilled alcohol from a wide variety of products—apples, peaches, plums, grapes—through the late nineteenth century. Foley, *History of Missouri*, 58 (1.11); Kochtitzky, *The Story of a Busy Life*, 19 (6.9). 740 gallons: Stoddard, *Sketches, Historical and Descriptive, of Louisiana*, 212 (1.17). Similar breadth of trading supplies: Foster, "First Years of American Justice," 138 (2.14).

62. "Grand Council": John B. Treat, Arkansa, to the Secretary of War, 20 May 1806, in Carter, *Territorial Papers*, 13:511–12 (2.16). Zone of passage: Militia Officers of New Madrid to Gen. William Clark, no date, 1809, in Carter, *Territorial Papers*, 14:270 (2.16). Ronda, *Lewis and Clark among the Indians*, chap. 4 (1.15).

63. See 2.8. Hunters, trappers: Arnold, *Colonial Arkansas*, 17–18 (2.12). "They pass from one side to the other": William Clark, St. Louis, to the Secretary of War, 30 April 1809, in Carter, *Territorial Papers*, 14:271 (2.16).

64. "A tract of 750 arpents of land": Statement by Joseph Michel, Administrator, of estates of Peter Saffrey and Francis Contelmy, 11 February 1808, Saffrey and Contelmy Probate, MJR (1.11). (Peter Saffray is noted as having "no personal property" in the same document.) In a statement from Joseph Michel to Probate Judge Andrew Wellson, 3 May 1806, Saffrey and Contelmy Probate, Michel refers to both men as "in Habitants of this place."

65. "Is no river at all": Kochtitzky, *The Story of a Busy Life*, 52–53 (6.9).

66. "St. Francois" as early boundary: Swem, "A Letter from New Madrid, 1789," 345 (2.2). Pierre Antoine Laforge, "Mémoire contenant des details sur les proprieties des habitans de la Nouvelle Madrid," 20 December 1804, in Carter, *Territorial Papers*, 13:405 (2.16); enclosed translation, 416; Quapaws: Berry, Beasley, and Clements, *The Forgotten Expedition*, 172 (1.15). Creation of Arkansas District: James Wilkinson, "A Proclamation," St. Louis, 1 January 1806, 540; subsequent divisions: James Wilkinson, "A Proclamation," St. Louis, 8 July 1806, 542; "a vast frontier": "Representation of the Inhabitants of New Madrid," to Maj. James Bruff, First Military Commandant of Upper Louisiana, 19 September 1804, 61, in Carter, *Territorial Papers*, vol. 13 (2.16).

67. Paleoseismology of the St. Francis: Guccione, "Late Pleistocene and Holocene Paleoseismology," 252 (2.21). Chain of bayous: Myers, "Cherokee Pioneers in Arkansas," 133 (2.12.1). "The economical and efficient means": Chambers, "River of Gray Gold," 104–5 (1.11). Stationing a force along the St. Francis: Maj. James Bruff, St. Louis, to Gov. Wilkinson, 28 May 1805, in Carter, *Territorial Papers*, 13:137 (2.16). But: pre-earthquake navigation: Berry, Beasley, and Clements, *The Forgotten Expedition*, 84 (1.15).

68. "As its current was much gentler": Flint, *The History and Geography of the Mississippi Valley*, 278–79; "Previous to the earthquakes": J. W. Foster, *The Mississippi Valley*, 21 (both in 1.17). Also Myers, "Cherokee Pioneers," 133 (2.12.1). transporting lead by wagon in 1819: Rafferty, ed., *Rude Pursuits and Rugged Peaks*, 19 (1.17).

69. Environmental disruption was surprisingly localized. Portions of the upper St. Francis basin were still "good rich Lands" in 1814: William Russell, St. Louis, to William Rector, Missouri land surveyor, in Carter, *Territorial Papers*, 14:754 (2.16). Near and below New Madrid, things were a different story. Drowned trader: Myers, "Cherokee Pioneers in Arkansas," 154 (2.12.1). "The St. Francis was at one time very low": "From the Columbian: Earthquake," *Independent American*, 3 March 1812, America's Historical Newspapers; Smith, *An Account of the Earthquakes*, 38 (1.2). "The traces of earth-cracks": Owen, Elderhorst, and Cox, *First Report of a Geological Reconnaissance*, 31 (1.17).

70. "Sunk country": Lyell, *A Second Visit to the United States*, 2:178–79 (5.16). See also 2.5.

71. Kochtitzky, *The Story of a Busy Life*, 51–52 (6.9).

72. Guccione, Van Arsdale, and Hehr, "Origin and Age of the Manila High and Associated Big Lake 'Sunklands' in the New Madrid Seismic Zone" (2.5); Guccione, "Late Pleistocene and Holocene Paleoseismology," 252, 259 (2.21). Prior uplifts occurred twice between 90 and 1640 CE. The quakes formed both of those sunklands as well as Reelfoot Lake and the now-drained Lake Obion: Mihills and Van Arsdale, "Late Wisconsin to Holocene Deformation in the New Madrid Seismic Zone," 1019 (2.21). Nineteenth-century accounts: Parker, *Missouri as It Is in 1867*; Wetmore, *Gazetteer of the State of Missouri*, 129. "Submerged ten and twenty feet": Foster, *The Mississippi Valley*, 21 (all in 1.17).

73. "A little river called Pemiscoe": James Fletcher, "Earthquake," *Western Sun*, Vincennes, IN, 15 February 1812; "A gentleman in attempting": "Earthquake," *Carthage Gazette*, Carthage, TN, 8 February 1812, both in CERI Compendium (1.2).

74. Downed timber and limited cypress growth: Guccione, "Late Pleistocene and Holocene Paleoseismology," 255, 258 (2.21). Surveyor Brown: King, "The Delimitation and Demarcation of the State Boundary," 136–37 (1.14). Cypress: Fuller, *The New Madrid Earthquake*, 70–71 (1.1); Van Arsdale et al., "Earthquake Signals in Tree-Ring Data," 255 (2.21).

75. "East branch of the St. Francis": Samuel G. Hopkins, Point Pleasant, to Major Richard Graham, Agent for Indian Affairs, St. Louis, 6 August 1825, Richard Graham Papers, Missouri Historical Society. Also Morrow, "New Madrid and Its Hinter-

land" (2.1). Court case: Feezor, "Fraud and Deceit in Dunklin County." "Old logs taken out of the bottom of the river": Testimony of Joseph Pelts, in *St. Francis Mill ... against William O. Sugg*, 280 (2.5).

76. "There is a great many places": testimony of Joseph Pelts, 282. "No straight channel": testimony of Brown Rosenbaum, 266. "All water," dugout canoes, and cordeling: testimony of G. T. Terrance, 181, 179, 173, 176. Could not canoe: testimony of W. F. Shelton, 310. Foot-and-a-half draft: testimony of E. C. Brandon, 191, 199, 202. Canoes and smartweed: testimony of Brown Rosenbaum, 260, 263–26; all from *St. Francis Mill ... against William O. Sugg* (2.5).

77. Testimony of J. W. Estep, in *St. Francis Mill ... against William O. Sugg*, 152, 186 (fishing and frogging, 155; tracking raccoons, 168; whiskey boat, 162). The point of this questioning was to establish that the Sugg boat that was supposed to have cleared that portion of the swamp had demonstrably not done so. Similar testimony: 280. Use of downed timber: 186 (2.5).

78. Shepard, *The New Madrid Earthquake*, 49–50, 56 (1.2); 1819 surveyor: Kochtitzky, *The Story of a Busy Life*, 50–51 (6.9).

79. Deposition of Henry Cassidy, 23 January 1813, in Carter, *Territorial Papers*, 14:623–24 (2.16).

80. Cassidy, 23 January 1813, in Carter, *Territorial Papers*, 14:623–24 (2.16).

81. "Beautiful and excellent lands" and timber railroads: Parker, *Missouri as It Is in 1867*, 81 (1.17). Roads: Kochtitzky, *The Story of a Busy Life*, 73–75. Railroads: Rhodes, *A Missouri Railroad Pioneer* (both 6.9).

82. "The face of the country": Silas Bent, St. Louis, to Hon. Jared Mansfield, Surveyor General, 11 August 1812, 590–91; "The country for some hundred miles": Silas Bent, St. Louis, to Josiah Meigs, 22 June 1813, 684; in Carter, *Territorial Papers*, vol. 14 (2.16). Confusion about the spelling of "Arkansas" would take years to settle.

83. King, "The Delimitation and Demarcation of the State Boundary," 133, 135 (1.14).

84. "Great Swamp": "Letters from Honorable L. F. Linn," 3 (1.17). Unhealthy: William Russell, St. Louis, to William Rector, Missouri land surveyor, in Carter, *Territorial Papers*, 14:756 (2.16). Parker, *Missouri as It Is in 1867*, 30 (1.17). "Ruins": Samuel G. Hopkins, New Madrid, to Richard Graham, Indian Agent, St. Louis, 29 January 1826, Richard Graham Papers, Missouri Historical Society.

85. King, "The Delimitation and Demarcation of the State Boundary," 140 and abstract (1.14); Geographer Bill Keegan confirmed this in working with early General Land Office maps in the course of this project.

86. "Letters from Honorable L. F. Linn and A. H. Sevier," 3–5 (1.17).

87. Later drainage work in southeast Missouri would make clear how wildly optimistic Linn's estimates were: see chapter 6. "Letters from Honorable L. F. Linn and A. H. Sevier," 3, 6 (1.17).

88. "Letters from Honorable L. F. Linn and A. H. Sevier," 6–7 (1.17); see 2.19.

89. Owen, Elderhorst, and Cox, *First Report of a Geological Reconnaissance*, 204; "Geographical Notices: Physical Geography of the Report on the Mississippi River," 227 (both in 1.17).

90. "Tennessee Stud," Jimmy Driftwood, 1958; Zac Cothren, "Jimmy Driftwood (1907–1998)," updated March 2012, *Encyclopedia of Arkansas History & Culture* (1.11). Also Evans, *A Pedestrious Tour* (1.17).

91. "The Indians are the only people": Morrow, "New Madrid and Its Hinterland," 248 (2.1). "Rather a nuisance": Col. D. Mooney, Monticello, Arkansas, to Gov. James Miller, 8 August 1820, in "L. C. Gulley Collection, 1819–1898," folder 2, no. 6, Arkansas History Commission. Mooney referred to these Native Americans as living "this side [of] white river near the Island." An 1850 map at the Arkansas History Commission shows a Big Island different from present Big Island, though both are formed by cutoffs near the mouth of the White River. Warren, *The Shawnees and Their Neighbors*, 79 (3.10).

92. Ancient sand blow debris: Wingerson, "In Search of Ancient Earthquakes," in (2.21).

93. Meigs account: Sugden, *Tecumseh*, 249 (3.10). Modern understanding: Tuttle, "The Use of Liquefaction Features in Paleoseismology," 362 (7.5.5); Tuttle et al., "Use of Archaeology to Date Liquefaction Features," 452 (7.5.5); Guccione, "Late Pleistocene and Holocene Paleoseismology," 245 (2.21).

94. *Biographical and Historical Memoirs of Northeast Arkansas*, 451 (1.17).

95. Myers, "Cherokee Pioneers in Arkansas," 153–57 (2.12.1); Bowes, *Exiles and Pioneers*, 41 (2.12). Lankford, "Shawnee Convergence," 398 (2.12), presents indirect evidence of this move; Warren, *The Shawnees and Their Neighbors*, 17 (3.10).

96. "A large body of Cherokees": "A Petition to the Secretary of War from the Inhabitants of Arkansas District," received 14 April 1812, in Carter, *Territorial Papers*, 14:544 (2.16). "The Cherokees have now left the ponds": Myers, "Cherokee Pioneers in Arkansas," 155; on emigration: 156 (2.12.1). It had not been two full years since the earthquakes; Tahlonteskee may have been counting winters or planting seasons.

97. Resource stress: Markman, "The Arkansas Cherokees," 36 (2.12.1); Key, "Indians and Ecological Conflict," 134 (2.12); see requests by Arkansas River Cherokees to accompany French and Quapaw tallow-gathering hunting expeditions: William L. Lovely to Governor Clark, 1 October 1813; and William L. Lovely to Governor Clark, 9 August 1814, in Carter, *Territorial Papers*, 14:51, 54 (2.16). "Remnants of tribes": Return J. Meigs to the Secretary of War, 17 February 1816, in Carter, *Territorial Papers*, 15:121–23 (2.16).

98. "The Cherokees who were exploring": Bolton, "Jeffersonian Indian Removal," 88 (1.11). "Many of the Indians": Ross, *Life and Times of Elder Reuben Ross*, 210 (1.17).

99. William Russell, St. Louis, to William Rector, Missouri land surveyor, in Carter, *Territorial Papers*, 14:755 (2.16). "The year the yerth shook so": Huddleston, "Some Indian Incidents along the White River," 40 (1.17). Traders along lower St. Francis: Arnold, *Colonial Arkansas*, 20, 186–87n3 (2.12).

100. Richard Searcy, White River, AR, to Gov. James Miller, 21 November 1820, folder 2, no. 9; Cherokee leaders, Point Pleasant, [AR], to Gov. James Miller, 18 June 1820, folder 2, no. 12; L. C. Gulley Collection, AHC; Morrow, "New Madrid and Its Hinterland," 248 (2.1); Myers, "Cherokee Pioneers in Arkansas," 156 (2.12.1); George Anne Draper, documentation from National Register of Historic Places

application for Mt. Hope Cemetery. Comparable processes elsewhere: O'Brien, *Firsting and Lasting* (2.3).

101. Connetoo resettlement: James Wilkinson, Governor of Louisiana, to H. Dearborn, Secretary of War, 22 September 1805, in Carter, *Territorial Papers*, 13:228–29 (2.16). "A certain Mᵣ Hunt": Report of James McFarland to Governor Lewis, Arkansa, 11 December 1808, in Carter, *Territorial Papers*, 14:266–68 (2.16).

102. Osage negotiations: Fausz, "Becoming 'a Nation of Quakers'"; Clark's lament: 37 (2.14); Treaty of Fort Osage: Foley, *History of Missouri*, 131–32 (1.11) Quapaw land claims: John B. Treat, Arkansa [Post], to the Secretary of War, 15 November 1805, in Carter, *Territorial Papers*, 13:277 (2.16). Cherokee newcomers: Usner, "American Indian Gateway," 49 (2.1); DuVal, *The Native Ground*, 151; Quapaw/Osage tension: chaps. 6, 7 (1.12).

103. Key, "Indians and Ecological Conflict," esp. 128–32 (2.12). "The Cherokees of St. Francis River": James Wilkinson, St. Louis, to the Secretary of War, 8 October 1805, in Carter, *Territorial Papers*, 13:235 (2.16); James Wilkinson, Governor of Louisiana, to H. Dearborn, Secretary of War, 22 September 1805, 228–29; efforts for peace: Secretary of War to Governor Wilkinson, 10 February 1806, 433. Continued conflict: James McFarland to Governor Lewis, Arkansa, 11 December 1808, in Carter, *Territorial Papers*, 14:266–68 (2.16). Depressed fur trade: Fausz, "Becoming 'a Nation of Quakers,'" 34 (2.14). Also: 2.17.

104. Key, "Indians and Ecological Conflict," 136 (2.12); Markman, "The Arkansas Cherokees," 33–34 (2.12.1). Cherokees leaving Nation for Arkansas: Bringier, "Notices," 30 (2.4); McLoughlin, *Cherokee Renascence*, chap. 11 (2.12.1); Bolton, "Jeffersonian Indian Removal," 88 (1.11). Cherokee numbers: McLoughlin, "Thomas Jefferson and the Beginning of Cherokee Nationalism," 109 (2.13). "Old Settlers": McKenney and Hall, *The Indian Tribes of North America*, 403–4 (1.17). See 2.12.2.

105. Settlement in northwest Arkansas: McLoughlin, *Cherokee Renascence*, 263 (2.12.1).

106. See 2.12.2 and 2.15.

107. Leaders of the "Cherokee Nation, on Arkansas," to Gov. James Miller, 22 April 1820, L. C. Gulley Collection, folder 2, no. 7 Arkansas History Commission; other documents by Walter Webber are in folder 2, no. 12; and folder 11, no. 99. Identity of Walter Webber: Hoig, *The Cherokees and Their Chiefs*, 135 (2.12.1).

108. Faragher, "'More Motley Than Mackinaw,'" 316, 321–23 (2.12); de Montulé, *Travels in America*, 105 (1.2).

109. Osage cessions: DuVal, *The Native Ground*, 224 (1.12); Baird, *The Osage People*, 31 (2.13). Treaty of 1828: Key, "Indians and Ecological Conflict," 144 (2.12).

110. Faragher, "'More Motley Than Mackinaw,'" 323 (2.12); Lankford, "Shawnee Convergence," 411–13 (2.12). Baird, "The Reduction of a People"; Key, "Outcasts upon the World," 100–104 (both in 2.14).

111. Stephen Warren does not see 1812 as crucial: Warren, *The Shawnees and Their Neighbors* (3.10).

112. Quakes did not change much: Hough and Bilham, *After the Earth Quakes*, 84 (1.6).

113. "You need not be alarmed about the Indians": Foley, *History of Missouri*, 162; populations: 166 (1.11). Post family and their emigrations: Valencius, *The Health of the Country*, 1–3, 13–16 (1.13).

114. See DuVal, *The Native Ground* (1.12).

115. "A hand carved in the wood": *Biographical and Historical Memoirs of Northeast Arkansas*, 452 (1.17). Carvings: Berry, Beasley, and Clements, *The Forgotten Expedition*, 70 (1.15). Widespread but little-studied hand symbolism: Thorne, *The Many Hands of My Relations*, 1 (1.11). Families forced off Reelfoot Lake: Miss Effie Cowan, interview with Mrs. William Price, Marlin, Texas, Library of Congress American Memory Project, American Life Histories: Manuscripts from the Federal Writers' Project, 1936–40. Displacement of other trapping, hunting families: Jacoby, *Crimes against Nature*; Warren, *The Hunter's Game* (both in 1.13).

116. *Biographical and Historical Memoirs of Northeast Arkansas*, 452 (1.17). Missionaries among the Cherokees had to post students to guard their orchards from surreptitious night-time picking. McClinton, ed., *The Moravian Springplace Mission*, 2:75 (3.17). Narrative strategies for effacing Native peoples from history: O'Brien, *Firsting and Lasting* (2.3).

117. Search for similarly disappeared communities: Harington, *Let Us Build Us a City* (6.6). Draper, personal communication, 2006.

118. Flint, *Recollections*, 220 (1.17).

119. Verhoff, *The Intractable Atom*, and "Album: Unnatural Nature," on the Rocky Mountain Proving Ground, Cronon, ed. *Uncommon Ground* (both in 1.13). Parallels with Lovely's Purchase: see 2.15.

120. "Muskrat, otter, mink, rackoon": Wetmore, *Gazetteer of the State of Missouri*, 129 (1.17). Wetmore estimated annual returns of $15,000 to $20,000 from this trade.

121. Isenberg, *Destruction of the Bison* (1.13). The Saline River in Arkansas was named for bison salt licks, but the animals had been hunted out by 1808: Key, "Indians and Ecological Conflict," 138; canebrakes as bison hideout, 135 (2.12). Cane as feed: Parker, *Missouri as It Is in 1867*, 34 (1.17). Lyell, *A Second Visit to the U.S.*, 179 (5.16). See marker for Buffalo Island, a plateau between Big Lake, Little River, and the St. Francis River, identified as home to bison until the 1811–12 earthquakes, but likely a haven after as well: Pat Lendennie, "Photo: Buffalo Island," *Encyclopedia of Arkansas*, accessed 19 June 2012 (1.11).

122. Bennett, "Big Lake National Wildlife Refuge" (2.5).

123. See citations on 2.5.

124. Lyell, *A Second Visit to the U.S.*, 179 (5.16).

Chapter 3

1. Speech of Gomo, Ninian Edwards papers, Sugden, *Tecumseh*, 266–67 (3.10).

2. "Shawnee Prophets": Sugden, *Tecumseh*, 251 (3.10).

3. "Creeks": "Muskogee" may be a better term, but "Creek" is more widely recognized: Martin, *Sacred Revolt*, 6–8 (3.19). "Houses shook": McClinton, ed., *Moravian Springplace Mission*, vol. 1, 16 December 1811, 460; continued tremors: 469, 470,

475, 478, 479, 483, 484, 486 (3.17). "The Great convultion of the Earth": M. Terrall, Natchez, Louisiana, to firm of Brown & Ives, Providence, RI, 3 January 1812, Brown family business papers, John Carter Brown Library, Brown University.

4. See sources on 3.1.1 and 3.17.

5. See 3.4.

6. See 3.2 and 3.1.1. Still untapped are records from British posts—the allies of Tecumseh and Tenskwatawa—as well as anthropological material from Plains Indians groups.

7. James, Long, and Say, "Account of an Expedition from Pittsburgh to the Rocky Mountaints," 57–58 (1.2). Hill, *Roads, Rails, and Waterways*, 157 (1.32). See 3.2. Also Skinner, "Traditions of the Iowa Indians," 497–98, and other sources listed in 3.1.1.

8. Hunter, *Memoirs of a Captivity*, 39 (3.9). See 3.8.

9. Martin, "Tecumseh," 9027 (3.10); Dowd, *A Spirited Resistance*, 110–12 (3.3.1).

10. Dowd, *A Spirited Resistance*, esp. 120, 181 (3.3.1); Martin, *Sacred Revolt*, 44–45 (3.19); McLoughlin, "Cherokee Anomie," 7, 17 (2.13); Fausz, "Becoming 'a Nation of Quakers'" (2.14); "There is no part of this country": Warren, *The Shawnees and Their Neighbors*, 38; Sugden, *Tecumseh*, esp. 107–8 (both in 3.10); processed fabric and food: Saunt, "Domestick . . . Quiet Being Broke" (3.19); McClinton, ed., *Moravian Springplace Mission*, 1:28 (3.17); Sugden, *Tecumseh*, 104–5 (3.10); see also 3.6.

11. See 3.14. Fears of British/Native cooperation: William Clark, St. Louis, to Henry Dearborn, Secretary of War, 29 April 1809, in Carter, ed., *Territorial Papers*, 14:266 (2.16); "General combination": William Henry Harrison, in Wood, *Empire of Liberty*, 675 (1.8); Drake, *Life of Tecumseh*, 133 (3.10). "It was known": Ross, *Life and Times of Elder Reuben Ross*, 198 (1.17).

12. Francis McHenry, "The Indian Prophet: From the Georgia Journal," *Halcyon Luminary, and Theological Repository* 1, no. 6 (1812): 275–77, Google Books.

13. See 3.13.

14. See 3.14 and 3.19. Historical reimaginings: DuVal, "Choosing Enemies" and "Could Louisiana Have Become an Hispano-Indian Republic?" (both in 2.14).

15. Pronunciation of "Tecumthé": J. G. Vore, Mission Agency, Muscogee, to Lyman C. Draper, Madison, WI, 29 November 1881, in Draper Manuscripts, vol. 4, 17; Sugden, *Tecumseh*, 23, 285 (both in 3.10). Varied versions and spelling: Ross, *Life and Times of Elder Reuben Ross*, 199 (1.17). Also see 3.10.

16. Dowd, *A Spirited Resistance*, 39–40, 126, 136–37 (3.3.1); Drake, *Life of Tecumseh*, 88–89; Edmunds, *The Shawnee Prophet*, 28–39, 42–46; William Henry Harrison to the Delawares, "Early in 1806," in Esarey, ed., *Governors Messages and Letters*, 1:184; Sugden, *Tecumseh*, 23–24, 113–17, 127, 208–9; Warren, *The Shawnees and Their Neighbors*, 24–28, 31–32, 37 (all at 3.10); Sugden, "Early Pan-Indianism: Tecumseh's Tour of the Indian Country," 114–15 (3.11); *Missouri: A Guide to the "Show Me" State*, 36 (1.11); "I found them burning their friends and relations": James McFarland, Arkansare Town, to William Clark, 20 February 1809, in Carter, ed., *Territorial Papers*, 14:269 (2.16).

17. Dowd, *A Spirited Resistance*, 128 (3.3.1); Edmunds, *Tecumseh and the Quest*, 84, 92;

Edmunds, *The Shawnee Prophet*, 34, 40; Sugden, *Tecumseh*, 10, 133, 147, 168; Warren, *The Shawnees and Their Neighbors*, 27 (all at 3.10).

18. See 3.6 and McClinton, ed., *Moravian Springplace Mission*, 1:15 (3.17).

19. "Napoleon of the west": Drake, *Life of Tecumseh*, 229; hunting story: 83 (3.10).
Retellings: for instance, Ross, *Life and Times of Elder Reuben Ross*, 199 (both at
1.17). Noble Tecumseh in modern histories: Feldman, *When the Mississippi Ran
Backwards*, 65–66 (1.1). Tenskwatawa as religious prophet versus Tecumseh as
charismatic war chief: Taylor, *The Civil War of 1812*, 126–27 (3.14). Complications
of brothers' roles: Dowd, *A Spirited Resistance*, 139, 146 (3.3.1); Edmunds, *The
Shawnee Prophet*, 92, 116; Edmunds, *Tecumseh and the Quest*, 111; Martin, "Tecum-
seh," 9028–29; Warren, *The Shawnees and Their Neighbors*, 32. "There were two
Tecumsehs": Editor's note, Esarey, ed. *Governors Messages and Letters* (all at 3.10).
Cherokee oral histories: W. H. Duncan, Cherokee Orphan Asylum, Cherokee Na-
tion, Indian Terr., to Lyman C. Draper, 12 January 1882, and William P. Doran,
Hempstead, Waller County, TX, to Lyman Draper, 27 November 1889, Draper
Manuscripts, vol. 4, pp. 72 and 92 (3.10). "This Warlike Spirit or Indian *Prophet*":
William Clark, St. Louis, to Henry Dearborn, Secretary of War, 29 April 1809, in
Carter, *Territorial Papers*, 14:265 (2.16); also Edmunds, *Tecumseh and the Quest*, 152
(all at 3.10); Stiggins Narrative, in Nunez, "Creek Nativism," 145 (3.19); Sugden,
"Early Pan-Indianism: Tecumseh's Tour of the Indian Country," 295 (3.11). "The
Prophets": Benjamin Hawkins, Creek Agency, 28 July 1813, *American State Papers:
Indian Affairs*, I (3.19).

20. "Turn over the earth": Dowd, *A Spirited Resistance*, 128 (3.3.1); see also 3.6.

21. Harrison to the Delawares, "Early in 1806," Esarey, ed. *Governors Messages and Letters*, 183 (3.10). Harrison knew his Bible: sun stands still, Joshua 10:13; also moon,
Habukkuk 3:11; dead rise, Acts 9:36–41; John 11:1–27.

22. Drake, *Life of Tecumseh*, 90–94; Edmunds, *The Shawnee Prophet*, 48–49; Edmunds,
Tecumseh and the Quest, 81–82 (all at 3.10); Benjamin Drake, "Anthony Shane's
Statements about Tecumseh," 28 November 1821, in Draper Manuscripts, vol. 12,
pp. 20, 54 (3.10). Scientific investigations: Greene, *American Science in the Age of
Jefferson*, 147–49 (5.3).

23. Indian league: Marshall, *The History of Kentucky*, 2:483 (1.17). "All the Indian tribes
on the continent" and "in common with the western tribes": Nunez, "Creek Nativism," 146, 14 (3.19). Challenges to Indian confederacy: DuVal, "Choosing Enemies"
(2.14); and DuVal, *The Native Ground* (1.12).

24. Sugden, *Tecumseh*, 80 (3.10); Aron, "Pigs and Hunters" (2.14); "kill the cattle":
Nunez, "Creek Nativism," 14 (3.19).

25. Edmunds, *Tecumseh and the Quest*, 141–47; Edmunds, *The Shawnee Prophet*, 100,
110–14. Impact of battle on Indian morale: "Statement of William Brigham, in
Company of Riflemen at Tippecanoe," Esarey, ed. *Governors Messages and Letters*,
706 (all at 3.10).

26. See 3.11. "You see him today on the Wabash": Sugden, *Tecumseh*, 215; also 179–82,
93–94, 205–17, 37–57, and 67; Edmunds, *Tecumseh and the Quest*, 203–4 (both at

3.10). Sugden, "Early Pan-Indianism: Tecumseh's Tour of the Indian Country," 272 (3.11).

27. Secrecy: William Henry Harrison to Secretary of War, 6 June 1811, Esarey, ed. *Governors Messages and Letters*, 5513 (3.10); Stiggins Narrative, in Nunez, "Creek Nativism," 147 (3.19). "Cultivate the ground": McKenney and Hall, *The Indian Tribes of North America*, 93–94 (1.17); "the way he came," John P. Sincecum to Lyman C. Draper, March 1882, Draper Manuscripts, vol. 4, p. 49 (3.10).

28. Groups traveled to hear and report back about Tecumseh: McLoughlin, *Cherokee Renascence*, 179 (2.12.1); Dowd, *A Spirited Resistance*, 145 (3.3.1). "Dance of the Lakes": Nunez, "Creek Nativism," 8 (3.19); A. W. Crain, Sasakwa, Seminole Nation, Indian Territory, to Lyman C. Draper, 11 January 1882, Draper Manuscripts, vol. 4, p. 16 (3.10); Benjamin Hawkins, Creek Agency, 28 July 1813, US Congress, *American State Papers: Indian Affairs*, 1:849 (3.19). "Sing the song": Nunez, "Creek Nativism," 14 (3.19); Martin, *Sacred Revolt*, 145–48 (3.19). Giving out red sticks: McKenney and Hall, *The Indian Tribes of North America*, 93 (2.14). Distribution of sticks as calendrical practice: Martin, *Sacred Revolt*, 187 (3.19).

29. Hunter, *Memoirs*, 43 (3.9); Dowd, *A Spirited Resistance*, 146 (3.3.1); Sugden, *Tecumseh*, 237 (3.10); Sugden, "Early Pan-Indianism: Tecumseh's Tour of the Indian Country," 279 (3.11).

30. "The fall of meteors": A. E. W. Robertson, Muscogee, Indian Territory, to Lyman C. Draper, 5 August 1884, Draper Manuscripts, vol. 4, p. 46. Edmunds, *Tecumseh and the Quest*, 237–51. Tecumseh's name: Sugden, *Tecumseh*, 4, 23 (all at 3.10); "fire panther": Mooney, "Myths of the Cherokee," 442 (3.16).

31. Nunez, "Creek Nativism," 42 (3.19); also Dowd, *A Spirited Resistance*, 156, 197 (3.3.1).

32. Stiggins Narrative, in Nunez, "Creek Nativism," 147–50 (3.19).

33. McKenney and Hall, *The Indian Tribes of North America*, 94–95 (2.14).

34. McKenney and Hall, *The Indian Tribes of North America*, 94–95 (2.14). Widely publicized in the nineteenth century: Drake, *Life of Tecumseh*, 144–45 (3.10).

35. Many stories: J. G. Vore, Wetunka, Indian Territory, to Lyman C. Draper, 16 August 1886; Statement of Tustenuckochee, 22 August 1883; and A. E. W. Robertson, Tullahassee, Indian Territory, to Lyman C. Draper, 8 December 1881; in Draper Manuscripts, vol. 4, pp. 1, 2, and 44 (3.10). "Cause the land": Will P. Ross, Fort Gibson, Ind. Terr., to Lyman C. Draper, 6 February 1882, in Draper Manuscripts, vol. 4, p. 61. "Strike the ground with his feet": A. E. W. Robertson, Tullahassee, Indian Territory, to Lyman C. Draper, 8 December 1881, Draper Manuscripts, vol. 4, p. 45.

36. "A man to make the world shake": Coleman Cole, Indian Territory, to Lyman C. Draper, 1 May 1884, Draper Manuscripts, vol. 4, p. 7 (3.10). Sugden, "Early Pan-Indianism: Tecumseh's Tour of the Indian Country," 290 (3.11).

37. Sugden, *Tecumseh*, 252–55 (3.10); and Sugden, "Early Pan-Indianism: Tecumseh's Tour of the Indian Country," 292–95 (3.11).

38. See 3.9.

39. Hunter, *Memoirs*, 47 (3.9). Placing this event: Sugden, "Early Pan-Indianism: Tecumseh's Tour of the Indian Country," 292–94 (3.11).

40. "Earthquakes," *Weekly Visitor*, 29 February 1812, NewsBank: America's Historical Newspapers.

41. For instance: "From the Columbian: Earthquake," *Independent American*, 3 March 1812; and "Earthquakes; Westward; Upper Louisiana; Little Praria; New Madrid," *Newburyport Herald*, 25 February 1812, both NewsBank: America's Historical Newspapers.

42. Edmunds, *The Shawnee Prophet*, 117; Edmunds, *Tecumseh and the Quest*, 148–49 (both in 3.10).

43. "Statement of William Brigham, in Company of Riflemen at Tippecanoe," in Esarey, ed. *Governors Messages and Letters*, 708 (3.10). Also Taylor, *The Civil War of 1812*, 129 (3.14).

44. McLoughlin, "Cherokee Anomie"; "Cherokee Ghost Dance Movement"; "Thomas Jefferson and the Beginning of Cherokee Nationalism" (all in 2.13); and *Cherokee Renascence* (2.12.1).

45. "We do not consider *you* as *white people*": McClinton, ed., *Moravian Springplace Mission*, 1:431; horse epidemic, 457 (3.17). McLoughlin, *Cherokee Renascence*, 179 (2.12.1); McLoughlin, "Cherokee Ghost Dance Movement," 118 (2.13).

46. See 3.5. "Communications from the Great Spirit": McLoughlin, "Cherokee Ghost Dance Movement," 136–37 (2.13).

47. McLoughlin, "Cherokee Anomie, 1794–1810," 11; "kill their cats": McLoughlin, "Cherokee Ghost Dance Movement," 136–37 (both in 2.13).

48. McLoughlin, "Cherokee Ghost Dance Movement," 142, 114 (2.13); Dowd, *A Spirited Resistance*, 175–76 (3.3.1).

49. "Fanatics who tell them": Col. Return J. Meigs, "Some Reflections on Cherokee Concerns," in National Archives Record Group 75, Roll M-208, "Records of the Cherokee Indian Agency in Tennessee, 1801–1835," reel 5, letter to Eustis of 9 March 1812. Conflicts over domestic animals and fowl: Aron, "Pigs and Hunters," 189–93 (2.14). Later attacks on whites' livestock as cultural proxy: McClinton, ed., *Moravian Springplace Mission*, 1:550 (3.17). "abandoned their bees": McLoughlin, "Cherokee Ghost Dance Movement," 138 (2.13).

50. Headgear: Sugden, *Tecumseh*, 145 (3.10); Dowd, *A Spirited Resistance*, 127, 179 (3.3.1). Meigs, "Some Reflections on Cherokee Concerns" (3.17).

51. McLoughlin, "Moravian Mission Dairy Excerpts," 341; also 3.17.

52. McClinton, ed., *Moravian Springplace Mission*, vol. 1, introduction, 23–24, 34–35, 149, 191, 420 (3.17). Mooney, "Cherokee Ball Play" (3.16).

53. "I have never heard anything bad about you": McLoughlin, "Moravian Mission Dairy Excerpts," 343 (3.17). Interpreters: McLoughlin, "Cherokee Ghost Dance Movement," 121 (2.13); McClinton, ed., *Moravian Springplace Mission*, 1:433 (3.17).

54. McClinton, ed., *Moravian Springplace Mission*, vol. 1: "Even old people": 462; "said that the earth is already very old," 460 (3.17). Creation story: "How the World Was Made," Mooney, "Myths of the Cherokee," 239 (3.16).

55. McClinton, ed., *Moravian Springplace Mission*, 1:460–61 (3.17).

56. McClinton, ed., *Moravian Springplace Mission*, 1:548 (3.17).

57. McClinton, ed., *Moravian Springplace Mission*, 1:461 (3.17).

58. McClinton, ed., *Moravian Springplace Mission*, 1:37, 461 (3.17). Hicks: McLoughlin, "The Beginning of Cherokee Nationalism," 85 (2.13); and Bolton, "Jeffersonian Indian Removal," 86 (1.11).

59. McClinton, ed., *Moravian Springplace Mission*, 1:27, 461 (3.17). Under world: Mooney, "Myths of the Cherokee," 240, 346–50 (3.16).

60. McClinton, ed., *Moravian Springplace Mission*, 1:461; on Vann: 37 (3.17). Snakes: Mooney, "Myths of the Cherokee," 251–52, 294–306 (3.16).

61. McClinton, ed., *Moravian Springplace Mission*, 1:462, 473 (3.17).

62. McClinton, ed., *Moravian Springplace Mission*, 1:469–70 (3.17). Solemnity marked important matters: Mooney, "The Cherokee Ball Play," 117 (3.16); Dowd, *A Spirited Resistance*, 173 (3.3.1).

63. McClinton, ed., *Moravian Springplace Mission*, 1:470, 473, 484–85 (3.17).

64. McClinton, ed., *Moravian Springplace Mission*, 1:470–72 (3.17).

65. Meigs, "Some Reflections" (3.17). Going to water: Mooney, "The Cherokee Ball Play," 120; Mooney, "The Cherokee River Cult"; Mooney, "The Sacred Formulas," 379 (3.16); Nabokov, *Where the Lightning Strikes*, 57 (3.3.2); McClinton, ed., *Moravian Springplace Mission*, 1:110 and 510n26 (3.17).

66. McClinton, ed., *Moravian Springplace Mission*, 1:475, 473 (3.17).

67. McClinton, ed., *Moravian Springplace Mission*, 1:639n12 (3.17).

68. Contemplation and prophecy: Dowd, *A Spirited Resistance*, 179 (3.3.1); McClinton, ed., *Moravian Springplace Mission*, 1:474 (3.17). Sacred sites: Mooney, "Myths of the Cherokee," 231, 404–19 (3.16); "Between River and Fire: Cherokee" in Nabokov, *Where the Lightning Strikes*, chap. 4 (3.3.2).

69. McClinton, ed., *Moravian Springplace Mission*, 1:474 (3.17).

70. Historical prophetic precedence: Dowd, *A Spirited Resistance*, 102, 179 (3.3.1). McClinton, ed., *Moravian Springplace Mission*, 1:474 (3.17). "The Book" as white knowledge and Indians as people without the book: Martin, *Sacred Revolt*, 115 (3.19).

71. McClinton, ed., *Moravian Springplace Mission*, 1:475–479 (3.17); McLoughlin, "Cherokee Ghost Dance Movement" (2.13).

72. McLoughlin, *Cherokee Renascence*, 183 (2.12.1); McClinton, ed., *Moravian Springplace Mission*, 1:479 (3.17); McLoughlin, "Cherokee Ghost Dance Movement," 328 (2.13).

73. McLoughlin, *Cherokee Renascence*, 184–85 (2.12.1); McLoughlin, "Cherokee Ghost Dance Movement" (2.13); McClinton, ed., *Moravian Springplace Mission*, 1:478–80, 387, 640n34 (3.17). Vann: Dowd, *A Spirited Resistance*, 178 (3.3.1).

74. McClinton, ed., *Moravian Springplace Mission*, 1:487 (3.17); McLoughlin, "Cherokee Ghost Dance Movement," 132 (2.13).

75. McLoughlin, "Cherokee Ghost Dance Movement," 134 (2.13). From 1828 to 1834, the *Cherokee Phoenix* carried several stories covering earthquakes, but all focused on Europe and South America. The paper made no mention of those felt in Cherokee country before the removal. For instance: "Earthquakes: Translated from a

Work Published in 1815," *Cherokee Phoenix*, 29 April 1829; and "Earthquakes (from the *Juvenile Encyclopedia*)," *Cherokee Phoenix*, 1 December 1832.

76. McClinton, ed., *Moravian Springplace Mission*, 1:474 (3.17).

77. Nunez, "Creek Nativism," 3 (3.19).

78. Relationship between Cherokee and Creek territory and spiritual movement: Dowd, *A Spirited Resistance*, 180 (3.3.1); McLoughlin, "Cherokee Ghost Dance Movement," 137 (2.13).

79. See 3.6 and 3.19. Deep roots of this movement: Dowd, *A Spirited Resistance*, 100–101 (3.3.1); Martin, *Sacred Revolt*, 123 (3.19). "Instead of beef and bacon": McLoughlin, "Cherokee Ghost Dance Movement," 138 (2.13). Sources of prophetic power: Nunez, "Creek Nativism," 8; and Stiggins Narrative, 149–52; Martin, *Sacred Revolt*, 122–24 (both at 3.19). Prophets' claims: Sugden, *Tecumseh*, 262 (3.10).

80. Witches: Sugden, *Tecumseh*, 262–63 (3.10). "Whenever such a person was found": Stiggins Narrative, in Nunez, "Creek Nativism," 149; "flying tales": 146; fatherless rumors, "ferment and agitation": 150; Big Warrior and schism: 154 (3.19). Prophetic frenzy: Hawkins, *Letters, Journals, and Writings*, 133 (3.19).

81. Splintering: Dowd, *A Spirited Resistance*, 156–58 (3.3.1). "Prophetic warriors" and prophetic rituals: Nunez, "Creek Nativism," 12, 167–68, 301. Attacks: Martin, *Sacred Revolt*, 130, 152; "Report of Alexander Cornells, Interpreter, Upper Creeks, to Colonel Hawkins"; I:846; "Destroy every thing": Benjamin Hawkins, Creek Agency, 28 July 1813, US Congress, *American State Papers: Indian Affairs*, 1:849; Hawkins, *Letters, Journals, and Writings*, 135, 42 (all at 3.19); Dowd, *A Spirited Resistance*, 170 (3.3.1).

82. "Had power to destroy them by an earthquake" and "render the earth quaggy": "Report of Alexander Cornells," 1:846. Martin, *Sacred Revolt*, 130 (both in 3.19); Dowd, *A Spirited Resistance*, 62 (3.3.1). See 3.19.1.

83. "By showing them they are both feeble and ignorant": "Report of Alexander Cornells," 1:846. "You may frighten one another": Martin, *Sacred Revolt*, 130 (both at 3.19).

84. See sources on 3.19.

85. Lack of coordination: Statement of Tustenuckochee, 22 August 1883, in Draper Manuscripts, vol. 4, p. 3 (3.10). Failed appeal for Cherokee alliance: Dowd, *A Spirited Resistance*, 157–58, 179–80 (3.3.1); McLoughlin, *Cherokee Renascence*, 191 (2.12.1); Martin, *Sacred Revolt*, 166 (3.19); Sugden, *Tecumseh*, 349–53 (3.10).

86. "Came near exterminating the Creeks": Statement of Tustenuckochee, 22 August 1883, Draper Manuscripts, vol. 4, p. 3 (3.10).

87. Cost to Cherokees: Howe, *What Hath God Wrought*, 74–76 (1.8); Dowd, *A Spirited Resistance*, 189 (3.3.1); and to Cherokee slaves: McLoughlin, *Cherokee Renascence*, 194–95 (2.12.1). "Unpleasant stories": McClinton, ed., *Moravian Springplace Mission*, 1:571 (3.17).

88. See 3.19. Also McLoughlin, *Cherokee Renascence*, 196 (2.12.1).

89. Sugden, *Tecumseh*, 271 (3.10).

90. Edmunds, *The Shawnee Prophet*, 61–63; Sugden, Tecumseh, 271 (both in 3.10); Dowd, *A Spirited Resistance*, 118–19 (3.3.1).

91. "White man found fighting": Sugden, *Tecumseh*, 284; also 291, 303 (3.10).

92. Sugden, *Tecumseh*, 303–9, 320–21 (3.10).

93. Wood, *Empire of Liberty* (1.8). Friendly fire incident: Edmunds, *Tecumseh and the Quest*, 160–61; Sugden, *Tecumseh*, 296, 357 (both at 3.10).

94. Debating Tecumseh's death: Drake, *Life of Tecumseh*, chap. 15; Sugden, *Tecumseh*, 375–76 (both at 3.10). Cheerful ditty: Shelley D. Rouse, "Colonel Dick John's Choctaw Academy: A Forgotten Educational Experiment," *Ohio Archeological and Historical Quarterly* 25, no. 2 (1916), Google Books, 91. Dowd, *A Spirited Resistance*, 184 (3.3.1). Sugden, *Tecumseh*, 391–93 (3.10).

95. Dowd, *A Spirited Resistance*, 184; Prophet Francis, 172, 189 (3.3.1). Sugden, *Tecumseh*, 108 (3.10); Martin, *Sacred Revolt*, 168, 96–107 (3.19).

96. War of 1812 separates Indian South and North: Dowd, *A Spirited Resistance*, 171, 181 (3.3.1). Debating the impact of Wars of 1812–15: Green, "The Expansion of European Colonization," 497 (3.6); Bowes, *Exiles and Pioneers*, 39 (2.12); Warren, *The Shawnees and Their Neighbors* (3.10).

97. Name "Tecumseh": postcard from A. E. W. Robertson, Tullahassee, Indian Terr., to Lyman C. Draper, 2 December 1881, Draper Manuscripts, vol. 4 (3.10); Dowd, *A Spirited Resistance*, 189 (3.3.1); Sugden, *Tecumseh*, 399 (3.10). Te-cu-wa-se, a Cherokee version of Tecumseh, was a common name among the Cherokees: W. H. Duncan, Cherokee Orphan Asylum, Cherokee Nation, Indian Terr., to Lyman C. Draper, 12 January 1882, Draper Manuscripts, vol. 4, p. 72 (3.10).

98. Calloway, *The Shawnees and the War for America*, 160; Edmunds, *The Shawnee Prophet*, 168–83, 187; Warren, *The Shawnees and Their Neighbors*, 13, 109–10 (all at 3.10).

Chapter 4

1. The author wrote from Jamaica, Long Island, which is now the borough of Queens. Smith, *An Account of the Earthquakes*, 45 (1.2). See also 4.3.

2. Smith, *An Account of the Earthquakes*, "strange sensation": 16; "sudden and deadly sickness": 44; "giddiness in their heads": 61; nausea: 9; trouble keeping feet: 13, 38; "dizziness and vertigo": 14; drunk or seasick: 51, 11; on a poise: 14 (1.2). Uneasy Canadians: "'Extract of a Letter from William Henry, 29th January, 1812," *Kingston Gazette* (Kingston, Upper Canada), 3 March 1812, University of Western Ontario archives. Nausea across eras: Davison, "The Feeling of Nausea Experienced during Earthquakes" (4.1); drunk or at sea: Pusey, "The New Madrid Earthquake," 286; Stevenson, "Autobiography of the Rev. William Stevenson," 41 (both in 1.2).

3. "Much wonder": Smith, *An Account of the Earthquakes*, 45 (1.2); "the earthquake which took place": Rush, *Medical Inquiries*, 326; quake of 1886: Porcher, "Influence of the Recent Earthquakes, 652, 651 (both at 4.1).

4. Smith, *An Account of the Earthquakes*, 16 (1.2).

5. Smith, *An Account of the Earthquakes*, 39–40 (1.2).

6. Montgomery letter to Barton (1.2).

7. See 4.1. "Summary [of News]," *Cherokee Phoenix*, 19 August 1829.

8. Webster, "On the Connections of Earthquakes with Epidemic Diseases," 341 (4.1); Norman K. Risjord, "Webster, Noah," in *American National Biography Online* (1.8).

9. Mitchill, "A Detailed Narrative," 304 (1.2).

10. Beaufort: Smith, *An Account of the Earthquakes*, 51; also 22 (1.2). Major Burlison: Griffith, ed. *Letters from Alabama*, 195 (1.17). Acclimation: Valencius, *The Health of the Country*, 22–34 (1.13).

11. Smith, *An Account of the Earthquakes*, 41, 43, 45 (1.2); Coen, *Earthquake Observers* (1.6).

12. See notes on 4.2—on aftershocks: Jenson, "I at Home," pt. 1, p. 43; Thomas's relapse, 39.

13. 20 August 1813: Jensen, "I at Home," pt. 1, p. 40 (4.2).

14. Smith, *An Account of the Earthquakes*, 29 (1.2); "Earthquake," *The Supporter* (1.2); Mitchill, "Description of the Volcano and Earthquake Which Happened in the Island of St. Vincents, on the 30th Day of April, 1812" (1.2).

15. "Bilious fever" and "Bowel complaint": 11 July and 21 June 1820; Mary Hempstead illness: 2, 7, 8 September 1820, in Jensen, "I at Home," pt. 5, pp. 44–47 (4.2).

16. 9 September 1820, Jensen, "I at Home," pt. 5, pp. 47–48 (4.2).

17. Smith, *An Account of the Earthquakes*, 60–61 (1.2). See also 4.5.

18. "Foreign News, London," *Green-Mountain Farmer*, 19 August 1811; "Several Shocks by Earthquakes," *Niles Register*, 4 January 1812.

19. See 4.7.

20. Smith, *An Account of the Earthquakes*, 39, 42, 13, 12, and 16 (1.2).

21. Smith, *An Account of the Earthquakes*, 48, 41, 12, and 42 (1.2).

22. Smith, *An Account of the Earthquakes*, 46 (1.2).

23. Smith, *An Account of the Earthquakes*, 44 (1.2); "Extract of a Letter from Coosawhatchie, Dated January 23," *The Times* (Charleston, SC), 25 January 1812; "The Earthquake Was Very Distinctly Felt," *Argus of Western America*, 1812, both in CERI Compendium (1.2).

24. Smith, *An Account of the Earthquakes*, 43 (1.2).

25. Sargent, "Account of the Several Shocks," 352; Smith, *An Account of the Earthquakes*, 59 (both at 1.2).

26. Drake, *Natural and Statistical View*, and other sources listed at 4.9.2 and 4.9.

27. Drake, *Natural and Statistical View*, 233 (4.9.2).

28. Drake, *Natural and Statistical View*, 235 (4.9.2).

29. Drake, *Natural and Statistical View*, 237 (4.9.2).

30. See 4.9.1.

31. See 4.9.

32. See 4.13; on naming this movement: esp. Boles, *The Great Revival*; and Hatch, *The Democratization of American Christianity*, 220–26.

33. See 4.16.

34. Thunderous noise: Montgomery letter to Barton (1.2); Flint, *Recollections*, 217, 222; "Thunder of hell": Wetmore, *Gazetteer of the State of Missouri*, 133 (both at 1.17). "I could but stand and wonder": Hudson, "A Ballad of the New Madrid Earthquake,"

NOTES TO PAGES 160–165 [361]

149; Downs, "Earth Quake, 1812," 266 (both at 4.16); "I heard a mighty drum": "Climbing up a Mountain" (4.15.1) See 4.16 and 1.4.

35. "Sulphureous smell": Smith, *An Account of the Earthquakes*, 27, 38 (1.2). Earthquakes fouling water: Montgomery letter to Barton (1.2); Starling, *History of Henderson County, Kentucky*, 134 (1.17). Altering wells: Smith, "On the Changes Which Have Taken Place in the Wells of Water Situated in Columbia, South-Carolina, since the Earthquakes of 1811–12" (5.6.1). Healing springs: Valencius, *The Health of the Country*, 152–58 (1.13). Mary Hempstead: Jensen, "I at Home," pt. 2, pp. 84–88 (4.2).

36. Sulfur stink in seismology: Jackson, *Earthquakes and Earthquake History of Arkansas*, 46 (7.3). In the Bible: Genesis 19:24, Psalm 11:5.

37. Flint, *Recollections of the Last Ten Years*, 218 (1.17); Revelation 20:2–3.

38. Downs, "Earth Quake, 1812," 265 (4.16).

39. See 1.30. Prince, "An Improvement of the Doctrine of Earthquakes" (5.24); Jesus's death: Matthew 27:46–54.

40. Hymn 134, Dupuy, *Hymns and Spiritual Songs*, 152; Terry, *Hymns and Spiritual Songs*, 155 (both at 4.15); Hudson, "A Ballad of the New Madrid Earthquake," 149 (4.16).

41. Johnson, *The Frontier Camp Meeting*, 104 (4.13).

42. Smith, *An Account of the Earthquakes*, 60–61 (1.2); Bruce, *And They All Sang Hallelujah*, 65 (4.13). See 4.11.

43. Smith, *An Account of the Earthquakes*, 51, 50 (1.2); see 4.17.

44. Finley, *Autobiography of Rev. James B. Finley*, 239 (4.14).

45. "Famous old circuit-rider": Ross, *Life and Times of Elder Reuben Ross*, 208 (1.17). Cartwright, *Autobiography of Peter Cartwright*, 126 (4.14).

46. Cartwright, *Autobiography of Peter Cartwright*, 126–27, 6, introduction (4.14).

47. Hymn 127, "Rejoicing in Christ," in Mudge, *The American Camp-Meeting Hymn Book*, 127 (4.15).

48. Milburn, *The Pioneers, Preachers, and People of the Mississippi Valley*, 356 (4.14); Hankins, *The Second Great Awakening and the Transcendentalists*, 154; spread esp. to South: Boles, *The Great Revival* (and on early movement, esp. chap 5) (both at 4.13).

49. Ross, *Life and Times of Elder Reuben Ross*, 236–40 (1.17); Weisberger, *They Gathered at the River*, 34–36 Godwin, *The Great Revivalists*, 155–56; Hankins, *The Second Great Awakening and the Transcendentalists*, 9 (all at 4.13). Shakers: Howe, *What Hath God Wrought*, 296–98 (1.8).

50. Mudge, *The American Camp-Meeting Hymn Book*, iii (4.15). "Awfully sublime": "Theophilis Arminius," "Religious and Missionary Intelligence," 273 (4.14). Also Ross, *Life and Times of Elder Reuben Ross*, 204–5 (1.17). Importance of "vernacular preaching": Hatch, *The Democratization of American Christianity*, 133–41 and other citations listed in 4.13.

51. "Slain of the Lord": Mudge, *The American Camp-Meeting Hymn Book*, iii (4.15); "glory box": Bruce, *And They All Sang Hallelujah*, 71–82 (4.13); "jerk by": Lorenzo Dow, 1804, quoted in Ross, *Life and Times of Elder Reuben Ross*, 238 (1.17).

52. "Bodily exercises": Ross, *Life and Times of Elder Reuben Ross*, 241 (1.17). Skeptics: Trollope, *Domestic Manners of the Americans*, chap. 15 (4.14). Revival pregnancy: Weisberger, *They Gathered at the River*, 36 (4.13).

53. Conkin, *Cane Ridge*, 122 (4.13); Finley, *Autobiography*, 240; Milburn, *The Pioneers, Preachers and People of the Mississippi Valley*, 376–78 (4.14).

54. Ross, *Life and Times of Elder Reuben Ross*, 239–40 (1.17); a woman "wrestled and prayed" to grace: Cartwright, *Autobiography of Peter Cartwright*, 89 (4.14).

55. "Shake your own war clubs": Hawkins, "Agent for Indian Affairs, Creek Nation, Letter to the Big Warrior" (3.19). "He shook hands with me": Sugden, *Tecumseh*, 262 (3.10). See also 3.7 and 3.20.

56. Hempstead: Jensen, "I at Home," pt. 1, p. 50 (4.2); Backsliding and "His wrath is come": Posey, "The Earthquake of 1811," 110 (4.12). Camp meetings in Missouri: Foley, *The Genesis of Missouri*, 273 (1.11).

57. "Brought many to grace": Bruce, *And They All Sang Hallelujah*, 65; Godwin, *The Great Revivalists*, 156 (both at 4.13). "The earthquake struck terror": Cartwright, *Autobiography of Peter Cartwright*, 126; Finley, *Autobiography*, 239–40 (4.14). New Madrid membership: Clark, *One Hundred Years of New Madrid Methodism*, 11 (2.2). "It was easy to preach to the people": Stevenson, "Autobiography of the Rev. William Stevenson," 42 (1.2). Baptists: Johnson, *The Frontier Camp Meeting*, 105 (4.13). "Had a great impact": McClinton, ed., *The Moravian Springplace Mission*, I, 468 (3.17). "Earthquake Christians": Ross, *Life and Times of Elder Reuben Ross*, 204 (1.17).

58. Downs, "Earth Quake, 1812," 265 (4.16).

59. Cartwright, *Autobiography of Peter Cartwright*, 87–88 (4.14). Dow's travels: Wood, *Empire of Liberty*, 610 (1.8).

60. See 4.18, 4.19, 4.20. "Like a flash" and "sudden and irregular storms": Delbourgo, *A Most Amazing Scene of Wonders*, 133–34 (4.18). "Powerful or pungent convictions" and "struck down": [Rev. Thomas Hindle], "Account of the Rise and Progress," 305; also Cartwright, *Autobiography of Peter Cartwright*, 104; Weisberger, *They Gathered at the River*, 32–34 (all at 4.14).

61. Animal deaths: Delbourgo, *A Most Amazing Scene of Wonders*, 46, 123; Shattering globes: Schiffer, *Draw the Lightning Down*, 53 (both at 4.18).

62. Yes, a dead ox's eyes really did open and close: Lucier, *Scientists and Swindlers*, 33 (5.3). Also: 4.21.

63. T. Gale: Delbourgo, "Electrical Humanitarianism in North America" and *A Most Amazing Scene of Wonders* (both at 4.18). See 4.20, 4.21.

64. Electrical erotica: Delbourgo, *A Most Amazing Scene of Wonders*, 155 (4.18). Bringier, "Notices," 45 (2.4). This is thus an even earlier American physical history of electricity than that outlined, for instance, in Herzig, "Subjected to the Current" (4.20).

65. "Universal blow": Delbourgo, *A Most Amazing Scene of Wonders*, 46 (4.18); "The shocks of the earthquake": Beck, *A Gazetteer*, 800 (1.2). Coen, *Earthquake Observers*, 33 (1.16). Bodily sensations told the truth of electricity just as they told the power of good preaching: see Conkin, *Cane Ridge*, 122 (4.13).

66. "The shocks seemed to produce": Mitchill, "A Detailed Narrative," 304 (1.2). Delbourgo identifies this as "sensational analogy": Delbourgo, *A Most Amazing Scene of Wonders*, 55, 179 (4.18).

67. "Strong and sudden jirks": Smith, *An Account of the Earthquakes*, 48 (1.2). Later Charlestonians' sensations: Porcher, "Influence of the Recent Earthquakes," 652–53 (4.1).

68. Guitéras, "Influence of the Recent Earthquakes" (4.1).

Chapter 5

1. Montgomery letter to Barton (1.2).

2. "From the American Daily Advertiser," reprinted letter from "A.F." to "Mr. Poulson," enclosing extracts from "Dr. Montgomery," Frankfort, Kentucky, 29 January 1812, in Smith, *An Account of the Earthquakes*, 47 (1.2). This article does not give Montgomery's first name, but strong textual parallels make likely that the two letters are by the same Alexander Montgomery.

3. See 5.10. William Leigh Pierce conducted similar trials with "some of the substances thrown up from the bowels of the Earth": "Earthquakes," *Georgia Journal* (Milledgeville), 25 March 2012.

4. See 5.3. Early American science is generally viewed as embryonic, frustrated, and—to the extent that it existed at all—based in cities: Cohen, "American Science"; more measured discussions in Greene, *American Science in the Age of Jefferson* (on urban-based scientific efforts, 107, 127); Daniels, *American Science in the Age of Jackson*. Newer research opens up countryside, lecture halls, exhibition circuits, engineering sites: see, esp. Lucier, *Scientists and Swindlers*; Pandora, "Popular Science in National and Transnational Perspective" (4.21); Delbourgo, *A Most Amazing Scene of Wonders* (4.18). Fruitless searches on Montgomery: databases of early American scientific and medical publishing; Kentucky Historical Society; Filson Historical Society. Subsequent history of Peale exhibits: see 5.10.

5. See citations at (5.2) and (5.3).

6. Kohlstedt, *The Formation of the American Scientific Community*, 6 (5.3); Laudan, *From Mineralogy to Geology*, 223; Rudwick, *Bursting the Limits of Time*, 130 (both at 5.11). "Lord, what is man," Capac, listed in 5.6.1 in the bibliography and drawing from Psalm 8:4.

7. Goldstein, "Yours for Science" (5.3); Smith, *An Account* (1.2). Also: 1.9, 5.21.

8. Montgomery letter to Barton (1.2). Smith, *An Account*, 47 (1.2).

9. Skill, not formal training: Pandora, "Popular Science," 355 (5.3). See sources under 5.21.

10. Pumice stone in Missouri River before New Madrid earthquakes: Meriwether Lewis, 14 April 1804, *The Journals of the Lewis and Clark Expedition*; Stoddard, *Sketches, Historical and Descriptive*, 240 (both at 1.17). Mysterious identity of the rocks: conversations with Scott Ausbrooks, Arkansas Geological Survey; Marticia Tuttle, M.Tuttle & Associates; and Jonathan McIntyre, Kentucky Geological Survey.

11. See 5.9. Austin: Chambers, "Land of Ores, Country of Minerals," chap. 4 (1.11); Schoolcraft, technical making-do: Rafferty, *Rude Pursuits and Rugged Peaks*, 31, 75 (1.17); Dunbar and Hunter: Berry, Beasley, and Clements, *The Forgotten Expedition*, for instance, 87, 123 (1.55); Tasting minerals: 11 April 1805; Sacagawea: 16 June 1805, *The Journals of the Lewis and Clark Expedition* (1.17); Kastor and Valencius, "Sacagawea's 'Cold'" (5.9).

12. Montgomery letter to Barton (1.2).

13. William L. Pierce, letter, *Savannah Republican*, March 1812, 3. Though Pierce does not say, it is likely he himself had supplied the specimens. "Paper proxies" stood in for specimens among European savants, much as Pierce makes his assertions stand in here: Rudwick, *Bursting the Limits of Time*, 387 (5.11).

14. See 5.18, 5.19. Specimens: Bringier, "Notices," 25—though his were lost overboard in the sinking of his boat (2.4); Lucier, *Scientists and Swindlers*, 29 (5.3). Clark: Foley, *A History of Missouri*, 189 (1.11). Twain, *Life on the Mississippi*, 230 (1.17). McMillan, "The Discovery of Fossil Vertebrates" (on Clark's museum, 37) (5.6). Importance of parlor as intellectual space: Kohlstedt, "Parlors, Primers, and Public Schooling," esp. 65–71 (1.10). Fascination with preserved creatures: Valencius, *The Health of the Country*, 65 (1.13); "Uncleaned bones": Porter, *The Eagle's Nest*, 37 (5.3).

15. Fossils spit out by the earthquakes: Drake, *Natural and Statistical View*, 237 (4.9.2); Charles Joseph Latrobe, *The Rambler in North America, 1832–1833*, 1:92 (1.31). Used as metaphor: Rush, *Medical Inquiries and Observations*, 154 (4.1). "Curious objects": de Montulé, *Travels in America*, 113 (1.2). Also: 5.10, 5.13, 5.19.

16. See 5.20.

17. See 5.12, 5.14, 5.15, as well as 2.18.

18. See 5.17. Werner and Hutton: Greene, *Geology in the Nineteenth Century*, chaps. 1 and 2 (5.11).

19. See 5.13.

20. Geological Society: Rudwick, *Bursting the Limits of Time*, esp. 464 (5.11). See 5.16.

21. McMillan, "The Discovery of Fossil Vertebrates"; and Rieppel, "Articulating the Past" (both at 5.6). Also: 4.21.

22. Fascination with North American landforms and geologic phenomena: Flores, *Journal of an Indian Trader* (1.17), 94–98; Greene, *Geology in the Nineteenth Century*, chap. 5 (5.11); Oldroyd, *Thinking About the Earth*, chap. 7 (1.6); Rudwick, *Bursting the Limits of Time*, 267–70; Taylor, "American Geological Investigations and the French" (both in 5.11); Rieppel, "Articulating the Past" (5.6).

23. "Marly appearance": Samuel Postlethwaite, 12 March 1813, "Memorandum on a Voyage from Natchez," Postlethwaite family papers, HM 28849 no. 2, Henry E. Huntington Library; background from Henry E. Huntington archives catalog. Geological knowledge a chief form of scientific expertise: Lucier, *Scientists and Swindlers*, 108; Goldstein, "Yours for Science," 581 (both in 5.3). See 5.6 and 5.7.

24. See 5.8; also White, "Early Geological Observations in the American Midwest" (5.6). Maclure and similar maps: Laudan, *From Mineralogy to Geology*, 106 (5.11); Lucier, *Scientists and Swindlers*, chap. 1 (5.3).

25. Hazen, "The Founding of Geology in America," 1827; Hendrickson, "Nineteenth-Century State Geological Surveys" (both in 5.6); Kohlstedt, *The Formation of the American Scientific Community*, 63; Lucier, *Scientists and Swindlers*, 33 (both in 5.3). Continuing mandate of private resource extraction: mission statement of the Arkansas Geological Survey: http://www.geology.ar.gov/home/index.htm (accessed 7 July 2012).

26. "Too long has it been fashionable": Greene, "Science and the Public in the Age of Jefferson," 207 (5.3). See 5.6 and 5.8.

27. Earthquake lecture: Greene, *American Science in the Age of Jefferson*, 116–17 (5.3). Drake may have given this lecture, as he did many of the talks. Drake, *Natural and Statistical View* (4.9.2). Mitchill, "A Detailed Narrative" (1.2). See 4.21 and 5.4. On the close relationship of science and arts, see Walls, *Passage to Cosmos* (5.11).

28. See 5.21.

29. See 3.2. Coen, *Earthquake Observers*, chap. 4 (1.6).

30. Barton: Greene, *American Science in the Age of Jefferson*, 47; Judd, *The Untilled Garden*, 122–24; Porter, *The Eagle's Nest*, 37 (all in 5.3).

31. Slow, halting, often fractured growth of early American institutions: Porter, *The Eagle's Nest* (5.3). Junto: Schiffer, *Draw the Lightning Down*, 22 (4.18). Drake: Greene, *American Science in the Age of Jefferson*, 116–17 (5.3). See 4.21 and 5.21.

32. See 1.16. Such interest is not simply "preprofessional" (Kohlstedt, "Parlors, Primers, and Public Schooling" (1.10)) but broadly integrative. The term "scientist" was coined in Britain in 1834, but met with general disdain; it came into use in in the United States following the Civil War as part of Gilded Age reform efforts: Lucier, "The Professional and the Scientist," 723–32 (5.3).

33. Oldroyd et al., "The Study of Earthquakes," 343–44 (1.6).

34. *An Account of the Great Earthquakes*; and Smith, *An Account of the Earthquakes*, 16 (both in 1.2).

35. Boorstin, *The Americans*, 20–25 (1.8). Examples: spread of McCormick reaper (Stoll, *Larding the Lean Earth*, 191 (5.7)), workshops that provided context for development of the telephone (Shulman, *The Telephone Gambit* (5.3)).

36. *An Account of the Great Earthquakes*, 9: letter from William Leigh Pierce (1.2).

37. Impulse to measure, quantify: Cannon, "Humboldtian Science" (5.11); "in a field 13 sink holes appeared": McLoughlin, "Excerpts from the Official Diary of the Moravian Mission," 341 (3.17). Maryland steeple: Smith, *An Account of the Earthquakes*, 44 (1.2); Finley, *Autobiography of Rev. James B. Finley*, 238 (4.14). 142° Fahrenheit: *An Account of the Great Earthquakes*, 15 (1.2).

38. Time zones: Carl-Henry Geschwind, *California Earthquakes*, 14 (1.6).

39. Sargent, "Account of the Several Shocks," 351 (1.2). Estimating duration: Smith, *An Account of the Earthquakes*, 10 (1.2). Precision in timing: "A Subscriber," "Earthquake!," *Poulson's American Daily Advertiser*, 23 December 1811, Infoweb Newsbank: America's Historical Newspapers. "I find by T. Parker's regulator": Smith, *An Account of the Earthquakes*, 55 (1.2). Precision in time: Canales, *A Tenth of a Second* (6.17). Coen, *Earthquake Observers*, 84 (1.6).

40. Smith, *An Account of the Earthquakes*, 12 (1.2).

41. Pierce, *An Account of the Great Earthquakes*, 12, 31 (1.2); Drake, *Natural and Statistical View* (4.9.2). Valencius, *The Health of the Country*, "Local Knowledge" (1.13).

42. Smith, *An Account of the Earthquakes*, 46; "occurred at 30 minutes past 2 o'clock": 23 (1.2).

43. "A Subscriber," "Earthquake!" *Poulson's Daily Advertiser*, 23 Dec 1811, Infoweb Newsbank: America's Historical Newspapers.

44. Smith, *An Account of the Earthquakes*; compass: 18; looking glass: 52; clocks: 40, 43; oscillating vials: 57 (1.2).

45. See 6.1.

46. Pandora, "The Children's Republic of Science," 89, 91 (1.10). Practicality as national virtue: Johnson, "Material Experiments" (5.8). "Statement of William Brigham, in Company of Riflemen at Tippecanoe," Esarey, *Governors Messages and Letters*, 707 (3.10). Smith, *An Account of the Earthquakes*; suspended scissors: 50; egg: 44 (1.2). Many people have asked "how?" Reasoning from family Easter decorations, I assume the egg had been blown out from a pinhole pricked at either end. Bowl-of-molasses seismoscopes: Geschwind, *California Earthquakes*, 14 (1.6).

47. Appendix to M'Murtrie, *Sketches of Louisville*, 240–45 (1.2).

48. Appendix to M'Murtrie, *Sketches of Louisville*, 234, 237 (1.2).

49. Drake, *Natural and Statistical View* (4.9.2); Table of Meteorological Observations from Charleston, SC, in Smith, *An Account of the Earthquakes*, 9; weather mentioned even when unremarkable: 40 (1.2).

50. "Warm as in the Indian summer": Sargent, "An Account of the Several Shocks," 356; see also 354 (1.2). "Indian summer" as theme: Foster, *The Mississippi Valley*, 19 (1.17). Similar to reports of other earthquakes: Smith, *An Account of the Earthquakes*, 76; hazy, dark weather: 9, 13, 29, 40; and Mitchill, "A Detailed Narrative," 288–89. "The atmosphere was filled": Broadhead, "The New Madrid Earthquake," 82 (all at 1.2). "Indian smoke" probably meant the thick smoke of prairie burning. See 5.23.

51. Audubon, *Audubon and His Journals*, 234–35 (1.18). Weather changes after quakes: Sargent, "An Account of the Several Shocks," 352, 356; Smith, *An Account of the Earthquakes*, 67; Pusey, "The New Madrid Earthquake," 186. "It began to rain transparent ice": Appendix to M'Murtrie, *Sketches of Louisville*, 236–37; "lowering appearances of the weather": *An Account of the Great Earthquakes*, 5. "From the looks of the weather today," Smith, *An Account of the Earthquakes*, 51 (all in 1.2).

52. Both in Sargent, "Account of the Several Shocks," 357. Puzzlingly good weather with earthquakes: Mitchill, "A Detailed Narrative," 290; Smith, *An Account of the Earthquakes*, 53 (all in 1.2). On "earthquake weather": Similarly: conversations with David Rabenau, fellow earthquake refugee, 1989. Coen, *Earthquake Observers*, 5 (1.6).

53. David Oldroyd and colleagues characterize preinstrumental earthquake science as "pre-paradigm," with collection of data and many competing theories: Oldroyd et al., "The Study of Earthquakes," 364 (1.6). Similarly: Rudwick, *Bursting the Limits of Time*, 80 (5.11). Perhaps vernacular accounts offer insights different from those of elite European literature: to dismiss New Madrid quakes compilations as

atheoretical or acausal is to miss the subtle nature of theorizing—to construct an account in a particular way is to make choices about what is significant or potentially causal. Coen, *Earthquake Observers*, 5 (1.6).

54. Janković, *Reading the Skies* (5.23). Horns of the comet: Bradbury, *Travels*, 209 (1.2).

55. See citations for 1.19 and 1.20.

56. "The shock was preceded": Montgomery letter to Barton (1.2). Flint, *Recollections*, 216–17 (1.17). Brookes: Appendix to M'Murtrie, *Sketches of Louisville*, 245 (1.2).

57. Directionality of sound: Bradbury, *Travels in the Interior of North America*, 206; Mitchill, "A Detailed Narrative," 282; Smith, *An Account of the Earthquakes*, 41, 56 (all in 1.2). Mallet: Geschwind, *California Earthquakes*, 10–11 (1.6).

58. Dickeson and Brown, "On the Cypress Timber of Mississippi and Louisiana" (1.14).

59. Noise causes rain: Ward, "How Far Can Man Control His Climate?" 9, 12 (5.23). Gettysburg: Opie, *Nature's Nation*, 358; "Try it": Worster, *Dust Bowl*, 39 (both at 1.13). Anti-hail device: "Eggers Hail Cannon Manufacturer—Anti Hail Device," http://www.hailcannon.com/index.html, home page and "How It Works" page, accessed 10 June 2012.

60. Bryan, letter from New Madrid, 346 (1.2).

61. French and New Madrid earthquakes on same newspaper page: "Severe Shocks of an Earthquake Were Felt in France," "Earthquake," *The Supporter* (1.2).

62. Mitchill, "A Detailed Narrative," Mt. Etna: Smith, *An Account of the Earthquakes*, 84 (both in 1.2). Local symptoms of global disturbances: Foster, *The Mississippi Valley*, 18–19 (1.17). Humboldt, "Earthquakes and Volcanoes in the Americas," in Mather and Mason, *Source Book in Geology*, 182–87 (5.11); "Foreign News, London," *Green-Mountain Farmer*, 19 August 1811.

63. "Some dreadful calamity": "Several Shocks by Earthquakes," *Niles' Register*, 4 January 1812. "Many and repeated shocks": *An Account of the Great Earthquakes* (1.2).

64. Interest in sequences of earthquakes: "Seismic Science: Is Number of Earthquakes on the Rise? Live Q&A," *Washington Post*, 9 March 2010. Remotely triggered earthquakes: Hough and Bilham, *After the Earth Quakes*, 132–33 (1.6). McCook, "Nature, God, and Nation in Revolutionary Venezuela" (1.7). New Madrid/Venezuela connections: Duden, *Report on a Journey*, 159 (1.17). "Continued glare of vivid flashes": Flint, *Recollections*, 217, using nineteenth-century spelling for the city (1.17).

65. Volcano: Fuller, *The New Madrid Earthquake*, 102–3 (1.1). "A distant convulsion": Smith, *An Account of the Earthquakes*, 28 (1.2).

66. "Within 30 miles of the great Osage village": *Louisiana Gazette* (St. Louis), 21 December 1811, CERI Compendium (1.2).

67. Smith, *An Account of the Earthquakes*, 83 (1.2).

68. "We may yet receive authentic information": Smith, *An Account of the Earthquakes*, 82 (1.2). Volcanoes in West: Flores, "Jefferson's Grand Expedition," 21–40, 271 (2.14). Kentucky volcanoes?: Marshall, *The History of Kentucky*, 2:13 (1.17).

69. Strange underground phenomena: "Jacques," "Notes of a Tourist," 396. "This unaccountable rumbling Sound": "unacountable" in the original. *The Journals of the Lewis and Clark Expedition*, 20 June 1805. Flint, *History and Geography*, 283 (all in

1.17). Flint: Valencius, *The Health of the Country*, 22–35 (1.13). "Loud explosions": Cutler, *A Topographical Description*, 112 (1.17).

70. Earthquakes and volcanoes together: Smith, *An Account of the Earthquakes*, 64–66 (1.2). Modern understanding: USGS, FAQ about Relationships between Earthquakes and Volcanic Eruptions, http://volcanoes.usgs.gov/about/faq/faqeq .php, updated 17 March 2011. More work needed on theoretical connections between volcanoes and earthquakes: Oldroyd, "The Earth Sciences," 114–15 (5.11).

71. Nummedal, "Kircher's Subterranean World," 40 (5.11).

72. Oldroyd et al., "Study of Earthquakes," 333–36; Carozzi, "Robert Hooke, Rudolf Erich Raspe, and the Concept of 'Earthquakes'" (both at 1.6); Laudan, *From Mineralogy to Geology*, 184–87 (5.11). Active theory after New Madrid earthquakes: for instance, "Another Conjecture of the Cause of the Earthquakes" (5.6.1).

73. Laudan, *From Mineralogy to Geology*, 187, 217 (5.11); Oldroyd et al., "Study of Earthquakes," 340–42, 346 (1.6).

74. "We regard Earthquakes": Capac, "Reflections concerning Earthquakes, No. 1"; Lea, "On Earthquakes" (both at 5.6.1); Smith, *An Account of the Earthquakes*, 68–69 (1.2).

75. "Thundering noise": "Earthquakes and Eruption in Ternate (Taken from the *Javasche Courant*)," *American Journal of Science and Arts* 5, new ser. (1848): 422. "Emission of flames": Smith, *An Account of the Earthquakes*, 66 (1.2).

76. Mitchill, "A Detailed Narrative," 311; Featherstonhaugh, *Excursion through the Slave States*, 326 (both in 1.2).

77. Dismissal of volcano/New Madrid connection: Beck, *A Gazetteer of the States of Illinois and Missouri*, 300 (1.2). One claim was energetically debunked: Smith, *An Account of the Earthquakes*, 6–7. "I shall not pretend to hazard": Sargent, "Account of the Several Shocks," 350. "There is not in any of these places": Austin, diary, 207–8 (all in 1.2).

78. Smith, *An Account of the Earthquakes*, 82 (1.2).

79. Volney, *A View of the Soil and Climate of the United States*, 53, 65, 79, 97–99; "subterranean fires": 99 (1.17).

80. Rebuttal of Volney occupied a whole appendix in Drake, *Natural and Statistical View* (4.9.2). Indignant rebuttal of European savants as a theme: Wood, *Empire of Liberty*, 393 (1.8). Volney's influence: Judd, *The Untilled Garden*, 137–39, Greene, *American Science in the Age of Jefferson*, 225–27 (both in 5.3).

81. Stoddard, *Sketches, Historical and Descriptive, of Louisiana*, 239–40 (1.17). Duden, *Report on a Journey*, 92 (1.17).

82. Mayor, "Suppression of Indigenous Fossil Knowledge" (2.3); McMillan, "The Discovery of Fossil Vertebrates," 33 (5.6).

83. *An Account of the Great Earthquakes*, 13–14 (1.2). Sargent, "An Account," 352 (1.2).

84. Montgomery letter to Barton (italics in the original) (1.2).

85. Volcanoes in Buncombe County: Mitchill, "A Detailed Narrative," 299; Smith, *An Account of the Earthquakes*, Buncombe County: 6–7; "An Earthquake was witnessed": 14 (both in 1.2). Witnessing: Foster, *The Mississippi Valley*, 19 (1.17). Epistemological centrality of witnessing: Shapin and Schaffer, *Leviathan and the Air-Pump*, 55–60 (4.8); Shapin, *Social History of Truth* (5.22).

86. Credibility: Sargent, "Account of the Several Shocks," 354; *An Account of the Great Earthquakes*, 15. Smith, *An Account of the Earthquakes*, "well known": 38; "personally familiar": 82; "from the same respectable individual": 49; "formerly resident": 60; Pusey, "The New Madrid Earthquake," 286 (all in 1.2). This is in some ways a reversal of Steven Shapin's *Social History of Truth* (5.22): rather than gentlemanly credibility giving events ontological status as "true," events grant observers status as credible. See Coen, *Earthquake Observers*, 92–99 (1.6).

87. Sand blow as volcano: Kendall Owens, "Major Fault Line Could Exist near Marianna: Researchers with UALR Say Sand Blows May Be Evidence of Major Earthquakes Originating in This Area," *Times-Herald/thnews.com*, 5 March 2007; Fuller, *The New Madrid Earthquake*, 82 (1.1). Diagram of volcano-looking sand blow: Tuttle, "The Use of Liquefaction Features in Paleoseismology," 362 (7.5.5).

88. "Ever and anon": "Letter from the Hon. L. F. Linn," Washington City, 1 February 1836, 139, in Wetmore, *Gazetteer of the State of Missouri* (1.17). Seismic shocks associated with lightning: "Extract of a letter to the Editors, dated West River [MD], January, 23, 1812," *Boston Independent Chronicle* 6 February 1812, CERI Compendium (1.2). "The thunder and lightning which prevail": Evans, *A Pedestrious Tour*, 194, 195 (1.17).

89. "In a direction due north": Smith, *An Account of the Earthquakes*, 21 (1.2). "Flashes such as result": Foster, *The Mississippi Valley*, 20 (1.17). Rumbling noises: Smith, *An Account of the Earthquakes*, 22; *An Account of the Great Earthquakes*, 7; Broadhead, "The New Madrid Earthquake," 79 (all in 1.2).

90. "Extract of a letter to the Editors, dated West River [MD], January, 23, 1812," *Boston Independent Chronicle* 6 February 1812, CERI Compendium (1.2). Also 1.10.

91. "Subterranean thunder," Wetmore, *Gazetteer of the State of Missouri*, 133 (1.17). On association of lightning with material below ground: Meredith, "Actic Smoke," 234 (5.23). Long associations with power of thunder: Rath, *How Early America Sounded* (5.20). Lightning, thunder, earthquakes in long historical association: Oldroyd et al., "Study of Earthquakes," 336 (1.6). Similar sounds of thunderstorm and volcanic eruption told many eighteenth-century researchers that they must be the same underlying process: Rappaport, *When Geologists Were Historians*, 182 (5.11). "The fulminating of the mountains": *An Account of the Great Earthquakes*, 15; Mitchill, "A Detailed Narrative," 286 (both at 1.2).

92. See 5.5. Other European scholars: Oldroyd et al., "Study of Earthquakes," 329, 36–37, 40 (1.6). Early theological interest: Van De Wetering, "Moralizing in Puritan Natural Science, 425 (1.30). Priestley: Delbourgo, *A Most Amazing Scene of Wonders*, 106 (4.18). Featherstonhaugh, *Excursion through the Slave States*, 326 (1.2). Electricity, earthquakes, epic poetry: Schoolcraft, *Transallegania* (1.4).

93. Oldroyd et al., "Study of Earthquakes," 329, 36–37, 40 (1.6). On Stukeley's drawing on Franklin, see Dean, "Benjamin Franklin and Earthquakes," 487 (5.5), Stukeley, "On the Causes of Earthquakes," 5.6.1.

94. The Lisbon earthquake and tsunami were a major event throughout European literature, featuring prominently, for example, in Voltaire's *Candide*. Cape Ann: see 1.7.1.

95. See 5.24.

96. "The Dome and the Lightning Rod," "The Maryland State House," state website, http://www.msa.md.gov/msa/mdstatehouse/html/dome_lightning.html, accessed 10 June 2012. See 5.5.

97. Electricity/earthquake link losing ground in Europe: Oldroyd et al., "Study of Earthquakes," 337 (1.6). But still lively in US: Lea, "On Earthquakes," 209–10 (5.6.1); Beck, *Gazetteer of Missouri*, 800; Smith, *An Account of the Earthquakes*, 68–69 (both in 1.2), explicitly rejects Stukeley's ideas.

98. See 5.25, "On the Cold of the Present Season" (5.6.1). On the climatic impact of historic volcanic eruptions, see the USGS publication, "This Dynamic Earth: Plate Tectonics and People," online at http://pubs.usgs.gov/gip/dynamic/tectonics.html, accessed 10 June 2012.

99. Skeen "The Year without a Summer," 60 (5.25); "On the Cold of the Present Season." Similar arguments in Capac, "Reflections concerning Earthquakes, No. 3" (both in 5.6.1) and "We Insert with Pleasure . . . ," *Augusta Herald*, 2 April 1812; and reply: J. E., "To the Editors of the Augusta Herald (Concluded from Our Last)," *Augusta Herald*, 9 April 1812, CERI Compendium (1.2).

100. Varied eruptions around the globe: Mitchill, "A Detailed Narrative," 322, 30. Also Smith, *An Account of the Earthquakes*, 62–63 (both at 1.2).

101. "Productive Cow," *Daily National Intelligencer*, 30 September 1816. Paltry nature of scientific publication has long been taken to indicate paltry American scientific life (for instance, Greene, "Science and the Public in the Age of Jefferson" [5.3]). Yet this analysis fails to see how science was interwoven with many other interests—including the productivity of Connecticut cows. See 1.9, 5.2, 5.4.

102. Howe, *What Hath God Wrought*, 187 (1.8).

Chapter 6

1. Rudwick, *Bursting the Limits of Time* (5.11).

2. See 2.3 and 6.6.

3. Continued tremors: McClinton, ed., *The Moravian Springplace Mission*, 151, 190 (3.17). Felt in Kentucky: Starling, *History of Henderson County, Kentucky*, 180 (1.17). "Paid little or no regard": Shaw, "Collections of the State Historical Society of Wisconsin," 203–4 (1.2).

4. Cave Burton: Joseph G. Baldwin, *The Flush Times of Alabama and Mississippi: A Series of Sketches* (San Francisco, CA: Bancroft-Whitney, 1901), 153–76. My thanks to S. Charles Bolton. Old-timers remember earthquakes: Parker, *Missouri as It Is in 1867*, 81 (1.17).

5. Parker, *Missouri as It Is in 1867* (section on geology and geography) ignores sand blows (1.17). "Kindo' throwed up": Testimony by J. B. Dial, in *St. Francis Mill . . . against William O. Sugg*, 361 (2.5). My great thanks to Lynn Morrow for this source. E. C. Hall et al., *Soil Survey of Mississippi County, Arkansas* (Washington, DC: Government Printing Office, for the US Department of Agriculture, Bureau of Soils,

1916). My thanks to Scott Ausbrooks, Arkansas Geological Survey, for pointing out Hall's silence on sand blows. Worster, *Dust Bowl*, 14 (1.13).

6. See **6.14**. "A question as to the truthfulness": "Discussion," 324.

7. "Lantern views": McGee, "A Fossil Earthquake," 414 (**6.14**); Houck, quoted in Kochtitzky, *The Story of a Busy Life*, 47 (**6.9**).

8. Ogilvie, "Development of the Southeast Missouri Lowlands," 18, 22–25 (**6.9**). "Hysteria": Penick, *The New Madrid Earthquakes*, 72 (1.1). St. Francis sunk lands not an earthquake feature: Saucier, "Origin of the St. Francis Sunk Lands" (2.6).

9. In-migration: Howe, *What Hath God Wrought*, 136–42 (1.8); Faragher, "More Motley Than Mackinaw," 316 (2.12). Land grants: Hendricks, "Sunken Lands" (2.5).

10. Hendricks, "Sunken Lands" (2.5). Van Arsdale et al., "Earthquake Signals in Tree-Ring Data," 517 (2.21).

11. Hendricks, "Sunken Lands" (2.5).

12. Hendricks, "Sunken Lands" (2.5).

13. March to the sea and earthquake reports: Nathan K. (Kent) Moran, phone interview, 7 August 2009, and presentation at Seismological Society of America annual meeting, April 2011.

14. Theaters of the war: DeBlack, *With Fire and Sword*, 35. See **6.12**.

15. McPherson, *Battle Cry of Freedom*, chap. 13. Island 10 no more: Swain, "Capture of Island No. 10" (both in **6.12**).

16. McPherson, *Battle Cry of Freedom*, 415–17 (**6.12**).

17. Cooling, "A People's War," 116–19; DeBlack, *With Fire and Sword*, 45–49, 71; Mackey, "Bushwackers, Provosts, and Tories," 171–75, 179, 183; Montgomery, "Battle of Prairie Grove"; Sutherland, "Guerillas," 133–36, 149; "I wanted to be my own General," 136; and Sutherland, "Introduction: The Desperate Side of War," 10–11 (all in **6.12**). Impact of river piracy on local communities: Bolsterli, *During Wind and Rain*, 11 (**6.11**). Jayhawker vs. bushwhacker: Sutherland, "Jayhawkers and Bushwhackers" (**6.12**).

18. Sutherland, "Guerillas," 137, 140, 150. Cooling, "A People's War," 114; Mackey, "Bushwackers, Provosts, and Tories," 182. James brothers: DeBlack, *With Fire and Sword*, 61, 125; and "Wild Dick," 142. Warrensburg violence: Fellman, "Inside Wars," 191 (all in **6.12**).

19. Dougan, "Missouri Bootheel" (**6.9**); Mobley, *Making Sense of the Civil War*, 58–59, 204; Sutherland, "Introduction: The Desperate Side of War," 14. "When we camp": Sutherland, "Guerillas," 136 (all in **6.12**).

20. Crowley, *Tennessee Cavalier in the Missouri Cavalry*, 99–101; DeBlack, *With Fire and Sword*, 86; Taylor, "Skirmish at Chalk Bluff" (all in **6.12**).

21. Taylor, "Skirmish at Chalk Bluff"; Crowley, *Tennessee Cavalier*, 100–101 (all in **6.12**).

22. Crowley, *Tennessee Cavalier*, 101; National Park Service, "Chalk Bluff"; Taylor, Skirmish at Chalk Bluff" (all in **6.12**).

23. "Town of several hundred inhabitants": George W. Clymans, "Diary of a Trip on the Mississippi and Missouri Rivers from Memphis, Tenn., to Omaha, Neb., and

by Land to the Vicinity of De Soto, Neb.," HM 19471, Henry E. Huntington Library, mid-May 1866, [10].

24. Overviews of these changes: Brattan, "The Geography of the St. Francis Basin" (2.7); and Snow, *From Missouri* (6.11.1). Estep: *St. Francis Mill ... against William Sugg*, 157–58, 169, 348–49 (2.5).

25. A few black families were able to claim homesteads under the Southern Homestead Act of 1866. DeBlack, *With Fire and Sword*, 154–55, 180–83, 189–90, 197–99, 208–9, 228 (6.12); Dougan, "Missouri Bootheel" (6.9).

26. Moore, *Antiquities of the St. Francis, White, and Black Rivers*, 256 (2.7). Sites still uninvestigated: Claudine Payne, "Middle-Period Mississippian in the St. Francis Basin: Unpublished Draft Paper" (2009); Claudine Payne, PhD, Station Archeologist, phone conversation, 30 January 2009; Tuttle et al., "Use of Archaeology to Date Liquefaction Features," 477 (7.5.5).

27. Snow, *From Missouri*, 2; Stepenoff, *Thad Snow*, 149, 173 (both in 6.11.1). Laymon, *Pfeiffer Country* (6.11).

28. Snow, *From Missouri*, 152, 155 (6.11.1); Dougan, "Missouri Bootheel" (6.9); Brattan, "The Geography of the St. Francis Basin," vii, 27, 48 (2.7); Cantor, "A Prologue to the Protest Movement," 805 (6.11); Blaney, "U.S. Cotton Almost Clear of Voracious Boll Weevil"; Giesen, "The Truth about the Boll Weevil," esp. 696; Sorenson, "The Boll Weevil in Missouri" (all in 6.8). Bluesman Lead Belly recorded versions of this traditional song beginning in 1934. His version is good, but I may have enjoyed Joy Harvey's more.

29. Snow, *From Missouri*, chap. 18, esp. 159; "timber workers" and "cotton Negroes": 141, "highly organized fashion": 139 (6.11.1); Wylie, "Race and Class Conflict," 184 (population numbers), 195 (6.11).

30. Dougan, "Missouri Bootheel" (6.9); Brattan, "The Geography of the St. Francis Basin," 27 (2.7); Cantor, "The Missouri Sharecropper Roadside Demonstration of 1939," 805–6; Wylie, "Race and Class Conflict," 185, 196 (both in 6.11); DeBlack, *With Fire and Sword*, 151 (6.12). Populations: Stepenoff, "The Last Tree Cut Down," 71 (6.9). Ratios: Kochtitzky, *The Story of a Busy Life*, 172 (6.9). Further south in the Delta, landowners got half the cotton and three-quarters of the corn raised by tenant farmers on their allotments. Bolsterli, *During Wind and Rain*, 89 (6.11). "I Never Picked Cotton," written by Roy Clark in 1969 and recorded by Johnny Cash for his 1996 Columbia Records album *Unchained*, where he discusses growing up as a dirt-poor sharecropper in the liner notes.

31. Flood in 1927: Barry, *Rising Tide* (1.13). The agricultural crisis in the Mississippi Valley was parallel to the Great Flood: in both cases local (white) elites wanted no intrusion of extralocal forces that might ultimately lead to the loss of their local labor force. Denial of subsidies: Snow, *From Missouri*, 199, 234; "Even Negroes lose color": 172. Snow had twenty sharecropping families on his own family's holdings. Stepenoff, *Thad Snow*, 4 (both in 6.11.1).

32. STFU: Cobb, "Southern Tenant Farmers Union"; Cantor, "The Missouri Sharecropper Roadside Demonstration of 1939," 809–10 (6.11); Dougan, "Missouri Bootheel" (6.9). Social tension, not seismic trouble: I appreciate commentary by

Harold Forsythe, Program in Agrarian Studies, Yale University, 19 January 2007. Snow, *From Missouri*, chaps. 27, 28; Snow estimated the 1,500 well-organized protesters had the knowledge and help of probably ten thousand more: 240–45, 276; "Let our camps do the talkin' for us": 242; "guarded and cared for": 281; "socially minded St. Louis people": 289 (6.11.1). KKK in Arkansas: DeBlack, *With Fire and Sword*, 180–228 (6.12). Cantor traces the history but fails to recognize the locally based nature of the organization and the centrality of local black leaders. At the time, many local white leaders pilloried Snow as the likely ringleader of all this protest. He pointed out the racism of their charges: things were so well organized, they assumed a white man had to be in charge. (African American leaders were careful to stay low-profile: as Whitfield explained, "A dead nigger's no use to nobody" [Snow, *From Missouri*, 245 (6.11.1)]) Snow then muddied the waters by publishing a increasingly complex and implausible "Confession" that somehow involved Trotsky and the Mexican revolutionaries. It was taken at face value by at least some readers, leading to lasting confusion about his actual level of involvement in a movement that was indeed extremely well hidden from the white majority in the region. Stepenoff, *Thad Snow*, 7 (6.11.1).

33. Snow, *From Missouri*, 288–89 (6.11.1); Cantor, "The Missouri Sharecropper Roadside Demonstration of 1939," 804, 820–21; Wylie, "Race and Class Conflict," 195 (6.11).

34. In 1988 Louis Dreyfus opened a rice mill at New Madrid (it was acquired by Riceland in 2002). Dougan, "Missouri Bootheel" (6.9)

35. See 6.13. "Where the land has been cleared and cultivated": Kochtitzky, *The Story of a Busy Life*, 57 (6.9).

36. Audience discussion with the lead singer of Creole Stomp, 24 July 2010, Ossippee Valley Bluegrass Festival, Ossippee, Maine.

37. Challenges of seismology in a floodplain: Guccione, "Late Pleistocene and Holocene Paleoseismology," 240, 261 (2.21).

38. Kochtitzky, *The Story of a Busy Life*, 90 (6.9); harvesting corn underwater: Keefe and Morrow, *The White River Chronicles of S. C. Turnbo*, 19–21 (6.11).

39. Snow, *From Missouri*; floods of 1912 and 1913: 96–104; flood of 1927: chap. 23 (New Madrid residents, 100); 1937: chap. 24 (dynamiting, 204) (6.11.1). Flood in 1927: Barry, *Rising Tide*; Daniel, *Deep'n as It Come*; Kelman, "Boundary Issues" (all in 1.13) "The Mississippi is no longer a river": "Needs of the Flood Sufferers," *New York Times*, 26 April 1927. New Madrid residents fleeing 1927 flood waters: "South in Terror as Flood Sweeps on toward the Gulf," *New York Times*, 22 April 1927.

40. Vegetation quickly covers all: Odum et al., "Near-Surface Structural Model for Deformation," 151 (7.5.3); Bolsterli, *During Wind and Rain*, 5 (6.11). Railroad tracks hidden: Rhodes, *A Missouri Railroad Pioneer*, 197 (6.9). Environmental challenges: Fuller, "Our Greatest Earthquakes," 79 (1.1). Ague: Brattan, "The Geography of the St. Francis Basin," 11 (2.7); Snow, *From Missouri*, 135–37 (6.11.1). Sick lists: W. H. Applewhite, *Appellants' Abstract of the Record*, Sugg case, 308 (2.5). Missouri Botanical Garden: B. F. Bush, "Notes on a List of Plants Collected in Southeastern Missouri in 1893," *Missouri Botanical Garden Annual Report* (St. Louis: Missouri Botanical Garden, 1894).

41. "Pole roads": Testimony of C. R. Parr, *St. Francis Mill . . . against William Sugg*, 281 (2.5). Two county seats: Brattan, "The Geography of the St. Francis Basin," 14 (2.7). Pemiscot: Rhodes, *A Missouri Railroad Pioneer*, 149 (6.9).

42. Most of the early roads in southeast Missouri ran north-south, along a series of sand ridges and terraces. East-west transportation was challenging. Ogilvie, "Development of the Southeast Missouri Lowlands," 3–4 (6.9); Brattan, "The Geography of the St. Francis Basin," 10; on common pattern of population centers separated by impassable low ground until connected by railroads and, later, state highways, 14 (2.7). Houck and railway quotations: Rhodes, *A Missouri Railroad Pioneer*, 149 (6.9).

43. Railroads: Dougan, "Missouri Bootheel" (6.9); Parker, "Missouri as It Is in 1867," 81 (1.17); connections with timber industry: Balogh, "Timber Industry" (6.9); and Brattan, "The Geography of the St. Francis Basin," 45 (2.7).

44. Rhodes, *A Missouri Railroad Pioneer*, esp. 127; battle with Goulds: 174; boast: 112, 198; railroad commissioners' disgust: 126, 155–56; also Kochtitzky, *The Story of a Busy Life*, 152 (both in 6.9). DeBlack, *With Fire and Sword*, 209–11 (6.12).

45. Railroading and swamp reclamation: Giesen, " 'The Truth about the Boll Weevil,' " esp. 688 (6.8). Houck a master: Rhodes, *A Missouri Railroad Pioneer*, 87, 91 (6.9); like many of Houck's endeavors, this ended in tangled and long-running litigation. Pisani, "Beyond the Hundredth Meridian," 477 (1.13).

46. Rail and timber personnel as sources of local knowledge: Shepard, "The New Madrid Earthquake," 59 (1.2); timber engineer C. B. Baily, for instance, worked with geologists Shepherd and Fuller. Parker, "Missouri as It Is in 1867," 30 (1.17). Timber executives and push for seismic network: Macelwane, "The Ozark Earthquake Investigation," 15 (6.4). Houck sending botanical specimens: Rhodes, *A Missouri Railroad Pioneer*, 197 (6.9). Railways promoting use of Great Plains: West, *The Contested Plains*, chap. 6 (1.13).

47. "Swamp railroading": Rhodes, *A Missouri Railroad Pioneer*, 90; scooping sand: 145; similar practice of piling downed trees: 155 (6.9). "Borrow pits" silent about origin as earthquake features: "Borrow Pit—Diversion Channel—Sta. 106" [3093 V4 085], Little River Drainage District Records. Earthquake sand: Fuller, *The New Madrid Earthquake*, plate 1, "Map of Earthquake Features of the New Madrid District" (1.1). Earthquake sand in modern foundations: Wingerson, "In Search of Ancient Earthquakes," 35 (2.21); Tish Tuttle, January 2011. Also: 6.13.

48. Hendricks, "Sunken Lands" (2.5); Rhodes, *A Missouri Railroad Pioneer*, 194–96; Jackson, "Manila (Mississippi County)" (6.9). American pattern of railroads building communities, hinterlands: Cronon, *Nature's Metropolis* (1.13). Panther: Snow, *From Missouri*, 148–49 (6.11.1). Howling wolves: Stepenoff, *Thad Snow*, 14 (6.11.1). Timbering ends trapping by 1890s: Z. T. Hicks, in *St. Francis Mill . . . against William Sugg*," 302, 306; testimony of W. F. Shelton, 310 (2.5).

49. Jackson, "Manila (Mississippi County)" (6.9); Hendricks, "Sunken Lands" (2.5). The New Madrid hinterland continued patterns present elsewhere: railroads allowed for the exploitation of once-abundant species such as bison (Isenberg, *The*

Destruction of the Bison) and passenger pigeons (Price, *Flight Maps*); game-rich territory was set off limits by well-to-do-hunting clubs (Jacoby, *Crimes against Nature*; Warren, *The Hunter's Game*) (all in 1.13).

50. Dougan, "Missouri Bootheel"; Jackson, "Manila (Mississippi County)"; Rhodes, *A Missouri Railroad Pioneer*, 262, 257 (all in 6.9); Snow *From Missouri* (6.11.1); Local railroads were the shorter-lived but roughly chronological rural parallel of the trolley lines that criss-crossed many American cities in the first half of the twentieth century and that were almost completely pulled up in the postwar heyday of the American automobile.

51. Dougan, "Missouri Bootheel" (6.9).

52. See 6.9.

53. Linn: "Letters from Honorable L. F. Linn," 5 (1 February 1836) (1.2). Later calls for swamp reclamation: Parker, "Missouri as It Is in 1867" (1.17). Swamp Land Acts and percentage wetland reductions: Pisani, "Beyond the Hundredth Meridian," 467 (1.13). Bootheel swamp reclamation: Feezor, "Fraud and Deceit in Dunklin County, 1865–1880" (6.11); Stepenoff, "Last Tree Cut Down"; Ogilvie, "The Development of the Southeast Missouri Lowlands" (both in 6.9); Decision, in *St. Francis Mill . . . against William Sugg*," 380 (2.5); and sources listed in 6.9.

54. Otto Kochtitzky, Jr., foreword and (presumably) postscript; Otto Kochtitzky [Sr.], author's foreword; in Kochtitzky, *The Story of a Busy Life* (6.9).

55. Kochtitzky, *The Story of a Busy Life*, 19, 21, 26, 32, 51–52, 74–80, 97, 119, 140 (6.9).

56. Little River Drainage District Records; Rhodes, *A Missouri Railroad Pioneer*, 262–63; Kochtitzky, *The Story of a Busy Life*, 144 (both in 6.9); Brattan, "The Geography of the St. Francis Basin," 4, 7 (2.7).

57. Brattan, "The Geography of the St. Francis Basin," 4–6 (2.7); Rhodes, *A Missouri Railroad Pioneer*, 263 (6.9).

58. Morris, "St. Francis River": Brattan, "The Geography of the St. Francis Basin," 7 (both in 2.7); Hendricks, "Sunken Lands" (2.5); Blackwell, "A Landscape Transformed" (6.9).

59. Kochtitzky, *The Story of a Busy Life*, 152–53; Rhodes, *A Missouri Railroad Pioneer*, 263 (both in 6.9).

60. Hendricks, "Sunken Lands" (2.5); Kochtitzky, *The Story of a Busy Life*, 149; Rhodes, *A Missouri Railroad Pioneer*, 264–66 (both in 6.9).

61. Kochtitzky, *The Story of a Busy Life*, 41, 49, 61 (6.9).

62. See 7.5.4. Kochtitzky, *The Story of a Busy Life*, 50–51 (6.9); Shepard, "The New Madrid Earthquake," 49–50, 56 (1.2); US Department of Agriculture Natural Resources Conservation Service, "Bald Cypress: Plant Fact Sheet," http://plants.usda.gov/factsheet/pdf/fs_tadi2.pdf.

63. Kochtitzky, *The Story of a Busy Life*, 47–51, 54 (6.9).

64. Kochtitzky, *The Story of a Busy Life*, 41, 46–47. Stepenoff, "The Last Tree Cut Down," 64 (both in 6.9). It is a mark of how successfully history of the region could be rewritten that even such a well-informed historian could view the environments that way.

65. Conversation with Scott Ausbrooks, Arkansas Geological Survey, July 2006. Snow, *From Missouri*, 3 (**6.11.1**); Dougan, "Missouri Bootheel" (**6.9**). "Hi": www .ratemybody.com profile, posted 25 May 2010.

66. Compilations: Hough and Bilham, *After the Earth Quakes*, 94 (**1.6**). Shaler, "Earthquakes of the Western United States" (**1.1**). Coen, *Earthquake Observers*, 114 (**1.6**).

67. Bolt, "The Development of Earthquake Seismology in the Western United States" and Geschwind, *California Earthquakes* (both listed in **1.6**).

68. Geschwind, *California Earthquakes*, 18 (**1.6**).

69. Geschwind, *California Earthquakes*, 17–18, 22 (quoting the *San Francisco Chronicle*, 1 April 1892; and *Sacramento Record-Union*, 20 April 1892) (**1.6**).

70. Geschwind, California Earthquakes, 22, 72–75 (quoting from the Pasadena *Star-News*, 3 July 1925 (**1.6**). Coen, *Earthquake Observers*, chaps. 9, 10 (**1.6**).

71. Geschwind, *California Earthquakes*, 22 (**1.6**).

72. Geschwind, *California Earthquakes*, 75 (**1.6**); L. Frank Baum, *The Wonderful Wizard of Oz* (Chicago: G. M. Hill, 1900); *The Wizard of Oz*, film, MGM Studios, 1939.

73. Seismological Society of America: Davison, *The Founders of Seismology*, 152–55; and similarly significant for international science—for instance, see 192 (**1.6**).

74. Davison, *The Founders of Seismology*, 177; also well summed in Coen, "Comparative Histories" (see **1.6**).

75. Davison, *The Founders of Seismology*, 70–71; similar experiments by John Milne: 187 (**1.6**).

76. Dewey and Byerly, "Early History of Seismometry," 184–98. Schweitzer, "Birth of Modern Seismology," 263 (both in **6.1**); Oldroyd et al., "The Study of Earthquakes," 346–50 (**1.6**). It's not entirely clear to what extent these devices were put into practice or remained theoretical proposals only, but certainly in the 1870s researchers in Europe and Japan began to actually build and experiment with devices to detect and record ground movement.

77. See **6.1**. Irony of discovery: Schweitzer, "Birth of Modern Seismology," 268–69.

78. International system of earthquake observatories: Davison, *The Founders of Seismology*, 195 (**1.6**); Dewey and Byerly, "Early History of Seismometry," 212; Schweitzer, "Birth of Modern Seismology," 270–72 (both at **6.1**); Geschwind, "Embracing Science and Research" (**6.2**). The international cooperation in geology in the 1880s operated in seismology: Greene, *Geology in the Nineteenth Century*, 204 (**5.11**). Challenges: Coen, *Earthquake Observers*, chap. 8 (**1.6**).

79. These changes in knowledge of the New Madrid earthquakes paralleled knowledge of other aspects of the natural world in the late nineteenth and early twentieth centuries; see Nash, "The Changing Experience of Nature," and other citations listed in **7.20**; though for important counterexamples see Coen, *The Earthquake Observers and "The Tongues of Seismology"* (**1.6**). Earthquake engineering: Geschwind, *California Earthquakes*, 9–15 (**1.6**).

80. Increasingly networked and institutionally robust American science: Goldstein, "Yours for Science"; and Lucier, *Scientists and Swindlers*, esp. 315–20; on Hopkins, 315–16 (**5.3**).

81. Geschwind, *California Earthquakes*, 82 (**1.6**). Davison, *Founders*, 152–55 (**1.6**).

82. Udías, "Jesuits' Studies of Earthquakes and Seismological Stations," 140 (6.2). Earth science did not, of course, become a visual science with the advent of seimo-grams: geology had a profoundly visual basis from its origins as a modern science: Rudwick, "The Emergence of a Visual Language for Geological Science" (5.11); and Rudwick, *Bursting the Limits of Time* (5.11). "That zig-zag line": Branagan, "Earth, Sky and Prayer in Harmony," 70 (6.2).

83. But see Coen, *The Earthquake Observers*, 227–34, "Tongues" and "Witness to Disaster: Comparative Histories"; on educability of an earthquake public, see Davison, *Founders of Seismology*, 184 (all in 1.6). Privileging of expert observation in epidemiology, for instance: Nash, *Inescapable Ecologies*, 105 (1.13). Kochtitzky, *The Story of a Busy Life*, 57–58 (6.9). Kochtitzky disputes an unnamed scientist's conclusions about both New Madrid and Charleston; this scientist was most likely Myron Fuller. Photo of Charleston earthquake: Stein, *Disaster Deferred*, 139 (7.11). The Charleston earthquake killed nearly a hundred people out of a population of 49,000. It caused $5–8 million in damages and rendered as many as seven out of eight homes in Charleston uninhabitable. Hough, *Earthshaking Science*, 211 (1.6).

84. Geschwind, *California Earthquakes*, 45 (1.6); Geschwind, "Embracing Science and Research," 47–48 (6.2).

85. Geschwind, "Embracing Science and Research," 47–52 (6.2); Coen, *Earthquake Observers*, chap. 10 (1.6).

86. Shepard, "The New Madrid Earthquake," 49 (1.2); Shapin, *Social History of Truth* (5.22).

87. Davison, *The Founders of Seismology*, 184; Coen, *Earthquake Observers* (both in 1.6) and "Tongues of Seismology" (4.8).

88. Hough, *Richter's Scale*, chap. 6, describes this camaraderie and close coordination of people and instruments in the CalTech Seismo Lab (later the Kresge Lab) from the 1920s to the 1960s (1.6).

89. Geschwind, "Embracing Science and Research"; Udías, "Jesuits' Studies of Earthquakes and Seismological Stations," 140; Udías and Stauder, "Jesuit Geophysical Observatories" (all in 6.2); SLU Department of Earth and Atmospheric Sciences, "Department History" (6.5).

90. Though the date of the school's founding is only six years after the New Madrid quakes, official histories of the founding of the school make no mention of the tremors. Meteorology and seismology: Geschwind, "Embracing Science and Research"; Udías, "Jesuits' Study of Earthquakes and Seismological Stations"; Udías and Stauder, "Jesuit Geophysical Observation"; Udías and Stauder, "The Jesuit Contribution to Seismology" (all in 6.2). An early term for the study of the interior of the earth was "endogenous meteorology." Dewey and Byerly, "Early History of Seismometry," 214 (6.1). SLU observers: Saint Louis University, "About SLU" (6.5). The names track changes in how scientific approaches have been categorized and defined—founded in 1925 as a program in geophysics, the program was renamed during the 1960s as part of a larger movement toward interdisciplinarity as a Department of Earth and Atmospheric Sciences (on movement toward "earth science" as indicative of interdisciplinarity: LeGrand, *Drifting Continents and*

Shifting Theories, 238-39 (1.6)). First church service: Foley, *A History of Missouri*, 1:184 (1.11). Rhetorical power of claiming of "first" events: O'Brien, "Firsting," in *Firsting and Lasting* (2.3).

91. Saint Louis University, "About SLU"; Saint Louis University Department of Earth and Atmospheric Sciences, "SLU Dept History" (6.5). This instrumentation was part of a wave of American earthquake investigation that followed the 1906 San Francisco earthquake, but it also followed local small earthquakes: Shepard, "The New Madrid Earthquake," 59 (1.2). SLU seismographs recorded their first local earthquake in 1909: Jackson, *Earthquakes and Earthquake History of Arkansas*, 54 (7.3); though see 49. Macelwane as scholar and teacher: Birkenhauer, "Father Macelwane"; Blum, "Sketch of the Life of James Bernard Macelwane"; Byerly and Stauder, "James B. Macelwane"; 261; Heinrich, "James B. Macelwane, S.J., Scholar," 14 (all in 6.3); Howell, *An Introduction to Seismological Research*, 131-32 (1.6).

92. Byerly and Stauder, "James B. Macelwane," 260 (6.3). California seismic study as model for the SLU approach: Macwelwane, "The Mississippi Valley Earthquake Problem," 96 (6.4).

93. Blum, "Sketch of the Life of James Bernard Macelwane, S.J.," 10; Byerly and Stauder, "James B. Macelwane, S.J."; Hodgson, "The Contribution of Father Macelwane" (all in 6.3); Macelwane, "Seismicity of the Mississippi Valley," 3 (6.4); Saint Louis University Department of Earth and Atmospheric Sciences, "SLU Dept History" (6.5); Udías, "Jesuits' Studies of Earthquakes and Seismological Stations," 140; Udías and Stauder, "Jesuit Geophysical Observatories," 189 (6.2).

94. Some work did happen in Arkansas in the same period, but it appears to have had even more limited public impact: Branner and Hansell, *Earthquake Risks in Arkansas* (7.2). Macelwane was a public figure in St. Louis and was noted for writing accessible introductions to esoteric scientific subjects. See 6.3. Writing in the late '40s, Macelwane referred to these activities as having occurred "only within the last few decades." Saint Louis University Department of Earth and Atmospheric Sciences, "SLU Dept History" (6.5); Byerly and Stauder, "James B. Macelwane," 262 (6.3). California array: Geschwind, *California Earthquakes*, 82 (1.6). Blum, "Sketch of the Life of James Bernard Macelwane"; Byerly and Stauder, "James B. Macelwane"; Hodgson, "The Contribution of Father Macelwane" (all in 6.3); Udías, "Jesuits' Study of Earthquakes and Seismological Stations," 140; Udías and Stauder "Jesuit Geophysical Observatories," 189 (6.2).

95. Macelwane, "The Mississippi Valley Earthquake Problem," 97 (6.4). This was part of his long-standing cooperation with the National Research Council, from the mid-'20s until midcentury: Byerly and Stauder, "James B. Macelwane, S.J.," 271 (6.3); Macelwane, "Progress in the Study of Earthquakes," 2, 4; Bradford and Macelwane, "A Preliminary Sketch"; Macelwane, "Can We Make St. Louis Earthquake Proof?"; "Earthquake Survey of Ozark Region Proposed by St. Louis Scientist" (6.4).

96. Macelwane, "Seismicity of the Mississippi Valley," 1; seismograph and technical frustration: Macelwane, "Grover, Missouri, Earthquake," 4 (both in 6.4).

97. Lucier, *Scientists and Swindlers* (5.3).
98. Udías and Stauder, "Jesuit Geophysical Observatories" (6.2); American Geophysical Union, "James B. Macelwane Medal" (6.3). Because Nuttli's report languished for several decades, I discuss it in the next chapter, rather than this one.
99. LeGrand, *Drifting Continents and Shifting Theories*, 171–75 (1.6); Tuttle et al. "The Earthquake Potential" (1.1). Massive expansion of science began in the 1920s, not simply postwar—see Galison and Hevly, *Big Science*, esp. Galison, "The Many Faces of Big Science," 5 (6.17)—but especially in seismology, changes amplified after the war.
100. Udías and Stauder, "Jesuit Geophysical Observatories," 140, 189; Udías, "Jesuits' Studies of Earthquakes and Seismological Stations" (6.2); LeGrand, *Drifting Continents and Shifting Theories*, 175 (1.6). Some variation in the information in these sources indicates how much still remains to be studied about these important developments. Johnston and Shedlock, "Overview of Research in the New Madrid Seismic Zone," 194 (7.3).
101. Udías, "Jesuits' Studies of Earthquakes and Seismological Stations," 135 (6.2).
102. Oreskes, "From Continental Drift to Plate Tectonics,"in *Plate Tectonics*, ed. Oreskes, chap. 1, overviews this process (also see 150–51); and Hough, *Earthshaking Science*, 1–23 (both in 1.6).
103. Oreskes, "From Continental Drift to Plate Tectonics," in *Plate Tectonics*,chap. 1; LeGrand, *Drifting Continents and Shifting Theories*, chaps. 4, 5, 6, 8, esp. 37–39, 82–83, 170 (both in 1.6).
104. Oreskes, *Plate Tectonics*, part 2; LeGrand, *Drifting Continents and Shifting Theories*, 170–78, 188, 196–99 (both in 1.6).
105. Oreskes, *Plate Tectonics*, part 2; LeGrand, *Drifting Continents and Shifting Theories*, 206, revision by Vine and Wilson, 8–9, 210–14 (both in 1.6). Robert Proctor emphasizes the intellectual cost of wartime secrecy, pointing out that these stripes were observed during the war, but were kept classified until the 1950s because they were useful for identifying enemy submarines. Such secrecy, he argues, thus delayed geological understanding: Proctor, *Agnotology*, 19 (6.6).
106. LeGrand, *Drifting Continents and Shifting Theories*, 215, 233–35 (1.6).
107. Stunning swiftness of this new model: Oreskes, *Plate Tectonics*, 171; LeGrand, *Drifting Continents and Shifting Theories*, 218, 229, 241–2 (both in 1.6).
108. Johnston, "A Major Earthquake Zone on the Mississippi," 63 (7.3).
109. False prediction: Geschwind, *California Earthquakes*, chap. 8; Bailey Willis and Santa Barbara: 77; Howell, *An Introduction to Seismological Research*, 53–56. Interest in earthquake prediction: Hough, *Predicting the Unpredictable* (all in 1.6); also Fan, "Collective Monitoring, Collective Defense" (5.2).
110. Loma Prieta: USGS Historic Earthquakes website, Loma Prieta, http://earthquake.usgs.gov/earthquakes/states/events/1989_10_18.php, posted October 2009; Michael Rosenberg, "Twenty Years Ago, an Earthquake Shook a Bay Area World Series," *SI.com (Sports Illustrated)*, 28 October 2009. Loma Prieta context for concerns about New Madrid seismicity: for instance, opening statement by Robert A. Roe, Chairman, 3, and testimony of Arch C. Johnston, director, Center for

Earthquake Research and Information, 55; *The Earthquake Threat in the Central United States* (6.16). Phone call: Coen, *Earthquake Observers*, 234 (1.6).

111. Browning: Farley, *Earthquake Fears, Predictions, and Preparations*, 45; Spence et al., *Responses to Iben Browning's Prediction* (6.15) Congressional hearing: *The Earthquake Threat in the Central United States*, 83 (6.16). Stewart was introduced by Rep. Emerson from Missouri, who according to Arch Johnston had unsuccessfully attempted to get Stewart introduced as a witness to the hearing but instead had to settle for introducing his written testimony. (Personal communication with Arch Johnston, August 2009).

112. Farley, *Earthquake Fears, Predictions, and Preparations*, 30–31; Stewart quotations: 45. The USGS put out a publication precisely to address the question of how Browning came to be granted such credibility: Spence et al., *Responses to Iben Browning's Prediction* (6.15). Manufacture of public uncertainty: Brandt, *The Cigarette Century*; and Oreskes and Conway, *Merchants of Doubt* (6.6).

113. Conversation with John David McFarland, retired geologist with the Arkansas Geological Survey, August 2007. Spence et al., *Responses to Iben Browning's Prediction*, 14–15. Farley, *Earthquake Fears, Predictions, and Preparations*, 44 (6.15). Johnston and Shedlock, "Overview of Research, 105 (7.3).

114. Spence et al., *Responses to Iben Browning's Prediction*, app. C, D, E, F; Steltzer, "On Shaky Ground" (6.15).

115. Spence et al., *Responses to Iben Browning's Prediction*, 20–22 (6.15); Hough, *Predicting the Unpredictable*, 185–88 (1.6). Coen, *Earthquake Observers*, 38–44 (1.6).

116. Similar episode of sudden public fear arising from nonexpert opinion: Seargent, *The Greatest Comets in History*, 191 (1.27). McFarland, conversation August 2007. The Arkansas Geological Survey was then named the Arkansas Geological Commission.

117. Abandoned and rusty boilers: Brattan, "The Geography of the St. Francis Basin," 46 (2.7). In Willa Cather's *My Ántonia* (book 5, chap. 3), wagon tracks barely visible in prairie grass are similar environmental *pentimenti*.

118. Subtlety of environmental signs: Guccione, "Late Pleistocene and Holocene Paleoseismology," 238, 48 (2.21). Electric plant: Dougan, "Missouri Bootheel" (6.9). Continuing challenges of seismic preparation in New Madrid: "Seismic Rehabilitation & Retrofit of Structures in the Central U.S.," presentation, EERI Annual Meeting and National Earthquake Conference, April 2012.

119. Jackson, *Earthquakes and Earthquake History of Arkansas*, 45 (7.3). The agency was at the time the Arkansas Geological Commission.

120. See 6.6, esp. Mathews, "Suppressing Fire and Memory"; and Fiege, *Irrigating Eden*. Also: 2.3.

121. Fuller, "Our Greatest Earthquakes," 79 (1.1).

Chapter 7

1. James Macelwane and colleagues worked on "old records" of the quakes, but apparently without publication: Macelwane, "Seismicity of the Mississippi Valley,"

3; Macelwane, "Progress in the Study of Earthquakes," 4; Bradford and Macelwane, "A Preliminary Sketch"; "Earthquake Survey of Ozark Region Proposed by St. Louis Scientist"; Macelwane, "The Ozark Investigation"; Macelwane, "The Ozark Program and Our Seismographs" (all in 6.4).

2. Davis, "Studies of Ground-Water Pollution"; Schrock, *Geology at MIT*, 39, M-7; Myron L. Fuller correspondence 1917 (all in 7.1.1). Fuller Craft Museum: Personal communication with Wyona Lynch-McWhite, Director, Fuller Craft Museum, Brockton, MA, 7 July 2010.

3. Fuller, *The New Madrid Earthquake*, 7 (1.1); Shepard, "The New Madrid Earthquake" (1.2).

4. "Up the old De Soto trail": "Fuller, *The New Madrid Earthquake*, 7–8 (1.1); Galloway, *The Hernando De Soto Expedition*" (2.11).

5. Shepard, "The New Madrid Earthquake," 50 (1.2); Fuller, *The New Madrid Earthquake*, 8–9 (1.1).

6. Other contemporary compilations: Broadhead, "The New Madrid Earthquake" (1.2); McGee, "Correspondence" (6.4) See 7.1.2.

7. Fuller, *The New Madrid Earthquake*, plate 7 (1.1); Davison, *The Founders of Seismology*, 154 (1.6).

8. Fuller, *The New Madrid Earthquake*, 7, 9 (1.1).

9. Fuller, *The New Madrid Earthquake*, 70 (1.1). See the section on baldcypress, chap. 2.

10. Fuller, *The New Madrid Earthquake*, 33–42 (1.1).

11. Fuller, *The New Madrid Earthquake*, 11–13, 109–10, plate 2; larger than Charleston and San Francisco: 76 (1.1); Fuller references: Lyell, *A Second Visit to the United States*, 2:180 (5.16).

12. Fuller, *The New Madrid Earthquake*, 109–10 (1.1); Macelwane, "Progress in the Study of Earthquakes," 2 (6.4); guide of 1979 summarizing Fuller: Jackson, *Earthquakes and Earthquake History of Arkansas*, 42–46 (7.3).

13. Brattan, "The Geography of the St. Francis Basin," 24 and 52 (2.7). Conversation with Scott Ausbrooks, Arkansas Geological Survey, summer 2007.

14. Researchers of the 1980s and '90s refer back to Fuller's insights with the surprise of rediscovery—a rediscovery that came about in part because of the republishing efforts of David Stewart's questionably creditable press. Johnston and Schweig, "The Enigma of the New Madrid Earthquakes," 339 (1.1).

15. See 7.1.3.

16. Disavowal of Stewart: Spence et al., *Responses to Iben Browning's Prediction*, 9–13 (6.15); see also discussion of Browning in that chapter. Fuller's report is available on the USGS website, but the Stewart version is still the most accessible hard copy.

17. See 7.15.

18. One compilation of earthquake reports was made in Arkansas in the 1930s, but apparently had little impact: Branner and Hansell, "Earthquake Risks in Arkansas" (7.2).

19. One 1936 survey of the field listed the New Madrid quakes under a chapter called "Description of Great Earthquakes," but sections on current research made clear

the distance between modern instrumentation and such anecdotally recorded tremors. Heck, *Earthquakes*, 130–31 (1.6). Records as medium of exchange—and of sale: Clancey, *Earthquake Nation*, 172 (1.6).

20. Better, more detection: Jackson, *Earthquakes and Earthquake History of Arkansas*, 49, 61; and on increasingly precise data, table 2 (7.3). Broader meanings of such precision: Canales, *A Tenth of a Second* (6.17).

21. See 7.3.1.

22. See 7.3.1.

23. Nuttli, "The Mississippi Valley Earthquakes of 1811 and 1812," 228 (7.3.1).

24. Nuttli, "The Mississippi Valley Earthquakes of 1811 and 1812," 228 (7.3.1).

25. Nuttli, "The Mississippi Valley Earthquakes of 1811 and 1812," 228 (7.3.1). See also Hough et al., "On the Modified Mercalli Intensities and Magnitudes," 23845 (7.5.2).

26. Nuttli, "The Mississippi Valley Earthquakes of 1811 and 1812," abstract (7.3.1).

27. CERI home page, http://www.ceri.memphis.edu/index.shtml, accessed 14 June 2012; "Dr. Chuck Langston Takes Helm at U of M's Earthquake Center," University of Memphis press release, 3 December 2008.

28. Arch Johnston, phone conversations, 10 August and 17 August 2009; "Inception" page on CERI website: http://www.ceri.memphis.edu/about_us/inception.html, accessed 14 June 2012; Jackson, *Earthquakes and Earthquake History of Arkansas*, 69 (7.3).

29. Arch Johnston, phone conversations 10 August and 17 August 2009; Hough, *Predicting the Unpredictable*, 184 (1.6). Funding came about largely because the Nuclear Regulatory Commission was still studying areas to site nuclear power plants—a priority to which the NRC returned in the early twenty-first century because of political concerns and concerns for global climate change.

30. Arch Johnston, phone conversations 10 August and 17 August 2009.

31. See 7.3.

32. Johnston called this "one of the tragedies of the profession." Johnston, conversation 17 August 2009; confirmed by Christine Harper, assistant archivist, Saint Louis University, e-mail, 29 July 2010.

33. Johnston, phone conversation, 10 August 2009; and Nathan K. (Kent) Moran, phone interview, 7 August 2009; and e-mail, 21 August 2009.

34. Conversations with Kent Moran, 7 August 2009 and October 2003; Hough and Bilham, *After the Earth Quakes* (1.6); Charles Langston, phone interview, 22 July 2010. I too use the CERI Compendium in exactly the same way.

35. When I asked about use of the compendium and user feedback, Johnston called it "underutilized." He was not sure if there had been any uptick because of the bicentennial. Researchers do contact CERI with questions because of the compendium and submit new material for it, but infrequently.

36. Chuck Langston, phone conversation, 22 July 2010.

37. Nuclear concerns: Johnston, "A Major Earthquake Zone on the Mississippi," 68 (7.3). Midcontinent network: Johnston and Shedlock, "Overview of Research in the New Madrid Seismic Zone," 196; network ceased: 200 (7.3). Testimony of Dal-

las L. Peck, director, US Geological Survey, Department of the Interior, *The Earthquake Threat in the Central United States*, 37 (6.16).

38. End of seismic monitoring: Johnston and Shedlock, "Overview of Research in the New Madrid Seismic Zone," 197 (7.3); written testimony by Johnston, *The Earthquake Threat in the Central United States*, 65 (quotation 57) (6.16).

39. See 7.4.

40. Geschwind, *California Earthquakes*, chap. 8; Hough, *Predicting the Unpredictable*, 106, 108 (both in 1.6). Report of 1974: Johnston, "A Major Earthquake Zone on the Mississippi," 67 (7.3).

41. Steltzer, "On Shaky Ground" (6.15); Hester, Satterfield, and Bauer, "Consortia Aim to Mitigate Potential Losses," [1] (7.1.1).

42. "If you have an earthquake in New Madrid": Steltzer, On Shaky Ground" (6.15); Van Arsdale, "Seismic Hazards," esp. 2 (1.1).

43. Increasing attention to seismic building codes: *The Earthquake Threat in the Central United States*, 72; pipelines, cables: 27 (6.16); Johnston and Shedlock, "Overview of Research in the New Madrid Seismic Zone," 202 (7.3). Pipelines and cables: Gambrell, "Scientist: Previously Unknown Earthquake Zone" (7.11); Steltzer, On Shaky Ground" (6.15); Zagier, "Earthquake Expert Warns of Economic Damage" (7.11).

44. Fréchet, Meghraoui, and Stucchi, *Historical Seismology*; Street and Green, "The Historical Seismicity of the United States"; and Street, "Contribution to the Documentation" (both in 7.7). Street et al., "NEHRP Soil Classifications" (7.5.12).

45. Geological structures: Johnston and Shedlock, "Overview of Research in the New Madrid Seismic Zone," 194 (7.3); Hamilton and McKeown, "Structure of the Blytheville Arch"; Gomberg, "Tectonic Deformation in the New Madrid Seismic Zone"; Howe and Thompson, "Tectonics, Sedimentation, and Hydrocarbon Potential." Glacial rebound: Grollimund and Zoback, "Did Glaciation Trigger Intraplate Seismicity?"; erosion: Calais et al., "Letter: Triggering of New Madrid Seismicity" (all in 7.5.3); Hough, "The Magnitude of the Problem" (7.5.2); Hough, *Earthshaking Science*, 175 (1.6). Hotspot: Cox and Van Arsdale, "Hotspot Origin of the Mississippi Embayment" (7.5.3).

46. See 7.6. My mother, Susan May, buys Crowley's Ridge peaches for visiting grandchildren.

47. See 7.5.4. Baldcypress: Stein, *Disaster Deferred*, chap. 6 (7.11). Lyell, *A Second Visit to the United States*, 2:176 (5.16). Large research teams: Galison, "The Many Faces of Science," introduction to Galison and Hevly, *Big Science* (6.17); Stein, *Disaster Deferred*, 7 (7.11).

48. See 7.5.5; recent work referenced succinctly in "NMSZ Expert Panel Report to NEPEC," 5 (7.5.1).

49. Tuttle, "The Use of Liquefaction Features in Paleoseismology," 361 (7.5.5). First efforts: Guccione, "Late Pleistocene and Holocene Paleoseismology," 247 (2.21); Tuttle and Schweig, "Archeological and Pedological Evidence," 253 (7.5.5).

50. Wingerson, "In Search of Ancient Earthquakes," 30, 35 (2.21).

51. "Read Fuller cover-to-cover": communication, 25 June 2012. Conventional wisdom: Saucier, "Evidence for Episodic Sand-Blow Activity" (1.22). Very long ago:

see Tuttle, "Search for and Study of Sand Blows," 19 (7.5.5). Good overview: Tuttle et al., "Dating of Liquefaction Features in the New Madrid Seismic Zone" (7.5.5).

52. Citations at 7.5.5, particularly Tuttle and Wolf, "Towards a Paleoearthquake Chronology," esp. 9–10, 15–16; and Tuttle, "Search for and Study of Sand Blows," 6.

53. Very old earthquakes: Tuttle, Al-Shukri, and Mahdi, "Very Large Earthquakes." Tuttle et al., "Use of Archaeology to Date Liquefaction Features," 451, 453, 468; Tuttle and Schweig, "Archeological and Pedological Evidence"; Tuttle et al., "Evidence for New Madrid Earthquakes in A.D. 300 and 2350 B.C." (all in 7.5.5); Johnston and Shedlock, "Overview of Research in the New Madrid Seismic Zone," 199 (7.3); Guccione, "Late Pleistocene and Holocene Paleoseismology," 247 (2.21); Tuttle, personal communications, 13 January 2011, 25 June 2012.

54. "Modest geologic goldrush": Hough, *Earthshaking Science*, 179 (1.6). Marianna: appendix E of CEUS Seismic Source Characterization, 10–13; Daytona Beach: Tuttle, Al-Shukri, and Mahdi, "Very Large Earthquakes," 756; Southeast Arkansas sand blows: Al-Shukri et al., "Spatial and Temporal Characteristics of Paleoseismic Features" (all in 7.5.5). "Science by backhoe": Shawna Vogel, *Naked Earth: The New Geophysics* (New York: Dutton, 1995), 162.

55. See 7.5.10 and 7.5.6. Wabash research summed in appendix E of CEUS Seismic Source Characterization, 15–16 (7.5.5). Earthquake of 1987: Hester, Satterfield, and Bauer, "Partnerships for the Central United States" (7.11.1)

56. Michetti, Audemard, and Marco, "Future Trends in Paleoseismology," section 2 (7.5.5). Multiple main shocks: Van Arsdale et al., "Faulting along the Southern Margin of Reelfoot Lake," 138 (1.25). New Madrid a main example: Guccione, "Late Pleistocene and Holocene Paleoseismology," 239 (2.21). New Madrid is often a referent in Stein and Mazzotti, eds., *Continental Intraplate Earthquakes* (7.8).

57. Blomeley, "Seismic Building Code Sparks Safety Debate"; Heard, "Geologist Urges Less Stringent Seismic Building Codes" (both in 7.11).

58. Problems of knowledge are, as historians remind us, also problems of social order: see Shapin and Schaffer, *Leviathan and the Air-Pump* (4.8).

59. Summation of magnitude debates (quite civilly): Hough, *Earthshaking Science*, 180–85 (1.6); Hough, "Large 19th Century Earthquakes," 352 (7.7).

60. Johnston and Schweig, "The Enigma of the New Madrid Earthquakes," 339 (1.1).

61. Johnston and Schweig, "The Enigma of the New Madrid Earthquakes," esp. 339 (1.1). Critiques: for instance, Hough, *Predicting the Unpredictable*, 183 (1.6); Stein, *Disaster Deferred*, 16 (7.11). Largest earthquakes: Hough, *Earthshaking Science*, 213 (1.6); US Geological Survey, "Historical Earthquakes: Prince William Sound, Alaska," http://earthquake.usgs.gov/earthquakes/states/events/1964_03_28.php; USGS, "Historic U.S. Earthquakes," http://earthquake.usgs.gov/earthquakes /states/historical_mag.php, updated 23 July 2012.

62. Sample of Memphis debates: Charlier, "Seismic Audit Is Jolt for Builders" (7.11). In a set of 2011 online teaching lectures, Stein engagingly illustrates his arguments about interacting faults and hazard assessment for migrating earthquake using the games Booby Trap and Whac-A-Mole.

NOTES TO PAGES 296-300 [385]

63. Landmark revision of magnitude estimates: Hough et al., "Modified Mercalli Intensities and Magnitudes" (7.5.2); further developed in Hough, "Large 19th Century Earthquakes," 361–62 (7.7). Popular writing: *Earthshaking science* (quotation 185); Hough, *Predicting the Unpredictable; Richter's Scale* (both in 1.6); Hough, *Finding Fault in California: An Earthquake Tourist's Guide* (Missoula, MT: Mountain Press, 1994).

64. Prior recognition of differential response: for instance, Jackson, *Earthquakes and Earthquake History of Arkansas*, 10–11 (7.3). Explained: Hough et al., "Modified Mercalli Intensities and Magnitudes" (7.5.2); Hough, "Large 19th Century Earthquakes," 361–62 (7.7).

65. Good review of intensity debate: Hough, "Scientific Overview and Historical Context" (7.5.1). Nuttli transcription error: Hough et al., "Modified Mercalli Intensities and Magnitudes," 23845–46 (7.5.2).

66. Hough, "Scientific Overview and Historical Context," 526 (7.5.1).

67. Mallet: Dewey and Byerly, "The Early History of Seismometry," 190 (6.1). Same impulse present in modern science: Kochkin and Crandell, "Survey of Historic Buildings" (7.5.12). Compare Clancey, *Earthquake Nation*, 71 (1.6).

68. Hough, "Large 19th Century Earthquakes," 351, 365 (7.7). See also Hough et al., "Modified Mercalli Intensities and Magnitudes of the New Madrid Earthquakes" (7.5.2).

69. Magnitude discussions summarized in NMSZ Expert Panel Report to NEPEC, 15–16, listed in 7.5.1. Also: 7.5.2.

70. USGS, Historic United States Earthquakes, 1994 Northridge, http://earthquake .usgs.gov/earthquakes/states/events/1994_01_17.php, updated 21 October 2009. Public debate: Steltzer, "On Shaky Ground" (6.15); "Sustainability Initiatives," Missouri Botanical Garden, http://www.missouribotanicalgarden.org/media /fact-pages/sustainability.aspx, accessed 15 June 2012.

71. Stein and Tomasello, "When Safety Costs Too Much" (7.11). Trashy movie: 7.13.

72. "You can build a school with steel": Heard, "Geologist Urges Less Stringent Seismic Building Codes." "And it's going to cost a million more dollars": UPI, "Mayor Fights Arkansas Earthquake Law" (both in 7.11).

73. Zoeller, "Earthquake Code Enforcement"; and Charlier, "Reeling Housing Industry Feels New Jolt" (both in 7.11); Greg Peterson, "Debating Earthquake Safety," *Geotimes Web* (2003); also 7.5.7.

74. Nuttli, "The Mississippi Valley Earthquakes of 1811 and 1812," 245 (7.3.1); Gomberg and Schweig, "Earthquake Hazard in the Heart of the Homeland," [4] (7.5.11).

75. See 3.2 and 7.9. "This is not a California earthquake": Associated Press, "New Madrid Zone Quake Will Be Very Different, Expert Warns" (7.11). See also the written testimony of Arch Johnston, *The Earthquake Threat in the Central United States* (6.16); Steltzer, "On Shaky Ground" (6.15); Johnson, "All Shook Up" (7.11). "Low probability, high consequence hazard": Crandell, "Policy Development and Uncertainty in Earthquake Risk" (7.8).

76. Stein, Tomasello, and Newman, "Should Memphis Build for California's Earthquakes?" (7.5.11); background: Stein, *Disaster Deferred*, 193 (7.11).

77. Frankel, "Forum Commentary"; and Hough, "Forum Commentary" on "Should Memphis Build for California's Earthquakes?" (7.5.11); see also 7.5.6.

78. New instrumental data: Tuttle, "New Madrid in Motion," 1038 (7.5.1). "The hazard posed by great earthquakes in the NMSZ": Newman et al., "Slow Deformation and Lower Seismic Hazard," 619 (7.5.11); see also other sources at 7.5.8.

79. Stein and Liu, "Long Aftershock Sequences"; Parsons "Earth Science." "Many of the earthquakes we see today in the Midwest": Fellman, "Earthquakes Actually 19th Century Aftershocks" (all in 7.5.8). Also Charlier, "Scientists Debate If New Madrid Fault Is Dead"; Van Arsdale, "Displacement History and Slip Rate on the Reelfoot Fault"; McKenna, Stein, and Stein, "Is the New Madrid Seismic Zone Hotter and Weaker?" (all in 7.5.8); Stein, Disaster Deferred, 76, 171 (7.11); Ebel, Bonjer, and Oncescu, "Paleoseismicity" (7.5.5).

80. "It just doesn't work that way": Charlier, "Scientists Debate If New Madrid Fault Is Dead" (7.5.8), and sources listed at 7.5.9.

81. Wide reporting: Stein, Disaster Deferred, 175; Britt, "Tiny Geologic Shifts"; Arkansas: Heard, "Scientists Divided on Quake Research" (all in 7.11).

82. Johnson, "All Shook Up" (7.11).

83. Nelson, "Why the Delta Matters" (7.16).

84. Frankel et al., "Earthquake Hazard in the New Madrid Seismic Zone" (7.5.11). Also Gomberg and Schweig, "Earthquake Hazard in the Heart of the Homeland" (7.5.11).

85. Earthquake Engineering Research Institute, "The Wenchuan, Sichuan Province, China Earthquake of May 12, 2008: EERI Special Earthquake Report," 2008. Frankel, "Earthquake Hazard" (7.5.11).

86. Elnashai et al., "Impact of New Madrid Seismic Zone Earthquakes" (7.11.1); Hemenway, "Midwest Earthquake Would Create Heavy Losses" (7.11).

87. Charlier, "Scientists Debate If New Madrid Fault Is Dead" (7.11); Gomberg and Schweig, "Earthquake Hazard in the Heart of the Homeland" (7.5.11); NMSZ Expert Panel Report to NEPEC, and "Central and Eastern Seismic Source Characterization for Nuclear Facilities" (7.5.1).

88. Protecting America, "New Madrid Fault: Is Little Rock Prepared?" Little Rock, AR, 17 July 2006; session on "Large, Vulnerable, and Critical Infrastructure," 2012 EERI Annual Meeting and National Earthquake Conference.

89. CUSEC, "Reducing the Risk," [n.d.], http://www.cusec.org/publications/other /prospectus.pdf; "2009 Earthquake Awareness Week Recap," CUSEC Journal 13, no. 2 (2009), 3 (7.11.2); New Madrid Bicentennial website (7.11.3); USGS, "Putting Down Roots in Earthquake Country"; FEMA and NEHRP, "The National Earthquake Hazards Reduction Program" (both in 7.11); The Great Central U.S. Shake-Out website (7.11.1).

90. The Great Central U.S. Shake-Out website (7.11.1).

91. USGS, "Putting Down Roots in Earthquake Country"; FEMA and NEHRP, "The National Earthquake Hazards Reduction Program" (both at 7.11).

92. "Katrina: Storm That Drowned a City," PBS, NOVA (7.14). In formal ways Katrina is a reference point for concerns about a New Madrid disaster: Dowd, "Conferees

Discuss Ways to Get Disaster Area Back to 'New Normal'" (7.11.1); Zoeller, "Earthquake Code Enforcement" (7.11). It is also a reference point for skepticism about FEMA and federal programs generally: Stein, *Disaster Deferred*, 42 (7.11). Informally, many involved in earthquake planning in the New Madrid seismic zone made reference to the hurricane throughout their conversations with me.

93. See citations for 7.14.1; quotations from "Legislators Gain Ice Storm Insight"; Claudine Payne, personal communication, February 2009; Revkin, "Disaster Awaits Cities in Earthquake Zones"; Dowd, "Businesses Are Urged to Be Earthquake-Ready" (both in 7.11.1).

94. Johnston and Schweig, "The Enigma of the New Madrid Earthquakes of 1811–1812," 379 (1.1).

95. See 7.8.2 and 7.8.3.

96. See 7.5.10 and 1.7.1. Not-so-quiet midcontinent zones: Hough, "Triggered Earthquakes," 1581 (7.5.6).

97. Witt and Loy, "Heartland Haiti" (7.11.4).

98. EERI, Learning from Earthquakes (1.7), "The M_w 7.0 Haiti Earthquake of January 12, 2010: Report #1," April 2010; effects, pre-existing Haitian challenges: 1; "The M_w 7.0 Haiti Earthquake of January 12, 2010: Report #2," May 2010; affected 15 percent of nation's population: 1; French building codes in Haitian engineering: 5. Total casualties: USGS, "Poster of the Haiti Earthquake of 12 January 2010" (1.7).

99. EERI, Chile Research Needs Workshop Report: largest instrumentally recorded earthquake: 2; standards modeled on U.S.: 1; hospital damage: 7; "natural laboratory": 6; casualties: 1, 7. Earthquake of 1960: USGS Historic World Earthquakes, "The Largest Earthquake in the World" (both in 1.7). Take-home message: CBS News, "Lessons Learned Limited Chile Quake Damage," posted 28 February 2010 to cbsnews.com.

100. EERI, Learning from Earthquakes (1.7), "The M_w 7.1 Darfield (Canterbury), New Zealand Earthquake"; EERI, "The 2010 Canterbury and 2011 Christchurch New Zealand Earthquakes"; USGS Significant Earthquake Archive (1.7), "Magnitude 6.1—South Island of New Zealand"; USGS Earthquake Summary Posters of South Island. Reaction of highway engineers: I talked with a variety of professionals at the 2012 EERI Annual Meeting and National Earthquake Conference, at which the New Zealand and Tohoku earthquakes—along with New Madrid seismicity—were main topics.

101. EERI, "The 2010 Canterbury and 2011 Christchurch New Zealand Earthquakes" (1.7); USGS, Earthquake Summary, "Magnitude 9.0—Near the East Coast of Honshu, Japan," and Poster of the Great Tohoku Earthquake (northeast of Honshu, Japan) of March 11, 2011—Magnitude 9.0, Earthquake Hazards Program website (7.5.7). Prior, unheeded warnings from paleoseismologists: K. Minoura, F. Inmamura, D. Sugawara, Y. Kono, and T. Iwashita, "The 869 Jōgan Tsunami Deposit," *Journal of Natural Disaster Science* 23, no. 2 (2001): 83–88.

102. See 7.9. EERI, Learning from Earthquakes (1.7), "The M_w 5.8 Virginia Earthquake of August 23, 2011"; USGS, Historic World Earthquakes website (1.7):

"Magnitude 7.0—Haiti Region," updated 12 May 2012 and USGS, "Poster of the Haiti Earthquake of 12 January 2010—Magnitude 7.0" (1.7).

103. Ten states: USGS, "Poster of the 2010–2011 Arkansas Earthquake Swarm" (7.9).

104. Chris Bury, "Are Arkansas' Natural Gas Injection Wells Causing Earthquakes?" *ABC Nightline*, posted on abcnews.go.com, 21 April 2011.

105. "If we get an earthquake in Greenbrier": conversation with Scott Ausbrooks, 3 July 2012; and citations listed in 7.12.

106. See 7.12.

107. USGS, "Science Features: Is the Recent Increase in Felt Earthquakes in the Central U.S. Natural or Manmade?" (7.12).

108. Johnston and Schweig, "The Enigma of the New Madrid Earthquakes of 1811–1812," 376 (1.1). "Extracting Facts from Folklore": Odum et al., "Near-Surface Structural Model for Deformation" (7.5.3). Skepticism about disruption to Mississippi: Stein, *Disaster Deferred*, 61 (7.11).

109. "Since the settlement": introductory note to *An Account of the Great Earthquakes* (1.2); "The Earthquake Was Very Distinctly Felt," *Argus of Western America* (Frankfort, KY), 1812, CERI Compendium (1.2). Insistence on earthquake history: Parker, *Missouri as It Is in 1867*, 24. "The fact is": Stoddard, *Sketches, Historical and Descriptive*, 240–41 (both at 1.17). See Schroeder, *Opening of the Ozarks*, 54 (1.14).

110. Lyell, *A Second Visit to the United States*, 2:180 (5.16); Featherstonhaugh, *Excursion through the Slave States*, 326 (1.2). Fuller, *The New Madrid Earthquake*, 11–12 (1.1). Jackson, *Earthquakes and Earthquake History of Arkansas*, 38 (likely from Fuller) (7.3).

111. M'Murtrie, *Sketches of Louisville*, 25 (1.2).

112. Piccardi, "Paleoseismic Evidence of Legendary Earthquakes," sec. 3.1 (4.10); Burnham, *Great Comets*, 213 (1.27). Also: Coen, *Earthquake Observers*, 209–210 (1.6).

113. Hough and Bilham, *After the Earth Quakes*, 37 (3.13); Tuttle, personal communication, January 2011; Van Arsdale, "Hazard in the Heartland," 18 (7.5.1). Jay K. Johnson, "From Chiefdom to Tribe in Northeast Mississippi: The Soto Expedition as a Window on a Culture in Transition," in Galloway, ed., *The Hernando De Soto Expedition* (2.11). Claudine Payne, "Middle-Period Mississippian in the St. Francis Basin" (manuscript, 2009). At the same time, archeologists are generally skeptical that one-time disasters cause fundamental and long-term cultural shifts. Claudine Payne, personal communication, January 2009.

114. Nuttli, "The Mississippi Valley Earthquakes of 1811 and 1812," 242 (7.3.1); Fuller, *The New Madrid Earthquake*, 18–36, 40–44 (1.1); Hough, "Scientific Overview and Historical Context," 531 (7.5.1); Drake, *Natural and Statistical View*, 233 (4.9.2).

115. Kochtitzky, *The Story of a Busy Life*, 55 (6.9).

116. "Seismic Science: Is Number of Earthquakes on the Rise?" (7.11). Lights, sounds, etc.: Fuller, *The New Madrid Earthquake*, 46 (1.1). Also: 1.19, 1:20.

117. Latrobe, *The First Steamboat Voyage on the Western Waters*, 21, 26–27 (1.31). Porcher, "Influence of the Recent Earthquake Shocks," 653 (4.1); Davison, *The Founders of Seismology*, 226 (1.6).

118. Hough, *Predicting the Unpredictable*, 56–68, 82–83, 161–62 (1.6). Sarah Jansen, an experienced animal handler as well as a historian of science, told me in March 2007 that individual cats and dogs have specific ways of registering electrical disturbances like weather, and certainly anecdotal conversation with many cat and dog owners bears out her perspective.

119. Hough, *Predicting the Unpredictable*, chap. 6 (1.6); see citations to 7.21.

120. See 7.17. Typical seismologist humor: Stein, *Disaster Deferred*, 106, 27, 83 (7.11).

121. See 7.18.

122. See 7.19.

123. On earthquakes: Hough, *Earthshaking Science*, 171 (1.6).

124. Discussions with CERI seismologists, 2004. See also 1.19.

125. "Earthquake Prediction by Tiempe," http://earthquakepredictionbytiempe .blogspot.com/, accessed 13 June 2012 and 27 February 2007 (then a Yahoo Group).

126. See 7.20, as well as 4.4, 4.8, 4.7.

127. USGS, "Did You Feel It?": Kathleen Tierney, "Societal Impacts," Lessons from Recent Earthquakes," 2012 EERI Annual Meeting and National Earthquake Conference; EERI Learning from Earthquakes (1.7), "The M_w 5.8 Virginia Earthquake of August 23, 2011," 1–2.

128. Tuttle et al., "The Earthquake Potential" (1.1); also Tuttle et al., "Use of Archaeology to Date Liquefaction Features" (7.5.5).

129. Marion Haynes, personal communication, January 2011 and January 2005.

130. Talks with Marion Haynes.

131. Conversations with Tish Tuttle, 2011, and Claudine Payne, 2011. Iron staining of old sand blows: Tuttle, Al-Shukri, and Mahdi, "Very Large Earthquakes," 762 (7.5.5).

132. Many geologists would now agree with Seth Stein that multiple causal mechanisms likely combine to create the intracontinental motion in the middle Mississippi Valley: Stein, *Disaster Deferred*, 156 (7.11).

133. Farley, "Scientists Seek Local Help" (7.11).

134. See 7.22.

135. Preston and Healy, "Small Earthquake Hits near Chicago"; Mackey, "Latest Updates on Chile's Earthquake" (both in 7.11).

136. Krupa, "Scientists Seek Clues to New Madrid Fault" (7.11).

137. See 7.23. Dowd, "Conferees Discuss Ways to Get Disaster Area Back to 'New Normal'" (7.11). Tony Fitzpatrick, "Seminar to Address Ways to Lessen Earthquake Damage," news release, Washington University in St. Louis, 11 Aug 2008; "Earthquakes: Disaster Can Strike Anywhere," *Amica Today Magazine*, Fall 2002. Heather Moyer, "States Push Quake Awareness," Disaster News Network, 30 Jan 2007.

138. Patterson, "Guest Column: Funding to Address Seismic Vulnerability"; OurAmazingPlanet Staff, "Doomsday Midwest Quake Predictions Overblown, Scientists Say"; Bailey and Dorman, "Guest Column: For Safer Schools—from Earthquakes" (all in 7.11).

139. Hough, "Large 19th Century Earthquakes," 352 (7.7); Chitwood, "Study Risk of Quake" (7.11).

Conclusion

1. Bradbury, *Travels in the Interior of America*, 245 (1.2).
2. I am grateful to Mike Finke and Elizabeth Oyler for this expedition and for gracious friendship and many invigorating conversations.
3. Marion Haynes, personal communication, January 2011.
4. Terrence Dickinson, *Nightwatch: A Practical Guide to Viewing the Universe*, rev. ed. (Buffalo, NY: Firefly Books, 1998), 47.
5. Tuttle, personal communication, 13 January 2011; Haynes, personal communication, January 2011; on land-leveling technology in the Delta: Bolsterli, *During Wind and Rain*, 119 (6.11). Also: 6.13.
6. Melissa Block, "In China, Quake Tourism Becoming Big Business," *All Things Considered* on National Public Radio, 6 May 2009, heard over WBUR-FM Boston; "[China] Clones Castrated Pig Who Survived 2008 Earthquake," BBC News Asia-Pacific, 18 September 2011, BBC online.
7. See 1.28.
8. Rudwick, *Bursting the Limits of Time* (5.11).
9. Montgomery letter to Barton (1.2).
10. Indeed, even what *historical* means. When I phoned Arch Johnston and asked about how he came to use historical accounts, he paused and clarified, "Now in terms of historical accounts, do you mean the paleoearthquake record?" I had to pause in turn to register his query, and to acknowledge my own assumptions—in phrasing the question, I had assumed a distinction between documents intentionally written down and the indirect evidence of much longer-ago people whose lives we know about only through a form of post hoc eavesdropping that they likely could not have anticipated. Johnston, phone interview, 10 August 2009.
11. Martitia Tuttle, personal communication on trenching expedition, October 2003.

BIBLIOGRAPHIC ESSAYS

Scripture quotations are from *Holy Bible*, New Living Translation (Wheaton, IL: Tyndale House, 1996); or *New Revised Standard Version Bible* (N.p.: National Council of the Churches, 1989).

Introduction: Earthquake Cracks

I.1 David Crockett: *David Crockett: A Narrative of the Life of David Crockett, by Himself [1834]* (Lincoln: University of Nebraska Press, 1987), 185–92; Howe, *What Hath God Wrought*, 663 (1.8). TV series: "Davy Crockett (TV Miniseries)," Wikipedia, http://en.wikipedia.org/wiki/Davy_Crockett_ (TV_miniseries) (last accessed 20 June 2012).

Chapter One: A Great Commotion

1.1 Overviews of New Madrid: Jay Feldman, *When the Mississippi Ran Backwards: Empire, Intrigue, Murder, and the New Madrid Earthquakes* (New York: Free Press, 2005); Myron L. Fuller, *The New Madrid Earthquake*, United States Geological Survey Bulletin 494 (Washington, DC: Government Printing Office, 1912), available through http://pubs.usgs.gov/bul/0494/report.pdf; Myron Leslie Fuller, "Our Greatest Earthquakes," *Popular Science Monthly* 69 (1906): 76–86, Google Books; Arch C. Johnston and Eugene S. Schweig, "The Enigma of the New Madrid Earthquakes of 1811–1812," *Annual Review of Earth and Planetary Sciences* 24 (1996): 339–84; James Lal Penick, Jr., *The New Madrid Earthquakes*, rev. ed. (Columbia: University of Missouri Press, 1981); "Report of the Independent Expert Panel on New Madrid Seismic Zone" (7.5.1); Nathaniel Southgate Shaler, "Earthquakes of the Western United States," *Atlantic Monthly* 24, no. 145 (1869): 549–60; Martitia P. Tuttle, Eugene S. Schweig, John D. Sims, Robert H. Lafferty, III, Lorraine W. Wolf, and Marion L. Haynes, "The Earthquake Potential of the New Madrid Seismic Zone," *Bulletin of the Seismological Society of America* 92, no. 6 (2002): 2080–89; Roy Van Arsdale, "Seismic Hazards of the Upper Mississippi Embayment," Waterways Experiment Station, US Army Corps of Engineers: Contract Report GL-9-1, 1998.

1.2 New Madrid Accounts: The accounts make clear both the felt extent of these quakes and the lack of clarity about precisely what happened when. We talk about three, perhaps four, major quakes—but that number had not yet been consolidated down even by 1814, when Samuel Mitchill presented his material at the Literary and Philosophical Society of New-York. His listing of the major shocks differs from his contemporaries' and from the consensus of modern seismology—giving us a sense for how unobvious even main facts about these quakes prove to be. *An Account of the Great Earthquakes, in the Western States, Particularly on the Mississippi River* . . . (New-buryport: Herald Office, 1812); Stephen F. Austin, "Diary, 17–19 May 1812 (?)," in *Annual Report of the American Historical Association*, ed. Eugene C. Barker (Washington, DC: Government Printing Office, 1924); Lewis C. Beck, *A Gazetteer of the States of Illinois and Missouri* . . . (Albany, NY: Charles R. and George Webster, 1823); John Bradbury, *Travels in the Interior of America: In the Years 1809, 1810, and 1811*, ed. Reuben Gold Thwaites, vol. 5, *Early Western Travels, 1748–1846* (Cleveland: A. H. Clark, 1904); Garland C. Broadhead, "The New Madrid Earthquake," *American Geologist* 30 (1902): 76–87; Eliza Bryan, letter from New Madrid, 22 March 1816, in Lorenzo Dow, *History of Cosmopolite: Or, the Four Volumes of Lorenzo Dow's Journal Concentrated in One*. . . . (Washington, Ohio: Joshua Martin, 1848): 344–46; Bringier, "Notices" (2.4); Center for Earthquake Research and Information New Madrid Compendium (CERI Compendium), http://www.ceri.memphis.edu/compendium/search/farfield.html, last accessed 29 May 2012; "Earthquake," *The Supporter* [Chillicothe, Ohio], 14 March 1812, Archive of Americana: America's Historical Newspapers; George William Featherstonhaugh, *Excursion through the Slave States*, vol. 1 (London: John Murray, 1844), Google Books; Edwin James, Stephen Harriman Long, and Thomas Say, "Account of an Expedition from Pittsburgh to the Rocky Mountains Performed in the Years 1819, 1820," in *Early Western Travels, 1748–1846: A Series of Annotated Reprints of Some of the Best and Rarest Contemporary Volumes of Travel*, vol. 15 (Cleveland: A. H. Clark, 1905); Lyell, *A Second Visit to the United States of North America* (5.16); H[enry] M'Murtrie, *Sketches of Louisville and Its Environs . . . To Which Is Added an Appendix, Containing an Accurate Account of the Earthquakes*, 1st ed. (Louisville: S. Penn, 1819); Samuel Latham Mitchill, "A Detailed Narrative of the Earthquakes Which Occurred on the 16th Day of December, 1811," *Transactions of the Literary and Philosophical Society of New-York* 1 (1815); Mitchill, "Description of the Volcano and Earthquake Which Happened in the Island of St. Vincents, on the 30th Day of April, 1812," *Transactions of the Literary and Philosophical Society of New-York* (1815): 1:315–23; Alexander Montgomery, letter, Frankfort, Kentucky, to Benjamin Smith Barton, [Philadelphia], 25 February 1812, American Philosophical Society; Edouard de Montulé, *Travels in America, 1816–1817*, trans. Edward D. Seeber (Bloomington: Indiana University Press, 1950); William Allen Pusey, "The New Madrid Earthquake: An Unpublished Contemporaneous Account," *Science* 71, no. 1837 (1930): 285–86; Winthrop Sargent, "Account of the Several Shocks of an Earthquake in the Southern and Western Parts of the United States," *Memoirs of the American Academy of Arts and Sciences* 3, ser. 1 (1815): 350–60; Col. John Shaw, "Collections of the State Historical Society of Wisconsin," *Wisconsin Historical Collections*, vol. 2 (1903), ed. Lyman Copeland Draper (Wisconsin Historical Society): 197–232; "Personal Narrative of Col. John Shaw, of Marquette County, Wisconsin,"

Missouri Historical Review 6, no. 2 (1912): 91–92; Edward M. Shepard, *The New Madrid Earthquake*, Pamphlet Reprinted from the *Journal of Geology*, 13, no. 1 (Jan-Feb 1905), Pamphlet Collection, Missouri Historical Society, St. Louis, MO; Robert Smith, *An Account of the Earthquakes Which Occurred in the United States, North America, on the 16th of December, 1811, the 23d of January, and the 7th of February, 1812 . . .* (Philadelphia: Robert Smith, 1812); William Stevenson, "Autobiography of the Rev. William Stevenson, Originally in the *New Orleans Christian Advocate* 1858, vol. 8, no. 2, and Compiled by the Rev. Walter N. Vernon, Jr.," Arkansas United Methodist Archives, Bailey Library, Hendrix College.

1.3 William Leigh Pierce, Publications: Pierce, "New-York, Feb. 11, Earthquake," *Hampshire Federalist*, 20 February 1812, NewsBank/Readex: America's Historical Newspapers. Transcription also available through the CERI Compendium, with a few minor errors. Pierce, "Earthquakes," *Georgia Journal*, 25 March 1812, p. 3, cols. 1–3, CERI Compendium (1.2); *An Account of the Great Earthquakes, in the Western States, Particularly on the Mississippi River; December 16–23, 1811, Collected from Facts* (Newburyport: Herald Office 1812); Smith, *An Account*, 31–37 (1.2); Pierce, Esq., *The Year: A Poem, in Three Cantoes* (New York: David Longworth, 1813).

1.3.1 Background: Robert M. Weir, "Pierce, William Leigh," in *American National Biography Online*, February 2000 (1.8).

1.4 New Madrid Earthquake Poetry: Pierce, *The Year* (1.3); Henry Rowe Schoolcraft, "Transallegania: Or, the Groans of Missouri," in Rafferty, ed., *Rude Pursuits and Rugged Peaks*, 126–41 (1.17); M.A.W., "Reflections on the Occurrences of the Last Year, Concluding with the Last Earthquake at Caraccas," *The Halcyon Luminary, and Theological Repository*, no. 6 (1812), Google Books. On Schoolcraft's poem: Penick, *The New Madrid Earthquake*, 126–27 (1.1). See also 4.16.

1.5 New Madrid Fiction: Emily Crofford, *When the River Ran Backward* (Minneapolis: Carolrhoda Books, 2000); Douglas C. Jones, *This Savage Race: A Novel* (New York: Henry Holt, 1993).

1.6 History of Earthquake Science: Bruce A. Bolt, "The Development of Earthquake Seismology in the Western United States," in *Geologists and Ideas: A History of North American Geology: Centennial Special Volume I*, ed. Ellen T. Drake and William M. Jordan (Boulder, CO: Geological Society of America, 1985); Albert V. Carozzi, "Robert Hooke, Rudolf Erich Raspe, and the Concept of 'Earthquakes,'" *Isis* 61, no. 1 (1970): 85–91; Gregory K. Clancey, *Earthquake Nation: The Cultural Politics of Japanese Seismicity, 1868–1930* (Berkeley: University of California Press, 2006); Deborah R. Coen, *The Earthquake Observers: Disaster Science from Lisbon to Richter* (Chicago: University of Chicago Press, 2013); Coen, ed., *Witness to Disaster: Earthquakes and Expertise in Comparative Perspective*, special issue of *Science in Context* 25, no. 1 (2012); Deborah R. Coen, "Introduction: Witness to Disaster: Comparative Histories of Earthquake Science and Response," in Coen, ed., *Witness to Disaster*: 1–15; Charles Davison, *The Founders of Seismology* (1927; repr. New York: Arno Press, 1978); Dennis R. Dean, "Benjamin Franklin and Earthquakes," *Annals of Science* 40 (1989): 481–94; Carl-Henry Geschwind, *California Earthquakes: Science, Risk, and the Politics of Hazard Mitigation*

(Baltimore: Johns Hopkins University Press, 2001); Geschwind, "Embracing Science
and Research: Early Twentieth-Century Jesuits and Seismology in the United States"
(6.2); Nicholas Hunter Heck, *Earthquakes*, 1936 facsimile ed. (1936; New York: Hafner
Publishing Co., 1965); Susan Elizabeth Hough and Roger G. Bilham, *After the Earth
Quakes: Elastic Rebound on an Urban Planet* (Oxford: Oxford University Press, 2006);
Susan Elizabeth Hough, *Earthshaking Science: What We Know (and Don't Know) about
Earthquakes* (Princeton, NJ: Princeton University Press, 2002); Susan Elizabeth Hough,
Predicting the Unpredictable: The Tumultuous Science of Earthquake Prediction (Princeton:
Princeton University Press, 2010); Susan Elizabeth Hough, *Richter's Scale: Measure of
an Earthquake, Measure of a Man* (Princeton, NJ: Princeton University Press, 2007);
Benjamin F. Howell, *An Introduction to Seismological Research: History and Development*
(Cambridge: Cambridge University Press, 1990); H. E. LeGrand, *Drifting Continents
and Shifting Theories: The Modern Revolution in Geology and Scientific Change* (Cam-
bridge:: Cambridge University Press, 1988); David Oldroyd, Filomena Amodor, Jan
Kozák, Ana Carneiro, and Manuel Pinto, "The Study of Earthquakes in the Hundred
Years Following the Lisbon Earthquake of 1755," *Earth Sciences History* 26, no. 2 (2007):
321–70; David R. Oldroyd, *Thinking about the Earth: A History of Ideas in Geology* (Cam-
bridge, MA: Harvard University Press, 1996); Naomi Oreskes, ed., with Homer Le
Grand, *Plate Tectonics: An Insider's History of the Modern Theory of the Earth* (Boulder,
CO: Westview Press, 2001); Udías, "Jesuits' Studies of Earthquakes and Seismological
Stations" (6.2).

1.7 **History of Earthquakes:** Earthquake Engineering Research Institute (EERI),
"Learning from Earthquakes: EERI Special Earthquake Reports," http://www.eeri
.org/site/projects/learning-from-earthquakes#; EERI, "The 27 February 2010 Central
South Chile Earthquake: Emerging Research Needs and Opportunities: Workshop Re-
port," Earthquake Engineering Research Institute, November 2010; EERI, "The 2010
Canterbury and 2011 Christchurch New Zealand Earthquakes and the 2011 Tohoku
Japan Earthquake: Emerging Research Needs and Opportunities: Workshop Report,"
November 2012; Stuart McCook, "Nature, God, and Nation in Revolutionary Venezu-
ela: The Holy Thursday Earthquake of 1812," in *Aftershocks: Earthquakes and Popular
Politics in Latin America*, ed. Jürgen Buchenau and Lyman L. Johnson (Albequerque:
University of New Mexico Press, 2010); USGS website on "Earthquake Summary Post-
ers," http://earthquake.usgs.gov/earthquakes/eqarchives/poster/; "Historic United
States Earthquakes," http://earthquake.usgs.gov/earthquakes/states/historical_state
.php; "Historic World Earthquakes," http://earthquake.usgs.gov/earthquakes/world
/historical.php; and "Significant Earthquake Archive," http://earthquake.usgs.gov
/earthquakes/eqinthenews/, accessed 2 July 2012; Simon Winchester, *A Crack in the
Edge of the World: America and the Great California Earthquake of 1906* (New York: Harper-
Collins, 2005).

1.7.1 **History of Earthquakes—New England:** William D. Andrews, "The Literature of the 1727 New England Earthquake," *Early American Literature* 7, no. 3 (1973):
281–94; William T. Brigham, "Catalogue of New England Earthquakes," *Boston Society
for Natural History Memoirs* 2 (1871): 1–28; John E. Ebel, "The Cape Ann, Massachusetts

Earthquake of 1755: A 250th Anniversary Perspective," *Seismological Research Letters* 77, no. 1 (2006): 74–86; John E. Ebel, "A Comparison of the 1981, 1982, 1986 and 1987–1988 Microearthquake Swarms at Moodus, Connecticut," *Seismological Research Letters* 60, no. 4 (1989): 177–84; John E. Ebel, "A Reanalysis of the 1727 Earthquake at Newbury, Massachusetts," *Seismological Research Letters* 71, no. 3 (2000): 364–74; John E. Ebel, "The Seventeenth Century Seismicity of Northeastern North America," *Seismological Research Letters* 67, no. 3 (1996): 51–68; Jeremy Miller, "Boston's Earthquake Problem: It's Not the Likelihood of a Major Earthquake That Makes Experts Tremble—Though It's Worse Than You Think; It's the Damage One Would Do—Because It's Much Worse Than You Think," *Boston Globe Magazine* (2006): 26–29. Also: 7.5.10, 7.9.

1.8 US History: *American National Biography Online*, Oxford University Press under the auspices of the American Council of Learned Societies: 2000–2008; Daniel J. Boorstin, *The Americans: The National Experience* (New York: Vintage Books/Random House, 1965); Daniel Walker Howe, *What Hath God Wrought: The Transformation of America, 1815–1848*, Oxford History of the United States (Oxford: Oxford University Press, 2007); Jane Kamensky, *The Exchange Artist: A Tale of High-Flying Speculation and America's First Banking Collapse* (New York: Viking, 2008); D. W. Meinig, *The Shaping of America: A Geographical Perspective on 500 Years of History*, vol. 2, *Continental America 1800–1867* (New Haven: Yale University Press, 1993); James M. McPherson, *Battle Cry of Freedom: The Civil War Era*, The Oxford History of the United States (Oxford: Oxford University Press, 1988); Michael O'Brien, *Conjectures of Order: Intellectual Life and the American South, 1810–1860*, 2 vols. (Chapel Hill: University of North Carolina Press, 2004); W. J. Rorabaugh, *The Alcoholic Republic: An American Tradition* (Oxford: Oxford University Press, 1981); Gordon S. Wood, *Empire of Liberty: A History of the Early Republic, 1789–1815*, The Oxford History of the United States (Oxford: Oxford University Press, 2009).

1.9 Newspapers and Transmission of Ideas: The liberal borrowing and republishing of interesting material such as William Leigh Pierce's earthquake accounts is a far different and more public model of idea transmission than what Martin Rudwick describes in the world of scholarly European intellectual exchange among savants, in which ideas gradually took form across a "gradient of privacy," from close colleagues out to publication—strong ideas got stronger across this gradient, whereas weaker ones often quietly dissipated (Rudwick, *Bursting the Limits of Time*, 44–46 (5.11). More private exchange took place between figures in the United States, as well, but the relative diversity of correspondents, the episodic nature of exchange and discussion, and the relative paucity of recognized figures suggest a far greater intellectual role for discussion in public media and public discursive space. Newspapers' important role in early American science: Pandora, "Popular Science in National and Transnational Perspective," 354–55 (5.3). Coen, *Earthquake Observers*, chap. 3 (1.6); O'Brien, *Conjectures of Order*, vol. 1, chap. 13 (1.8). Newspapers in the early US: Howe, *What Hath God Wrought*, 226–27, 232; Wood, *Empire of Liberty*, 478 (both at 1.8). See also 5.4, 5.21.

1.10 American Women as Observers and Authors: A few New Madrid accounts are explicitly noted as written by women—for instance, Smith, *An Account of the Earthquakes*, 50 (1.2). In analyzing scientific work of the second half of the nineteenth century, Daniel Goldstein noted the difficulty of estimating how many women were directly involved in scientific collecting, observation, and discussion, because many of those who engaged in such work conducted their formal correspondence through male relatives: Goldstein, "Yours for Science" (5.3), esp. 581. Philosophical discussions and demonstration did engage and involve both women and men, providing a socially acceptable meeting place and outlet for scientifically curious women: Delbourgo, *A Most Amazing Scene of Wonders*, 110–15, 4.18. Science education involving women and girls: Sally Gregory Kohlstedt, "Parlors, Primers, and Public Schooling: Education for Science in Nineteenth-Century America," *Isis* 81 (1990): 425–45; Katherine Pandora, "The Children's Republic of Science in the Antebellum Literature of Samuel Griswold Goodrich and Jacob Abbott," in *Osiris: National Identity; The Role of Science and Technology*, ed. Carol E. Harrison and Ann Johnson, vol. 24 (2009): 75–98, esp. 90; Kim Tolley, "Science for Ladies, Classics for Gentlemen: A Comparative Analysis of Scientific Subjects in the Curricula of Boys' and Girls' Secondary Schools in the United States, 1794–1850," *History of Education Quarterly* 36, no. 2 (1996): 129–53.

1.11 **History of the Middle Mississippi Valley:** S. Charles Bolton, *Territorial Ambition: Land and Society in Arkansas, 1800–1840* (Fayetteville: University of Arkansas Press, 1993); Carter, ed., *The Territorial Papers of the United States* (2.16); Mark M. Chambers, "River of Gray Gold: Cultural and Material Changes in the Land of Ores, Country of Minerals, 1719–1839" (PhD diss.: Department of History, Stony Brook University, 2012); Carl J. Ekberg, *François Vallé and His World: Upper Louisiana before Lewis and Clark* (Columbia: University of Missouri Press, 2002); Carl J. Ekberg, *French Roots in the Illinois Country: The Mississippi Frontier in Colonial Times* (Urbana: University of Illinois Press, 1998); *Encyclopedia of Arkansas History & Culture*, Butler Center for Arkansas Studies, Central Arkansas Library System, http://www.encyclopediaofarkansas .net/encyclopedia, accessed June 2012; William E. Foley, *The Genesis of Missouri: From Wilderness Outpost to Statehood* (Columbia: University of Missouri Press, 1989); Foley, *A History of Missouri*, vol. 1, *1673 to 1820* (Columbia: University of Missouri Press, 1971); *Missouri: A Guide to the "Show Me" State*, American Guide Series (1941; repr. New York: Hastings House, 1954); Missouri's Judicial Records (MJR), collection within the online archive Missouri's Digital Heritage, accessible via http://www.sos.mo.gov/archives /mojudicial/; Morrow, "New Madrid and Its Hinterland" (2.1); Tanis C. Thorne, *The Many Hands of My Relations: French and Indians on the Lower Missouri* (Columbia: University of Missouri Press, 1996); Julie Winch, *The Clamorgans: One Family's History of Race in America* (New York: Hill & Wang, 2011); Patrick G. Williams, S. Charles Bolton, and Jeanne M. Whayne, eds., *A Whole Country in Commotion: The Louisiana Purchase and the American Southwest* (Fayetteville: University of Arkansas Press, 2005), esp. S. Charles Bolton, "Jeffersonian Indian Removal and the Emergence of Arkansas Territory"; Charles F. Robinson, III, "The Louisiana Purchase and the Black Experience"; and Jeannie M. Whayne, "A Shifting Middle Ground: Arkansas's Frontier Exchange Economy and the Louisiana Purchase." Also: 6.11.

1.12 **Middle Mississippi Valley and Regional History:** This project comprehends the middle Mississippi Valley not as constituent states but as a geographic region. Shining parallel examples: Stephen Aron, *American Confluence: The Missouri Frontier* *from Borderland to Border State*, A History of the Trans-Appalachian Frontier, ed. Walter Nugent and Malcolm Rohrbough (Bloomington: Indiana University Press, 2006); Kathleen DuVal, *The Native Ground: Indians and Colonists in the Heart of the Continent*, ed. Daniel K. Richter and Kathleen M. Brown, Early American Studies (Philadelphia: University of Pennsylvania Press, 2006); Peter J. Kastor, *The Nation's Crucible: The Louisiana Purchase and the Creation of America* (New Haven, CT: Yale University Press, 2004); Daniel H. Usner Jr., *Indians, Settlers, and Slaves in a Frontier Exchange Economy: The Lower Mississippi Valley before 1783* (Chapel Hill: Institute of Early American History and Culture/University of North Carolina Press, 1992); West, *The Contested Plains* (1.13); Richard White, *The Middle Ground: Indians, Empires, and Republics in the Great Lakes Region, 1650–1815*, Cambridge Studies in North American Indian History (Cambridge: Cambridge University Press, 1991). Also see 2.1 and 2.8.

1.13 US Environmental History: John M. Barry, *Rising Tide: The Great Mississippi Flood of 1927 and How It Changed America* (New York: Simon & Schuster, 1997); William Cronon, *Nature's Metropolis: Chicago and the Great West* (London: W. W. Norton, 1991); William Cronon, ed. *Uncommon Ground: Rethinking the Human Place in Nature* (New York: W.W. Norton, 1996 [1995]); Pete Daniel, *Deep'n as It Come: The 1927 Mississippi River Flood* (New York: Oxford University Press, 1977); Andrew C. Isenberg, *The Destruction of the Bison: An Environmental History, 1750–1920* (Cambridge: Cambridge University Press, 2000); Karl Jacoby, *Crimes against Nature: Squatters, Poachers, Thieves, and the Hidden History of American Conservation* (Berkeley: University of California Press, 2001); Ari Kelman, "Boundary Issues: Clarifying New Orleans's Murky Edges," *Journal of American History* 94 (2007): 695–703; Linda Nash, "The Changing Experience of Nature: Historical Encounters with a Northwest River," *Journal of American History* 86, no. 4 (2000): 1600–1629; Nash, *Inescapable Ecologies: A History of Environment, Disease, and Knowledge* (Berkeley: University of California Press, 2006); Gregory H. Nobles, "Straight Lines and Stability: Mapping the Political Order of the Anglo-American Frontier," *Journal of American History* 80, no. 1 (1993): 9–35; John Opie, *Nature's Nation: An Environmental History of the United States* (Fort Worth: Harcourt Brace College Publishers, 1998); Donald J. Pisani, "Beyond the Hundredth Meridian: Nationalizing the History of Water in the United States," *Environmental History* 5, no. 4 (2000): 466–82; Jennifer Price, *Flight Maps: Adventures with Nature in Modern America* (New York: Basic Books, 1999); Ted Steinberg, *Down to Earth: Nature's Role in American History*, 2nd ed. (New York: Oxford University Press, 2009); Mart A. Stewart, "Rice, Water, and Power: Landscapes of Domination and Resistance in the Lowcountry, 1790–1880," *Environmental History Review* 15, no. 3 (1991): 47–64; Stewart, *"What Nature Suffers to Groe": Life, Labor, and Landscape on the Georgia Coast, 1680–1920*, Wormsloe Foundation Publications 19 (Athens: University of Georgia Press, 1996); Conevery Bolton Valencius, *The Health of the Country: How American Settlers Understood Themselves and Their Land* (New York: Basic Books, 2002); Gwendolyn Verhoff, "The Intractable Atom: The Challenge of Radiation and Radioactive Waste in American Life, 1942 to Present" (PhD thesis, Washington University in St. Louis, 2008); Louis S. Warren, *The Hunter's Game: Poachers and Conservationists in Twentieth-Century America*, Yale Historical Publications (New Haven: Yale University Press, 1997); Elliott West, *The Contested Plains: Indians,*

Goldseekers, and the Rush to Colorado (Lawrence: University Press of Kansas, 1998); Donald Worster, *Dust Bowl: The Southern Plains in the 1930s* (Oxford: Oxford University Press, 1979). Also see 1.14 and 6.10.

1.14 **Environmental History of the Middle Mississippi Valley:** Virgil H. Holder, "Historical Geography of the Lower White River," *Arkansas Historical Quarterly* 27, no. 2 (1968): 130–45; Ogilvie, "The Development of the Southeast Missouri Lowlands" (6.9); Andrew Hurley, ed., *Common Fields: An Environmental History of St. Louis* (St. Louis: Missouri Historical Society Press, 1997), esp. Patricia Cleary, "Contested Terrain: Environmental Agendas and Settlement Choices in Colonial St. Louis"; and William R. Iseminger, "Culture and Environment in the American Bottom: The Rise and Fall of Cahokia Mounds"; Guy Lancaster, "Red River," updated May 2011, *Encyclopedia of Arkansas History & Culture* (1.11); Brian Aaron King, "The Delimitation and Demarcation of the State Boundary of Missouri" (MA thesis, University of Missouri–Columbia, 1995); Montroville W. Dickeson, M.D., and Andrew Brown, "On the Cypress Timber of Mississippi and Louisiana," *American Journal of Science and Arts* 5, ser. 2 (1848): 19–20; Walter A. Schroeder, *Opening the Ozarks: A Historical Geography of Missouri's Ste. Genevieve District, 1760–1830* (Columbia: University of Missouri Press, 2002); Valencius, *The Health of the Country* (1.13). Also: 2.5, 2.7, 2.19, 6.8, and 6.9.

1.15 **Americans Explore and Encounter:** Trey Berry, Pam Beasley, and Jeanne Clements, eds., *The Forgotten Expedition, 1804–1805: The Louisiana Purchase Journals of Dunbar and Hunter* (Baton Rouge: Louisiana State University Press, 2006); Dan L. Flores, "The Ecology of the Red River in 1806: Peter Custis and Early Southwestern Natural History," *Southwestern Historical Quarterly* 88 (1984–85): 1–42; Robert J. Malone, "Everyday Science, Surveying, and Politics in the Old Southeast: William Dunbar and the Influence of Place on Natural Philosophy" (PhD, Department of History, University of Florida, 1996); James Ronda, *Lewis and Clark among the Indians* (Lincoln: University of Nebraska Press, 1984); Dan Flores, "Jefferson's Grand Expedition and the Mystery of the Red River"; and Elliott West, "Lewis and Clark: Kidnappers"; in Williams, Bolton, and Whayne, eds., *A Whole Country in Commotion* (1.11).

1.16 **Slow Mail:** A description of the quakes written by a trader on 3 January 1812 from Natchez, Louisiana, to a merchant firm in Providence, Rhode Island, did not arrive until 11 February 1812. M. Terrell, letter, Natchez, Louisiana, to firm of Brown & Ives, Providence, RI, 3 January 1812, Brown family business papers, John Carter Brown Library, Brown University. Mail generally took about three weeks to travel from New Orleans to Washington, DC, though bad weather could delay it further: Howe, *What Hath God Wrought*, 63 (1.8); officials' frustration: Foley, *A History of Missouri*, 1:136 (1.11); on the ironies of slow communication at the beginning and the ending of the War of 1812, 16, 71; on radical decrease in travel time for news even by the early 1820s, 222–24. Piecing together acts of Congress from scraps: Whayne, "A Shifting Middle Ground," 73 (1.11). Judge unable to resign because resignation letter could not reach capital: John Coburt, Maysville, KY, to James Monroe, Secretary of State, 24 January 1813, in Carter, *Territorial Papers*, 14:625 (2.16). Endlessly frustrating to those with scientific interests: Greene, "Science and the Public in the Age of Jefferson," 206–7 (5.3).

1.17 **Early Observers of the Middle Mississippi Valley and Western Territories:** A. H. Abney, *Life and Adventures of L.D. Lafferty* . . . (New York: Garland Publishing, 1976); *Biographical and Historical Memoirs of Northeast Arkansas* . . . (Chicago: Goodspeed Publishing Co., 1889); [Jervis Cutler], *A Topographical Description of the State of Ohio, Indiana Territory, and Louisiana* . . . (Boston: Charles Williams, 1812); Gottfried Duden, *Report on a Journey to the Western States of North America and a Stay of Several Years Along the Missouri (During the Years 1824, '25, '26, and 1827)* (1829; repr. Columbia: The State Historical Society of Missouri and University of Missouri Press, 1980); Estwick Evans, *A Pedestrious Tour, of Four Thousand Miles: Through the Western States and Territories, During the Winter and Spring of 1818* (Concord, NH: Printed by Joseph C. Spear, 1819); Timothy Flint, *The History and Geography of the Mississippi Valley*, 2nd ed., vol. 1 (Cincinnati: E. H. Flint and L. R. Lincoln, 1832); Flint, *Recollections of the Last Ten Years in the Valley of the Mississippi*, ed. E. Hartley Grattan (1826; repr. New York: Alfred A. Knopf, 1932); Dan L. Flores, ed., *Journal of an Indian Trader: Anthony Glass and the Texas Trading Frontier, 1790–1810* (College Station: Texas A&M University Press, 1985); J. W. Foster, *The Mississippi Valley: Its Physical Geography* (Chicago: S. C. Griggs, 1869); "Geographical Notices: Physical Geography of the Report on the Mississippi River, by Humphreys and Abbot," *American Journal of Science and Arts* 35, ser. 2 (1863): 223–35; Lucille Griffith, ed., *Letters from Alabama, 1817–1822*, by Anne Newport Royall, Southern Historical Publications 14 (Tuscaloosa: University of Alabama Press, 1969); Duane Huddleston, "Some Indian Incidents along the White River, 1813–1822," *Independence County Chronicle* 15, no. 14 (1974): 36–46; "Jacques," "Notes of a Tourist through the Upper Missouri Region (from the *Missouri Saturday News*, 1838)" *Missouri Historical Society Bulletin* 22, no. 4 (1966): 393–409; Dorsey D. Jones, "An Original Letter from Pork Bayou, Arkansas, April 16, 1815," *Mississippi Valley Historical Review* 32, no. 1 (1945): 89–94; *The Journals of the Lewis and Clark Expedition*, University of Nebraska Press Center and University of Nebraska-Lincoln Libraries, University of Nevada, http://lewisandclarkjournals.unl.edu, accessed 12 June 2012; James F. Keefe and Lynn Morrow, eds., *A Connecticut Yankee in the Frontier Ozarks: The Writings of Theodore Pease Russell* (Columbia: University of Missouri Press, 1988); "Letters from Honorable L. F. Linn, of the Senate of the United States, and Honorable A. H. Sevier, Delegate in Congress from Arkansas . . ." (Washington, DC: Government Printing Office (printed for the 24th Cong., 1st sess.), 1836); Lyell, *A Second Visit* (5.16); Thomas Loraine McKenney and James Hall, *The Indian Tribes of North America: With Biographical Sketches and Anecdotes of the Principal Chiefs*, ed. Frederick Webb Hodge, vol. 3 (1837–44; Edinburgh: J. Grant, 1933); Edouard de Montulé, *Travels in America, 1816–1817*, trans. Edward D. Seeber (Bloomington: Indiana University Press, 1950); David Dale Owen, William Elderhorst, and Edward Travers Cox, *First Report of a Geological Reconnaissance of the Northern Counties of Arkansas: Made During the Years 1857 and 1858 [for the Arkansas Geological Survey]* (Little Rock: Johnson & Yerkes, state printers, 1858); Humphrey Marshall, *The History of Kentucky* (Frankfort [KY]: Henry Gore, 1812), 2 vols.; Nathan Howe Parker, *Missouri as It Is in 1867: An Illustrated Historical Gazetteer* . . . (Philadelphia: J.B. Lippincott & Co., 1867), Google Books; Milton D. Rafferty, ed., *Rude Pursuits and Rugged Peaks: Schoolcraft's Ozark Journal, 1818–1819* . . . (Fayetteville:

University of Arkansas Press, 1996); James Ross, *Life and Times of Elder Reuben Ross by His Son, James Ross* (Philadelphia: Grant, Faires & Rodgers, 1882); Edmund L. Starling, *History of Henderson County, Kentucky* (Henderson, KY, 1887); Amos Stoddard, *Sketches, Historical and Descriptive, of Louisiana* (Philadelphia: Mathew Carey, 1812); Mark Twain, *Life on the Mississippi* (1883; New York: Signet, 1980); C. F. Volney, *A View of the Soil and Climate of the United States of America: With Supplementary Remarks . . .* , trans. C. B. Brown (Philadelphia: J. Conrad & Co., 1804); Alphonso Wetmore, *Gazetteer of the State of Missouri . . .* (St. Louis: C. Keemle, 1837).

1.18 Audubon's Earthquake Account: Maria R. Audubon, ed. *Audubon and His Journals*, vol. 2 (New York: Charles Scribner's Sons, 1897): 234–37; Myron Leslie Fuller, "Audubon's Account of the New Madrid Earthquake," *Science* (1905): 748–49. Audubon remembered the shock as occurring on an afternoon in November. While he is apparently mistaken about the month, aspects of his account—including emphasis on unseasonal, strange, or foreboding weather—are consistent with how many people experienced the quakes.

1.19 Earthquake Sounds: Accounts from the New Madrid earthquakes consistently stress how loud the earthquakes were: Smith, *An Account of the Earthquakes*; Mitchill, "A Detailed Narrative of the Earthquakes" (both in 1.2). To experience, order "Earthquake Sounds," 21 recordings, available from the Seismological Society of America, http://www.seismosoc.org/publications/EQ-Sounds.php. Part of early seismological research: Fuller, *The New Madrid Earthquake*, 46 (1.1); Davison, *The Founders of Seismology*, 226 (1.6). Modern scientific understanding: "Earthquake Booms, Seneca Guns, and Other Sounds," USGS: Earthquake Topics for Education, updated 25 June 2012, http://earthquake.usgs.gov/learn/topics/booms.php. People who feel earthquake sounds are probably experiencing seismic waves that are just below the level of audibility—like the deepest notes on a powerful organ when heard from right near the pipes, or the deep calls of elephants' long-distance seismic communication, usually inaudible to humans, which some observant elephant keepers can feel as a tingling in the eardrums. Gregory Mone, "Earth Speaks in an Inaudible Voice," *Discover Magazine* (August 2007), posted 2 August 2007 to http://discovermagazine.com; Mark Schwartz, "Elephants Pick up Good Vibrations—Through Their Feet," news release, Stanford University, Stanford News Service, 5 March 2001, http://news.stanford.edu /pr/01/elephants37.html. Coen, *Earthquake Observers*, 18, 27–28, 222 (1.6).

1.20 History of Sound: Peter A. Coates, "The Strange Stillness of the Past: Toward an Environmental History of Sound and Noise," *Environmental History* 10 (October 2005); Sarah Keyes, "'Like a Roaring Lion': The Overland Trail as a Sonic Conquest," *Journal of American History* 96, no. 1 (2009): 19–43; Richard Cullen Rath, *How Early America Sounded* (Ithaca: Cornell University Press, 2003), earthquake noises, 26–28; Mark M. Smith, *Listening to Nineteenth-Century America* (Chapel Hill: University of North Carolina Press, 2001).

1.21 Earthquake Lights: John S. Derr, "Earthquake Lights: A Review of Observations and Present Theories," *Bulletin of the Seismological Society of America* 63, no. 6 (1973): 2177–87.

1.22 **Sand Blows and Liquefaction Features:** Overall: Sudhir K. Jain, William R. Lettis, C. V. R. Murty, and Jean-Pierre Bardet, "Bhuj, India Earthquake of January 26, 2001, Reconnaissance Report," *Earthquake Spectra* 18 (2002), suppl. A, chap. 7, "Liquefaction." Characteristic only of large earthquakes, magnitude 5 or greater: Martitia P. Tuttle, personal communication, 13 January 2011. New Madrid seismic zone: Appendix E of CEUS Seismic Source Characterization (7.5.5); Fuller, *The New Madrid Earthquake*, 79–83 (1.1); Tuttle, "Late Holocene Earthquakes, Use of Liquefaction Features in Paleoseismology" (7.5.5); in NMSZ, commonly a meter to a meter and half thick and ten to thirty meters in diameter: Johnston and Schweig, "The Enigma of the New Madrid Earthquakes of 1811–1812," 369 (1.1). A sand blow so large it is nicknamed "The Beach": Steltzer, "On Shaky Ground" (6.15). Size of sand blows: Tuttle, "The Use of Liquefaction Features in Paleoseismology," 366 (7.5.5); Roger T. Saucier, "Evidence for Episodic Sand-Blow Activity during the 1811–1812 New Madrid (Missouri) Earthquake Series," *Geology* 17 (February 1989): 103–6; similarly, Haydar J. Al-Shukri et al., "Spatial and Temporal Characteristics of Paleoseismic Features in the Southern Terminus of the New Madrid Seismic Zone in Eastern Arkansas," *Seismological Research Letters* 76, no. 4 (2005); Van Arsdale, "Seismic Hazards of the Upper Missouri Embayment," 27 (1.1). Tabletop liquefaction: Wendy Gerstel, "Earth Connections: Resources for Teaching Earth Science; Experiment 1, Shake, Rattle, and Liquefy," *Washington Geology* 27, nos. 2–4 (1999), Washington State Department of Natural Resources, available through http://www.dnr.wa.gov/Publications/ger_washington_geology_1999_v27 _no2-3–4.pdf. See also 7.5.5.

1.23 **Earthquake Cracks:** Fuller, *The New Madrid Earthquake*, 47–58 (1.1); J. E. Ebel, "The Cape Ann, Massachusetts, Earthquake of 1755: A 250th Anniversary Perspective," *Seismological Research Letters* 77, no. 1 (2006): 74–86, esp. 76.

1.24 **Earthquake Lakes:** Bringier, "Notices," 21–22 (2.4); Featherstonhaugh, *Excursion through the Slave States*, 78 (1.2); Flint, *Recollections of the Last Ten Years*, 215 (1.7); Bennett, "Big Lake National Wildlife Refuge" (2.5); Ogilvie, "The Development of the Southeast Missouri Lowlands," 417n52 (6.9). Van Arsdale, "Seismic Hazards of the Upper Mississippi Embayment," 19–20 (1.1).

1.25 **Reelfoot Lake:** Foster, *The Mississippi Valley*, 22 (1.17). Widely noted in early geologies of region: David Dale Owens, *Report of the Geological Survey in Kentucky* (Frankfort, KY: A. G. Hodges, State Printer, 1856), illustrated plate of the lake; Seismic uplift: Fuller, *The New Madrid Earthquake*, 62–64 (1.1); relation to Reelfoot Lake: Van Arsdale, "Earthquake Signals in Tree-Ring Data from the New Madrid Seismic Zone," 516 (2.21); Roy B. Van Arsdale, Jodi Purser, William Stephenson, and Jack Odum, "Faulting along the Southern Margin of Reelfoot Lake, Tennessee," *Bulletin of the Seismological Society of America* 88 (1998): 322–23. Seemingly unimpressive, the uplift causes in places 11 meters of surface relief—substantial topographic change to an otherwise flat floodplain. Guccione, "Late Pleistocene and Holocene Paleoseismology" (2.21). Arkansas friends tell me that because of the dead cypress trunks still spiking up toward the surface, fishers in Reelfoot can't use fiberglass jon boats that would be ripped up by the stumps. They either use old-fashioned V-shaped pirogues or finish the bottom of their boats in graphite to slide more easily over the stumps. Reelfoot boating techniques: Acorn Point Lodge, FAQ, http://www.acornpointlodge.com/faq .htm#f5; "Old Sparky's Forum," Reelfoot Lake Boat thread, http://www.southernpaddler .com/phpBB3/viewtopic.php?f=3&t=4007, both accessed 22 June 2012. Reelfoot is a shallow "flooded forest" requiring slow boating speeds but offering great views of bald eagles as well as excellent fishing: Tennessee State Parks Department, Tennessee Department of Environment and Conservation, "Reelfoot Lake," http://www.tn.gov /environment/parks/ReelfootLake/activities/, accessed 22 June 2012.

1.26 Wonder and Analysis: Richard Holmes, *The Age of Wonder: How the Romantic Generation Discovered the Beauty and the Terror of Science* (New York: Vintage Books/ Random House, 2010).

1.27 Comet of 1811: The comet was observed by naked eye March 1811–January 1812 (and last glimpsed by an observer with a telescope in Russia in August 1812), giving it the longest period in which any comet had been kept in view until Comet Hale-Bopp in 1997. A comet had blazed after the assassination of Julius Caesar—as later evoked by William Shakespeare in his play (Calpurnia warns of the ill omen in act 2). Many Americans who saw the comet in 1811 could well have been familiar with both the classical and the Shakespearian references. Not all meanings were negative: one set of explorers in the northwest reported being cordially received because the comet was a positive omen. David A. Seargent, *The Greatest Comets in History: Broom Stars and Celestial Scimitars*, Astronomer's Universe (New York: Springer, 2009), 128, 130. As late as 1833, comet and meteor showers were taken as portents—memorialized as "the years the stars fell." Johnson, *The Frontier Camp Meeting*, 105–6 (4.13). The 1811 comet as a promising omen of individual destiny: Leo Tolstoy, *War and Peace*, trans. Constance Garnett (New York: T. Y. Crowell Co., 1898), 3:100–101; 293–94; 5:142–43; Google Books. As an ominous harbinger: Robert Burnham, *Great Comets* (New York: Cambridge University Press, 2000), 166. Latrobe, *The Rambler in North America*, 90 (1.31); Pusey, "The New Madrid Earthquake," 286 (1.2); Ross, *Life and Times of Elder Reuben Ross*, 201 (1.17). Comets and epidemics: Daniel Drake, "Travelling Editorials: Voyage Up the Mississippi," *Western Journal of Medicine and Surgery* 8, no. 2 (1843). Symbolism: Penick, *The New Madrid Earthquakes*, 11–14, 118–20 (1.1); Sugden, "Early Pan-Indianism: Tecumseh's Tour of the Indian Country," 289 (3.11). Scientific interest: Greene, *American Science in the Age of Jefferson*, 152 (5.3).

1.28 New Madrid Death Toll: Travelers noted evidence of death on river: "Earthquake," *The Supporter* (1.2). Charles Lyell: "the loss of life was considerable"; he was also taken to see what locals called "the sink-hole where the negro was drowned." Lyell, *A Second Visit*, 174, 176 (5.16). Drawing on ample first-person accounts, Myron Fuller made clear a small (by earthquake standards) but significant death toll: Fuller, *The New Madrid Earthquake*, 42–43; also Penick, *The New Madrid Earthquakes*, 34, 109–11; and Feldman, *When the River Ran Backward* (all 1.1). Yet: "As far as we know, no one has ever died in an earthquake in the New Madrid zone": Stein and Tomasello, "When Safety Costs Too Much"; softer assertion that "we don't know": Stein, *Disaster Deferred*, 18, 58 (both in 7.11).

1.29 New Madrid Relief Act and New Madrid Claims: Fuller, *The New Madrid Earthquake*, 44; Penick, *The New Madrid Earthquakes*, 46–47 (both 1.1); Foley, *History of Missouri*, 170–73 (1.11); Schroeder, *Opening the Ozarks*, 149–52 (1.14). Failed claim for Hot Springs: Berry, Beasley, and Clements, *The Forgotten Expedition*, 191n56 (1.15); other desirable land: James W. Sherby, *From New Madrid to Claverach: How an Earthquake Spawned a St. Louis Suburb* (St. Louis, MO: Virginia Publishing, 2009); successful claims for Little Rock: Bolton, *Territorial Ambition*, 64–66 (1.11). The court case was *Lessieur v. Price*, decided 17 December 1851: e-mail from Lynn Morrow, director of local records, Office of the Secretary of State, Missouri, 17 December 2008. One indication of the logistical problems created by these claims is the sheer length of the index entry for "New Madrid claims" in Carter, ed, *The Territorial Papers of the United States*, 15:799 (2.16). Land claimed under the New Madrid relief act was still being resurveyed in the 1840s: see for instance a letter from Silas Reed dated 8 December 1842, 1CFD vol. 47, p. 54, Missouri State Archives.

1.29.1 History of US Disaster Response: Gareth Davies, "The Nineteenth Century Origin of Federal Disaster Relief," work-in-progress, 2012; Michele Landis Dauber, "The Sympathetic State," *Law and History Review* 23, no. 2 (2005), text markers 9, 61, and 64.

1.30 Earthquakes and Theology: Maxine Van de Wetering, "Moralizing in Puritan Natural Science: Mysteriousness in Earthquake Sermons," *Journal of the History of Ideas* 43, no. 3 (1982): 417–38.

1.31 Voyage of the *New Orleans*: Mary Helen Dohan, *Mr. Roosevelt's Steamboat* (New York: Dodd, Mead, 1981); Feldman, *When the Mississippi Ran Backwards*, 16–19, 104–6, 122–30, 148–50 (1.1); Ellsworth S. Grant, "Roosevelt, Nicholas J.," *American National Biography Online* (1.8); Charles Joseph Latrobe, *The Rambler in North America, 1832–1833*, vol. 1 (New York: Harper & Brothers, 1835); J[ohn] H[azlehurst] B[oneval] Latrobe, *The First Steamboat Voyage on the Western Waters*, Maryland Historical Society Fund Publication 6 (Baltimore: Maryland Historical Society, 1871); Penick, *The New Madrid Earthquakes*, 76–82 (1.1). Transformation of American life with new commerce and communication: Howe, *What Hath God Wrought* (1.8). Wreck of the New Orleans: Ben Casseday, *The History of Louisville, from Its Earliest Settlement Till the Year 1852* (Louisville, KY: Hull and Brother, 1852): 128.

1.31.1 Latrobe and Roosevelt Families: Pamela Scott, "Latrobe, Benjamin Henry"; Grant, "Roosevelt, Nicholas J."; C. M. Harris, "Fulton, Robert"; David Hochfelder, "Latrobe, John Hazlehurst Boneval"; Mary M. Thomas, "Latrobe, Benjamin Henry"; all in *American National Biography Online* (1.8); Dohan, *Mr. Roosevelt's Steamboat*, 17 (1.31); Latrobe, *The First Steamboat Voyage on the Western Waters* (1.31); Jill Eastwood, "La Trobe, Charles Joseph (1801–1875)," in *Australian Dictionary of Biography, Online Edition* (Australian National University, 2006). Fulton: Wood, *Empire of Liberty*, 483–84 (1.8).

1.32 Early American Transportation: Forest G. Hill, *Roads, Rails, and Waterways: The Army Engineers and Early Transportation* (Norman: University of Oklahoma Press, 1957).

1.33 **Steamboats and the West:** Central to nineteenth-century revolution in commerce and communication: Howe, *What Hath God Wrought* 5, 214–15; and Meinig, *The Shaping of America: Continental America 1800–1867*, 319 (both 1.8). Early generation of steamboats, impact on smaller rivers and St. Louis: Foley, *History of Missouri*, 167–68 (1.11); Holder, "Historical Geography of the Lower White River" (1.14); Leslie C. Stewart-Abernathy, "Steamboats," in *Encyclopedia of Arkansas History & Culture*, updated December 2011 (1.11). Deforestation: F. Terry Norris, "Where Did the Villages Go? Steamboats, Deforestation, and Archaeological Loss in the Mississippi Valley," in Hurley, ed., *Common Field* (1.14). Abraham Lincoln, young steamboat inventor: Howe, *What Hath God Wrought*, 596 (1.8).

1.34 **History of Disasters:** Steven Biel, *American Disasters* (New York: New York University Press, 2001); Alessa Johns, ed., *Dreadful Visitations: Confronting Natural Catastrophe in the Age of Enlightenment* (New York: Routledge, 1999); Anthony N. Penna and Jennifer S Rivers, *Natural Disasters in a Global Environment* (Malden, MA: Wiley-Blackwell, 2013). Also: 7.14.

Chapter Two: Earthquakes and the End of the New Madrid Hinterland

2.1 **New Madrid Hinterland:** I borrow from Lynn Morrow: "New Madrid and Its Hinterland: 1783–1826," *Missouri Historical Society Bulletin* 36, no. 4, pt. 1 (1980): 241–50, to indicate the area from the Apple Creek Shawnee and Delaware villages near Cape Girardeau south through the Cherokee and mixed-tribe settlements near Arkansas Post. Daniel Usner regards St. Louis down through the lower St. Francis as a zone of interest (Daniel H. Usner, Jr., "An American Indian Gateway: Some Thoughts on the Migration and Settlement of Eastern Indians around Early St. Louis," *Gateway Heritage* 11, no. 3 [1990–91]: 42–51), while John Bowes treats St. Louis through New Madrid as one region (Bowes, *Exiles and Pioneers*, 37 (2.12). Because I am interested in the effects of the quakes on settlement, I am less focused on the northern reaches up toward St. Louis, where the quakes were alarming but not terribly destructive. Such definitions are by nature fuzzy, but sometimes fuzziness is historically accurate: Stephen Warren observes that keeping a flexible definition of region often served the Native American emigrants into the middle Mississippi: Warren, *The Shawnees and Their Neighbors*, 73 (3.10). The New Madrid hinterland included the communities of Little Prairie and Point Pleasant, villages whose importance in early Indian affairs have been insufficiently developed. For a beginning: Lynn Morrow, "Trader William Gillis and Delaware Migration in Southern Missouri," *Missouri Historical Review* 75 (1981): 147–67, esp. 150, 155; Morrow, "New Madrid and Its Hinterland," esp. 248. On environmental, commercial hinterlands broadly, see Cronon, *Nature's Metropolis* (1.13). Also see 1.12 and 2.8 and 2.20.

2.2 **Early New Madrid:** New Madrid multiethnic in 1780s: Rev. Elmer Talmage Clark, *One Hundred Years of New Madrid Methodism, a History of the Methodist Episcopal Church, South, in New Madrid, Missouri, 1812–1912* (New Madrid: 1912), 5; Parker, *Missouri as It Is in 1867*, 46 (1.17); E. G. Swem, "A Letter from New Madrid, 1789," *Mississippi Valley Historical Review* 5, no. 3 (1918): 342–46. esp. 343; Ethnic tensions and violence: Sugden, *Tecumseh*, 55 (3.10). New Madrid and American and European rivalries: Brattan, "The Geography of the St. Francis Basin," 2 (2.7); Penick *The New Madrid Earthquakes*, 15–29 (1.1). Narrative power of stories of non-Indian town foundings: O'Brien, "Firsting," in *Firsting and Lasting* (2.3). New Madrid from point of view of eastern Indians' ambitions for western lands: Morrow, "New Madrid and Its Hinterland," 241–42 (2.1); Sugden, *Tecumseh*, 54–55; Warren, *The Shawnees and Their Neighbors*, 79 (both at 3.10); Bowes, *Exiles and Pioneers*, 22–27 (2.12).

2.3 **Efforts to Efface Native History:** Philip J. Deloria, *Indians in Unexpected Places* (Lawrence: University Press of Kansas, 2004), esp. introduction; Adrienne Mayor, "Suppression of Indigenous Fossil Knowledge: From Claverack, New York, 1705 to Agate Springs, Nebraska, 2005," in Proctor and Schiebinger, *Agnotology* (6.6); Jean M. O'Brien, *Firsting and Lasting: Writing Indians out of Existence in New England* (Minneapolis: University of Minnesota Press, 2010). Also: **6.6.**

2.4 **Louis Bringier:** Louis Bringier, "Notices of the Geology, Mineralogy, Topography, Productions, and Aboriginal Inhabitants of the Regions around the Mississippi and Its Confluent Waters," *American Journal of Science* 3 (1821): 15–46; W. D. Williams and Louis Bringier, "Louis Bringier and His Description of Arkansas in 1812," *Arkansas Historical Quarterly* 48, no. 2 (1989): 108–36.

2.5 **Sunk Lands:** Excellent summaries: Margaret J. Guccione, Roy B. Van Arsdale, and Lynne H. Hehr, "Origin and Age of the Manila High and Associated Big Lake 'Sunklands' in the New Madrid Seismic Zone, Northeastern Arkansas," *Geological Society of America Bulletin* 112, no. 4 (2000): 579–90; and Nancy Hendricks, "Sunken Lands," updated May 2009, *Encyclopedia of Arkansas History & Culture* (1.11). Further: Arkansas Game & Fish Commission, "St. Francis Sunken Lands," http://www.agfc.com/hunting /Pages/wmaDetails.aspx?show=600, accessed 19 June 2012; Julie Bennett, "Big Lake National Wildlife Refuge," updated May 2011, *Encyclopedia of Arkansas History & Culture* (1.11); Fuller, *The New Madrid Earthquake*, 64–75 (1.1); Parker, *Missouri as It Is in 1867*, 30–38 (1.17); W. J. McGee, "Correspondence: The New Madrid Earthquake," *American Geologist* 30 (1902): 93–96; USDA Forest Service, "Ozark-St. Francis National Forests," http://www.fs.usda.gov/osfnf/ accessed 1 July 2012; *St. Francis Mill Company et al., Respondents, against William O. Sugg et al., Appellants,* in the Supreme Court of the State of Missouri, April Term, 1902, *Appellants' abstract of the Record,* 1902; Mel White, "The Mississippi Flyway," updated April 2010, *Encyclopedia of Arkansas History & Culture* (1.11). See also **6.9, 6.10.**

2.6 **Debates over Creation:** Guccione, Van Arsdale, and Hehr, "Origin and Age of the Manila High," 579 (2.5); Roger T. Saucier, "Origin of the St. Francis Sunk Lands, Arkansas and Missouri," *Geological Society of America Bulletin* 81 (1970): 2847–54; Van Arsdale et al., "Earthquake Signals in Tree-Ring Data," 516 (2.21). Also see **2.7, 6.9.**

2.7 **St. Francis River:** Arkansas Department of Parks & Tourism, "St. Francis River," http://www.arkansas.com/places-to-go/lakes-rivers/river.aspx?id=18, updated 2012; Samuel Tilden Brattan, "The Geography of the St. Francis Basin," *University of Missouri Studies* 1, no. 3 (1926): vi–vii, 1–54; Clarence B. Moore, *Antiquities of the St. Francis, White, and Black Rivers, Arkansas* (Philadelphia: P. C. Stockhausen, 1910). Jodi Morris, "St. Francis River," updated September 2011, *Encyclopedia of Arkansas*

(1.11), provides an account of earthquake damage and Cherokee settlement consistent with this analysis and unlike most earlier histories. Geomorphology of lower St. Francis region: Tuttle, "Late Holocene Earthquakes," 22–26 (7.5.5); Tuttle et al., "Use of Archaeology to Date Liquefaction Features and Seismic Events in the New Madrid Seismic Zone" (7.5.5); "Report of the Independent Expert Panel on New Madrid Seismic Zone," 6 (7.5.1). Also see 2.5 and 6.9.

2.8 **Midcontinent History Based around Waterways:** River-based history reveals a new regional understanding: Usner, "American Indian Gateway" (2.1); DuVal, *The Native Ground* (1.12). In US environmental history broadly: Pisani, "Beyond the Hundredth Meridian," 467–74. River travel in colonial and early American Missouri: Foley, *History of Missouri*, 50–51 (1.11). Importance of rivers as conduits and corridors in understanding Native American history: Bowes, *Exiles and Pioneers*, chap. 1, esp. p. 21 (2.12); Lisa Tanya Brooks, *The Common Pot: The Recovery of Native Space in the Northeast* (Minneapolis: University of Minnesota Press, 2008), xli. Cultural role of rivermen in connecting peoples of the eighteenth-century lower Mississippi Valley: Usner, *Indians, Settlers, and Slaves*, chap. 7, esp. p. 220 (1.12).

2.9 **Indigenous Middle Mississippi Valley before 1780:** DuVal, *The Native Ground*; and Usner, *Indians, Settlers, and Slaves* (both 1.12); Joseph Patrick Key, "The Calumet and the Cross: Religious Encounters in the Lower Mississippi River," *Arkansas Historical Quarterly* 61, no. 2 (2002): 151–68; Robert C. Mainfort, Jr., "The Late Prehistoric and Protohistoric Periods in the Central Mississippi Valley," in *Societies in Eclipse: Archaeology of the Eastern Woodlands Indians, A.D. 1400–1700*, ed. David S. Brose, C. Wesley Cowan, and Robert C. Mainfort (Washington, DC: Smithsonian Institution Press, 2001); Dan F. Morse, "Protohistoric Hunting Sites in Northeastern Arkansas," in Dye and Brister, eds. *The Protohistoric Period in the Mid-South.*

2.10 **Cahokia:** Glenn Hodges, "America's Forgotten City," *National Geographic* (January 2011), through http://ngm.nationalgeographic.com; William R. Iseminger, "Culture and Environment in the American Bottom: The Rise and Fall of Cahokia Mounds," in Hurley, ed. *Common Fields* (1.14); "Timeline: Cahokia Mounds State Historic Site," http://cahokiamounds.org/explore/timeline, updated 2008.

2.11 **Soto and the "Corn Chiefdoms":** DuVal, *The Native Ground*, chap. 2 (1.12); Patricia Galloway, *The Hernando De Soto Expedition: History, Historiography, and "Discovery" in the Southeast* (Lincoln: University of Nebraska Press, 1997).

2.12 **Native Resettlement of the Middle Mississippi Valley:** John Bowes, *Exiles and Pioneers: Eastern Indians in the Trans-Mississippi West*, Studies in North American Indian History, ed. Frederick Hoxie and Neal Salisbury (Cambridge: Cambridge University Press, 2007); John Mack Faragher, "'More Motley Than Mackinaw': From Ethnic Mixing to Ethnic Cleansing on the Frontier of the Lower Missouri, 1783–1833," in Cayton and Teute, eds., *Contact Points* (2.14); Joseph Patrick Key, "Indians and Ecological Conflict in Territorial Arkansas," *Arkansas Historical Quarterly* 59, no. 2 (2000): 127–46 (Key's argument centers on parallel environmental use by later waves of Cherokees and Americans but is also useful for understanding early Cherokee, American, and French emigration, which centered on trading and hunting); George E. Lankford,

"Shawnee Convergence: Immigrant Indians in the Ozarks," *Arkansas Historical Quarterly* 58, no. 4 (1999): 390–413; Lynn Morrow, "New Madrid and Its Hinterland"; Morrow, "Trader William Gillis and Delaware Migration in Southern Missouri"; Usner, Jr., "An American Indian Gateway" (all at 2.1). Warren, *The Shawnees and Their Neighbors* (3.10), emphasizes the Shawnee vision of trans-Mississippi emigration as an alternative to acculturation or violent resistance, but (unlike this argument) does not see 1812 as a major break point. Recognition of waves of emigration is starting to influence archeological research: "Cherokees in Arkansas: UA-WRI Research Station Current Projects Homepage," Arkansas Archeological Survey, http://www.uark.edu/campus-resources/archinfo/atucherokees.html, revised 2007.

2.12.1 **Cherokees:** Foundational argument: Robert A. Myers, "Cherokee Pioneers in Arkansas: The St. Francis Years, 1785–1813," *Arkansas Historical Quarterly* 56, no. 2 (1997): 127, 52–57. Further: Ann M. Early, "Duwali (1756–1839)," updated March 2009, *Encyclopedia of Arkansas* (1.11); Stan Hoig, *The Cherokees and Their Chiefs: In the Wake of Empire* (Fayetteville: University of Arkansas Press, 1998), 102–7 and chap. 8; William G. McLoughlin, *Cherokee Renascence in the New Republic* (Princeton: Princeton University Press, 1986), 56; Robert Paul Markman, "The Arkansas Cherokees: 1817–1828" (PhD diss., University of Oklahoma, 1972). Many accounts cite the leadership of Duwali (also Diwali, Bowl, or the Bowles), in early emigration: Usner, "American Indian Gateway," 47 (2.1); Markman, "The Arkansas Cherokees," 7–9. Robert Myers contests, arguing that Diwali did not emigrate until about fifteen years later: Myers, "Cherokee Pioneers in Arkansas," 137. Early, "Duwali," asserts Duwali's emigration around 1810. Evidence for Duwali's later emigration is convincing, but he may have settled or traveled back and forth (see Hoig, *The Cherokees and Their Chiefs*, 182).

2.12.2 **Cherokee/Osage War:** Guerilla warfare far from governmental officials is messily hard to periodize. Hostilities and violence continued throughout the early nineteenth century, periodically flaring into larger-scale war. McLoughlin, *Cherokee Renascence*, 218 (2.12.1); Bringier, "Notices," 33 (2.4); DuVal, *Native Ground*, chap. 7 (1.12); Markman, "The Arkansas Cherokees," 15–16. Violence of 1815: Jones, "An Original Letter from Polk Bayou," 92 (**Early Observers of the New Madrid Hinterland**, **1.17**); McLoughlin, *Cherokee Renascence*, 263 (2.12.1), reports hostilities through 1824, but see report of Col. Matthew Arbuckle, Headquarters, 7th Infantry, Cantonment Gibson, 13 July 1826, in "L. C. Gulley Collection, 1819–1898," folder 8, no. 84. This was a war of immigrants: Bowes, *Exiles and Pioneers*, 43 (2.12).

2.13 **Native American History Broadly:** W. David Baird, *The Osage People*, ed. John I. Griffin, Indian Tribal Series (Phoenix, 1972); William G. McLoughlin, "Cherokee Anomie, 1794–1810: New Roles for Red Men, Red Women, and Black Slaves"; "The Cherokee Ghost Dance Movement, 1811–1813"; and "Thomas Jefferson and the Beginning of Cherokee Nationalism, 1806–1809," in *The Cherokee Ghost Dance: Essays on the Southeastern Indians, 1789–1861*, ed. William G. McLoughlin, Walter H. Conser, Jr., and Virginia Duffy McLoughlin ([Macon, GA]: Mercer University Press, 1984); Sugden, "Early Pan-Indianism: Tecumseh's Tour of the Indian Country" (3.11).

2.14 **American, European, Indigenous Encounters:** W. David Baird, "The

Reduction of a People: The Quapaw Removal, 1824–1834," *Red River Valley Historical Review* 1, no. 1 (1974): 21–36; Andrew R. L. Cayton and Fredrika J. Teute, eds., *Contact Points: American Frontiers from the Mohawk Valley to the Mississippi, 1750–1830* (Chapel Hill: University of North Carolina Press, 1998), esp. Stephen Aron, "Pigs and Hunters: 'Rights in the Woods' on the Trans-Appalachian Frontier"; Dowd, *A Spirited Resistance* (3.3.1); Kathleen DuVal, "Choosing Enemies: The Prospects for an Anti-American Alliance in the Louisiana Territory," *Arkansas Historical Quarterly* 62, no. 3 (2003): 233–52; DuVal, "Could Louisiana Have Become an Hispano-Indian Republic?" in Williams, Bolton, and Whayne, eds., *A Whole Country in Commotion*, 41–58 (1.11); Frederick J. Fausz, "Becoming 'a Nation of Quakers': The Removal of the Osage Indians from Missouri," *Gateway Heritage* (2000): 28–39; C. J. Miller, "Lovely County," updated May 2008, *Encyclopedia of Arkansas History and Culture* (1.11); Nunez, "Creek Nativism" (3.19); Williams, Bolton, and Whayne, eds., *A Whole Country in Commotion* (1.11), esp. Dan Flores, "Jefferson's Grand Expedition and the Mystery of the Red River"; Lynn Foster, "The First Years of American Justice: Courts and Lawyers on the Arkansas Frontier"; Joseph Patrick Key, "'Outcasts Upon the World': The Louisiana Purchase and the Quapaws"; Bolton, "Jeffersonian Indian Removal and the Emergence of Arkansas Territory" (1.11) and Elliot West, "Lewis and Clark: Kidnappers" (1.15).

2.15 **Lovely's Purchase:** Key, "Indians and Ecological Conflict," 136–37 (2.12); Miller, "Lovely County" (2.14); Bolton, *Territorial Ambition*, 26 (1.11). Lovely's Purchase thus represents, like Kentucky's fought-over Bloody Ground, an earlier cognate of the nature reserves created by late twentieth-century weapons plants and proving grounds: "Album: Unnatural Nature," on the Rocky Mountain Proving Ground, in Cronon, ed., *Uncommon Ground*, 57–66 (1.13); Faragher, *Daniel Boone*, on ample game in tribally contested territories of the eighteenth century.

2.16 **Colonial and Early American Mississippi Valley:** Morris S. Arnold, *Colonial Arkansas, 1686–1804: A Social and Cultural History* (Fayetteville: University of Arkansas Press, 1991); Aron, *American Confluence* (1.2); Clarence Edwin Carter, ed., *The Territorial Papers of the United States* (Washington, DC: United States Government Printing Office, 1934–1962), vol. 9, *The Territory of Orleans 1803–1812*; vol. 13: *The Territory of Louisiana-Missouri, 1803–1806*; vol. 14: *The Territory of Louisiana-Missouri, 1806–1814*; vol. 15: *The Territory of Louisiana-Missouri, 1815–1821*; vol. 19: *The Territory of Arkansas, 1819–1825*; vol. 20: *The Territory of Arkansas, 1825–1829*; vol. 21: *The Territory of Arkansas, 1829–1836*; Ekberg, *French Roots in the Illinois Country*; Ekberg, *François Vallé and His World* (both at 1.11).

2.17 **James Wilkinson**, though governor of Louisiana Territory, manifestly opposed many federal policies and worked as a paid agent of Spain. His double loyalties make evaluating his statements about the middle Mississippi Valley a challenging exercise: Flores, "Jefferson's Grand Expedition," 33 (2.14); Foley, *History of Missouri*, chap. 6 (1.11); West, "Lewis and Clark: Kidnappers," 11 (1.15); Paul David Nelson, "Wilkinson, James," *American National Biography Online* (1.8).

2.18 **Early Mining in the Middle Mississippi Valley:** Chambers, "River of Gray Gold"; Foley, *The Genesis of Missouri*, 14–18; Foley, *A History of Missouri*, 1:8–10 (all in

1.11); Schroeder, *Opening of the Ozarks*, 70–71 (1.14); Swem, "A Letter from New Madrid, 1789"; Williams and Bringier, "Louis Bringier and His Description of Arkansas" (2.4). One example of burgeoning American interest: William Russell, St. Louis, to William Rector, Missouri land surveyor, in Carter, ed., *Territorial Papers*, 15:752–59, esp. 754 (2.16). See also 5.14.

2.19 **Red River Raft:** Berry, Beasley, and Clements, *The Forgotten Expedition*, 194; Flores, "The Ecology of the Red River in 1806," 19–22 (both in 1.15); Hill, *Roads, Rails, and Waterways*, 166 (1.32); Lancaster, "Red River" (1.14); Valencius, *The Health of the Country*, 101–2, 141, 144–45 (1.13).

2.20 **Missouri Probate Records:** New Madrid County probate records (and those of many counties) are available through Missouri's Judicial Records (MJR), online database of the Missouri Secretary of State's office, through the Missouri Digital Heritage Initiative, http://www.sos.mo.gov/archives/mojudicial/.

2.21 **Modern Interpretations of Earthquake Evidence:** Margaret J. Guccione, "Late Pleistocene and Holocene Paleoseismology of an Intraplate Seismic Zone in a Large Alluvial Valley, the New Madrid Seismic Zone, Central USA," in *Paleoseismology: Integrated Study of the Quaternary Geological Record for Earthquake Deformation and Faulting*, ed. Alessandro M. Michetti, Franck A. Audemard M., and Schmuel Marco, special issue, *Tectonophysics* 408, nos. 1–4 (2005): 237–64; Robin K. Mihills and Roy B. Van Arsdale, "Late Wisconsin to Holocene Deformation in the New Madrid Seismic Zone," *Bulletin of the Seismological Society of America* 89, no. 4 (1999): 1019–24; Odum et al., "Near-Surface Structural Model for Deformation Associated with the February 7, 1812, New Madrid, Missouri Earthquake" (7.5.3); Roy B. Van Arsdale, David W. Stahle, Malcolm K. Cleaveland, and Margaret J. Guccione, "Earthquake Signals in Tree-Ring Data from the New Madrid Seismic Zone and Implications for Paleoseismicity," *Geology* 26 (1998): 515–18; Lois Wingerson, "In Search of Ancient Earthquakes: Can Archeologists Help Predict the Next Big One?" *Archaeology* (2006): 30–35.

Chapter Three: Revival and Resistance

3.1 **Indigenous Knowledge:** David M. Gordon and Shepard Krech, III, "Introduction: Indigenous Knowledge and the Environment," in Gordon and Krech, eds., *Indigenous Knowledge and the Environment in Africa and North America* (Athens: Ohio University Press, 2012); Mayor, "Suppression of Indigenous Fossil Knowledge" (2.3); Helen Tilley, "Global Histories, Vernacular Science, and African Genealogies; Or, Is the History of Science Ready for the World?" *Isis* 101, no. 1 (2010): 110–19.

3.1.1 **Indigenous Knowledge of Earthquakes:** Alan D. McMillan and Ian Hutchison, "When the Mountain Dwarfs Danced: Aboriginal Traditions of Paleoseismic Events along the Cascadia Subduction Zone of Western North America," *Ethnohistory* 49, no. 1 (2001): 41–68; Coll Thrush with Ruth S. Ludwin, "Finding Fault: Indigenous Seismology, Colonial Science, and the Rediscovery of Earthquakes and Tsunamis in Cascadia," *American Indian Culture and Research Journal* 31, no. 4 (2007): 1–24; Krajick, "Future Shocks" (7.10); Hough, *Earthshaking Science*, 170–72 (1.6); "Hare Kills the

U´Yê," in Alanson Skinner, "Traditions of the Iowa Indians," *Journal of American Folklore* 38, no. 150 (1925): 497–98.

3.2 New Madrid Earthquakes as Felt in Eastern versus Western North America: Seismic waves travel better in old, cold, eastern North American crust than in the relatively hotter, younger crust of the western United States. Gomberg and Schweig, "Earthquake Hazard in the Heart of the Homeland," [4] (7.11); Hough et al., "On the Modified Mercalli Intensities and Magnitudes of the 1811–1812 New Madrid, Central United States Earthquakes" (7.5.2); Johnston and Shedlock, "Overview of Research in the New Madrid Seismic Zone," 205–6 (7.3). Researchers typically conclude that little was felt of the New Madrid earthquakes much west of St. Louis, yet little research has been done in Spanish or British military records or Native oral history sources to investigate western reports or lack of reports, and Edwin James's report of upper Missouri Indian conversation about the tremors suggests they were indeed felt and discussed in the West as well as in the East (James, Long, and Say, "Account of an Expedition from Pittsburgh to the Rocky Mountains," 57–58 (1.2). Evidence that a later Mississippi Valley quake was felt as far west as Nebraska and New Mexico: Margaret H. Hopper and S. T. Algermissen, "An Evaluation of the Effects of the October 31, 1895, Charleston, Missouri, Earthquake," ed. United States Department of the Interior Geological Survey (Open-File Report 80–778 [preliminary], 1980), 6. Geologist and earthquake historian Susan E. Hough observes that in geological terms the New Madrid quakes were much more an event of the American East than of the West, tying the Mississippi Valley to eastern networks of information. Hough, *Earthshaking Science*, 91 (1.6). This useful insight with respect to American culture suggests also the question of how the quakes might have served as an event giving western peoples more information about the midcontinent. Also: 7.9.

3.3 Native American Spirituality: 3.4–3.8, 3.16, 3.19.1.

3.3.1 Native Resistance to Foreign Encroachment: Gregory Evans Dowd, *A Spirited Resistance: The North American Indian Struggle for Unity, 1745–1815*, Johns Hopkins University Studies in Historical and Political Science; ser. 109, 4 (Baltimore, MD: Johns Hopkins University Press, 1992); William G. McLoughlin, "Ghost Dance Movements: Some Thoughts on Definitions Based on Cherokee History," *Ethnohistory* 37, no. 1 (1990): 25–44.

3.3.2 and American Environments: Peter Nabokov, *Where the Lightning Strikes: The Lives of American Indian Sacred Places* (New York: Viking, 2006).

3.4 Naming Divine Being: Bringier, "Notices," 31 (2.4); McLoughlin, "The Cherokee Ghost Dance Movement," 122 (2.13); Nunez, "Creek Nativism," 9, 151 (3.19); Sugden, *Tecumseh*, 8 (3.10).

3.5 Cherokee Religious Revival: Dowd, *A Spirited Resistance*, 173–81 (3.3.1); McLoughlin, "The Cherokee Ghost Dance Movement" (2.13); McLoughlin, "Ghost Dance Movements" (3.3.1); William G. McLoughlin, "New Angles of Vision on the Cherokee Ghost Dance Movement of 1811–1812," *American Indian Quarterly* 5, no. 4 (1979): 317–45.

3.6 Prophetic Spiritual Leaders: Michael D. Green, "The Expansion of European Colonization to the Mississippi Valley, 1780–1880," in *The Cambridge History of the Native*

Peoples of North America, ed. Bruce G. Trigger and Wilcomb E. Washburn, vol. 1 (Cambridge: Cambridge University Press, 1996); Sugden, *Tecumseh*, 119–20, 440 (3.10).

3.7 Native American Religious Rituals Parallel with Those of American Christian Revival: connections between Protestant spiritualism and the Prophet: Warren, *The Shawnees and Their Neighbors*, 24, 29–31 (3.10); parallels esp. in physical movement: Dowd, *A Spirited Resistance*, 128, 169 (3.3.1); Martin, *Sacred Revolt*, 123 (3.19); Sugden, *Tecumseh*, 262 (3.10). Also see 3.20.

3.8 Osage Spirituality: Cutler, *A Topographical Description of the State of Ohio, Indiana Territory, and Louisiana*, 119 (1.17); Willard H. Rollings, *The Osage: An Ethnohistorical Study of Hegemony on the Prairie-Plains* (Columbia: University of Missouri Press, 1992), chap. 1.

3.9 Osage Reaction to Earthquakes: John Dunn Hunter, *Memoirs of a Captivity among the Indians of North America*, 3rd ed. (1823; repr. London: Longmans, Hurst & Co., 1824), 39. Hunter was attacked as an anti-American imposter in the early nineteenth century, but recent historians have established the value of his account: Sugden, "Early Pan-Indianism: Tecumseh's Tour of the Indian Country," 292–94 (3.11).

3.10 Tecumseh and Tenskwatawa: Colin G. Calloway, *The Shawnees and the War for America*, Penguin Library of American Indian History (New York: Viking, 2007); Benjamin Drake, *Life of Tecumseh, and of His Brother the Prophet; with a Historical Sketch of the Shawanoe Indians* (1841; repr. Cincinnati: Queen City Publishing House, 1855); Draper Manuscripts, Tecumseh Papers, Series YY, Wisconsin Historical Society; R. David Edmunds, *The Shawnee Prophet* (Lincoln: University of Nebraska Press, 1983); R. David Edmunds, *Tecumseh and the Quest for Indian Leadership*, 2nd ed., Library of American Biography, ed. Mark C. Carnes (New York: Pearson Longman, 2007); Logan Esarey, ed. *Governors Messages and Letters*, vol. 1: *Messages and Letters of William Henry Harrison, 1800–1811*, Indiana Historical Collections (Indianapolis: Indiana Historical Commission, 1922); Joel W. Martin, "Tecumseh," in *Encyclopedia of Religion*, ed. Lindsay Jones (Detroit: Macmillan Reference USA, 2005), 9027–29, Gale Group; John Sugden, *Tecumseh: A Life* (New York: Henry Holt, 1997); Stephen Warren, *The Shawnees and Their Neighbors, 1795–1870* (Urbana: University of Illinois Press, 2005).

3.10.1 Family Emigration to New Madrid Hinterland: The brothers' mother, Methotaske (or Methoataaskee), may have moved to the St. Francis after 1779, leaving her children to be raised by an aunt: Warren, *The Shawnees and Their Neighbors*, 17 (3.10). She may have lived among southeastern Cherokees in the early 1790s: Sugden, *Tecumseh*, 77 (3.10). She may have done both: Methotaske may have emigrated west along with the Cherokees. Details are hard to pin down, but the brothers apparently had family networks among the New Madrid hinterland: they likely had a sister, possibly a niece in a Shawnee/French family in Missouri: Sugden, *Tecumseh*, 210–11 (3.10). Wisconsin historian Lyman Draper attempted to track down evidence of a daughter of Tecumseh's living along the Arkansas: Draper Manuscripts, 4YY, no. 47 (3.10).

3.11 Tecumseh's Journey of Recruitment: John Sugden, "Early Pan-Indianism: Tecumseh's Tour of the Indian Country, 1811–1812," *American Indian Quarterly* 10, no. 4 (1986): 273–304; Sugden, *Tecumseh*, 37–57, 67, 93–94, 179–82, 205–17 (3.10); Edmunds, *Tecumseh and the Quest for Indian Leadership*, 203–4 (3.10). Also 3.12.

3.12 Tecumseh Sending Emissaries to Other Groups: Thomas Loraine Mc-Kenney and James Hall, *The Indian Tribes of North America: With Biographical Sketches and Anecdotes of the Principal Chiefs*, ed. Frederick Webb Hodge, vol. 3 (1837–44; repr. Edinburgh: J. Grant, 1933), 192n2.

3.13 Tecumseh's Earthquake Prophecy: Feldman, *When the Mississippi Ran Backwards*, chaps. 4, 5, 11, 12, 13, esp. pp. 160–61 and 195–211 (1.1); Susan Elizabeth Hough and Roger G. Bilham, *After the Earth Quakes: Elastic Rebound on an Urban Planet* (Oxford: Oxford University Press, 2006), 53–57; Sugden, "Early Pan-Indianism: Tecumseh's Tour of the Indian Country," 290 (3.11). Studiously downplayed: Edmunds, *Tecumseh*, 134–39 (3.10).

3.14 War of 1812: Alan Taylor, *The Civil War of 1812: American Citizens, British Subjects, Irish Rebels, & Indian Allies* (New York: Alfred A. Knopf, 2010); Wood, *Empire of Liberty* (1.8), chap. 18; Lisa Brooks understands Tecumseh and Tenskwatawa's movement as an expansion of the United Indian Nations that opposed the early United States: *The Common Pot*, chap. 3, esp. 121–27 (2.18). The war forgotten: Tony Horwitz, "Remember the Raisin!" *Smithsonian* 43, no. 3 (2012): 28–30, 35. See also 3.10.

3.15 Cherokee Nation in the Early Nineteenth Century: William G. McLoughlin, *After the Trail of Tears: The Cherokees' Struggle for Sovereignty, 1839–1880* (Chapel Hill: University of North Carolina Press, 1993); William G. McLoughlin, "Cherokee Anomie," and "Thomas Jefferson and the Beginning of Cherokee Nationalism" (2.13); McLoughlin, *Cherokee Renascence in the New Republic* (2.12.1).

3.16 Cherokee Practices and Beliefs: James Mooney, "The Cherokee Ball Play," *American Anthropologist* 3, no. 2 (1890): 105–32; Mooney, "The Cherokee River Cult," *Journal of Cherokee History* 7, no. 1 (1982): 30–36; Mooney, "Myths of the Cherokee" and "The Sacred Formulas of the Cherokee," in *James Mooney's History, Myths, and Sacred Formulas of the Cherokees*, ed. George Ellison (Fairview, NC: Historical Images, 1992).

3.17 Springplace Mission and Cherokee Reaction: Moravian missionaries recorded discussions with Cherokee neighbors who came to visit their school and settlement, which were on main trails, near a major ball field, and adjacent to an often-traveled river. Though with clear cultural bias, they were careful, observant, and centrally located: see 34–35; maps xxiii–xxvii, of Rowena McClinton, ed., Anna Rosina Kliest Gambold and John Gambold, *The Moravian Springplace Mission to the Cherokees*, vol. 1, *1805–1813*; and vol. 2, *1814–1821*, *Indians of the Southeast*, ed. Michael D. Green and Theda Perdue (Lincoln: University of Nebraska Press, 2007); shorter excerpt: William G. McLoughlin, "Appendix: Excerpts from the Official Diary of the Moravian Mission at Springplace, Georgia," in "New Angles of Vision on the Cherokee Ghost Dance Movement of 1811–1812," *American Indian Quarterly* 5, no. 4 (1979): 339–45. Still: on problems of mistaking missionaries' and other bureaucrats' writings for the perspectives of Indians themselves: Warren, *The Shawnees and Their Neighbors*, introduction, esp. 1–6 (3.10). The journal was mostly written by Anna, though John and others filled in: it was explicitly a collective, rather than individual, record. The Moravians established their school on a small scale in 1804. Presbyterians also conducted larger mission schools in the Cherokee Nation from 1803 to 1810. Missions and their schools were central to the cultural debates convulsing the nation. Meigs, "Some Reflections,"

National Archives RG 75, Roll M-208, 9 March 1812. McLoughlin, "The Cherokee Ghost Dance Movement," 117; McLoughlin, "Cherokee Anomie, 1794–1810," 13 (2.13).

3.18 Moravians: F. Ernest Stoeffler, "Pietism," 7141–44; and David A. Schattschneider, "Moravians," 6190–92; in *Encyclopedia of Religion*, ed. Lindsay Jones (Detroit: Macmillan Reference USA, 2005).

3.19 Creek War with the United States (Red Stick War): US Congress, *American State Papers: Indian Affairs*, vol. 1 (Washington, DC: Gales & Seaton, 1832), esp. "Report of Alexander Cornells, Interpreter, Upper Creeks, to Colonel Hawkins," 22 June 1813, 1:846; and Benjamin Hawkins, "Agent for Indian Affairs, Creek Nation, Letter to the Big Warrior, Little Prince, and Other Chiefs of the Creek Nation, 16 June 1814," 4:845; Dowd, *A Spirited Resistance*, chaps. 8 and 9 (3.3.1); Benjamin Hawkins, *Letters, Journals, and Writings*, ed. C. L. Grant (Savannah, GA: Beehive, 1980); Correspondence of Benjamin Hawkins, US agent to the Creeks, in *American State Papers: Documents, Legislative and Executive . . . March 4, 1789 to June 15, 1834*, ed. Walter Lowrie and Matthew St. Clair Clarke, vol. 4 (Washington, DC: Gales and Seaton, 1832); Howe, *What Hath God Wrought* (1.8), 75–76; Theron A. Nunez, Jr., "Creek Nativism and the Creek War of 1813–1814," *Ethnohistory* 5 (1958): 1–47, 131–75, 292–301; Claudio Saunt, "'Domestick . . . Quiet Being Broke': Gender Conflict among Creek Indians in the Eighteenth Century," in Cayton and Teute, eds., *Contact Points* (2.14); Joel W. Martin, *Sacred Revolt: The Muskogees' Struggle for a New World* (Boston: Beacon Press, 1991).

3.19.1 Creek Prophets: Nunez, "Creek Nativism," 8–14 (3.19); Edmunds, *Tecumseh*, 137 (3.10).

3.20 Connections between Protestant Spiritualism and the Prophet, see Warren, *The Shawnees and Their Neighbors*, 24, 29–31 (3.10). Also: 3.7.

Chapter Four: The Quaking Body

4.1 Earthquakes and Health: Charles Davison, "The Feeling of Nausea Experienced during Earthquakes," *The Lancet* 1 (1906): 706; John Guitéras, "Influence of the Recent Earthquakes in Charleston upon Health," *Medical News* (Philadelphia) 50 (1887): 37; Peyre F. Porcher, "Influence of the Recent Earthquakes in Charleston upon Health," *Medical News* (Philadelphia) 49 (1886): 651–53; Benjamin Rush, *Medical Inquiries and Observations upon the Diseases of the Mind*, History of Medicine Series 15 (1812; facs. New York: Library of the New York Academy of Medicine, 1962), 39 (insanity), 326, 172; Noah Webster, "On the Connection of Earthquakes with Epidemic Diseases, and on the Succession of Epidemics," *Medical Repository* 4 (1801); 340–44; Disease follows many earthquakes: Smith, *An Account of the Earthquakes*, 79–81; but also 64 (1.2); Jamaica earthquake caused illness: Prince, "An Improvement of the Doctrine of Earthquakes" 12 (5.24); Rush, *Medical Inquiries and Observations*, 106.

4.2 Stephen Hempstead and Hempstead Family: Mrs. Dana O. Jensen, "I at Home: The Diary of a Yankee Farmer in Missouri," *Missouri Historical Society Bulletin*: pt. 1, vol. 13, no. 1 (1956); pt. 1, vol. 13, no. 3 (1957); pt. 3, vol. 14, no. 1 (1957); pt. 4, vol. 14, no. 3 (1958); pt. 5, vol. 15, no. 1 (1958); pt. 6, vol. 15, no. 3 (1959); pt. 7, vol. 21, no. 1 (1965); pt. 7, vol. 22, no. 1 (1966); pt. 9, vol. 22, no. 4, pt. 1 (1966). Background:

Valencius, *The Health of the Country*, 5, 21–22, 38, 65–69, 105–7, 110–12, 140, 153–54, 207, 252–54 (1.13).

4.3 Scientific and Religious Truth in the Early US: Howe, *What Hath God Wrought*, esp. 3 and 186–87 (1.8).

4.4 Health, Environments, Bodily Knowledge: Gregg Mitman, "Hay Fever Holiday: Health, Leisure, and Place in Gilded-Age America," *Bulletin of the History of Medicine* 77 (2003): 600–635; Gregg Mitman, Michelle Murphy, and Chris Sellers, eds., "Landscapes of Exposure: Knowledge and Illness in Modern Environments," *Osiris* 19 (2004); Michelle Murphy, *Sick Building Syndrome and the Problem of Uncertainty: Environmental Politics, Technoscience, and Women Workers* (Durham: Duke University Press, 2006); Linda Nash, "The Changing Experience of Nature: Historical Encounters with a Northwest River," *Journal of American History* 86, no. 4 (2000): 1600–29; Nash, *Inescapable Ecologies* (1.13); Joy Parr, *Sensing Changes: Technologies, Environments and the Everyday, 1953–2003* (Vancouver: University of British Columbia Press, 2009); Victoria Sweet, "Hildegard of Bingen and the Greening of Medieval Medicine," *Bulletin of the History of Medicine* 73, no. 3 (1999); Valencius, *The Health of the Country* and Verhoff, "The Intractable Atom" (both at 1.13). Also: 4.7, 4.8 and 7.20.

4.5 Healing and Narrative in Eighteenth and Nineteenth Centuries: Barbara Duden, *The Woman beneath the Skin: A Doctor's Patients in Eighteenth-Century Germany*, trans. Thomas Dunlap (Cambridge, MA: Harvard University Press, 1991); Michael Sappol, "The Odd Case of Charles Knowlton: Anatomical Performance, Medical Narrative, and Identity in Antebellum America," *Bulletin of the History of Medicine* 83 (2009); Steven M. Stowe, *Doctoring the South: Southern Physicians and Everyday Medicine in the Mid-Nineteenth Century* (Chapel Hill: University of North Carolina Press, 2004) (on deathbed stories, esp. 225–26); Stowe, "Seeing Themselves at Work: Physicians and the Case Narrative in the Mid-Nineteenth-Century American South," *American Historical Review* 101, no. 1 (1996).

4.6 Nineteenth-Century Notions of the Body: Charles E. Rosenberg, "The Therapeutic Revolution: Medicine, Meaning, and Social Change in Nineteenth-Century America," in *Explaining Epidemics and Other Studies in the History of Medicine* (Cambridge: Cambridge University Press, 1992); "Airs" and "Body" in Valencius, *The Health of the Country* (1.13); John Harley Warner, *The Therapeutic Perspective: Medical Practice, Knowledge, and Identity in America, 1820–1885* (Cambridge, MA: Harvard University Press, 1986).

4.7 Geological Understanding of Earth as Body: Oldroyd, *Thinking about the Earth*, chap. 1; Oldroyd et al., "The Study of Earthquakes in the Hundred Years Following the Lisbon Earthquake," 325 (both at 1.6); Viermij, "Subterranean Fire" (5.11).

4.7.1 Shift away from Bodily Model: Laudan, *From Mineralogy to Geology*, chap. 7, esp. 138 (5.11).

4.7.2. Yet Bodily Language Persisted: Lucier, *Scientists and Swindlers*, 215, 222 (5.3).

4.8 Bodily Experience and Scientific Knowledge: Bertucci, "The Electrical Body of Knowledge" (4.20); Coen, *The Earthquake Observers* (1.6); Deborah R. Coen,

"The Tongues of Seismology in Nineteenth-Century Switzerland," in Coen, ed., *Witness to Disaster* (1.6); Delbourgo, *A Most Amazing Scene of Wonders* (4.18); Michelle Murphy, "The 'Elsewhere within Here' and Environmental Illness; or, How to Build Yourself a Body in a Safe Space," *Configurations* 8 (2000); Murphy, "Toxicity in the Details: The History of the Women's Office Worker Movement and Occupational Health in the Late-Capitalist Office," *Labor History* 41, no. 2 (2000); Gregg Mitman, Michelle Murphy, and Chris Sellers, "Introduction: A Cloud over History," in Mitman, Murphy, and Sellers, eds., *Osiris: Landscapes of Exposure*, 1–17 (4.4); Nash, *Inescapable Ecologies* (1.13); Simon Schaffer, "Self Evidence" and "Gestures in Question," in *Questions of Evidence: Proof, Practice, and Persuasion across the Disciplines*, ed. James Chandler, Arnold I. Davidson, and Harry Harootunian (Chicago: University of Chicago Press, 1994); Steven Shapin and Simon Schaffer, *Leviathan and the Air-Pump: Hobbes, Boyle, and the Experimental Life* (1985; repr. Princeton, NJ: Princeton University Press, 1989); Grace Yen Shen, "Taking to the Field: Geological Fieldwork and National Identity in Republican China," in *Osiris*, vol. 24 (2009), *National Identity: The Role of Science and Technology*, ed. Carol E. Harrison and Ann Johnson; Pamela H. Smith, *The Body of the Artisan: Art and Experience in the Scientific Revolution* (Chicago: University of Chicago Press, 2004). Also see: 4.4 and 7.20.

4.9 Daniel Drake: Frank A. Barrett, "Daniel Drake's Medical Geography," *Social Science and Medicine* 42, no. 6 (1996): 791–800; Michael L. Dorn, " (In)Temperate Zones: Daniel Drake's Medico-Moral Geographies of Urban Life in the Trans-Appalachian American West," *Journal of the History of Medicine and Allied Sciences* 55, no. 3 (2000): 256–91; Greene, *American Science in the Age of Jefferson*, 116–17 (5.3); Henry D. Shapiro, "Drake, Daniel," in *American National Biography Online* (1.8); Stephen Charles Szaraz, "History, Character, and Prospects: Daniel Drake and the Life of the Mind in the Ohio Valley, 1785–1852" (PhD diss., Harvard University, 1993); Valencius, *The Health of the Country*, esp. 165 (1.13).

4.9.1 Use of Drake in Contemporary Seismological Analysis: Hough and Bilham, *After the Earth Quakes*, 62, 85, 114 (1.6).

4.9.2 Publications: Daniel Drake, *Natural and Statistical View, or Picture of Cincinnati and the Miami Country* (Cincinnati: Looker & Wallace, 1815); Drake, *A Systematic Treatise, Historical, Etiological, and Practical, on the Principal Diseases*, 2 vols. (Philadelphia: Grigg, Elliot & Co.; New York: Mason & Law, 1850, 1854).

4.10 Earthquakes and Christian Theology: William D. Andrews, "The Literature of the 1727 New England Earthquake," *Early American Literature* 7, no. 3 (1973): 281–94; Van De Wetering, "Moralizing in Puritan Natural Science," esp. 436 (1.30); Luigi Piccardi, "Paleoseismic Evidence of Legendary Earthquakes: The Apparition of Archangel Michael at Monte Sant'Angelo (Italy)," *Tectonophysics* 408, nos. 1–4 (2005): 113–28, sect. 3.1.

4.11 End-Times and American Culture: Paul S. Boyer, *When Time Shall Be No More: Prophecy Belief in Modern American Culture* (Cambridge, MA: Belknap Press of Harvard University Press, 1992), esp. part 1; Wood, *Empire of Liberty*, 617–19 (1.8).

4.12 New Madrid Quakes and Religion: Clark, *One Hundred Years of New Madrid Methodism* (2.2); Penick, *The Madrid Earthquakes*, 115–26 (1.1); Walter Brownlow Posey,

"The Earthquake of 1811 and Its Influence on Evangelistic Methods in the Churches of the Old South," *Tennessee Historical Magazine* 1, ser. 2 (1931): 107–14.

4.13 Great Revival: Sydney E. Ahlstrom, *A Religious History of the American People* (New Haven: Yale University Press, 1972), chaps. 26, 27; John B. Boles, *The Great Revival: Beginnings of the Bible Belt* (Lexington: University Press of Kentucky, 1996); Dickson D. Bruce, Jr., *And They All Sang Hallelujah: Plain-Folk Camp-Meeting Religion, 1800–1845* (Knoxville: University of Tennessee Press, 1974); Paul Keith Conkin, *Cane Ridge: America's Pentacost* (Madison: University of Wisconsin Press, 1990); George Stanley Godwin, *The Great Revivalists* (Boston: Beacon Press, 1950); Nathan O. Hatch, *The Democratization of American Christianity* (New Haven: Yale University Press, 1989); Barry Hankins, *The Second Great Awakening and the Transcendentalists* (Westport, CT: Greenwood Press, 2004); Howe, *What Hath God Wrought*, chap. 5 (1.8); Charles A. Johnson, *The Frontier Camp Meeting; Religion's Harvest Time* (Dallas: Southern Methodist University Press, 1955); Bernard A. Weisberger, *They Gathered at the River: The Story of the Great Revivalists and Their Impact upon Religion in America* (Boston: Little, Brown, 1958), chap. 1; Wood, *Empire of Liberty*, chap. 16, esp. 595–619 (1.8).

4.14 Revivals and Revivalists: Peter Cartwright, *Autobiography of Peter Cartwright*, ed. Charles Langworthy Wallis (New York: Abingdon Press, 1956), James B. Finley, *Autobiography of Rev. James B. Finley, or, Pioneer Life in the West Edited by W. P. Strickland* (Cincinnati: Methodist Book Concern, 1856); William Henry Milburn, *The Pioneers, Preachers and People of the Mississippi Valley* (New York: Derby & Jackson, 1860); Ross, *Life and Times of Elder Reuben Ross* (1.17); "Theophilis Arminius" [Rev. Thomas Hindle], "Religious and Missionary Intelligence," *Methodist Magazine* 2 (1819): 273; for a characteristically ascerbic view: Frances Milton Trollope, *Domestic Manners of the Americans* (London: Whittaker, Treacher, 1832), chap. 15.

4.15 Revival Hymns: importance of widely shared song: Boles, *The Great Revival*, 121–24; Hatch, *The Democratization of American Christianity*, 128, 146–61 (both at 4.13). Produced as part of an explosion of religious printing: Paul C. Gutjahr, *An American Bible: A History of the Good Book in the United States, 1777–1880* (Stanford, CA: Stanford University Press, 1999), app. 1–6. Typical early hymnals: Starke Dupuy, *Hymns and Spiritual Songs, Original and Selected* (Frankfort [KY]: Printed for Butler and Wood by Kendall and Russells, 1818); Evangelical Lutheran Church, *A Collection of Hymns; and a Liturgy* (Philadelphia: G. & D. Billmeyer, 1814); Enoch Mudge, *The American Camp-Meeting Hymn Book* (Boston: Joseph Burdakin, 1818); Ezekiel Terry, *Hymns and Spiritual Songs* (Palmer [MA]: E. Terry, 1816). See also 5.4.

4.15.1 Spiritual Songs Draw on Natural Signs: "Climbing Up the Mountain," traditional Southern spiritual, arr. Doyle Lawson, Doyle Lawson & Quicksilver, in *Sacred Voices: An A Capella Gospel Collection* (Sugar Durham, NC: Hill Records, 1999); "My Lord, What a Morning," traditional spiritual, transcribed as hymn 13 in Episcopal Commission for Black Ministries, *Lift Every Voice and Sing II* (New York: Church Publishing Incorporated, 1993). See also 1.4.

4.16 New Madrid Hymns: William Downs, "Earth Quake, 1812—Reflections on the Same," in *A New Kentucky Composition of Hymns and Spiritual Songs: Together with* *a Few Odes, Poems, Elegies, &c.* (Frankfort, KY: Gerard & Berry, 1816), ReadEx/News-Bank; Arthur Palmer Hudson, "A Ballad of the New Madrid Earthquake," *Journal of American Folklore* 60, no. 236 (1947): 147–50. See also 1.4.

4.17 Nineteenth-Century American Religious Response to Civic Crises: Charles E. Rosenberg, *The Cholera Years: The United States in 1832, 1849, and 1866* (Chicago: University of Chicago Press, 1962).

4.18 Electricity in Early America: "Ben Franklin's Science," exhibit at the Collection of Historical Scientific Instruments, Harvard University, summer 2006; Joyce E. Chaplin, *The First Scientific American: Benjamin Franklin and the Pursuit of Genius* (New York: Basic Books, 2006); James Delburgo, "Electrical Humanitarianism in North America: Dr. T. Gale's *Electricity, or Ethereal Fire, Considered* (1802) in Historical Context," in Bertucci and Pancaldi, eds., *Electric Bodies*, 117–56 (4.20); James Delbourgo, *A Most Amazing Scene of Wonders: Electricity and Enlightenment in Early America* (Cambridge, MA: Harvard University Press, 2006); Michael Brian Schiffer, *Draw the Lightning Down: Benjamin Franklin and Electrical Technology in the Age of Enlightenment* (Berkeley: University of California Press, 2003). See also 4.20 and 5.5.

4.19 Electricity and Revival: Delbourgo, "Electrical Humanitarianism," 119, 122–25; Delbourgo, *A Most Amazing Scene of Wonders*, 133, 212–19 (both 4.18).

4.20 Early Electrical Science, Electrical Medicine, Electrical Physical Experience: Paola Bertucci and Guiliano Pancaldi, eds., *Electric Bodies: Episodes in the History of Medical Electricity* (Bologna: Università di Bologna, Dipartimento di Filosofia, Centro Internazionale per la Storia delle Università e della Scienza, 2001), esp. Bertucci, "The Electrical Body of Knowledge: Medical Electricity and Experimental Philosophy in the Mid-Eighteenth Century," 43–68; Marco Bresadola, "Early Galvanism as Technique and Medical Practice," 157–80; and Oliver Hochadel, "'My Patient Told Me How to Do It': The Practice of Medical Electricity in the German Enlightenment," 69–90; Rebecca Herzig, "Subjected to the Current: Batteries, Bodies, and the Early History of Electrification in the United States," *Journal of Social History* 41, no. 4 (2008); Schaffer, "Self Evidence" (4.8). Also: 5.5.

4.21 Lectures, Demonstrations, Entrepreneurial Science: Lectures and demonstrations were crucial sites of scientific engagement: Lucier, "The Professional and the Scientist in Nineteenth-Century America," 713; Pandora, "Popular Science," 354–55; Wright, "Scientific Interest and Observation," 235–36, though see Greene, *American Science in the Age of Jefferson*, 20 (all in 5.3). Central to the epistemological arguments of early electrical science: Schaffer, "Self Evidence" (4.8). See also 5.3.

4.21.1 Electrical Demonstration Kits: Schiffer, *Draw the Lightning Down*, 78 (4.18); and "Ben Franklin's Science" (4.18).

4.21.2 A Broad Audience: Chaplin, *The First Scientific American*, 104–14; Delbourgo, *A Most Amazing Scene of Wonders*, 87–128, 201–2 (both at 4.18); Hindle, *The Pursuit of Science in Revolutionary America*, 74–75 (5.3).

4.21.3 A Good Living: Lucier, *Scientists and Swindlers*, 32–33 (5.3).

4.21.4 Not a Good Living: Greene, "Science and the Public in the Age of Jefferson," 201–13 (5.3) and Schiffer, *Draw the Lightning Down* (4.18).

4.21.5 Intellectual and Public Impact of European Science: Laudan, *From Mineralogy to Geology*, 187 and Rudwick, *Bursting the Limits of Time*, 444 (both at 5.11).

Chapter Five: Vernacular Science

5.1 **Daniel Drake:** see 4.9, 4.9.1, and 4.9.2.

5.2 **Vernacular Science:** Pamela Smith describes ways of knowing and working developed by those skilled with manual crafts: Smith, *The Body of the Artisan* (4.8). I expand to include textually-based expertise as well as craft and physical aspects of knowledge. Vernacular science works in and through the everyday, accessible and revisable by those of many geographic or social places. Similarly: Katherine Pandora, "Knowledge Held in Common: Tales of Luther Burbank and Science in the American Vernacular," *Isis* 92 (2001): 484–516; Katherine Pandora and Karen A. Rader, "Science in the Everyday World: Why Perspectives from the History of Science Matter," *Isis* 99, no. 2 (2008): 350–64. Helen Tilley contrasts "vernacular science" of Western experts, with "vernacular knowledge" of local communities: Tilley, "Global Histories, Vernacular Science, and African Genealogies" (3.1). "Local knowledge" emphasizes place-based aspects of medical and environmental practice: Valencius, *The Health of the Country* (1.13). "Residential science" contrasts more cosmopolitan field sciences: Robert E. Kohler, "History of Field Science: Trends and Prospects," in *Knowing Global Environments: New Historical Perspectives on the Field Sciences*, ed. Jeremy Vetter, Studies in Modern Science, Technology, and the Environment (New Brunswick, NJ: Rutgers University Press, 2011), 214–16; "Indigenous" knowledge describes early modern European natural history: Alix Cooper, *Inventing the Indigenous: Local Knowledge and Natural History in Early Modern Europe* (Cambridge: Cambridge University Press, 2007), but see 3.1. "Vernacular dialogue" in early twentieth century: Deborah R. Coen, "The Tongues of Seismology in Nineteenth-Century Switzerland," 73–102. "Citizen science": Fa-ti Fan, "'Collective Monitoring, Collective Defense': Science, Earthquakes, and Politics in Communist China," in Coen, ed., *Witness to Disaster* (1.6 and 7.22). "Lay science": Jeremy Vetter, "Introduction to Special Issue on Lay Participation in the History of Scientific Observation," *Science in Context* 24, no. 2 (2011): 127–41; Jeremy Vetter, introduction to *Knowing Global Environments: New Historical Perspectives on the Field Sciences*, ed. Jeremy Vetter, Studies in Modern Science, Technology, and the Environment (New Brunswick, NJ: Rutgers University Press, 2011), 11–13. Broad questions also explored in: Conference on "Lay Participation in Scientific Observation," Max Planck Institute for the History of Science, Berlin, June 2007.

5.3 **Early American Science:** Stephen Case, "'Insufferably Stupid or Miserably Out of Place': F. A. P. Barnard and His Scientific Instrument Collection in the Antebellum South," *Historical Studies on the Natural Sciences* 39, no. 4 (2009): 418–43; I. Bernard Cohen, "Science in America: The Nineteenth Century," in *Paths of American Thought*, ed. Arthur M. Schlesinger Jr. and Morton White (Boston: Houghton Mifflin Co., 1963); George H. Daniels, *American Science in the Age of Jackson*, History of American Science and Technology Series, ed. Lester D. Stephens (Tuscaloosa: University of Alabama Press, 1968); Delbourgo, *A Most Amazing Scene of Wonders* (4.18); Daniel

Goldstein, "'Yours for Science': The Smithsonian Institution's Correspondents and the Shape of Scientific Community in Nineteenth-Century America," *Isis* 85 (1994): 573–99; John C. Greene, *American Science in the Age of Jefferson* (Ames: Iowa State University Press, 1984); Greene, "Science and the Public in the Age of Jefferson," in *Early American Science*, ed. Brooke Hindle, History of Science Selections from Isis (New York: Science History Publications, 1976); Hindle, *The Pursuit of Science in Revolutionary America, 1735–1789* (Chapel Hill: Institute of Early American History and Culture and University of North Carolina Press, 1956); Richard William Judd, *The Untilled Garden: Natural History and the Spirit of Conservation in America, 1740–1840*, Studies in Environment and History (New York: Cambridge University Press, 2009); Judd, "A 'Wonderfull Order and Ballance': Natural History and the Beginnings of Forest Conservation in America, 1730–1830," *Environmental History* 11 (2006): 8–36; Sally Gregory Kohlstedt, *The Formation of the American Scientific Community: The American Association for the Advancement of Science, 1848–1860* (Urbana: University of Illinois Press, 1976); Kohlstedt, "Parlors, Primers, and Public Schooling" (1.10); Paul Lucier, "The Professional and the Scientist in Nineteenth-Century America," *Isis* 100, no. 4 (2009): 699–732; Lucier, *Scientists and Swindlers: Consulting on Coal and Oil in America, 1820–1890*, ed. Merritt Roe Smith, Johns Hopkins Studies in the History of Technology (Baltimore: Johns Hopkins University Press, 2008); Judith Magee, *The Art and Science of William Bartram* (University Park: Pennsylvania State University Press, 2007); Malone, "Everyday Science, Surveying, and Politics in the Old Southeast" (1.15); Pandora, "The Children's Republic of Science" (1.10); Katherine Pandora, "Popular Science in National and Transnational Perspective: Suggestions from the American Context," *Isis* 100, no. 2 (2009): 346–58; Charlotte M. Porter, *The Eagle's Nest: Natural History and American Ideas, 1812–1842*, History of American Science and Technology Series, ed. Lester D. Stephens (Tuscaloosa: University of Alabama Press, 1986); Seth Shulman, *The Telephone Gambit: Chasing Alexander Graham Bell's Secret* (New York: W. W. Norton, 2008); Stoll, *Larding the Lean Earth* (5.7); Valencius, *The Health of the Country*, esp. "Local Knowledge" (1.13); Louis B. Wright, "Scientific Interest and Observation," in *The Cultural Life of the American Colonies, 1607–1763* (New York: Harper and Brothers, 1957). Also see 4.21, 5.4, 5.5, 5.6, 5.9.

5.4 **Early American Scientific Publishing**: often dismissed as limited, if not vestigial: Daniels, *American Science in the Age of Jackson*, 18, app. 2; Greene, *American Science in the Age of Jefferson*, 10–11, 21–24 (both at 5.3). Less dismissive approach to international publishing of colonial American authors: Frank R. Freemon, "American Colonial Scientists Who Published in the 'Philosophical Transactions' of the Royal Society," *Notes and Records of the Royal Society of London* 39, no. 2 (1985): 191–206. Yet American regional publishing flourished in the first half of the nineteenth century, churning out hymnals, Bibles, and do-it-yourself medical manuals. These were part of an explosion of knowledge and information that also included vernacular science and must be reckoned in our histories of early American science as well. Gutjahr, *An American Bible*, in (4.15); Hatch, *The Democratization of American Christianity*, 144 (4.13); Charles E. Rosenberg, "John Gunn: Everyman's Physician," in *Explaining Epidemics and Other Studies in the History of Medicine* (Cambridge: Cambridge University

Press, 1992). European context: Alex Csiczar, "Broken Pieces of Fact: The Rise of the Scientific Journal in the Nineteenth Century" (PhD thesis, Harvard University, 2010). See also 4.15, 5.3.

5.5 Franklin and Electricity: "Ben Franklin's Science"; Chaplin, *The First Scientific American*, 138; and Schiffer, *Draw the Lightning Down* (all at 4.18); Dean, "Benjamin Franklin and Earthquakes," esp. 487 (1.6); Philip Dray, *Stealing God's Thunder: Benjamin Franklin's Lightning Rod and the Invention of America* (New York: Random House, 2005); Hindle, *The Pursuit of Science in Revolutionary America*, 78–79, 188–89 (5.3) . See also 4.18, 4.20, 5.3.

5.6 Early American Earth Sciences: James X. Corgan, ed., *The Geological Sciences in the Antebellum South* (Tuscaloosa: University of Alabama Press, 1982); Greene, *American Science in the Age of Jefferson*, chap. 9 (5.3); Robert M. Hazen, "The Founding of Geology in America: 1771 to 1818," *Geological Society of America Bulletin* 85 (1974): 1827–34; Walter B. Hendrickson, "Nineteenth-Century State Geological Surveys: Early Government Support of Science," *Isis* 52, no. 3 (1961): 357–71; Judd, *The Untilled Garden*, chap. 4 (5.3); R. Bruce McMillan, "The Discovery of Fossil Vertebrates on Missouri's Western Frontier," *Earth Sciences History* 29, no. 1 (2010): 26–51; Lukas Rieppel, "Articulating the Past: A Cultural History of Paleontology, 1870–1930" (PhD diss., Harvard University Department of the History of Science, 2012); David Spanagel, "The Geological Imperative: Using New York's Natural History to Build an Empire State, 1810–1840" (manuscript in progress); Spanagel, "Great Convulsions and Parallel Scratches: The Era of Romantic Geology in Upstate New York," *Northeastern Geology and Environmental Sciences* 17, no. 2 (1995): 179–82; George W. White, "Early Geological Observations in the American Midwest," in *Toward a History of Geology*, ed. Cecil J. Schneer (Cambridge, MA: MIT Press, 1969). See also 5.3.

5.6.1 Early American Earthquake Study: "Another Conjecture of the Cause of the Earthquakes," *National Intelligencer* (Washington, DC), 28 March 1812; Capac, "Reflections concerning Earthquakes, No. 1," *Georgia Journal*, 19 Feb 1812; Capac, "Reflections concerning Earthquakes, No. 3," *Georgia Journal*, 11 March 1812; both in CERI Compendium (1.2) (presumably, Capac also wrote a no. 2); Isaac Lea, "On Earthquakes—Their Causes and Effects," *American Journal of Science and Arts* 9, no. 2 (1825): 209–15; "On the Cold of the Present Season" (Originally from the *Petersburg Intelligencer*), *Daily National Intelligencer*, 30 Sept 1816; Edward Darrell Smith, "On the Changes Which Have Taken Place in the Wells of Water Situated in Columbia, South-Carolina, since the Earthquakes of 1811–12," *American Journal of Science* 1 (1819): 93–95; many drew upon William Stukeley, "On the Causes of Earthquakes," *Philosophical Transactions* [of the Royal Society of London] 46 (1749): 641–46, 657–69.

5.7 Agricultural Improvement and Geological Interests in Early US: Emily Pawley, "The Balance-Sheet of Nature: Calculating the New York Farm, 1820–1860" (PhD thesis, University of Pennsylvania, 2009); Richard C. Sheridan, "Mineral Fertilizers in Southern Agriculture," in *The Geological Sciences in the Antebellum South*, ed. Corgan; Steven Stoll, *Larding the Lean Earth: Soil and Society in Nineteenth-Century America* (New York: Hill and Wang, 2002), esp. 150–52 and 90.

5.8 Nationalism of Early American Earth Science: Boorstein, *The Americans* (1.8); Greene, "Science and the Public in the Age of Jefferson," 211; Hindle, *The Pursuit of Science in Revolutionary America*, 317; Judd, *The Untilled Garden*, esp. 96, 170 (all in 5.3); Ann Johnson, "Material Experiments: Environment and Engineering Institutions in the Early American Republic," in *National Identity: The Role of Science and Technology*, ed. Carol E. Harrison and Ann Johnson, *Osiris* 24 (2009): 53–74; Peter A. Shulman, "Ships, Security, and the Politics of Trees: The Maritime Origins of American Forest Conservation" (work in progress, Boston Environmental History Seminar, Massachusetts Historical Society); Spanagel, *The Geological Imperative* (5.6).

5.9 Experimentation in Early American Medicine and Science: Chaplin, *The First Scientific American*, chap. 4 (4.18); Chambers, "Land of Ores, Country of Minerals," chap. 4 (1.11); Alexa Green, "Working Ethics: William Beaumont, Alexis St. Martin, and Medical Research in Antebellum America," *Bulletin of the History of Medicine* 84, no. 2 (2010): 193–216; Peter J. Kastor and Conevery Bolton Valencius, "Sacagawea's 'Cold': Pregnancy and the Written Record of the Lewis and Clark Expedition," *Bulletin of the History of Medicine* 82 (2008): 276–310; Ronald L. Numbers and William J. Orr, Jr., "William Beaumont's Reception at Home and Abroad," *Isis* 72, no. 264 (1981): 590–612; Stowe, "Borrowing, Experimenting, and Violence," 156–62 in *Doctoring the South* (4.5); John Harley Warner, "From Specificity to Universalism in Medical Therapeutics: Transformation in the 19th-Century United States," in *Sickness and Health in America: Readings in the History of Medicine and Public Health*, ed. Judith Walzer Leavitt and Ronald L. Numbers (Madison: University of Wisconsin Press, 1997); John Harley Warner, *The Therapeutic Perspective: Medical Practice, Knowledge, and Identity in America, 1820–1885* (Cambridge, MA: Harvard University Press, 1986). See also 4.21, 5.3.

5.10 Charles Willson Peale's Museum: Keith Tony Beutler, "The Memory Revolution in America and Memory of the American Revolution, 1790–1840" (PhD thesis, Department of History, Washington University in St. Louis, 2005); Walter Faxon, "Relics of Peale's Museum," *Bulletin of the Museum of Comparative Zoology* 59 (1915): 117–48; Otto Friedrich, "The Peales: America's First Family of Art," *National Geographic* 178, no. 6 (1990); Greene, *American Science in the Age of Jefferson*, 26–27, 52–57; Greene, "Science and the Public in the Age of Jefferson," 210; Judd, *The Untilled Garden*, 100; Porter, *The Eagle's Nest*, 27–40 (all in 5.3); Robert E. Schofield, "The Science Education of an Enlightened Entrepreneur: Charles Willson Peale and His Philadelphia Museum, 1784–1827," *American Studies* 30, no. 2 (1989): 21–40; Wood, *Empire of Liberty* (1.8), 555–57. On museums and history of American science, see Pandora and Rader, "Science in the Everyday" (5.2).

5.11 History of Earth Sciences through the Nineteenth Century: Susan Faye Cannon, "Humboldtian Science," in *Science in Culture: The Early Victorian Period*, (New York: Dawson and Science History Publications, 1978), chap. 3; Mott T. Greene, *Geology in the Nineteenth Century: Changing Views of a Changing World* (Ithaca: Cornell University Press, 1982); Nicolas Jardine, James A. Secord, and Emma C. Spary, eds., *The Cultures of Natural History* (Cambridge: Cambridge University Press, 1996); Rachel Laudan, *From Mineralogy to Geology: The Foundations of a Science, 1650–1830*, Science and Its Conceptual Foundations, ed. David L. Hull (Chicago: University of Chicago Press, 1987); Tara E. Nummedal, "Kircher's Subterranean World and the Dignity of the Geocosm," in *The Great Art of Knowing: The Baroque Encyclopedia of Athanasius Kircher*, ed. Daniel Stolzenberg (Stanford, CA: Stanford University Libraries, 2001), 37–47; Kirtley F. Mather and Shirley L. Mason, eds., *A Source Book in Geology*, 1st ed. (New York: McGraw-Hill, 1939); David R. Oldroyd, "The Earth Sciences," in *From Natural Philosophy to the Sciences: Writing the History of Nineteenth-Century Science*, ed. David Cahan (Chicago: University of Chicago Press, 2003): 88–128; Oldroyd, *Thinking about the Earth* (1.6); Rhoda Rappaport, *When Geologists Were Historians, 1665–1750* (Ithaca, NY: Cornell University Press, 1997); Martin J. S. Rudwick, *Bursting the Limits of Time: The Reconstruction of Geohistory in the Age of Revolution* (Chicago: University of Chicago Press, 2005); Rudwick, "The Emergence of a Visual Language for Geological Science, 1760–1840," *History of Science* 14 (1976): 149–95; Rudwick, *The Great Devonian Controversy: The Shaping of Scientific Knowledge among Gentlemanly Specialists* (Chicago: University of Chicago Press, 1985); Rudwick, *Worlds before Adam: The Reconstruction of Geohistory in the Age of Reform* (Chicago: University of Chicago Press, 2008); Kenneth L. Taylor, "American Geological Investigations and the French, 1750–1850," *Earth Sciences History* 9, no. 2 (1990): 118–25; Rienk Vermij, "Subterranean Fire: Changing Theories of the Earth during the Renaissance," *Early Science and Medicine* 3, no. 4 (1998): 323–47; Laura Dassow Walls, *Passage to Cosmos: Alexander Von Humboldt and the Shaping of America* (Chicago: University of Chicago Press, 2009) Also see 1.6, 4.7, 5.6, 5.12.

5.12 Late Eighteenth- and Early Nineteenth-Century Earth Science: Period of intellectual foment and development in Europe: Rudwick, *Bursting the Limits of Time*, 9; and *Worlds before Adam*, 3; Laudan, *From Mineralogy to Geology*. Quiet period for geological theory in US: Greene, *Geology in the Nineteenth Century*, 144–45 (all at 5.11).

5.13 Emergence of Historical, Fossil-Inspired Geology: Laudan, *From Mineralogy to Geology*, esp. 146; Rudwick, "Minerals, Strata, and Fossils" in Jardine, Secord, and Spary, eds., *The Cultures of Natural History*; Rudwick, *Bursting the Limits of Time*, esp. chap. 9; Rudwick, *The Great Devonian Controversy*; Rudwick, *Worlds before Adam*, chap. 1 (all at 5.11).

5.14 Importance of Mining and Commercial Interests in European Earth Sciences: Laudan, *From Mineralogy to Geology*, chap. 3 and Rudwick, *Bursting the Limits of Time*, chap. 2 (both at 5.11); Andre Wakefield, *The Disordered Police State: German Cameralism as Science and Practice* (Chicago: University of Chicago Press, 2009). See also 2.18.

5.15 Importance of Voyages of Discovery in European Earth Sciences: Michael Dettelbach, "Humboltian Science," and Janet Browne, "Biogeography and Empire," in Jardine, Secord, and Spary, eds., *The Cultures of Natural History* (5.11).

5.16 Lyell and New Madrid: Charles Lyell, *A Second Visit to the United States of North America* (New York: Harper & Brothers, 1849), esp. 2:33; Leonard G. Wilson, *Lyell in America: Transatlantic Geology, 1841–1853* (Baltimore: Johns Hopkins University Press, 1998); Laudan, *From Mineralogy to Geology*, chap. 9; and Rudwick, *Worlds before Adam*, esp. 15 (both at 5.11).

5.17 Debates over Formation and Composition of the Earth: Stephen G. Brush, "Nineteenth-Century Debates about the Inside of the Earth: Solid, Liquid or Gas?," *Annals of Science* 36 (1979): 225–54; Hazen, "The Founding of Geology in America," 1829 (5.6); Greene, *Geology in the Nineteenth Century*, chaps. 1 and 2; and Laudan, *From Mineralogy to Geology*, chap. 2 (5.11); Oldroyd, *Thinking about the Earth*, chap. 4 (1.6).

5.18 Artifact Collections: Natural History and Mineral Cabinets: Rudwick, "Minerals, Strata, and Fossils," in *Cultures of Natural History*, ed. Jardine, Secord, and Spary (5.13); Spanagel, *The Geological Imperative*, introduction (5.6).

5.19 Collection of Fossils Shaped Changing Nineteenth-Century Geological Ideas, Involved Broad Range of Participants: Rudwick, *Worlds before Adam*, chap. 2; Rudwick, *Bursting the Limits of Time*, 385–86 (both at 5.11).

5.20 Musk Ox Skull: Academy of Natural Sciences of Philadelphia (now Academy of Natural Sciences of Drexel University), "Harlan's Musk Ox (*Bootherium Bombifrons*)," Online Exhibition: Thomas Jefferson Fossil Collection, accessed 24 May 2010 at http://www.ansp.org/museum/jefferson/otherFossils/bootherium.php; Foster, *The Mississippi Valley*, 22 (1.17) Fuller, *The New Madrid Earthquake*, 77 (1.1); de Montulé, *Travels in America*, 113 (1.2); "Parts of Animals Brought to View by the Late Earthquakes near New Madrid, Mo. Letter to Jasper Lynch, Esq. From Samuel L. Mitchill," *Daily National Intelligencer*, 4 July 1820. The names change—*Ovibos cavifrons* Leidy in 1902; now *Symbos cavifrons*. But the 1828 skull is still held to be the first. J. B. Hatcher, "Discovery of a Musk Ox Skull (*Ovibos Cavifrons* Leidy), in West Virginia, near Steubenville, Ohio," *Science* 16, new ser. 409 (1902); and J. M. McDonald and K. C. Corkum, "A Woodland Musk Ox, *Symbos Cavifrons* (Artiodactyla: Bovidae), from Bayou Sara, Louisiana," *Southwestern Naturalist* 32, no. 1 (1987). Jerry N. McDonald and Clayton E. Ray, "The Autochthonous North American Musk Oxen Bootherium, Symbos, and Gidleya," *Smithsonian Contributions to Paleobiology* 66 (1989); conversations with Claudine Payne and Marion Haynes, Blytheville Archeological Complex, Arkansas Archeological Complex, 23 April 2008, 24 January 2005, 8 June 2012.

5.21 Networks of Scientific Correspondence: Involved many people, recognized and conferred expertise: James Delbourgo, *A Most Amazing Scene of Wonders*, esp. 201 (4.18); Geschwind, "Early Twentieth-Century Jesuits and Seismology in the United States," esp. 45–48 (6.2); Goldstein, "Yours for Science"; Judd, *The Untilled Garden*, 96–97; Lucier, *Scientists and Swindlers*, 20–21, 359n162 (all in 5.3); Rudwick, *Bursting the Limits of Time*, 31–32, 46, 289; and *The Great Devonian Controversy* (both at 5.11); Valencius, *The Health of the Country*, 183–90 (1.13); Kariann Yokota, "'To Pursue the Stream to Its Fountain': Race, Inequality, and the Post-Colonial Exchange of Knowledge across the Atlantic," *Explorations in Early American Culture* 5 (2001): 173–229. Also: 1.9.

5.22 Making Facts: Shapin and Schaffer, *Leviathan and the Air-Pump* (4.8); Steven Shapin, *A Social History of Truth: Civility and Science in Seventeenth-Century England* (Chicago: University of Chicago Press, 1994).

5.23 History of Weather: Vladimir Janković, *Reading the Skies: A Cultural History of English Weather, 1650–1820* (Manchester: Manchester University Press, 2000); Mamie J. Meredith, "'Arctic Smoke,' 'Lightning Nest,' and Other Weather Terms,"

American Speech 26, no. 3 (1951): 232–37; Geschwind, "Early Twentieth-Century Jesuits and Seismology in the United States" (6.2); Robert Ward, "How Far Can Man Control His Climate?" *Scientific Monthly* 30 (1930): 5–18. James Rodger Fleming, Vladimir Janković, and Deborah R. Coen, *Intimate Universality: Local and Global Themes in the History of Weather and Climate* (Sagamore Beach, MA: Science History Publications, 2006).

5.24 **Prince/Winthrop Debate:** Dean, "Benjamin Franklin and Earthquakes," 488 (1.6); Delbourgo, *A Most Amazing Scene of Wonders*, 1, 65–72, 107 (4.18); Dray, *Stealing God's Thunder*, chap. 4 (5.5); Zoltan Haraszti, "Young John Adams on Franklin's Iron Points," *Isis* 41, no. 1 (1950): 11–14; Hindle, *The Pursuit of Science in Revolutionary America*, 94–96 (5.3); Theodore Hornberger, "The Science of Thomas Prince," *New England Quarterly* 9, no. 1 (1936): 26–42; Thomas Prince, *An Improvement of the Doctrine of Earthquakes, Being the Works of God, and Tokens of His Just Displeasure . . .* (Boston, 1755), Early American Imprints, ser. 1, Evans 7550, American Antiquarian Society, ReadEx/NewsBank; Eleanor M. Tilton, "Lighting-Rods and the Earthquake of 1755," *New England Quarterly* 13, no. 1 (1940): 85–97; John E. Van de Wetering, "God, Science, and the Puritan Dilemma," *New England Quarterly* 38, no. 4 (1965): 494–507; John Winthrop, *A Letter to the Publishers of the Boston Gazette . . .* (Boston: Edes and Gill[?], 1756), Early American Imprints, ser. 1, Evans, no. 7820, Readex Digital Collections.

5.25 **"Year without a Summer"** (1816): C. Edward Skeen, "'The Year without a Summer': A Historical View," *Journal of the Early Republic* 1, no. 1 (1981): 51–67; Steinberg, *Down to Earth*, 47 (1.13).

Chapter Six: Sunk Lands and Submerged Knowledge

6.1 **History of Seismic Instrumentation:** James Dewey and Perry Byerly, "The Early History of Seismometry (to 1900)," *Bulletin of the Seismological Society of America* 59, no. 1 (1969): 194–98; Johannes Schweitzer, "The Birth of Modern Seismology in the Nineteenth and Twentieth Centuries," *Earth Sciences History* 26, no. 2 (2007): 263–80; Clancey, *Earthquake Nation*, chap. 10; Davison, *The Founders of Seismology* ; Geschwind, *California Earthquakes*, esp. chaps. 2, 4; Oldroyd, *Thinking about the Earth*, chap. 10; Oldroyd et al., "Study of Earthquakes in the Hundred Years Following the Lisbon Earthquake," esp. 346–50 (all at 1.6); Geschwind, "Embracing Science and Research" (6.2). In the context of human observations: Coen, *The Earthquake Observers*, esp. chap.1 (1.6).

6.2 **Jesuit Seismology:** David Branagan, "Earth, Sky and Prayer in Harmony: Aspects of the Interesting Life of Father Edward Pigot . . . Part I," *Earth Sciences History* 28, no. 1 (2010): 69–99; Carl-Henry Geschwind, "Embracing Science and Research: Early Twentieth-Century Jesuits and Seismology in the United States," *Isis* 89, no. 1 (1998): 27–49; Agustín Udías, "Jesuits' Studies of Earthquakes and Seismological Stations," in *Geology and Religion: A History of Harmony and Hostility*, ed. Martina Kölbl-Ebert, Geological Society Special Publication 310 (London: Geological Society, 2009); Agustín Udías and William Stauder, "The Jesuit Contribution to Seismology," *Seismological Research Letters* 67 (1996): 10–19; Agustín Udías and William Stauder,

"Jesuit Geophysical Observatories," *Eos: Transactions, American Geophysical Union* 72, no. 16 (1991): 88–89, 185.

6.3 **James B. Macelwane, SJ:** American Geophysical Union, "James B. Macelwane Medal," http://sites.agu.org/honors/medals-awards/james-b-macelwane /?sub=recipients, accessed 12 June 2012; Henry F. Birkenhauer, SJ, "Father Macelwane and the Jesuit Seismological Association," *Earthquake Notes* 27, no. 2 (1956): 12–13; Victor J. Blum, SJ, "Sketch of the Life of James Bernard Macelwane, SJ" *Earthquake Notes* 27, no. 2 (1956): 9–11; Perry Byerly and William V. Stauder, SJ, "James B. Macelwane, SJ," *National Academy of Science Biographical Memoirs* 31 (1958): 252–81; Ross B. Heinrich, "James B. Macelwane, SJ, Scholar," *Earthquake Notes* 27, no. 2 (1956): 13–15; Earnest A. Hodgson, "The Contribution of Father Macelwane to the Founding of the Eastern Section, Seismological Society of America," *Earthquake Notes* 27, no. 2 (1956): 11–12; Brian J. Mitchell, "James B. Macelwane (1883–1956)," website of the American Geophysical Union, http://sites.agu.org/honors/james-b-macelwane-1883%E2%80%931956/, accessed 12 June 2012.

6.4 **Macelwane Publications:** D. C. Bradford and James B. Macelwane, SJ, "A Preliminary Sketch of the Seismic History of Missouri," *Earthquake Notes* 7 (1935): 17; James B. Macelwane Manuscript Collection (DOC MSS 1), Saint Louis University Archives, St. Louis, MO, esp. (roughly chronologically): "The Geology of St. Louis," *The Fleur de Lis* (St. Louis University) 19, no. 2 (January 1918): 78–91; "The Ozark Earthquake Investigation," *Physics Bulletin* 5, no. 2 (1925): 15–16; "Earthquake Survey of Ozark Region Proposed by St. Louis Scientist," *St. Louis Post-Dispatch*, 30 September 1925, 21, 24; "Can We Make St. Louis Earthquake Proof?," address before the St. Louis Chamber of Commerce, 1925; "The Ozark Program and Our Seismographs," *Physics Bulletin* 7, no. 1, (1927): 12–13; "The Mississippi Valley Earthquake Problem," *Bulletin of the Seismological Society of America* 20, no. 2 (1930): 95–98; "Grover, Missouri, Earthquake November 16, 1933," *Earthquake Notes* 5, no. 3 (1934): 3–4; "Progress in the Study of Earthquakes in the New Madrid Region," typescript of address delivered at the meeting of the Missouri Academy of Science, Columbia, Missouri, 7 December 1934; "Seismicity of the Mississippi Valley," apparently unpublished typescript manuscript, 1948[?].

6.5 **Saint Louis University (SLU):** Saint Louis University, "About SLU," http:// www.slu.edu/x5525.xml; SLU Department of Earth and Atmospheric Sciences, "Department History," http://www.slu.edu/x36005.xml, accessed 12 June 2012.

6.6 **Forgetting, Denying Scientific, Environmental Events:** Allan M. Brandt, *The Cigarette Century: The Rise, Fall, and Deadly Persistence of the Product That Defined America* (New York: Basic Books, 2007); Geschwind, *California Earthquakes*, chap. 1 (1.6); Donald Harington, *Let Us Build Us a City: Eleven Lost Towns* (San Diego: Harcourt Brace, 1986); Alan MacEachern, "A History of the Miramichi Fire: Or, Spotting Fire in the Archives & the Field," work-in-progress presented at the MIT Seminar on Environmental and Agricultural History, 2010; Andrew Salvador Mathews, "Suppressing Fire and Memory: Environmental Degradation and Political Restoration in the Sierra Juárez of Oaxaca, 1887–2001," *Environmental History* 8, no. 1 (2003): 76–108; Roger Turner, "Where Is the Dust Bowl in 20th Century American Meteorology? Atmospheric Physicists' 'Flight' from an Integrative Climatology," discussion paper for

Climate and Cultural Anxiety Conference, Colby College, April 2009; Naomi Oreskes and Erik M. Conway, *Merchants of Doubt: How a Handful of Scientists Obscured the Truth on Issues from Tobacco Smoke to Global Warming* (New York: Bloomsbury Press, 2010); Robert N. Proctor and Londa L. Schiebinger, *Agnotology: The Making and Unmaking of Ignorance* (Stanford, CA.: Stanford University Press, 2008). Also: 2.3.

6.7 Environments of the New Madrid Seismic Zone: See 1.14, 2.5, 2.7, 2.19, 6.9 and 6.11.

6.8 Boll Weevil: Betsy Blaney, "U.S. Cotton Almost Clear of Voracious Boll Weevil: Farmers Prevail over U.S.'s Most Expensive Pest; Boll Weevil Gone from 98 Percent of U.S. Cotton," Associated Press, through abcnews.go.com, posted 30 October 2009; James C. Giesen, " 'The Truth about the Boll Weevil': The Nature of Planter Power in the Mississippi Delta," *Environmental History* 14 (2009): 683–704; Clyde E. Sorenson, "The Boll Weevil in Missouri: History, Biology and Management (Publication G4255)," University of Missouri Extension, http://extension.missouri.edu/publications /DisplayPub.aspx?P=g4255, posted April 1995.

6.9 Railroads, Lumbering, Swamp Drainage: George W. Balogh, "Timber Industry," posted July 2008; Michael Dougan, "Missouri Bootheel," posted April 2010; and Donna Brewer Jackson, "Manila (Mississippi County)," posted May 2009, both in *Encyclopedia of Arkansas History & Culture* (1.11); Sam Blackwell, "A Landscape Transformed by the Little River Drainage District," *Southeast Missourian*, 4 November 2007, published via southeastmissourian.com; Otto Kochtitzky, *Otto Kochtitzky: The Story of a Busy Life* (Cape Girardeau, MO: Ramfre Press, 1957); Little River Drainage District Records, digital collection and associated website, Special Collections and Archives, Kent Library, Southeast Missouri State University, accessed through http://library .semo.edu/archives/, updated November 2011; Leon Parker Ogilvie, "The Development of the Southeast Missouri Lowlands," PhD thesis, University of Missouri, 1967; Joel P. Rhodes, *A Missouri Railroad Pioneer: The Life of Louis Houck* (Columbia: University of Missouri Press, 2008); Bonnie Stepenoff, " 'The Last Tree Cut Down': The End of the Bootheel Frontier," *Missouri Historical Review* (October 1995): 61–78.

6.10 History of Swamp Reclamation: Shannon Stunden Bower, "Watersheds: Conceptualizing Manitoba's Drained Landscape, 1895–1950," *Environmental History* 12 (2007): 796–819; Nash, *Inescapable Ecologies*, esp. 108; Pisani, "Beyond the Hundredth Meridian," 474–79 (both in 1.13).

6.11 Culture and History of the New Madrid Seismic Zone: Margaret Jones Bolsterli, *During Wind and Rain: The Jones Family Farm in the Arkansas Delta, 1848–2006* (Fayetteville: University of Arkansas Press, 2008); Brattan, "The Geography of the St. Francis Basin" (2.7); Louis Cantor, "A Prologue to the Protest Movement: The Missouri Sharecropper Roadside Demonstration of 1939," *Journal of American History* 55, no. 4 (1969): 804–22; William E. Cobb, "Southern Tenant Farmers Union," *Encyclopedia of Arkansas History & Culture*, posted June 2010 (1.11); Faragher, " 'More Motley Than Mackinaw' " (2.14); Joan Tinsley Feezor, "Fraud and Deceit in Dunklin County, 1865– 1880," presentation at Missouri Conference on History, 1996; James F. Keefe and Lynn Morrow, *The White River Chronicles of S. C. Turnbo* (Fayetteville: University of Arkansas

Press, 1994); Sherry Laymon, *Pfeiffer Country: The Tenant Farms and Business Activities of Paul Pfeiffer in Clay County, Arkansas, 1902–1954* (Little Rock, AR: Butler Center Books, 2009); Irvin G. Wylie, "Race and Class Conflict on Missouri's Cotton Frontier," *Journal of Southern History* 20, no. 2 (1954): 183–96.

6.11.1 **Thad Snow:** Thad Snow, *From Missouri* (Boston: Houghton Mifflin Company/Riverside Press, 1954); Bonnie Stepenoff, *Thad Snow: A Life of Social Reform in the Missouri Bootheel*, ed. William E. Foley, Missouri Biography Series (Columbia: University of Missouri Press, 2003).

6.12 **Civil War in the New Madrid Seismic Zone:** William J. Crowley, *Tennessee Cavalier in the Missouri Cavalry: Major Henry Ewing, C.S.A.* (Homewood, IL: Crowley, 1978); Thomas A. DeBlack, *With Fire and Sword: Arkansas, 1861–1874*, Histories of Arkansas (Fayetteville: University of Arkansas Press, 2003); John Fiske, *The Mississippi Valley in the Civil War* (Boston: Houghton, Mifflin, 1900); McPherson, *Battle Cry of Freedom* (1.8); Freeman K. Mobley, *Making Sense of the Civil War in Batesville-Jacksonport and Northeast Arkansas, 1861–1874* (Batesville, AR: pivately published, 2005); National Park Service, "Chalk Bluff," CWSAC Battle Summary, American Battlefield Protection Program, accessed 11 June 2012; Daniel E. Sutherland, "Guerillas: The Real War in Arkansas," in *Civil War Arkansas: Beyond Battles and Leaders*, The Civil War in the West, ed. Anne J. Bailey and Daniel E. Sutherland (Fayetteville: University of Arkansas Press, 2000); Craig Swain, "Capture of Island No. 10," HMdb.org, the Historical Marker Database, http://www.hmdb.org/marker.asp?marker=18187, accessed 11 June 2012; Donald E. Sutherland, ed., *Guerrillas, Unionists, and Violence on the Confederate Home Front* (Fayetteville: University of Arkansas Press, 1999), esp. B. Franklin Cooling, "A People's War: Partisan Conflict in Tennessee and Kentucky"; Michael Fellman, "Inside Wars: The Cultural Crisis of Warfare and the Values of Ordinary People"; Robert R. Mackey, "Bushwackers, Provosts, and Tories: The Guerrilla War in Arkansas"; and Sutherland, "Introduction: The Desperate Side of War"; Dan Montgomery, "Battle of Prairie Grove," updated May 2010; Daniel E. Sutherland, "Jayhawkers and Bushwhackers," updated March 2009; and Michael Taylor, "Skirmish at Chalk Bluff," updated July 2009, *Encyclopedia of Arkansas History & Culture* (1.11).

6.13 **Agriculture and Earthquake Evidence:** Plowing destroys archeological sites: Tuttle et al., "Use of Archaeology to Date Liquefaction Features," 477 (7.5.5); Van Arsdale, "Seismic Hazards of the Upper Mississippi Embayment," 27 (1.1). The sheer size of some New Madrid sand blows made them less vulnerable to plowing: Johnston and Schweig, "The Enigma of the New Madrid Earthquakes of 1811–1812," 369 (1.1). Artificially cleared land hard to read for earthquake evidence: Odum et al., "Near-Surface Structural Model for Deformation Associated with the February 7, 1812, New Madrid, Missouri Earthquake" (7.5.3).

6.14 **1880s, '90s Debate over Earthquake:** James MacFarlane, "The "Earthquake" at New Madrid, Mo., in 1811, Probably Not an Earthquake," abstract, *Proceedings of the American Association for the Advancement of Science, Thirty-Second Meeting, 1883*; and subsequent "Discussion: The 'Earthquake' at New Madrid, Mo. In 1811, Probably Not an Earthquake," *Science* 2, no. 31 (1883): 324; W. J. McGee, "A Fossil Earthquake,"

Bulletin of the Geological Society of America 4 (1892): 411–15; also W. J. McGee, "Correspondence: The New Madrid Earthquake," *American Geologist* 30 (1902); Penick *The New Madrid Earthquake*, 71–75 (1.1).

6.15 Predicting New Madrid: John E. Farley, *Earthquake Fears, Predictions, and Preparations in Mid-America* (Carbondale: Southern Illinois University Press, 1998); William Spence, Robert B. Herrmann, Arch C. Johnston, and Glen Reagor, *Responses to Iben Browning's Prediction of a 1990 New Madrid, Missouri, Earthquake*, US Geological Survey Circular 1083 (Washington, DC: United States Government Printing Office, 1993); C. D. Steltzer, "On Shaky Ground: The Earthquake Hazard Here May Be Greater Than You Think, And Planning for It Is Less Than You'd Expect," *Riverfront Times Online* (St. Louis, MO), 15 December 1999.

6.16 New Madrid Preparation: Subcommittee on Investigations and Oversight and Subcommittee on Science, Space, and Technology of the Committee on Science, Space, and Technology, US House of Representatives, *The Earthquake Threat in the Central United States: Are We Prepared?*, 101st Congress, 1st Sess., 17 November 1989.

6.17 Modern Science: Jimena Canales, *A Tenth of a Second: A History* (Chicago: University of Chicago Press, 2009); Peter Galison and Bruce Hevly, eds., *Big Science: The Growth of Large-Scale Research* (Stanford, CA: Stanford University Press, 1992).

Chapter Seven: The Science of Deep History

7.1 Myron L. Fuller

7.1.1 Background: Stanley N. Davis, "Studies of Ground-Water Pollution, 1899–1945," paper 36–14, Abstracts, Geological Society of America, 2004 Annual Meeting, http://www.geosociety.org/meetings/searchabstracts.htm; Myron L. Fuller Correspondence, 1917, Record Unit 45—Office of the Secretary 1890–1929, Box 23, Folder 23, Smithsonian Institution Archives, Washington, DC; Robert Rakes Schrock, *Geology at MIT, 1865–1965: A History of the First Hundred Years of Geology at Massachusetts Institute of Technology* (Boston[?]: Murray Publishing, 1982), Google Books. Conversation with Wyona Lynch McWhite, director of the Fuller Craft Museum, Brockton, Massachusetts, 7 July 2010.

7.1.2 Publications: The most relevant of Fuller's many publications are Fuller, *The New Madrid Earthquake*; and "Our Greatest Earthquakes" (1.1); "Audubon's Account of the New Madrid Earthquake" (1.18); "Cause and Periods of Earthquakes in the New Madrid Area, Missouri, and Arkansas," *Science* (1905): 349–50; "Comparative Intensities of the New Madrid, Charleston, and San Francisco Earthquakes," *Science* 23 (1906): 917–18; "Earthquakes and the Forest," *Forestry and Irrigation* 12 (1906): 261–67; "Notes on the Jamaica Earthquake," *Journal of Geology* (1907): 696–721.

7.1.3 Reprintings of *The New Madrid Earthquake*: Myron L. Fuller, *The New Madrid Earthquake*, 4th ed. (Marble Hill, MO: Gutenberg-Richter, 1995). Dates are not clear for editions 1–3, but the print runs were apparently small: the 1995 edition is the only one generally available. Prior reprints had similarly small print runs: Cape Girardeau, MO: Ramfre Press, 1958; Central US Earthquake Consortium (David Stewart was director of CUSEC before it moved to Memphis in 1988), 1988; Stewart's Center for Earthquake Studies at Southeastern Missouri State University in Cape Girardeau, MO, 1989.

7.2 Early Twentieth-Century Study of New Madrid: George C. Branner and J. M. Hansell, *Earthquake Risks in Arkansas: A Statistical Study Covering the Period from 1811 to 1931* (1932; repr. Little Rock: Arkansas Geological Survey, 1937); Broadhead, "The New Madrid Earthquake" (1.2); W. J. McGee, "Correspondence: The New Madrid Earthquake," *American Geologist* 30 (1902): 200–201; F. A. Sampson, "The New Madrid and Other Earthquakes of Missouri," *Missouri Historical Review* 8, no. 4 (July 1913): 179–99. Also: **6.3, 6.4.**

7.3 Early Resurgence of New Madrid Study: Kern C. Jackson, *Earthquakes and Earthquake History of Arkansas*, Information Circular 26 (Little Rock: Arkansas Geological Commission, 1979); Johnston and Schweig, "The Enigma of the New Madrid Earthquakes" (1.1); Arch C. Johnston, "A Major Earthquake Zone on the Mississippi," *Scientific American* 246 (1982): 60–68; Arch C. Johnston and Kaye M. Shedlock, "Overview of Research in the New Madrid Seismic Zone," *Seismological Research Letters* 63, no. 3 (1992): 193–208; Penick, *The New Madrid Earthquakes* (1.1); Margaret Ross, "The New Madrid Earthquake," *Arkansas Historical Quarterly* 27, no. 2 (1968): 83–104.

7.3.1 Otto Nuttli: Howell, *An Introduction to Seismological Research*, 103–4, 127; Hough, *Predicting the Unpredictable*, 181–84 (both 1.6); Johnston and Shedlock, "Overview of Research in the New Madrid Seismic Zone," 194 (7.3); Brian J. Mitchell, "Memorial: Otto W. Nuttli (1926–1988)," *Bulletin of the Seismological Society of America* 78, no. 3 (1988): 1387–89; Otto Nuttli, "The Mississippi Valley Earthquakes of 1811 and 1812: Intensities, Ground Motion, and Magnitudes," *Bulletin of the Seismological Society of America* 63, no. 1 (1973): 227–48; and accompanying microfiche.

7.4 Seismic Monitoring: Arkansas Geological Survey Arkansas Seismic Network information website, http://www.geology.ar.gov/geohazards/ark_seismic_network .htm, accessed 27 June 2012; CERI Seismic Networks Station List, http://www.ceri .memphis.edu/seismic/stations/index.html, accessed 3 July 2012; Saint Louis University Earthquake Center, Seismic Network web page, http://www.eas.slu.edu/eqc /eqcnetwork.html, accessed 3 July 2012; University of Arkansas at Little Rock Earthquake Center Arkansas Seismic Observatory web page, http://quake.ualr.edu/ASO .htm, updated June 2012; USGS, "ANSS—Advanced National Seismic System—Accomplishments," updated 7 June 2012, at http://earthquake.usgs.gov/monitoring /anss/milestones.php. Digital equipment: USGS, "EQ Monitoring Today," updated 8 June 2012, at http://earthquake.usgs.gov/learn/eqmonitoring/eq-mon-9.php. Good brief review of often confusing sources of instrumental data for seismicity in the central United States is in the header notes for the New Madrid catalog, CERI: http:// www.ceri.memphis.edu/seismic/catalogs/cat_nm_help.html, accessed 3 July 2012. In 1984 IRIS, the Incorporated Research Institutions for Seismology, began to coordinate the activities of virtually all the universities with research programs in seismology. IRIS now administers the Global Seismographic Network: see http://www.iris .edu/hq/programs/gsn.

7.5 Recent Study of New Madrid: Recent decades have seen voluminous work. These sources are indicative.

7.5.1 Overviews: (chronologically): Roy B. Van Arsdale, "Hazard in the Heartland: The New Madrid Seismic Zone," *Geotimes* 42, no. 5 (1997): 16–19; S. E Hough, "Scientific Overview and Historical Context of the 1811–1812 New Madrid Earthquakes," *Annals of Geophysics* 47 (2004): 523–38; Martitia P. Tuttle, "New Madrid in Motion," *Nature* 435 (2005): 1037–39; Hough and Bilham, "Tecumseh's Legacy: The Enduring Enigma of the New Madrid Earthquake," in Hough and Bilham, *After the Earth Quakes*, chap. 4 (1.6); NMSZ Expert Panel Report to NEPEC, "Report of the Independent Expert Panel on New Madrid Seismic Zone Earthquake Hazards," National Earthquake Prediction Evaluation Council, US Geological Survey, 16 April 2011; US Nuclear Regulatory Commission, US Department of Energy, and Electric Power Research Institute, "Central and Eastern United States Seismic Source Characterization for Nuclear Facilities," NUREG-2115, January 2012, available through the NRC website: http://www.nrc.gov/reading-rm /doc-collections/nuregs/staff/sr2115/ (CEUS Seismic Source Characterization).

7.5.2 Magnitude: Susan E. Hough, Jon G. Armbruster, Leonardo Seeber, and Jerry F. Hough, "On the Modified Mercalli Intensities and Magnitudes of the 1811– 1812 New Madrid, Central United States Earthquakes," *Journal of Geophysical Research* 105, no. 23 (2000): 23839–64; S. E. Hough, "Scientific Overview and Historical Context of the 1811–1812 New Madrid Earthquakes," *Annals of Geophysics* 47 (2004): 523–38; Susan E. Hough, "The Magnitude of the Problem," *Seismological Research Letters* 82, no. 2 (2011): 167–69; "Report of the Independent Expert Panel on New Madrid Seismic Zone," 6, 7, 15, 16 (7.5.1). For accessible explanations of magnitude scales and other earthquake science, visit the USGS website, under "Earthquake Topics for Education" and "Earthquake Glossary." Hough, *Richter's Scale*; Coen, *Earthquake Observers*, 259– 260 (both 1.6).

7.5.3 Structures, Tectonics, Glaciation: E. Calais, A. M. Freed, R. Van Arsdale, and S. Stein, "Letter: Triggering of New Madrid Seismicity by Late-Pleistocene Erosion," *Nature* 466 (2010): 608–11; R. T. Cox, and R. B. Van Arsdale, "Hotspot Origin of the Mississippi Embayment and Its Possible Impact on Contemporary Seismology," *Engineering Geology* 46 (1997): 201–16; Randel Tom Cox and Roy B. Van Arsdale, "Neotectonics of the Southeastern Reelfoot Rift Zone Margin, Central United States, and Implications for Regional Strain Accommodation," *Geology* 29, no. 5 (2001): 419–22; Joan Gomberg, "Tectonic Deformation in the New Madrid Seismic Zone: Inferences from Map View and Cross-Sectional Boundary Element Models," *Journal of Geophysical Research* 98, no. B4 (1993): 6639–64; Balz Grollimund and Mark D. Zoback, "Did Deglaciation Trigger Intraplate Seismicity in the New Madrid Seismic Zone?" *Geology* 29, no. 2 (2001): 175–78; R. M. Hamilton, and F. A. McKeown. "Structure of the Blytheville Arch in the New Madrid Seismic Zone," *Seismological Research Letters* 59, no. 4 (1988): 117–21; James R. Howe and Thomas L. Thompson, "Tectonics, Sedimentation, and Hydrocarbon Potential of the Reelfoot Rift," *Oil & Gas Journal* (1984): 179–90; Jack K. Odum, William J. Stephenson, Kaye M. Shedlock, and Thomas L. Pratt, "Near-Surface Structural Model for Deformation Associated with the February 7, 1812, New Madrid, Missouri Earthquake," *GSA Bulletin* 110, no. 2 (1998): 149–62; Van Arsdale, "Faulting along the Southern Margin of Reelfoot Lake, Tennessee" (1.25).

7.5.4 **Dendrochronology:** Guccione, "Late Pleistocene and Holocene Paleoseismology," 255, 258; Van Arsdale et al., "Earthquake Signals in Tree-Ring Data" (both in 2.21).

7.5.5 **Paleoseismology:** "CEUS Paleoliquefaction Database, Uncertainties Associated with Paleoliquefaction Data, and Guidance for Seismic Source Characterization," Appendix E of CEUS Seismic Source Characterization (7.5.1), prepared by M. Tuttle & Associates; Haydar J. Al-Shukri, Robert E. Lemmer, Hanan H. Mahdi, and Jeffrey B. Connelly, "Spatial and Temporal Characteristics of Paleoseismic Features in the Southern Terminus of the New Madrid Seismic Zone in Eastern Arkansas," *Seismological Research Letters* 76, no. 4 (2005): 502–11; John E. Ebel, Klaus-Peter Bonjer, and Mihnea C. Oncescu, "Paleoseismicity: Seismicity Evidence for Past Large Earthquakes," *Seismological Research Letters* 71, no. 2 (2000): 283–94; Guccione, "Late Pleistocene and Holocene Paleoseismology" (2.21); Alessandro M. Michetti, Franck A. Audemard M., Shmuel Marco, eds., *Paleoseismology: Integrated Study of the Quaternary Geological Record for Earthquake Deformation and Faulting*, special issue, *Tectonophysics* 408, nos. 1–4 (2005), esp. Michetti, Audemard M., and Marco, "Future Trends in Paleoseismology: Integrated Study of the Seismic Landscape as a Vital Tool in Seismic Hazard Analyses," 3–21; and Luigi Piccardi, "Paleoseismic Evidence of Legendary Earthquakes," 113–28 (4.10); Saucier, "Evidence for Episodic Sand-Blow Activity during the 1811–1812 New Madrid (Missouri) Earthquake Series" (1.22); Martitia P. Tuttle and Eugene S. Schweig; "Archeological and Pedological Evidence for Large Prehistoric Earthquakes in the New Madrid Seismic Zone, Central United States," *Geology* 23, no. 3 (1995): 253–56; Martitia P. Tuttle, Robert H. Lafferty, III, Margaret J. Guccione, Eugene S. Schweig, III, Neal Lopinot, Robert F. Cande, Kathleen Dyer-Williams, and Marion Haynes, "Use of Archaeology to Date Liquefaction Features, Central United States," *Geoarchaeology* 11 (1996): 453–55; Tuttle, "Late Holocene Earthquakes and Their Implications for Earthquake Potential of the New Madrid Seismic Zone, Central United States" (PhD diss., Dept. of Geology, University of Maryland, 1999); M. P. Tuttle, J. Collier, L. W. Wolf, and R. H. Lafferty, III, "New Evidence for a Large Earthquake in the New Madrid Seismic Zone between A.D. 1400 and 1670," *Geology* 27, no. 9 (1999): 771–74; Martitia P. Tuttle, J. D. Sims, Kathleen Dyer-Williams, Robert H. Lafferty, III, and Eugene S. Schweig, III, "Dating of Liquefaction Features in the New Madrid Seismic Zone," US Nuclear Regulatory Commission, NUREG/GR-0018 (2000); M. P. Tuttle, "The Use of Liquefaction Features in Paleoseismology: Lessons Learned in the New Madrid Seismic Zone, Central United States," *Journal of Seismology* 5 (2001): 351–80; Martitia P. Tuttle and Lorraine W. Wolf, "Towards a Paleoearthquake Chronology of the New Madrid Seismic Zone," US Geological Survey, Earthquake Hazards Program, Progress Report 01HQGR0164 (2003); Martitia P. Tuttle, Eugene S. Schweig, III, Janice Campbell, Prentice M. Thomas, John D. Sims, and Robert H. Lafferty, III, "Evidence for New Madrid Earthquakes in A.D. 300 and 2350 B.C.," *Seismological Research Letters* 76, no. 4 (2005): 489–501; Martitia P. Tuttle, Haydar Al-Shukri, and Hanan Mahdi, "Very Large Earthquakes Centered Southwest of the New Madrid Seismic Zone 5,000–7,000 Years Ago," *Seismological Research Letters* 77, no. 6 (2006): 755–70; Martitia P. Tuttle, "Search for and Study of Sand Blows at Distant Sites Resulting from Prehistoric and Historic New Madrid Earthquakes: Collaborative Research, M. Tuttle & Associates and Central Region Hazards Team," US Geological Survey, Final Technical Report 02HQGR0097 (2010). See also 1.22.

7.5.6 Triggered Earthquakes: Susan E. Hough, "Remotely Triggered Earthquakes Following Moderate Main Shocks," in Stein and Mazzotti., eds., *Continental Intraplate Earthquakes* (7.8), "Forum Commentary on "Should Memphis Build for California's Earthquakes?" (7.4.11); Hough, "Triggered Earthquakes and the 1811–1812 New Madrid, Central United States, Earthquake Sequence," *Bulletin of the Seismological Society of America* 91, no. 6 (2001): 1574–81; stress triggering more broadly: Hough, *Earthshaking Science*, 68–71 (1.6).

7.5.7 Hazard Mapping: USGS, Earthquake Hazards Program website, http://earthquake.usgs.gov/hazards/, provides access to the 2008, 1996, and earlier seismic hazards maps (see "Lower 48 States"); for an overview, see USGS "2008 National Seismic Hazard Maps," http://pubs.usgs.gov/fs/2008/3018/pdf/FS08–3018_508.pdf.

7.5.8 System Shutting Down: Tom Charlier, "Scientists Debate If New Madrid Fault Is Dead or Just Sleeping," *ScrippsNews* (Scripps Howard News Service), 28 January 2010; Megan Fellman, "Earthquakes Actually 19th Century Aftershocks: Repercussions of 1811 and 1812 New Madrid Quakes Continue to Be Felt," 4 November 2009, Northwestern University news release, http://www.northwestern.edu/newscenter/stories/2009/11/quakes.html; Jason McKenna, Seth Stein, and Carol A. Stein, "Is the New Madrid Seismic Zone Hotter and Weaker Than Its Surroundings?" in Stein and Mazzotti, eds., *Continental Intraplate Earthquakes* (7.8) "New Madrid Fault System, U.S., May Be Shutting Down," *ScienceDaily*, 13 March 2009, http://www.sciencedaily.com /releases/2009/03/090313145956.htm; Tom Parsons, "Earth Science: Lasting Earthquake Legacy (News and Views)," *Nature.com* 462, no. 7269 (2009): 42–43; Seth Stein and Mian Liu, "Long Aftershock Sequences within Continents and Implications for Earthquake Hazard Assessment," *Nature.com* 462, no. 7269 (2009): 87–89; Roy B. Van Arsdale, "Displacement History and Slip Rate on the Reelfoot Fault of the New Madrid Seismic Zone: Opinion Paper," *Engineering Geology* 55 (2000): 219–26.

7.5.9 System Not Shutting Down: R. Smalley, Jr., M. A. Ellis, J. Paul, and R. B. Van Arsdale, "Letters: Space Geodetic Evidence for Rapid Strain Rates in the New Madrid Seismic Zone of Central USA," *Nature* 435 (2005): 1088–90; Tuttle, "New Madrid in Motion," 1038 (7.5.1).

7.5.10 Expanding Seismic Zones of Midcontinent and Eastern US: Appendix E of CEUS Seismic Source Characterization (7.5.1); Karl Mueller, Susan E. Hough, and Roger Bilham, "Analysing the 1811–1812 New Madrid Earthquakes with Recent Instrumentally Recorded Aftershocks," *Nature* 429, no. 6989 (2004): 284–88; "New Madrid Not at Fault, Scientists Say: Wabash Blamed in 1800s Quakes," *Columbia Daily Tribune*, 8 June 2010; Susan E. Hough, Roger Bilham, Karl Mueller, William Stephenson, Robert Williams, and Jack Odum, "Wagon Loads of Sand Blows in White County, Illinois," *Seismological Research Letters: Eastern Section* 76, no. 3 (2005): 373–86; Hough, *Earthshaking Science*, 210 (1.6); Kevin Krajick, "Future Shocks: Modern Science, Ancient Catastrophes and the Endless Quest to Predict Earthquakes," *Smithsonian* (2005): 38–46. Also: 1.7.1 and 7.9.

7.5.11 Debating Risk of New Madrid Seismic Zone: A. D. Frankel, D. Applegate, M. P. Tuttle, and R. A. Williams, "Earthquake Hazard in the New Madrid Seismic Zone Remains a Concern," US Geological Survey Fact Sheet, 2009–3071 (2009); Joan Gomberg and Eugene Schweig, "Earthquake Hazard in the Heart of the Homeland," USGS Fact Sheet 4 (2007); Andrew Newman, Seth Stein, John Weber, Joseph Engeln, Ailin Mao, and Timothy Dixon, "Slow Deformation and Lower Seismic Hazard at the New Madrid Seismic Zone," *Science* 284 (1999): 619–21; Seth Stein, Joseph Tomasello, and Andrew Newman, "Should Memphis Build for California's Earthquakes?" *Eos: Transactions, American Geophysical Union* 84, no. 19 (2003): 177, 184–85; A. D. Frankel and S. E. Hough, "Forum Commentary," 271–73; Stein, Tomasello, and Newman, "Response to Forum Commentary," 273.

7.5.12 Various Contributions: Alan L. Kafka, "Does Seismicity Delineate Zones Where Future Large Earthquakes Are Likely to Occur in Intraplate Environments?," in Stein and Mazzotti, eds., *Continental Intraplate Earthquakes* (7.8),; Vladimir G. Kochkin and Jay H. Crandell, "Survey of Historic Buildings Predating the 1811–1812 New Madrid Earthquakes and Magnitude Estimation Based on Structural Fragility," *Seismological Research Letters* 75 (January/February 2004): 22–35, doi:10.1785/gssrl.75.1.22; Ron Street, Edward W. Woolery, Zhenming Wang, and James B. Harris, "NEHRP Soil Classifications for Estimating Site-Dependent Seismic Coefficients in the Upper Mississippi Embayment," *Engineering Geology* 62 (2001): 123–35.

7.6 Crowley's Ridge: "Formation of the Crowley's Ridge Natural Division," in Thomas Foti, "Geography and Geology," updated December 2011, *Encyclopedia of Arkansas History & Culture* (1.11); Roy B. Van Arsdale, Robert A. Williams, Eugene S. Schweig, Kaye M. Shedlock, Jack K. Odum, and Kenneth W. King, "The Origin of Crowley's Ridge, Northeastern Arkansas: Erosional Remnant or Tectonic Uplift?" *Bulletin of the Seismological Society of America* 85, no. 4 (1995): 963–85. Broader context: Schweig and Van Arsdale, "Neotectonics of the Upper Mississippi Embayment"; Tuttle, "Late Holocene Earthquakes," 26–29 (7.5.5).

7.7 Historical Seismology: Julien Fréchet, Mustapha Meghraoui, and Massimiliano Stucchi, *Historical Seismology: Interdisciplinary Studies of Past and Recent Earthquakes*, Modern Approaches in Solid Earth Sciences, vol. 2 (Dordrecht, Netherlands: Springer, 2008), esp. Susan E. Hough, "Large 19th Century Earthquakes in Eastern/ Central North America: A Comparative Analysis"; Ronald Street, "A Contribution to the Documentation of the 1811–1812 Mississippi Valley Earthquake Sequence," *Earthquake Notes* 53, no. 2 (1982): 39–52; Ronald L. Street and Robert F. Green, "The Historical Seismicity of the United States: 1811–1928," contract report, US Geological Survey (Contract 14-08-0001-21251, 1984).

7.8 Intraplate Earthquakes: Seth Stein and Stephane Mazzotti, eds., *Continental Intraplate Earthquakes: Science, Hazard, and Policy Issues*, special paper, *Geological Society of America* 425 (2007), esp. Jay H. Crandell, "Policy Development and Uncertainty in Earthquake Risk in the New Madrid Seismic Zone"; and Seth Stein, "Approaches to Continental Intraplate Earthquake Issues."

7.8.1 Bhuj: Sudhir K. Jain, William R. Lettis, C. V. R. Murty, and Jean-Pierre Bardet, eds., "Supplement A to Vol 18: Bhuj, India, Earthquake of January 26, 2001, Reconnaissance Report," *Earthquake Spectra* 18 (2002). But see Qingsong Li, Mian Liu, and Youqing Yang, "The 01/26/2001 Bhuj, India, Earthquake: Intraplate or Interplate?" in *Plate Boundary Zones*, ed. Seth Stein and Jeffrey T. Freymueller (AGU Geodynamics series, 2002).

7.8.2 China: Rhett Butler, Gordon S. Stewart, and Hiroo Kanamori, "The July 27, 1976, Tangshan, China Earthquake: A Complex Sequence of Intraplate Events," *Bulletin of the Seismological Society of America* 69, no. 1 (1979): 207–20; Mian Liu, Youqiing Yang, Zhengkang Shen, Shimin Wang, Min Wang, and Yongge Wan, "Active Tectonics and Intracontinental Earthquakes in China: The Kinematics and Geodynamics," in Stein and Mazzotti, eds., *Continental Intraplate Earthquakes* (7.8); "NSF Awards MU $2.16 Million for Intraplate Earthquake Studies," press release, University of Missouri–Columbia, 21 November 2007.

7.8.3 Australia: Mark Leonard, David Robinson, Trevor Allen, John Schneider, Dan Clark, Trevor Dhu, and David Burbridge, "Toward a Better Model of Earthquake Hazard in Australia," in Stein and Mazzotti, eds., *Continental Intraplate Earthquakes* (7.8).

7.9 Eastern North American Earthquakes Area of Impact: USGS, "Poster of the 2010–2011 Arkansas Earthquake Swarm," updated April 2011, and "Poster of the Virginia Earthquake of 23 August 2011—Magnitude 5.8," updated Oct 2012, both available through http://earthquake.usgs.gov. Also see: 3.2.

7.10 Recent Seismology: Kevin Krajick, "Future Shocks: Modern Science, Ancient Catastrophes and the Endless Quest to Predict Earthquakes," *Smithsonian* (2005): 38–46.

7.11 Public Awareness and Debate: Associated Press, "New Madrid Zone Quake Will Be Very Different, Expert Warns," *Kansas City* (Missouri) *Star*, 18 June 2006, Kansas City.com; Nancy Bailey and Jim Dorman, "Guest Column: For Safer Schools—from Earthquakes" *Memphis Commercial Appeal*, www.commercialappeal.com, 14 May 2010; Clif Chitwood, "Study Risk of Quake," *Arkansas Democrat-Gazette*, 19 January 2008; Kathie Bassett, "Scientists Debate Area Earthquake Hazard," *Telegraph* (Alton, IL), 7 March 2010, TheTelegraph.com; Seth Blomeley, "Seismic Building Code Sparks Safety Debate: It'd Snag Development, Legislators Hear," *Arkansas Democrat-Gazette*, 4 January 2008, LexisNexis; Keith Boles, "Earthquake Seminar Being Held in Jonesboro," KAIT8 Jonesboro (Arkansas), kait8.com, 2 February 2010; Robert Roy Britt, "Tiny Geologic Shifts Confirm Quake Risk in Heart of America," *MSNBC News*, 22 June 2005; Tom Charlier, "Reeling Housing Industry Feels New Jolt from Expected Quake-Proof Code," *Memphis Commercial Appeal*, www.commercialappeal.com, 25 January 2008; Charlier, "Seismic Audit Is Jolt for Builders: Costly Shakeup Feared If Quake Code Updated," *Commercial Appeal*, 21 April 2009; Donna Farley, "Scientists Seek Local Help in Studying New Madrid Fault," *Southeast Missourian*, seMissourian.com, 14 July 2009; Mary Jo Feldstein, "It's Not How You React to an Earthquake, It's How You Prepare for One," *St. Louis Post-Dispatch*, 10 February 2004; Jon Gambrell, "Scientist: Previously Unknown Earthquake Zone Puts Arkansas Gas Pipeline on Shaky Ground,"

Los Angeles Times, 22 January 2009, on-line edition; Joan Gomberg and Eugene Sch-weig, "Earthquake Hazard in the Heart of the Homeland," USGS Fact Sheet FS-131–02 (October 2002); Kenneth Heard, "Geologist Urges Less Stringent Seismic Building Codes," *Arkansas Democrat-Gazette*, 29 January 2008, LexisNexis; Heard, "Scientists Divided on Quake Research," *Arkansas Democrat Gazette*, 3 July 2005; Chad Hemenway, "Midwest Earthquake Would Create Heavy Losses, Difficult Recovery," *P&C National Underwriter*, 30 June 2010, property-casualty.com; Kristen Gunderson Hunt, "Earth-quake Risk Looms in Midwest; Shakes Leave Little Damage, Raise Concerns," *Business Insurance*, 28 April 2008, LexisNexis; Halley Johnson, "All Shook Up: Earthquake Study Predicts Extensive Damage for Tennessee," *Memphis Flyer*, 7 July 2010, memphisflyer. com; Erin Karmon, "Discovery of Surface Fault Offers View of Quake Zone: New Fault in Bootheel Area Is Not Seismically Active," *St. Louis Post-Dispatch*, 13 December 2002; John Krupa, "Scientists Seek Clues to New Madrid Fault," *Arkansas Democrat Gazette*, 15 June 2008; Robert Mackey, "Latest Updates on Chile's Earthquake," *The Lede, New York Times* News Blog, nytimes.com, 1 March 2010; Missouri Department of Natural Resources, "Earthquakes in Missouri" (Rolla and Jefferson City, MO: Missouri Depart-ment of Natural Resources, Division of Geology and Land Survey in cooperation with Missouri Department of Public Safety), n.d. (post-1985 and likely 1990s); FEMA and NHRP, "The National Earthquake Hazards Reduction Program: FEMA Accomplish-ments in Fiscal Year 2011," 2012; Tim O'Neil, "Study Aims to Find Quake Danger Zones: Geologists Are Looking at Soil in Effort to Help Avert Damage," *St. Louis Post-Dispatch*, 30 December 2003; OurAmazingPlanet Staff, "Doomsday Midwest Quake Predictions Overblown, Scientists Say," Livescience.com, 22 October 2010, http://news.yahoo .com/s/livescience; Gary Patterson, "Guest Column: Funding to Address Seismic Vul-nerability," *Memphis Commercial Appeal*, 4 December 2009; Jennifer Preston and Jack Healy, "Small Earthquake Hits near Chicago," *New York Times*, nytimes.com, 10 Febru-ary 2010; "Seismic Science: Is Number of Earthquakes on the Rise? Live Q&A," 9 March 2010, *Washington Post*, washingtonpost.com; Steltzer, "On Shaky Ground" (6.15); Seth Stein, *Disaster Deferred: How New Science Is Changing Our View of Earthquake Hazards in the Midwest* (New York: Columbia University Press, 2010); Seth Stein and Joseph To-masello, "When Safety Costs Too Much," editorial, *New York Times*, 10 January 2004; *The Earthquake Threat in the Central United States* (6.16); United Press International, "Mayor Fights Arkansas Earthquake Law," 24 January 2008; US Department of the In-terior and US Geological Survey, "Putting Down Roots in Earthquake Country: Your Handbook for the Central United States," General Information Product 119 (Reston, VA: US Geological Survey, 2011); Alan Scher Zagier, "Earthquake Expert Warns of Economic Damage," Associated Press State & *Local Wire*, 14 August 2008, LexisNexis; Zachary Zoeller, "Earthquake Code Enforcement Better Late Than Never—But It Will Cost You," 6 November 2006, *Memphis Daily News*, www.memphisdailynews.com.

    7.11.1 New Madrid Disaster Planning: James Dowd, "Businesses Are Urged to Be Earthquake-Ready," *Memphis Commercial Appeal*, 11 Feb 2010; Amir S. Elnashai, Lisa J. Cleveland, Theresa Jefferson, and John Harrald, "Impact of New Madrid Seismic Zone Earthquakes on the Central USA, vol. 1 and 2," *FEMA New Madrid Catastrophic Planning Project Phase Two Report, MAE Center Report 09–03* (Mid-America Earthquake

Center, University of Illinois, 2009); James Dowd, "Conferees Discuss Ways to Get Disaster Area Back to 'New Normal,'" 11 September 2009, *Memphis Commercial Appeal*, commercialappeal.com; Great Central US Shake-Out website: http://www.shakeout .org/centralus/, updated 2012; Norman C. Hester, Ira Satterfield, and Robert A. Bauer, "Partnerships for the Central United States: Consortia Aim to Mitigate Potential Losses in the Central United States If Another New Madrid Hits," *Geotimes* (1999); Andrew Revkin, "Disaster Awaits Cities in Earthquake Zones" *New York Times*, 24 Feb 2010.

7.11.2 CUSEC: The *CUSEC Journal* chronicles the Central US Earthquake Consortium's efforts to coordinate earthquake preparation since 1993: see http://www.cusec .org/publications/cusec-newsletter.html.

7.11.3 New Madrid Bicentennial: Commemorations pulled together a wide variety of regional planning and disaster preparation organizations: http:// newmadrid2011.org/, updated 2012.

7.11.4 Global Lessons: James Lee Witt and Adm. James M. Loy, "Heartland Haiti: Prepare for Earthquakes before They Strike," *Washington Times*, 11 February 2010.

7.12 Earthquake Clusters, Fracking Fears: Conversations with Scott Ausbrooks, Arkansas Geological Survey, 19 January and 3 July 2012; Erica Doerr, "Earthquakes," updated March 2011, *Encyclopedia of Arkansas History & Culture* (1.11); "Earthquake Town Meeting Set in Dyersburg: Citizen Concern, Awareness up after Series of Quakes in Region," Center for Earthquake Research and Information (CERI) website, 14 June 2005; David J. Hayes, Deputy Secretary, US Department of the Interior, "Is the Recent Increase in Felt Earthquakes in the Central U.S. Natural or Manmade?" US Department of the Interior, DOI News, 11 April 2012, http://www.doi.gov/news /doinews/Is-the-Recent-Increase-in-Felt-Earthquakes-in-the-Central-US-Natural -or-Manmade.cfm; National Academy of Sciences, "Induced Seismicity Potential in Energy Technologies: Report in Brief," June 2012, accessed 3 July 2012 at http://dels .nas.edu/Materials/Report-In-Brief/4298-Induced-Seismicity; Campbell Robertson, "A Dot on the Map, until the Earth Started Shaking," *New York Times*, 5 February 2011, newyorktimes.com; Rainer Sabin, "Experts Say Pattern of Recent Earthquakes Unusual," Associated Press State & Local Wire, 10 May 2005, LexisNexis; USGS, "Poster of 2010–11 Arkansas Earthquake Swarm" (7.9); USGS, "Science Features: Is the Recent Increase in Felt Earthquakes in the Central U.S. Natural or Manmade?" posted 11 April 2012 to http://www.usgs.gov/blogs/features/usgs_top_story/is-the-recent-increase -in-felt-earthquakes-in-the-central-us-natural-or-manmade/. Popular concern for relationship between fracking-related earthquakes and New Madrid earthquakes: for instance, the Fracking Arkansas advocacy group, http://frackingarkansas.wordpress .com/new-madrid-seismic-zone/, accessed 3 July 2012. Calls for permanent moratorium on fracking: http://www.stoparkansasfracking.org/, accessed 27 June 2012.

7.13 Trashy Movie Earthquake Apocalypse: "Apocalypse 10.5," Wikipedia, http:// en.wikipedia.org/wiki/10.5:_Apocalypse, updated 12 May 2012; Rick Wilson, "California Geological Survey—EarthquakeDOC—10.5 Apocalypse Review," California Department of Conservation, 2007, http://www.conservation.ca.gov/cgs/earthquakedoc /eq-movie_reviews/Pages/10.aspx.

7.14 American Disasters: "Katrina: Storm That Drowned a City," PBS, *Nova* (2005), www.pbs.org/wgbh/nova/earth/storm-that-drowned-city.html. Also: 1:34.

7.14.1 2009 Ice Storm: "January 2009 Central Plains and Midwest Ice Storm," http://en.wikipedia.org/wiki/January_2009_Central_Plains_and_Midwest_ice _storm; Staff report, "Legislators Gain Ice Storm Insight," *Times Leader* (Princeton, KY), 12 October 2009, timesleader.net.

7.15 New Madrid Field Guides: David Stewart and Ray Knox, *The Earthquake That Never Went Away: The Shaking Stopped in 1812, But the Impact Goes On* (Marble Hill, MO: Gutenberg-Richter Publications, 1993); David Stewart, *The Earthquake That Never Went Away* (Marble Hill, MO: Gutenberg-Richter Publications, 1995); Ray Knox and David Stewart, *New Madrid Fault Finders Guide: A Set of Self-Guided Field Tours in the "World's Greatest Outdoor Earthquake Laboratory," The New Madrid Fault Zone* (Marble Hill, MO: Gutenberg-Richter Publications, 1995).

7.16 Recent Cultural History of the New Madrid Seismic Zone: Rex Nelson, "Why the Delta Matters," *Arkansas Democrat-Gazette*, 12 July 2009. Also: 6:11

7.17 Animals and Earthquakes: "Anxious Animals as Quake Precursors," *Science News* 113, no. 17 (1978): 278; Joseph L. Kirschvink, "Earthquake Prediction by Animals: Evolution and Sensory Perception," *Bulletin of the Seismological Society of America* 90, no. 2 (2000): 312–23; John R. B. Lighton and Frances D. Duncan, "Shaken, Not Stirred: A Serendipitous Study of Ants and Earthquakes," *Journal of Experimental Biology* 208, no. 16 (2005): 3103–7; Huang Zan, abstract, "The History and Future of the Abnormal Animal Behaviors Used as Earthquake Precursor," *Dizhen Dici Guance yu Yanjiu/Seismological and Geomagnetic Observation and Research* 27, no. 4 (2006): 44–51, accessed through Georef, 22 February 2010. Coen, *Earthquake Observers*, 270–272 (1.6).

7.18 Earthquake-Panicked Squirrels? Latrobe, *The Rambler in North America*, 86 (1.31); quoted and circulated by cousin J. H. B. Latrobe: Latrobe, *The First Steamboat Voyage on the Western Waters*, 22 (1.31). Cited in: Ross, "The New Madrid Earthquake," 83–84 (7.3), but generally dismissed as tall-tale: Penick, *The New Madrid Earthquakes*, 12–13 (1.1).

7.19 Documented Squirrel Migrations: Ernest Thompson Seton, "Migrations of the Graysquirrel (*Sciurus Carolinensis*)," *Journal of Mammalogy* 1, no. 2 (1920): 56–57; Vagn Flyger, "The 1968 Squirrel 'Migration' in the Eastern United States," paper presented at the Northeast Fish and Wildlife Conference, White Sulphur Springs, West Virginia, February 1969, accessed through http://www.myoutbox.net/flyger.htm, 22 February 2010 (this informal republishing seems consistent with content referenced in other peer-reviewed literature).

7.20 Bodily Sensation, Scientific Data: Coen, *Earthquake Observers* (1.6); and "Tongues of Seismology" (4.8); Christopher Lawrence and Steven Shapin, eds., *Science Incarnate: Historical Embodiments of Natural Knowledge* (Chicago: University of Chicago Press, 1998); Nash, "The Changing Experience of Nature" (1.13); USGS "Did You Feel It?" website: http://Earthquake.usgs.gov/Earthquakes/dyfi/, last accessed 16 June 2012. Some of the most interesting recent work in the history of ecological sciences and environmental activism has dealt with the conflict between expert data gatherers and nonspecialists insisting on the validity and usefulness of their subjective experience of their bodies (bad smells, loud noises, seeing patterns in their own miscarriages): see 4.4 and 4.8. Twitter feeds and the USGS "Did You Feel It?" website are excellent resources on what earthquakes feel like to modern human beings.

7.21 **Chinese Science:** Fa-ti Fan, "Redrawing the Map: Science in Twentieth-Century China," *Isis* 98 (2007): 524–38; Fan, "'Collective Monitoring, Collective Defense" (5.2); Shen, "Taking to the Field" (4.8).

7.22 **Contemporary "Citizen Science":** Ira Flatow, "NPR Science Friday: Citizen Science for the New Year," heard over WBUR-FM 90.9, Boston, MA, 7 January 2010; one major online project is "Zooniverse: Real Science Online," http://www.zooniverse .org/home, last accessed 16 June 2012, when 652,464 people were online contributing scientific data; The Quake-Catcher Network: http://qcn.stanford.edu/, last accessed 16 June 2012; Stein, *Disaster Deferred*, 83 (7.11).

7.23 **Tennessee Bridge Retrofit:** Tennessee Department of Transportation, "Major Closures on I-40 Bridge over Mississippi River Begin August 3," TNnewsroom.gov, posted 16 July 2010; Tennessee Department of Transportation, "I-40 Hernando Desoto Bridge," TN.gov, posted 22 October 2010; and "I-40/Hernando DeSoto Seismic Bridge Work to Be Complete This Weekend," posted 4 November 2010; TRC Companies, "Projects: I-40 Mississippi River Bridge Seismic Retrofit," http://www.trcsolutions .com, posted 2010; "Large, Vulnerable, and Critical Infrastructure," session, 2012 EERI Annual Meeting and National Earthquake Conference.

ACKNOWLEDGMENTS

I began and ended this book as a faculty member in a history department. In between, I stepped out of conventional academic work to raise young children as a free-range academic. I am therefore fortunate to be able to express my gratitude to the people and institutions who made possible Virginia Woolf's formula for a woman writing: a place of one's own and financial support.

Washington University in St. Louis, a fabulous and lively place, launched this book with a New Faculty Research Grant. Many years later, a New Faculty Research Fund at the University of Massachusetts Boston—also a fabulous and lively place, in completely different ways—allowed me to bring it to completion. In between, I was sustained by a research fellowship from the National Endowment for the Humanities, by a Senior Fellowship at the Dibner Institute for the History of Science, and by formal affiliation as an Associate of the Department of the History of Science at Harvard University, whose people and communities, not to mention libraries, made possible my continued academic work.

This project has been enabled by the generosity of geologists, seismologists, and other public scientists whose intellectual tenacity is matched only by their enthusiasm for the questions they research. I am humbled by the work and patience of the scientists at the US Geological Survey (USGS) and its allied projects, especially the Center for Earthquake Research and Information (CERI), as well as at state surveys and public universities and institutes, particularly the Arkansas Geological Survey (AGS). I single out a number of individuals here, but I want to foreground my deep respect for a craft that is difficult and often underfunded, but which is also both vital and exciting. Thank you all.

This book has a historical godfather, Lynn Morrow, for many years the director of local records for Missouri, whose encouragement to look further for sources set me on the path of this project and whose interest kept patience with long writing until the end. Lynn Morrow has been for years—

decades—a fine historian in his own right, an administrator making possible innovative public history, and a mighty encourager of others' projects. Lynn's questions, suggestions, and manila envelopes packed with offbeat and promising sources have inspired my writing throughout. I would call him a midwife except that he might glower at me, and that would be enough to quell a braver historian than I. Lynn: congratulations on your retirement, and thank you.

Many people working on the scientific, historical, and policy questions surrounding the New Madrid earthquakes gave their time, often great chunks of time, to help a historian understand and write about these events. I am grateful to Martitia (Tish) Tuttle, of M. Tuttle & Associates; Haydar J. Al-Shukri, of the Center for Earthquake Education and Technology Transfer and the program in applied sciences at the University of Arkansas at Little Rock (UALR); and graduate student Marilyn Egan for allowing me to follow them around on a trenching expedition. My enduring thanks for graciously allowing me to interview, visit, and generally pester go to: Tish Tuttle, director and state geologist Bekki White, retired geologist John David McFarland, William Prior, and the rest of the staff at AGS, and especially geohazards point person Scott Ausbrooks, who answered countless questions over many years with unfailing good humor and astounding energy; the enormously accommodating staff at CERI/USGS, especially CERI founding director Arch Johnston and historian Nathan K. (Kent) Moran, as well as Joan Gomberg, Gary Patterson, Paul Bodin, director Chuck Langston, Eugene (Buddy) Schweig, and Roy B. Van Arsdale; James M. Wilkinson, Jr., director of the Central US Earthquake Consortium (CUSEC); Claudine Payne and Marion Haynes of the Arkansas Archeological Survey; James E. Price at the Ozark National Scenic Riverways; John E. Ebel of the Weston Observatory of Boston College; and Veronica Villolobos-Pogue and Edwin Lyons of the Arkansas Department of Emergency Management.

Two writing groups shaped my craft and this history. I appreciate the commentary of a long-running group which included Steven Biel, Seth Rockman, Jona Hansen, Jane Kamensky, John Plotz, Jennifer Roberts, and Michael Willrich. I am proud to be a founding member of IWSS, the Independent Women Scholars Salon, a group whose conversation and critique, encouragement and admonition buttress creative writing and creative life-choices. My enduring thanks to Lara Friedenfelds, Joy Harvey, Susan Lanzoni, Rachael Rosner, Kara Swanson, and Nadine Weidman, as well as former members Deborah Levine, Monique Tello, and Deborah Weinstein. When I said I wasn't sure the Lead Belly boll weevil song really belonged in a book about earthquakes, Joy insisted, then sang a marvelous version. When I went onstage to accept an award in the history of women and science, I looked into the audience to see an entire row of my writing group cheering and giving thumbs-up. Camaraderie this powerful moves mountains: it has certainly gotten books written—with many more to come!

Generous colleagues, friends, and family read and questioned, improving my understanding and this book. My humble and hearty thanks flow to S. Charles Bolton, Deborah R. Coen, Jon Christensen, Julia Hansford, David Jones, Deborah Levine, Benjamin Looker, Susan Hughes May, Laura Owens, Emily Pawley, Joshua L. Reid, David Spanagel, Corinna Treitel, Martitia Tuttle, Matthew G. Valencius, Paul Valencius, and Michael Willrich for commenting on proposals, chapters, and whole book sections. Two anonymous reviewers from the Press insisted on revisions: I'm glad now they did. At Wash U, Gar Allen, Iver Bernstein, Hillel Kieval, and Derek Hirst helped create this project with their queries and interest. Allan M. Brandt, Charles Rosenberg, and Elliott West all touched base with me and this project throughout, ensuring that both continued to thrive. As a colleague and as an editor, Lara Friedenfelds helped shape this book with grace and care.

This project was shaped by conversations and presentations at many venues. I am grateful to Jay Berger of the Earthquake Engineering Research Institute (EERI), and Jim Wilkinson, CUSEC, for the invitation to discuss my work at a plenary at the 2012 EERI/National Earthquake Conference, and to Nancy Gwinn and Lilla Vekerdy for the invitation to present this project at the Smithsonian Institution's "Age of Wonder" conference in 2010. Comments by Emily Pawley, Chris Jones, David Spanagel, and Jeremy Vetter at the 2011 American Society for Environmental History conference helped sharpen key sections; on an impromptu field trip, Emily, Chris, Jeremy, and I saw the desert bloom, and Chris came up with this book's title. Conversations at the "Witness to Disaster" workshop, Barnard College, October 2009, especially with organizer Deborah R. Coen, Andrea Westermann, Peder Anker, and Fa-ti Fan, sharpened this book, as did comments from Debbie on the subsequent *Science in Context* special issue she edited in 2012. Creating a plenary at the 2007 American Association for the History of Medicine annual meeting with Phil Teigen and Susan D. Jones gave me a new perspective and this book a new chapter. My thanks to those who commented on work-progress presentations at CERI; the STS Circle of the Harvard University Kennedy School of Government; the Humanities Center, Department of History, and Department of Geosciences of Stony Brook University; the Junior Faculty Colloquium at UMass Boston; the Center for Culture, History, and Environment and the Department of the History of Science at University of Wisconsin–Madison, especially Nancy Langston, Gregg Mitmann, and Anna Zeide; a session at the 2009 History of Science Society annual meeting, especially commentator Mott Greene; the Brown

University Thursday Lunch Colloquium; a 2008 "Under the West" conference at the Huntington Library, especially Jon Christensen, Brian Frehner, and organizer William Deverell; MIT's Program in Science, Technology, and Society and Seminar on Environmental and Agricultural History, especially Deborah Fitzgerald, Chihyung Jeon, David Jones, Jamie Pietruska, Harriet Ritvo, Natalie Schüll, Rosalind Williams, and Rebecca Woods; the Northeastern University Department of History, especially Tony Penna; the 2007 conference on "Lay Participation in Scientific Observation" at the Max-Planck Institute for the History of Science, especially Jeremy Vetter, Lorraine Daston, Brian Frehner, and Fa-ti Fan; the Department of the History and Sociology of Science at the University of Pennsylvania, especially David Barnes, Ruth Schwartz Cohen, Chris Jones, Rob Kohler, Susan Lindee, Beth Linker, Roger Turner, and Jeremy Vetter; the Program in Agrarian Studies at Yale University, especially Valentine Cadieux, Harold Forsythe, Richard Kernaghan, Erika Olbricht, Jennifer Leigh Smith, and John F. Varty; the Dibner Institute for the History of Science, especially David Cahan and Sarah Wermeil; Washington University in St. Louis Special Collections; the Boston Environmental History Symposium; and undergraduates at WashU and UMB for their discussions.

Numerous people answered questions or sent me sources, from lovely little tidbits to whalloping pieces of evidence. My thanks to Scott Ausbrooks, S. Charles Bolton, Nancy Britton, Mark M. Chambers, Thomas Danisi, George Anne Draper, Xaq Frohlich, Lynn Foster, Michael Friedlander, Monica Garcia, Maggie Haller, David Jones, Jane Kamensky, David Konig, George Lankford, Deborah Levine, Jen Light, Jessica Martucci, Adrienne Naylor, David Pantalony, Seth Rockman, Charles Rosenberg, George Smith, Seth Stein, and Tish Tuttle. Retired AGS geologist John David McFarland shared his parody "The Night before the Big One," showing that even in the midst of the Iben Browning fuss, earth scientists retained a sense of humor. David Rabenau discussed earthquake weather as we sat in a San Francisco city park, fellow refugees from the Loma Prieta earthquake. Arkansas songwriter Charley Sandage even sent me a CD!

Terrific people at fine institutions helped my research. I am grateful to Joan Feezor, archivist, Local Records Division, Missouri State Archives, now retired, who accompanied me in 2004 to the New Madrid Courthouse and whose subsequent dedicated work in the Missouri Department of Local Records projects has brought into use myriad sources from early New Madrid. My great thanks also go to Fred Burchsted, research librarian, Widener Library; Michael Leach, head of collection development, and Reed Lowrie, science reference and cartographic librarian, both in the Cabot Science Library, Harvard University; multiple staffers at the New Madrid County

Courthouse, especially Anne Evans Copeland, registrar of deeds, for generosity and forbearance in sharing valuable sources; J. Michael Flowers, Steve McLoughlin, Darrell Pratte, James Palmer, and Bruce Wilson of the Missouri Department of Natural Resources; Harry Miller, reference archivist at the Wisconsin Historical Society; Kim Nusco, reference and manuscript librarian at the John Carter Brown Library of Brown University; archivist Christine Harper at Saint Louis University Pious XII Memorial Library; archivist Erin Davis of Washington University in St. Louis Special Collections; Jonathan McIntyre at the Kentucky Geological Survey; and librarians and archivists at AGS, at the University of Western Ontario, and at the incomparable Missouri Historical Society.

I am grateful for the images facilitated and provided by Tish Tuttle; by the Southeast Missouri State University Special Collections and Archives, with heroic help from Amber Miranda; by the State Historical Society of Missouri, facilitated by Sara Przybilski; by USGS, thanks to Rob Williams and Natasha McCallister; by SLU's archives and Earthquake Center, with special assistance from Chris Harper; by AGS, thanks to David Johnson; by the Missouri State Archives, thanks to Laura Jolley; and by the Huntington Library and the Newberry Library.

I am grateful to Ike Williams for bringing me aboard and Katherine Flynn, literary agent and editorial manager, for steering me well at the Kneerim, Williams & Bloom Agency. At the University of Chicago Press I benefited enormously from the skillful professionalism of Karen Merikangas Darling, senior editor; and Carol Fisher Saller, senior manuscript editor; as well as Abby Collier. Research by Wash U students Julie Bucy, Mark M. Chambers, Jeremy M. Mikecz, and Sarah E. Mullen helped get this book started. Geographical insight and maps from Bill Keegan helped complete it.

Raising children takes a village—often many villages. So does writing books. Friends and colleagues Mike Finke and Elizabeth Oyler took me flying over the New Madrid seismic zone on a day that reminded me of how incredibly fun my job is. Don and Audrey Evans provided a writing retreat in the Ozarks the memory of which daily sustains me. At a key moment, a scholarship from the Adams Street Early Learning Center gave my two sons a wonderful community and gave me the space to keep writing. Our local library's first-grader story hour with Ms. Amanda inspired me not to leave out the stories and caused me to bump into a book on astronomy that appears in the conclusion. On numerous occasions, Cleve and Susie May loaned their car for research trips; provided space, love, humor, and excellent meals; and swooped in to manage household and grandchildren while I holed up in libraries or went to conferences. Friends Jim and Joanne

Fox, Ashish Karamchandani and Vibha Krishnamurthy, Laura Owens, Todd and Claire Prono, and Corinna Treitel gave me encouragement and perspective and the occasional much-needed gin and tonic. They and Charlie Bolton, Paul and Marilyn Valencius, and all the families and communities in which my small household is lucky to be embedded made possible our family, my life, and this book.

Ilan, Casimir, and Zara John Valencius are a joy and blessing beyond measure. As I sat at my desk, preschooler Zara would sometimes come up and ask, "Mama, are you writing your book?" When I'd nod "yes," she would sit beside me and busy herself at her own desk, intently drawing, writing, and stapling small sheets together. "Here you go!" she'd announce after a while, with her usual cheerful energy. "Here's a book I wrote for you!" I am so very glad finally to be able to reciprocate.

More than ten years ago, Matthew G. Valencius asked, "Hey, has anyone written a book on the New Madrid earthquakes? Why don't you write a short, quick book about those?" Since then, he has shipped boxes upon boxes of books and files, moved my office while I was on bed rest, rearranged his own work schedules to make possible my research, witnessed innumerable soybean fields ("See, there's a sand blow!"), visited obscure and terribly hot and humid Civil War battlefields, manhandled staggering amounts of laundry, refused to put real life on hold for footnotes, and listened with admirable patience to intricate details of early nineteenth-century small town probate cases, all to help me write a book that proved neither short nor quick. He may roll his eyes frequently and eloquently—but he did *not* flee to the territories when I announced, not once but twice, that a book that I had declared done in fact required another rewrite. At a final stage, when I said I needed a block of time to finish, Matt said he would take the kids off Saturday mornings to a fencing class. I now have that book about the earthquakes—and a house full of people who wield swords. Thank you, partner.

INDEX

Italic page numbers indicate illustrations.